Sourdough Innovations

Sourdough fermentation was probably one of the first microbial processes employed by mankind for the production and preservation of food. This practice is still widely used worldwide due to the distinct sensorial and health properties attributed to these products. Traditional sourdough bread is achieved by spontaneous fermentations, leading to natural selections of microorganisms (mainly yeast and lactic acid bacteria) with health benefits for the consumers' microbiota. However, multiple opportunities are currently underexploited through the entire sourdough value chain. *Sourdough Innovations: Novel Uses of Metabolites, Enzymes, and Microbiota from Sourdough Processing* summarizes the latest scientific knowledge and current opportunities of sourdough technology at the biomass, microbiota, and enzymatic levels described in three distinctive sections.

Section I covers the fermentation process of cereals and non-cereals to produce sourdough-containing compounds with health-enhancement benefits. Section II includes novel advances in sourdough enzymology, and lastly, Section III explores various applications of sourdough microbiota as antimicrobial and probiotic microorganisms and opportunities to be included in both food and non-food applications.

Key Features:

- Includes extensive information on the use of innovative or emerging technologies aiming to promote circular exploitation systems.
- Promotes the full use of the cereal and non-cereal sourdough metabolites.
- Covers the functionality of sourdough microorganisms and functional compounds and future exploitation of some of them in the field of nutraceuticals or functional foods.

Sourdough Innovations is unique in its examination of the health beneficial compounds through the downstream processing of sourdough from cereals, microbiota, and enzymes. It is also a great resource for academic staff and scientists within the broad area of food science who are researching, lecturing, or developing their professional careers in food microbiology, food chemistry, food processing, and food technology, including bio-process engineers interested in the development of novel technological improvements in sourdough processing.

Sourdough Innovations

Novel Uses of Metabolites, Enzymes, and Microbiota from Sourdough Processing

Edited by
Marco Garcia-Vaquero
and João Miguel F. Rocha

CRC Press is an imprint of the
Taylor & Francis Group, an **informa** business

First edition published 2024
by CRC Press
6000 Broken Sound Parkway NW, Suite 300, Boca Raton, FL 33487-2742

and by CRC Press
4 Park Square, Milton Park, Abingdon, Oxon, OX14 4RN

CRC Press is an imprint of Taylor & Francis Group, LLC

© 2024 selection and editorial matter, Marco Garcia-Vaquero and João Miguel F. Rocha; individual chapters, the contributors

Reasonable efforts have been made to publish reliable data and information, but the author and publisher cannot assume responsibility for the validity of all materials or the consequences of their use. The authors and publishers have attempted to trace the copyright holders of all material reproduced in this publication and apologize to copyright holders if permission to publish in this form has not been obtained. If any copyright material has not been acknowledged please write and let us know so we may rectify in any future reprint.

Except as permitted under U.S. Copyright Law, no part of this book may be reprinted, reproduced, transmitted, or utilized in any form by any electronic, mechanical, or other means, now known or hereafter invented, including photocopying, microfilming, and recording, or in any information storage or retrieval system, without written permission from the publishers.

For permission to photocopy or use material electronically from this work, access www.copyright.com or contact the Copyright Clearance Center, Inc. (CCC), 222 Rosewood Drive, Danvers, MA 01923, 978-750-8400. For works that are not available on CCC please contact mpkbookspermissions@tandf.co.uk

Trademark notice: Product or corporate names may be trademarks or registered trademarks and are used only for identification and explanation without intent to infringe.

Library of Congress Cataloging-in-Publication Data
Names: Garcia-Vaquero, Marco, editor. | Rocha, João Miguel, editor.
Title: Sourdough innovations : novel uses of metabolites, enzymes, and microbiota from sourdough processing / edited by Marco Garcia-Vaquero, João Miguel Rocha.
Description: First edition. | Boca Raton : CRC Press, 2023. | Includes bibliographical references and index. | Summary: "Sourdough fermentation is a widely used to produce and preserve bread, while increasing the health benefits due to the activity of several microorganisms. This book provides an exploration of health beneficial compounds through the downstream processing of sourdough from cereals, microbiota and enzymes"-- Provided by publisher.
Identifiers: LCCN 2022060710 (print) | LCCN 2022060711 (ebook) | ISBN 9780367674977 (hbk) | ISBN 9780367692599 (pbk) | ISBN 9781003141143 (ebk)
Subjects: LCSH: Yeast. | Bioactive compounds. | Sourdough starter--Health aspects. | Dough. | Functional foods. | Food--Microbiology. | Food industry and trade--By-products.
Classification: LCC TP580 .S68 2023 (print) | LCC TP580 (ebook) | DDC 664/.08--dc23/eng/20230427
LC record available at https://lccn.loc.gov/2022060710
LC ebook record available at https://lccn.loc.gov/2022060711

ISBN: 978-0-367-67497-7 (hbk)
ISBN: 978-0-367-69259-9 (pbk)
ISBN: 978-1-003-14114-3 (ebk)

DOI: 10.1201/9781003141143

Typeset in Times
by Deanta Global Publishing Services, Chennai, India

Contents

Preface ix
About the Editors xiii
Contributors xv

SECTION I Cereal and Non-cereal Sourdough Metabolites

Chapter 1 Sourdough Fermentation as a Way to Improve Health Benefits and the Sensory Properties of Bakery Products 3

Sedef Nehir El, Yesim Elmaci, Sibel Karakaya, and Alexandrina Sirbu

Chapter 2 Cereals and Cereal Sourdoughs as a Source of Functional and Bioactive Compounds 31

Monica Trif, Claudia Terezia Socol, Sneh Punia Bangar, and Alexandru Vasile Rusu

Chapter 3 Non-cereal and Legume Based Sourdough Metabolites 63

Maria Aspri, Nikolina Čukelj Mustač, and Dimitrios Tsaltas

Chapter 4 Innovative Technologies to Extract High-Value Compounds 87

Anca C. Farcas, Sonia A. Socaci, Dubravka Novotni, and Marco Garcia-Vaquero

SECTION II Enzymes from Sourdough Cultures

Chapter 5 Introduction to Sourdough Enzymology 117

Bogdan Păcularu-Burada, Marina Pihurov, Mihaela Cotârleț, Elena Enachi, and Gabriela-Elena Bahrim

Chapter 6 Major Classes of Sourdough Enzymes 147

Maria Aspri and Dimitrios Tsaltas

Chapter 7 Discovery, Characterization, and Databases of Enzymes from Sourdough 161

Zeynep Agirbasli and Sebnem Harsa

Chapter 8 Enzyme Production from Sourdough 199

Mensure Elvan and Sebnem Harsa

Chapter 9 Biotechnological Applications of Sourdough Lactic Acid Bacteria: A Source for Vitamins Fortification and Exopolysaccharides Improvement 231

Pasquale Russo, Kenza Zarour,
María Goretti Llamas-Arriba, Norhane Besrour-Aouam,
Vittorio Capozzi, Nicola De Simone, Paloma López, and
Giuseppe Spano

SECTION III Innovative Applications of Sourdough Microbiota

Chapter 10 Sourdough as a Source of Technological, Antimicrobial, and Probiotic Microorganisms 265

Vera Fraberger, Görkem Özülkü, Penka Petrova,
Knežević Nada, Kaloyan Petrov,
Domig Konrad Johann, and João Miguel F. Rocha

Chapter 11 Isolation, Technological Functionalization, and Immobilization Techniques Applied to Cereals and Cereal-Based Products and Sourdough Microorganisms 311

Zlatina A. Genisheva, Pedro Ferreira-Santos, and José A. Teixeira

Chapter 12 Sourdough Microorganisms in Food Applications 337

Mensure Elvan and Sebnem Harsa

Chapter 13 Production of Nutraceuticals or Functional Foods Using
Sourdough Microorganisms 369

*Armin Mirzapour-Kouhdasht, Samaneh Shaghaghian, and
Marco Garcia-Vaquero*

Chapter 14 Applications of Sourdough in Animal Feed 391

*Kristina Kljak, Miona Belović, Marija Duvnjak, Vanja Jurišić,
and Miloš Radosavljević*

Index 435

Preface

There is evidence of the crucial role that grains, including cereals, pseudocereals, and non-cereal grains, have played in the development of multiple civilizations over the millennia. Bread, in both its leavened and unleavened forms, appears to be one of the most primitive known food biotechnological processes developed by humankind. There are archaeological remains that relate to sourdough bread production and consumption dating back several millennia in multiple countries around the world, including preserved cereal grains and advanced tools that were developed to process grain for the purposes of fermentation and bread making, including sieves and mills to process grains into flour, vessels with ancient residues of primitive fermentation, and multiple varieties of high-temperature ovens developed for bread making.

Sourdough fermentation is a practice still used worldwide due to the distinct sensorial and health properties of bread, including improvements in texture, flavor, shelf life, and nutritional properties compared to other non-sourdough alternatives. Moreover, traditional sourdough bread was achieved by spontaneous fermentations, leading to natural selections of microorganisms (mainly yeast and lactic acid bacteria) that also exert health benefits for the consumers' microbiota. Consistent efforts have been made through the years aiming to understand the intertwined and complex biological, chemical, and physical reactions needed for the production of sourdough and the interactions between these processes and the final health benefits and sensorial characteristics of sourdough bread. This book aims to provide a revised vision of the multiple opportunities and research gaps that are currently being researched through the entire sourdough value chain, including research performed to understand the raw materials, sourdough microbiota, and sourdough enzymes, as well as new opportunities and technologies applied at various stages of production. Sourdough biotechnology is reviewed in detail in this book in three distinct sections.

Section I. Cereal and Non-cereal Sourdough Metabolites. This section contains four chapters covering multiple aspects of the sourdough fermentation process of cereal and non-cereal products and the main health benefits achieved during fermentation processes, as well as the potential of the use of the raw materials themselves as a source of compounds that could exert health benefits (i.e. proteins, peptides, fibers, and phenolic compounds) and the influence of sourdough fermentation on the chemical profile of these products. Moreover, this section also covers one of the main trends in the food industry aiming to improve the utilization of crops and grains by the use of innovative technologies (i.e. ultrasounds and microwaves) to develop a "green sourdough biotechnology" exploitation model. These new technologies aim to generate new food ingredients and improve the sustainability and circularity of crop production by the generation of additional value from either low-value products, by-products, or waste generated during the production and utilization of the raw materials.

Section II. Enzymes from Sourdough Cultures. This section includes five chapters exploring the main sourdough microbiota involved in the production of enzymes for the bioconversion of several macromolecules during the process of sourdough fermentation, as well as those enzymes inherently present in the grains. The influence that these enzymes have when improving the quality and shelf life of bread and also when developing new food products is also explored in detail. The section also draws attention to the role of the different enzymatic processes as biocatalysts of multiple reactions, challenges, research, development, and innovation of these products for the food and feed sectors. The section covers the main enzymes, including those for the conversion of carbohydrates and fiber, lipids, proteins, phenolic compounds, and phytates, as well as the main strategies for the identification of sourdough enzymes together with the compilation of this information in databases for a wide variety of applications. The future of these enzymes as a source of new compounds with additional health benefits, and the role of these compounds for the fortification of foods or development of new food products and nutraceuticals, is also highlighted in this section.

Section III. Innovative Applications of Sourdough Microbiota. This section contains five chapters dedicated exclusively to the wide variety of microorganisms discovered and studied in sourdough to date that influenced the development of a wide variety of sourdough breads. This includes a description of the main characteristics and implications in sourdough development of bacteria (heterofermentative or homofermentative lactic acid bacteria species) and yeast (~80 different species) discovered to date in multiple sourdough varieties. The main advantages of these microorganisms, as well as the advances for their immobilization or confinement of cells in an area while preserving the catalytical activity that they are needed for, are covered in detail together with the main advantages and disadvantages of the methods developed and insights into the future direction of research in the field. This section also highlights the relevance of sourdough microorganisms affecting the nutritional, technological, and sensorial properties of food and the possibilities to exploit them as a source of nutraceuticals for the food and/or feed industries, as these fermentation processes may improve the digestibility of nutrients and provide additional health benefits for both humans and animals.

Overall, this book covers a wide range of topics aiming to integrate current agricultural developments related to cereal, pseudocereal, and non-cereal production and processing affecting the quality of sourdough, as well as the main microorganisms and enzymes produced at various stages and in multiple sourdough types, influencing the highly appreciated sensorial attributes of this bread. The book also provides extensive and detailed information on the use of these sourdough products beyond that of bread production, emphasizing the current research and future of sourdough as a source of microorganisms and enzymes for multiple applications, tailoring the information to mainly food and feed applications, as well as the current status and future direction of sourdough biotechnology, such as the inclusion of innovative technologies for the development of a "green sourdough biotechnology" concept by focusing on improving the efficiency and sustainability of the sourdough value chain.

I hope the reading of this book inspires new generations to dig deeper into the complexity of these "alive" food products, and that it will spark the curiosity of new scientists willing to explore and develop sourdough biotechnology further, developing new fields, exploring current and new research opportunities and challenges in sourdough production, and developing new food and feed products that will benefit future generations of consumers.

Dr. Marco Garcia-Vaquero
Dublin, Ireland

About the Editors

Marco Garcia-Vaquero is Assistant Professor in the section of Food and Nutrition at the School of Agriculture and Food Science at University College Dublin (UCD, Ireland). He has broad research experience in food science research, with working experiences in Spain (University of Santiago de Compostela) and Ireland (Teagasc and UCD), where he continues his research in the development of functional foods or nutraceuticals.

Dr. Garcia-Vaquero's primary research interest focuses on the extraction, purification, and characterization of different molecules (i.e. polysaccharides, proteins, peptides, and phenols) with potential applications in the food industry from multiple agri-food products, by-products, and wastes. This research involves the use and optimization of the extraction of multiple compounds using innovative technologies (ultrasounds, microwave, pulsed electric fields, supercritical fluid extraction, and others) and bioactivity–purification guided approaches for the isolation of functional molecules with potential to be sold as nutraceuticals, pharmaceuticals, or cosmetics and health claims (anti-hypertensive, anti-oxidant, anticancer, antimicrobial, and immunomodulatory). The elucidation of structure–function relationships of the isolated molecules that could influence their use within certain food and feed matrices are also key foci of his research together with the safety (chemical hazards, such as heavy metals) and bio-accessibility of these compounds *in vitro*.

Within this area of expertise, Dr. Garcia-Vaquero has published more than 80 articles and chapters, editing multiple books and special issues in high impact factor journals; contributing also to more than 50 conferences in the areas of functional foods, food safety, and innovative technologies. His expertise in the field is also recognized by his position as an editorial board member of several high impact factor journals, an advisor for policymakers, and a reviewer for European and international funding agencies. Dr. Garcia-Vaquero has provided guidance and supervision within the area of food science and technology to multiple masters, PhD, and postdoctoral researchers, and he currently manages his own laboratory and research group within UCD, developing further research in this area through multiple projects funded by several European and Irish funding agencies.

His expertise in the field of food science, chemistry, and technology has also been recognized by his election as a MIFST (member of the Institute of Food Science and Technology), a member of the Institute of Food and Health at UCD, as well as management committee member and group leader in multiple European research networks, leading European and international researchers, and guiding the future development and direction of this research field.

João Miguel F. Rocha holds a five-year degree in Biological Engineering, specialization in Chemical and Food Technology (University of Minho), and a PhD in Food Science and Engineering (ISA-University of Lisbon). Among complementary professional qualifications are the Life Cycle Assessment (LCA) and HACCP certifications. He is the Chair and Grant Holder Scientific Representative of the COST Action SOURDOMICS (CA18101). Currently, he is also a Secretary of the General Assembly board of the National Association of Science and Technology Researchers (ANICT). His research activities were developed between Portugal and Finland (University of Helsinki) in the fields of microbiology, molecular biology, biochemistry, chemistry, chromatography, mass spectrometry, and engineering. One of his first contacts with sourdough science was during his PhD, with the microbial and lipid characterization of baking raw materials (maize, rye and wheat flours, mother-doughs, and sourdoughs) and broa – a traditional sourdough bread made of regional maize and rye flours by farmers in Northern Portugal. Previous and current research activities and interests related with cereal science and sourdough biotechnology encompass: phenotypic and genotypic screening and characterization of microorganisms; design and development of fermentation processes and starter cultures; production, extraction, and purification of functional microbial metabolites; design of innovative food products and processes and valorization of by-products and food wastes; food safety and ascertaining the health-promoting effects and environmental sustainability of food. Among others, his research activities also involved: polycyclic aromatic compounds (PAHs) in rubber crumbs from recycled tires for synthetic turf pitch infills; dietary polyphenols as prophylactic agents in food allergies; development of methodologies for the extraction, separation, and characterization of food constituents by gas and liquid chromatographic and mass spectrometry, e.g., biogenic amines, lipids, and polyphenols; characterization of biopolymers, formulation and characterization of films, edible coatings and biodegradable packaging; batch and semi-continuous fermentations to yield baker's yeast, lipids, and hydrogen; life cycle assessment of products and processes; and protein crystallization in droplet-based multiphase microfluidic bioreactors. His professional career as a researcher has always been filled with other intensive activities: Assistant Professor of several courses, *viz.* General Microbiology, Microbiology, Molecular Biology of the Cell (Genetics), General Chemistry I (Inorganic Chemistry) and II (Physical Chemistry and Kinetic Physics), Toxicology, Pharmacotoxicology, Chemistry, and Structure of Foods, Applied Biochemistry (Enzyme Kinetics), Introduction to Industrial Processes and General Biochemistry I, in the College of Biotechnology, Portuguese Catholic University (ESB-UCP) and University of Minho; and as a Trainer in courses of Environmental Chemistry, Analytical Chemistry, Laboratories of Chemistry and Basic Food Chemistry. He was also co-founder, administrator, and managing partner of a biotech start-up, the core-business of which is the processing of dairy by-products and isolation of compounds with high added value.

Contributors

Zeynep Agirbasli
Izmir Institute of Technology
Izmir, Turkey

Maria Aspri
Cyprus University of Technology
Lemesos, Cyprus

Gabriela-Elena Bahrim
Dunărea de Jos University of Galaţi
Galaţi, Romania

Sneh Punia Bangar
Clemson University
Clemson, SC USA

Miona Belović
University of Novi Sad
Novi Sad, Siberia

Norhane Besrour-Aouam
Université Tunis El Manar
Tunis, Tunisia

Vittorio Capozzi
Institute of Sciences of Food Production
National Research Council of Italy (CNR)
Foggia, Italy

Mihaela Cotârleţ
Dunărea de Jos University of Galaţi
Galaţi, Romania

Nicola De Simone
University of Foggia
Foggia, Italy

Marija Duvnjak
University of Zagreb
Zagreb, Croatia

Sedef Nehir El
Ege University
Izmir, Turkey

Yesim Elmaci
Ege University
Izmir, Turkey

Mensure Elvan
Izmir Institute of Technology
Izmir, Turkey

Elena Enachi
Dunărea de Jos University of Galaţi
Galaţi, Romania

Anca C. Farcas
University of Agricultural Sciences and
    Veterinary Medicine Cluj-Napoca
Cluj-Napoca, Romania

Pedro Ferreira-Santos
University of Minho
Braga, Portugal

Vera Fraberger
BOKU – University of Natural
    Resources and Life Sciences Vienna
Vienna, Austria

Marco Garcia-Vaquero
University College Dublin
Dublin, Ireland

Zlatina A. Genisheva
University of Minho
Braga, Portugal

Sebnem Harsa
Izmir Institute of Technology
Izmir, Turkey

Contributors

Domig Konrad Johann
BOKU – University of Natural Resources and Life Sciences Vienna
Vienna, Austria

Vanja Jurišić
University of Zagreb
Zagreb, Croatia

Sibel Karakaya
Ege University
Izmir, Turkey

Kristina Kljak
University of Zagreb
Zagreb, Croatia

María Goretti Llamas-Arriba
University of Basque Country
San Sebastián, Spain

Paloma López
Margarita Salas Biological Research Centre
Madrid, Spain

Armin Mirzapour-Kouhdasht
University College Dublin
Dublin, Ireland

Nikolina Čukelj Mustač
University of Zagreb
Zagreb, Croatia

Knežević Nada
Podravka d.d. Research and Development
Koprivnica, Croatia

Dubravka Novotni
University of Zagreb
Zagreb, Croatia

Görkem Özülkü
University of Yildiz Technical
Esenler-Istanbul, Turkey

Bogdan Păcularu-Burada
Dunărea de Jos University of Galaţi
Galaţi, Romania

Kaloyan Petrov
Institute of Chemical Engineering
Bulgarian Academy of Sciences
Sofia, Bulgaria

Penka Petrova
Institute of Microbiology
Bulgarian Academy of Sciences
Sofia, Bulgaria

Marina Pihurov
Dunărea de Jos University of Galaţi
Galaţi, Romania

Miloš Radosavljević
University of Novi Sad
Novi Sad, Siberia

João M. F. Rocha
Universidade Católica Portuguesa
Porto, Portugal

Pasquale Russo
University of Milan
Milan, Italy

Alexandru Vasile Rusu
University of Agricultural Sciences and Veterinary Medicine Cluj-Napoca
Cluj-Napoca, Romania
University of Animal Sciences and Veterinary Medicine Cluj-Napoca
Cluj-Napoca, Romania

Samaneh Shaghaghian
Food Laboratory of Pars Pajhouhan
Shiraz, Iran

Alexandrina Sirbu
Constantin Brancoveanu University of Pitesti
Valcea, Romania

Contributors

Sonia A. Socaci
University of Agricultural Sciences and Veterinary Medicine Cluj-Napoca
Cluj-Napoca, Romania

Giuseppe Spano
University of Foggia
Foggia, Italy

José A. Teixeira
University of Minho
Braga, Portugal

Claudia Terezia Socol
University of Oradea
Oradea, Romania

Monica Trif
Centre for Innovative Process Engineering (CENTIV) GmbH
Syke, Germany

Dimitrios Tsaltas
Cyprus University of Technology
Lemesos, Cyprus

Kenza Zarour
Université Oran 1
Oran, Algeria

Section I

Cereal and Non-cereal Sourdough Metabolites

1 Sourdough Fermentation as a Way to Improve Health Benefits and the Sensory Properties of Bakery Products

Sedef Nehir El, Yesim Elmaci, Sibel Karakaya, and Alexandrina Sirbu

CONTENTS

1.1 Introduction 3
1.2 Nutritional Properties of Sourdough Bakery Products.................................5
    1.2.1 Starch Digestibility 5
    1.2.2 Protein Digestibility..6
    1.2.3 Gluten Degradation...6
    1.2.4 Dietary Fiber...7
    1.2.5 Total Energy 8
    1.2.6 Bioactive Compounds, Bioactive Peptides, and Vitamins................8
    1.2.7 Undesirable Compounds...9
    1.2.8 Anti-nutritional Compounds...10
    1.2.9 Bioactivities of Sourdough Bakery Products..................................11
    1.2.10 Phytase Activity..11
    1.2.11 Exopolysaccharides 12
1.3 Sensory Properties of Sourdough Bakery Products 12
    1.3.1 Effect of Grain Type and Flour Composition.................................12
    1.3.2 Acid Production 17
    1.3.3 Sensory Attributes of Sourdough Bread...18
References..21

1.1 INTRODUCTION

During sourdough fermentation, spontaneous or inoculated selected lactic acid bacteria produce organic acids, such as lactic and acetic acids, and yeasts produce carbon dioxide and ethanol through their metabolic activity under moderate temperatures over 24 hours (Catzeddu 2019). In addition, the metabolic activity of sourdough microorganisms greatly affects the technological performance of dough and the nutritional properties, aroma profile, shelf life and overall quality of bread. Microbial metabolites, as a result of interactions between yeast and lactic acid bacteria, are responsible for changes in the activation of enzymes. Sourdough fermentation of cereals causes a significant change in their nutritional properties with both enzyme-induced changes and/or metabolites of metabolic activity (Arora et al. 2021, Gobetti et al. 2019). All these changes during sourdough fermentation can affect nutritional properties by decreasing or increasing the amounts or functionality of some compounds and by enhancing or retarding their biological use (Canesin et al. 2021). The knowledge of the nutritional effects of sourdough cereal fermentation has been reviewed in detail under the topics of reduction of starch digestibility, release of bioactive peptides from parent proteins, gluten degradation, dietary fiber solubilization, enhanced mineral bioavailability, change of levels and/or bioaccessibility of phenolics, sterols and vitamins, inhibition of anti-nutritional compounds, etc. (Papadimitriou et al. 2019).

Together with nutritional properties (Figure 1.1), the sensory properties of the sourdough affect consumers' choices. Grain type, flour composition, and acid production change the characteristics of sourdough, which affects the sensory characteristics of the bakery products. This is particularly significant as the interest of

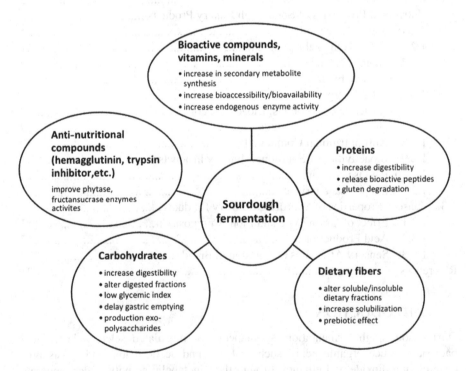

FIGURE 1.1 Nutritional aspects of sourdough fermentation.

consumers in more flavorful and healthier breads over the last decade increases day by day. Today, various new sourdough products, such as sweet baked goods, crackers, pizza, gluten-free products, and pasta, have been developed; and different flours, such as those from legumes are currently being explored. With the development of sourdough technology equipment, production in industrial bakeries has become widespread. Whether sourdough fermentation is produced with traditional or nonconventional flours (legumes and pseudo-grains) it is still a biological activity worth exploring in many new directions (Gobbetti et al. 1995, Mastilovic and Popov 2002, Vuyst et al. 2014, Ganzle 2014).

1.2 NUTRITIONAL PROPERTIES OF SOURDOUGH BAKERY PRODUCTS

1.2.1 STARCH DIGESTIBILITY

Many studies based on empirical and *in vitro* studies concluded that starch digestibility is increased by sourdough fermentation due to the presence of slow digestible starch (SDS) and rapid digestible starch (RDS) fractions and their resistance to digestive enzymes (Fois et al. 2018, Çetin-Babaoglu et al. 2020). Studies have demonstrated that sourdough fermentation has the capacity to shift the glycemic index (GI) of bread from a high to medium level (Scazzina et al. 2009, Novotni et al. 2012).

Recent reports showed that starch fractions (SDS, RDS, and resistant starch (RS)) of sourdough reduce GI. The biological acidification of bread as well as the release of peptides and free fatty acid content, and the presence of phenolic compounds and soluble dietary fiber are all factors that alter starch digestibility and lower the GI of the bread (Rizzello et al. 2019, Çetin-Babaoglu et al. 2020). In particular, acetic acid has been reported to cause a delay in gastric emptying, while the accumulation of lactic acid in the fermentation medium induces interactions between starch and gluten, during dough baking, and reduces starch digestibility. Novotni et al. (2012) investigated the influence of a 20% sourdough addition to gluten-free bread. In fact, the addition of sourdough reduces the viscous consistency of gluten-free bread and accelerates the enzymatic hydrolysis of starch, thereby decreasing GI.

The starch–protein matrix interactions, particularly those found in sorghum, result in lower digestibility than starches usually obtained from other grains, such as rice and corn. However, *in vitro* starch digestibility (IVSD) increased in sourdough bread prepared from sorghum flours. Lactic or acetic acid, which are produced during fermentation, especially can significantly lower the GI of bread. Macro and microstructure of cereal foods, gastric emptying, α-amylase binding to substrates and digestive enzyme properties are different factors affecting starch digestibility (Liljeberg and Bjorck 1996, Olojede et al. 2020). Scazzina et al. (2009) evaluated the possible effects of sourdough fermentation in whole grain bread on postprandial blood glucose levels in healthy subjects. They concluded that the presence of fiber did not influence the glycemic potential of bread, and organic acids produced during sourdough fermentation delayed gastric emptying, in contrast, without starch hydrolysis (Novotni et al. 2012, Shumoy et al. 2018, Olojede et al. 2020).

Many studies have reported a positive correlation between gastric emptying rate and perception of satiety (Rizello et al. 2019). These studies conclude that sourdough bakery products modulate the perception of hunger and satiety rather than mechanical satiety in some sourdough products. The conclusion of these studies is that sourdough bakery products modulate the perception of hunger and satiety rather than a mechanical satiety. Satiety perception can be simulated by hormonal responses such as glucagon-like peptide (GLP)-1, oxyntomodulin, pancreatic polypeptides, and glucose-dependent insulin trophic polypeptide (GIP) through metabolically active compounds released during sourdough fermentation. It has also been reported that the type and number of proteins affect gastric emptying time, satiety, and appetite perceptions. Several studies have revealed the effects of protein hydrolysis products, such as bioactive peptides, on the perception of satiety (Rizello et al. 2019, Catzeddu 2019, Canesin and Cazarin 2021).

1.2.2 Protein Digestibility

The proteolysis during sourdough fermentation starts with the action of the cereals' endogenous enzymes activated by the low pH of the fermentation medium. The role of sourdough microorganisms in proteolysis during fermentation is related to further hydrolysis of peptides into amino acids by intracellular peptidases of the microorganisms (Moroni et al. 2009, Fernández-Peláez et al. 2020). In particular, degradation of water-soluble proteins by sourdough fermentation was reported (Elkhalifa et al. 2006). In addition to proteolysis during fermentation, the digestibility of proteins in the gastrointestinal system was also reported. Generally, improved protein digestibility by sourdough fermentation was detected in many studies (Bartkiene et al. 2011, Rizzello et al. 2014). While the improved protein digestibility of products obtained from sourdough fermentation is not fully understood, possible mechanisms include biological acidification on the indirect activation of endogenous proteases of raw material, secondary proteolysis by peptidases released from lactobacilli, and, most probably, the modification of the gluten network, which becomes more susceptible to digestive enzymes (Arora et al. 2021).

1.2.3 Gluten Degradation

Ingestion of gliadins and secalins, the prolamins found in wheat and rye, respectively, caused damage to the small intestinal mucosa in people who suffer from celiac disease. This is especially true of peptides rich in proline and glutamine, which cause an autoimmune response in celiac patients and are released during endoluminal digestion of gliadins (Gänzle et al. 2008). While the cereal industry produces gluten-free (GF) products, the search for new technologies and formulations to improve the sensory and nutritional properties of gluten-free products continues. Sourdough technology has been used to produce GF products over the past decades (Olojede et al. 2020). Heterofermentative lactobacilli cause acidification and the degradation of disulfide bonds of gluten during sourdough fermentation, which promotes the activation of endogenous enzymes in cereals (Gobbetti et al. 2014, Fernández-Peláez et al. 2020). Sourdough fermentation can be convenient for the withdrawal of traces of gluten in cross-contaminated, gluten-free products and the partial hydrolysis of gluten (Rizzello et al. 2014). In addition, sourdough fermentation can degrade gluten, which is appropriate for providing gluten-free products' properties, as reported by Rizzello et al. (2007), who used fungal proteases obtained from *Aspergillus oryzae* and *Aspergillus niger* and sourdough lactobacilli selected on their peptidase systems. They observed that gluten content reduced from 74,592 to 12 ppm during long-time fermentation (48 h at 37°C). The lactobacilli used in this study were selected *Lactobacillus sanfranciscensis* strains with peptidase activity toward proline-rich peptides. The mechanism is explained as the primary proteolysis by fungal proteases and liberation of polypeptides. Then, specific peptidases, prolyl endopeptidases of microbial origin which hydrolyze peptides containing proline residues, may cleave proline-rich immune-stimulatory gluten peptides (Nionelli and Rizzello 2016). In conclusion, studies on the effect of sourdough fermentation on gluten degradation showed that sourdough technology is useful in the elimination of traces of gluten, the partial hydrolysis of gluten, renders flours free of gluten that are naturally rich in this protein network, and enhances protein digestibility due to degradation of proteins during fermentation both by proteases produced by microorganisms and endogenous enzymes of the raw materials.

1.2.4 Dietary Fiber

Today, many studies have proven the physiologically positive effects of grains on health and most of these effects have been attributed to their dietary fiber (e.g., hemicellulose, arabinoxylans, β-glucans and fructans). There is scientific evidence for dietary fiber decreasing the risk of diabetes, cardiovascular diseases, some cancers (such as colorectal), and metabolic syndrome (Kopec et al. 2011, Arora et al. 2021). Recently, Arora et al. (2021) reviewed 60 articles on the effects of sourdough fermentation on individual amount and ratio between soluble and insoluble dietary fiber fractions in various grains. They reported that sourdough fermentation increased the water-soluble dietary fraction. At the same time, although grains and pseudocereals are good sources of dietary fiber, when enriched with bran in varying proportions (5–20%), sourdough fermentation could use these bran-fortified matrices and increase the soluble dietary fraction. During sourdough fermentation, the composition and concentration of dietary fiber can change with a few intrinsic enzymes that are activated by flour hydration, as well as some enzymes produced by lactic acid bacteria and yeast metabolic activity. Mihhalevski et al. (2013) observed an increase in soluble dietary fiber content in rye sourdough bread and explained this increase as the partial transformation of soluble fiber to insoluble fiber form by activating the enzymes in the grain (α-amylase, α-arabinofuranosidase, β-glucanase, β-xylosidase, endo-xylanase, and cinnamoyl esterase) during fermentation.

On the other hand, depending on genotype, kernel maturation stage, and thermal treatments, sourdough fermentation has different effects on total dietary fiber content or the ratio between soluble–insoluble dietary fiber of breads (Saa et al. 2017). The differences between the results of the studies during sourdough fermentation could be explained by the variety of endogenous enzymes in grains, their acidification and activation under processing conditions, whole flours of milky grain-rich caryopsis in outer layers, and low degradation of insoluble dietary fiber by bacteria due to the lower release of fermentable sugar by endogenous enzymes (Gobetti et al. 2014, Saa et al. 2017, Peicz et al. 2017, Saa et al. 2018).

1.2.5 Total Energy

One of the main approaches to reduce sugar in bakeries is *in-situ* production of sweet polyols and exopolysaccharides (EPS) in a sourdough system. A reduction in caloric content is achieved by converting monosaccharides into polyols, and the remaining monosaccharides were then reduced by polymerization to long-chain carbohydrates. In particular, certain sourdough lactobacilli and yeast strains can produce polyols such as mannitol, xylitol, and erythritol, and/or EPS naturally. Polyols and EPS produced by sourdough microorganisms can be a potential novel approach to overcoming some adverse effects on the quality of foods – depending on sugar reduction – including volume, texture, structure, starch gelatinization, etc. (Sahin et al. 2019).

1.2.6 Bioactive Compounds, Bioactive Peptides, and Vitamins

Phytochemicals in cereals and pseudocereals (e.g., wheat, rye, barley, buckwheat) are flavonoids, alkylresorcinols, lignans, phenolic acids, tocols, benzoxazinoids, betaines, folates, phytates, and sterols (Gobetti et al. 2019). Many of these chemicals have important functions, such as antioxidant activity for the grain. The health benefits of these compounds in humans such as antimicrobial, antidiabetic, and anti-inflammatory activity depend on their quantity, bioaccessibility, conjugation, and food matrix. The levels and bioavailability/bioaccessibility of these compounds increase up to seven and ten times during fermentation. These effects can be explained by several mechanisms. Lactic acid bacteria show high tolerance to the antimicrobial effect of phenolic compounds. This resistance refers to metabolic activity that increases the capacity of lactic acids to convert phenolic acids into their metabolites. Secondary metabolites are produced by complex metabolic pathways used by bacteria during fermentation. Also, phenolic compounds exist in soluble form within the cytoplasm or bind covalently to the plant cell wall. Glycosyl hydrolases, produced as a result of the metabolic activity of lactic acid bacteria, release phenolic compounds from the food matrix by breaking down the plant cell wall and depolymerizing the high molecular weight phenolics (Gobetti et al. 2019, Ripari et al. 2019). This determined activity or released level of phytochemicals in a chemical system does not show high bioaccessibility in digestion systems. There have been reports on the limiting or enhancing factors of sourdough fermentation on *in vitro* phenolic bioaccessibility. Ferulic and sinapic acids were found to be more bioaccessible *in vitro* in sourdough bread compared to the control bread (Irakli et al. 2019), while sourdough breads with high fiber and protein content inhibited the activity of the digestive enzymes and impaired the bioaccessibility of bounded phenolics (Saa et al. 2017). Sourdough fermentation together with thermal processing affected the transformation of phenolic compounds. Luksic et al. (2016) reported that the levels of rutin in buckwheat were broken down and transformed during the process of sourdough fermentation into quercetin. The rutin molecule consists of quercetin and sugar parts, where the bacteria use sugars as an energy source after the enzymatic degradation of rutin. During the degradation of rutin and other substances by bacteria, the concentration of rutin decreased and quercetin (free and bounded to rutin) increased. Mihhalevski et al. (2013) determined the stability of B-complex vitamins (thiamin, riboflavin, nicotinic acid, nicotinamide, pantothenic acid, pyridoxine, and pyridoxal) in rye sourdough bread. The retention of vitamins in sourdough products varied, depending on the acidic or alkaline conditions of bacterial activity, the contribution of endogenous yeast vitamins, hydrolyzes the degree of chemically bound forms such as nicotinic acid and baking parameters. It has been reported that sourdough fermentation increases the content of many vitamins in grains. However, the baking process can affect the overall retention of vitamins. The thiamine content in wheat flour increased, especially after prolonged sourdough fermentation, but decreased even during the short baking process. Losses in tocopherol and tocotrienol contents were observed as a result of contact with air and during cooking. In addition, many studies have reported that yeast fermentation increases the folate content more than three times. When comparing the ability of different yeasts and lactic acid bacteria to affect the folate content in rye sourdough, it was reported that the effects of sourdough bacteria were minimal (Poutanen et al. 2009).

Lactobacilli can produce bioactive compounds including amino acid derivatives (e.g., γ-amino butyric acid), prebiotic exopolysaccharides, and bioactive peptides. Studies have reported that the formation of antioxidant peptides (Coda et al., 2012) and antihypertensive peptides (Rizzello et al. 2008, Zhao et al. 2013) takes place during sourdough fermentation of various cereal flours. Diowksz et al. (2020) reported the ACE inhibitory activity of sourdough bread before and after *in vitro* digestion as 93% and 59%, respectively. An anti-tumoral peptide, lunasin, and lunasin-like polypeptides were also determined after the fermentation of cereals flours and different legume species by sourdough lactic acid bacteria, respectively (Rizzello et al. 2012, Rizzello et al. 2015).

1.2.7 Undesirable Compounds

The most important Maillard reaction products that occur during food processing are furosine, 5-hydroxymethylfurfural, acrylamide, heterocyclic amines, advanced glycation end products, and melanoidins. The formation of the Maillard reaction products is desirable for the occurrence of flavor and aroma compounds. However, harmful effects of acrylamide, heterocyclic amines, and 5-hydroxymethylfurfural on human health (especially acrylamide) have been well documented (ALjahdali and Carbonero 2019). Studies reported that sourdough fermentation may prevent the formation of these compounds. Different lactic acid bacteria (LAB) combinations are able to reduce the formation of acrylamide in sourdough breads most probably due to direct and/or indirect effects of proteolysis and acidification on both sugar and amino acid composition, and accordingly, the inhibition of acrylamide. Most of the LAB strains are able to ferment L-arabinose, D-ribose, D-galactose, D-fructose, and D-maltose, which are the substrates of the acrylamide reaction (Bartkiene et al. 2017). However, Diana et al. (2014) reported that the acrylamide content of 12 artisan sourdough breads was higher than those of commercial breads. Acrylamide formation has been associated with free amino acids released during sourdough fermentation. Ertop and Sarıkaya (2017) demonstrated that the hydroxymethylfurfural (HMF) content of sourdough breads made with wheat flour and bran (156.95 mg/kg), wheat flour (105.74 mg/kg), and whole wheat flour (132.94 mg/kg) was higher than bread fermented with bakery yeast (23.74 mg/kg). The crust of bread especially contains the highest amount of advanced Maillard reaction products (Michalska et al. 2008).

Although LAB fermentation is generally considered non-toxic, biogenic amines (BA) produced during fermentation are an important safety concern (EFSA 2011). Both increase and decrease in BA levels during sourdough fermentation were reported (Bartkiene et al. 2011, Diana et al. 2014). The different BA levels have been associated with the raw material and microorganisms used in the fermentation process. When raw material rich in protein is used for fermentation, the level of BA increases at the end of the fermentation compared to the raw material with low protein content. The effect of LAB on BA levels is most probably related to their decarboxylase activity (Bartkiene et al. 2017). Bartkiene et al. (2011) indicated that the use of decarboxylase-negative LAB can reduce BA levels in fermented foods.

1.2.8 Anti-nutritional Compounds

Sourdough fermentation increases the amount or bioavailability of many components and decreases anti-nutritional compounds, such as trypsin inhibitors, raffinose, condensed tannin, and phytic acid, responsible for reducing the digestibility of most nutrients and the absorption minerals (Montemurro et al. 2019, Arora et al. 2021). Curiel et al. (2015) evaluated the potential of 19 legumes in sourdough fermentation. They recorded that lactic acid bacterial fermentation enzymes improved the nutritional value of the samples by decreasing the amount of condensed tannins and raffinose. This partial degradation by enzymes, such as fructansucrase activity of lactic acid bacteria of raffinose – which are not digested in the human intestine – is important for the control of the dose-dependent beneficial and adverse effects of raffinose in foods. The bonding of phytic acid – which is naturally found in grains – with dietary minerals, starch, and proteins reduces the bioaccessibility of these important nutrients in terms of nutrition. Also, trypsin inhibitors are found in high amounts in many grains and legumes and impair protein absorption by inhibiting the enzymes responsible for protein digestion. Sourdough fermentation is an effective process such as soaking, germination, and cooking to reduce these anti-nutritional compounds. The breakdown of phytate by increasing endogenous phytase enzyme activity of the grain and the role of lactic acid bacteria within its microbial population of sourdough fermentation have been reported with many studies (Gobbetti et al. 2014). It has been shown that phytase activity is more effective than sourdough bacteria in the reduction of phytic acid, but the decrease in pH during the fermentative metabolism of sour yeast bacteria provides favorable conditions for the endogenous phytase activity of the grain (Canesin and Cazarin 2021).

1.2.9 Bioactivities of Sourdough Bakery Products

It has been reported in many studies that sourdough fermentation of grains increases antioxidant activity (Curiel et al. 2015). These studies explain that antioxidant properties are in the form of inactivating reactive oxygen species or preventing oxidative stress. The reasons for this increase in antioxidant activity during sourdough fermentation are the release of more soluble phenolic compounds from the food matrix due to enzymatic activity and acidification of LAB and a decrease of soluble proteins to low molecular weight peptides due to the proteolytic enzyme activity. Also, heat treatment applied after fermentation causes denaturation and aggregation in soluble proteins; thus smaller peptides (1–5kDa) and free amino acids with a strong antioxidant activity are released (Nissen et al. 2020, Luti et al. 2020). The activities of these peptides are not limited by their antioxidant capacity. Many studies show that bioactive peptides produced by enzymatic hydrolysis of parent proteins have different biological roles in human health, including antimicrobial, anti-proliferative, cholesterol and blood pressure lowering, antithrombotic, and enhancing mineral bioaccessibility (Galli et al. 2018, Luti et al. 2020). However, there is no consensus in databases and literature on the sequence of peptides showing various bioactivities, and a structure/function relationship with many peptides of different molecular lengths released after fermentation has not been outlined (Luti et al. 2020).

1.2.10 Phytase Activity

Phytases, such as phosphatases, hydrolyze phytate into phosphoric acid and myo-inositol phosphates. Phytases are particularly crucial in human nutrition since they can degrade phytate during food processing and also gastrointestinal digestion (Haros et al. 2005, Sharma et al. 2020). Sourdough technology can eliminate the phytate of whole wheat bread. Phytase-positive LAB are preferred notably in whole wheat bread production due to their ability to increase mineral bioavailability (Cizeikiene et al. 2020). In the last decade, studies on the determination of residual content of phytic acid or the bioavailability of minerals in doughs and breads fermented by sourdough microorganisms were reported (Arora et al. 2021). Rodriguez-Ramiro et al. (2017) demonstrated that the bioaccessibility of iron in sourdough bread was 12% higher than that reported in conventional bread (1.3–1.4%). In addition to LAB and yeast phytase activity, acidification during sourdough fermentation stimulates endogenous grain phytase (Fekri et al. 2020). Although the majority of LAB produce intracellular phytases, reports of LAB expressing extracellular phytase have been recently published (Karaman et al. 2018, Cizeikiene et al. 2020). Furthermore, LAB isolated from sourdough were more effective in terms of extracellular phytase activity compared to yeast isolates (Karaman et al. 2018). Cizeikiene et al. (2020) reported that *Lactobacillus delbrueckii* ssp. *bulgaricus* MI was suitable for manufacturing bread with reduced phytate content. In conclusion, LABs are "Generally Recognized as Safe" (GRAS) for humans and their health benefits are well documented. Therefore, phytase-producing LAB can be safely used for combating micronutrient deficiencies in plant-based food. Recently, reports on LAB producing extracellular phytases have increased their significance in the production of phytate-free or phytate-reduced foods. Moreover, further studies on the isolation of efficient phytase-producing LAB and their commercial availability are needed.

1.2.11 EXOPOLYSACCHARIDES

Exopolysaccharides (EPS) are metabolic products produced by sourdough LAB during dough fermentation. The most outstanding EPS-producing LAB are *Lactobacillus, Lactococcus, Bifidobacterium, Leuconostoc, Pediococcus, Streptococcus, Enterococcus*, and *Weissella sp.* (Angelin and Kavitha 2020). Although sourdough LAB can produce homo- and heteroexopolysaccharides, the amount of synthesized heteroexopolysaccharides is insignificant (mg/ml) compared to the synthesized amount of homoexopolysaccharides (g/L). While heteroexopolysaccharides consist of different monosaccharides, such as glucose, fructose, galactose, and rhamnose, homoexopolysaccharides are either glucan or fructan polymers. Glucans produced by glucansucrases from LAB are dextran, mutan, alternan, and reuteran; and the fructans produced by fructansucrases from LAB are levan and inulin (Tieking and Gänzle 2005, Bounaix et al. 2009, Riaz Rajoka et al. 2020). Lactobacilli EPSs can reduce cholesterol levels, interfere with pathogen adhesion, and possess anti-tumor, anti-HIV, and immunomodulatory activities. The pro-inflammatory and anti-inflammatory properties of the EPSs are strongly related to their structure. For instance, the EPSs with negative charges most likely have more pro-inflammatory activity than the neutral EPSs. An anti-inflammatory effect has been shown for EPSs comprising phosphate groups, sulfate groups, and uronic acid, as well as for high molecular weight EPSs (Riaz-Rajoka et al. 2020). One of the health promoting impacts of EPS has been associated with their prebiotic effect. Exopolysaccharides, such as glucan, fructans, gluco- and fructo-oligosaccharides, produced by LAB can be metabolized by gut microorganisms. The end products of this metabolism are acetate, butyrate, and propionate. Propionate has several beneficial effects including a reduction in cholesterol and triglyceride levels and an increase in insulin sensitivity (Galle and Arendt 2014, Fernández-Peláez et al. 2020). The beneficial effect of EPS is strongly related to their composition such as the type of monosaccharides they contain, glycosidic linkage, chemical modification, etc. (Angelin and Kavitha 2020). In conclusion, EPS and oligosaccharide formation during sourdough fermentation have favorable advantages concerning industrial as well as therapeutic applications in terms of antioxidant, antimicrobial, and immunomodulatory activities.

1.3 SENSORY PROPERTIES OF SOURDOUGH BAKERY PRODUCTS

1.3.1 EFFECT OF GRAIN TYPE AND FLOUR COMPOSITION

Due to increased consumer demands for healthier foodstuffs, a huge diversification of bakery products has arisen and increased the interest for a combination of recipes, fermentation methods, and enlargement of the raw material base. The variety of cereals and flour types are changing constantly, and new ingredients are used especially for dietetic and functional baked foods.

The sourdough method is an ancient, but still insufficiently explored, method used in bakery-making, to improve nutritional and sensory characteristics of baked products as well as their safety and shelf life. Different kinds of liquid or/and dried sourdough, the mix of starter cultures, associations of lactobacilli with or without yeast, etc., are used for sourdough bread preparation. In order to analyze microbiota as a whole (lactic acid bacteria competition, or the effect of the combination of LAB and yeasts), various studies took into account different raw materials as substrates and different intrinsic properties or conditions. Nowadays, bakery formulations are tailored using either one type of flours or combinations of cereals, pseudocereals and/or legumes. When discussed on grain type and flour composition in sourdough bread, there are two distinct stages: the first addresses sourdough formulation, where flour is the specific substrate for spontaneous microflora or/and after microbial inoculation for microbiota growth (e.g., LAB) and sourdough fermentation, the second refers to the typical ingredient as the raw material in a recipe for bread-making. For instance, bread is made from rice flour, but the substrate for a specific microbiota of sourdough is ensured by wheat flour. That also means the sourdough fermentation makes its mark on dough properties (e.g., rheology, sensory) at all stages of bread-making, during the preparation of sourdough itself, and in the next technological stages of processing the bread dough with the addition of sourdough.

Regularly, sourdough bread is made of wheat and rye flours because wheat and rye have bread-making properties different from other grains. However, cereals used in bread-making are wheat, rye, rice, maize, barley, oat, sorghum, millet, emmer, and teff; while the most known pseudocereals are buckwheat, amaranth, and quinoa (Sirbu 2009, Coda et al. 2014). Besides cereal and pseudocereal flours, sourdough bread can be prepared with an addition of milling by-products (e.g., germ and bran), starchy tubers (e.g., potatoes, Jerusalem artichoke), cassava roots, soy derivatives, beans, lentils, etc. In this regard, various recipes take into account white, dark, or whole grain flour; soft or durum wheat flours, non-wheat cereal and pseudocereal flours; flour blends of cereals, pseudocereals, and/or legumes, etc., for sourdough bread-making (Salovaara and Valjakka 1987, Katina et al. 2006b, Schober et al. 2007, Vogelmann et al. 2009, Poutanen et al. 2009, Zannini et al. 2009, Huttner et al. 2010, Rieder et al. 2012, Baye et al. 2013, Rizzello et al. 2014, Rinaldi et al. 2015, Rizzello et al. 2016, Omedi and Huang 2016, Bender et al. 2018, Gobbetti et al. 2020, Teleky et al. 2020).

Concerning the substrate, grain type, flour composition, and quality all influence cereal fermentation and microbiota development with effects on the quality of baked end products. The role of cereal flours in the fermentation process, during their mixing with water and dough development, is already recognized. Dough composition (i.e., moisture content, mono- and disaccharides, starch, soluble and insoluble non-starch polysaccharides composition, amino acids, peptides and protein content, levels of vitamins and minerals, endogenous enzymes, etc.) offers suitable conditions that can support microbiological growth and fermentation processes.

Other environmental factors are related to water activity, pH, and different chemical compounds, such as preservatives, salts, etc.

In fact, besides the chemical composition of flours related to their source or nature (i.e., cereals, pseudocereals, or other origins), another factor that influences the quality of sourdough bread is flour extraction rate. The milling process destroys the grain integrity and changes the ratio between anatomical layers of the kernels. By modifying the flour extraction rate, the protein content and gluten/protein quality (with consequences on rheological and technological behavior of dough, too), water absorption, carbohydrates (starch, fiber, and sugar ratio), lipids (incl. fatty acids), vitamins, mineral matter (ash content), and other compounds (e.g., acids, flavonoids, tannins, etc.) are changed. Consequently, the crushing degree influences the biochemical composition of the flour as well as the bioavailability of its nutrients, making flour a more susceptible substrate for enzymes, including microbial ones.

Wheat flours remain the most appreciated ingredients for bread-making from the sensorial standpoint. However, to balance or improve the nutritional composition, the addition of alternative cereals and pseudocereals has been accomplished (Rollan et al. 2019). Although pseudocereals and other cereals have a better nutritional profile in terms of protein content and composition (including essential amino acids), their use is limited to a certain extent because of poor baking quality (as dough rheology, technological properties, and sensory quality of end products). Regarding protein fractions, some cereals (e.g., rice) and pseudocereals (i.e., buckwheat, amaranth, and quinoa) lack gluten allergen factors and are successfully used in gluten-free product formulations. Some alternative raw materials other than wheat flours (e.g., teff, super sorghum, pseudocereals) also come with higher levels of minerals.

Meroth et al. (2004) observed that the microbiota composition of the sourdough is strongly influenced by the fermentation process and starter addition. Fermentation parameters and flour characteristics exercise various effects on microbiota activity. During fermentation, many biochemical and microbiological changes occur due to the action of endogenous enzymes of flours and the microbial enzymes in sourdough.

The conversion of the flour compounds (i.e., carbohydrates, proteins, lipids, and others) by enzymes during dough processing determines the bread quality. For example, starch modifications contribute to the texture of the bread "crumb" and staling of bread; generation of oligosaccharides and other reducing sugars can secure flavor precursors; changes in dietary fibers influence their prebiotic effect. Flour protein conversion influences the rheological behavior of dough and its machinability, bread loaf volume, bread taste and flavor, and nutritional value. Lipid transformation in dough contributes to bread crumb texture, taste, flavor, and shelf life. Other bioactive compounds have different roles, as antioxidants, against anti-nutritive factors, or selective antimicrobial effects.

The activity of endogenous enzymes of flours is interdependent with the metabolism of sourdough microbiota in the conversion of flour compounds (Ganzle 2014). Carbohydrates play an essential role, as substrate, during the sourdough fermentation process because of active enzymes that break the linkages and release fermentable sugar (e.g., glucose and maltose) from polysaccharides and later from maltodextrin. The type of flour (i.e., its composition) and process conditions, dough hydration, dough yield, and fermentation time influence enzymatic hydrolysis (Gobbetti et al. 1995, Mastilovic and Popov 2002). Also, fermentation temperature, redox potential (oxygen), water activity, pH, and sodium chloride level are factors controlling the enzymatic activity in the fermentation process (Hammes and Ganzle 1998, Vuyst et al. 2014). Ganzle (2014) commented that oxygen consumption, acidification, and thiol accumulation by LAB metabolism modulate the solubility of substrates and activity of flour enzymes. At the same time, conversion of flour constituents under actions of the endogenous enzymes of cereal flours results in mono- and oligosaccharides, peptides, amino acids, and others (e.g., short-chain aldehydes) that provide carbon and nitrogen sources as well as regeneration cofactors for LAB growth and ensure metabolic energy (Ganzle et al. 2007).

According to cereal type and flour extraction rate, starch and carbohydrate metabolism follows different pathways due to flours' composition in starch and other saccharides, as well as endogenous amylases and glucoamylase activities. As Ganzle (2014) pointed out, maize, sorghum, cassava, potatoes, and pearl millet compared with wheat and rye have lower amylase activity, and starch degradation and conversion of maltodextrin in sourdough depend on extracellular amylases of LAB. As maltose is the main carbon source for many sourdough lactobacilli (e.g., *L. sanfranciscensis*, *Lactobacillus fermentum*, etc.), they are highly adapted for maltose metabolism (e.g., dextran glucosidase (DexB), which contributes to carbohydrate conversion and is widespread in genomes of lactobacilli). Degradation of arabinoxylans in wheat and rye sourdough seems to be secured by cereal enzymes. However, EPSs are produced by cereal-associated microbiota, the polymerization process being strain-dependent – for example, lactobacilli were screened for their different abilities to produce EPS, namely, either fructans (e.g., *L. sanfranciscensis*, *Lactobacillus frumenti*) or glucans (e.g., *Lactobacillus reuteri*) (Sahin et al. 2019). The formation of EPS by LAB in sourdough contributes to bread quality in terms of its loaf volume, crumb texture, dietary fiber content, sugar reduction, and anti-staling effect (Ganzle 2014, Torrieri et al. 2014, Sahin et al. 2019).

Protein modification is another key factor that influences the quality of sourdough during fermentation since the proteins (e.g., gluten proteins) determine the rheological behavior of dough and affect the overall quality of baked end products. The protein substrate is defined by the type, content and quality of the cereal proteins and their metabolites. Protein structure, associations between proteins, and their interactions with non-protein constituents during dough formation strongly direct the protein digestibility. Biochemical conversion of protein fractions is initiated through proteolysis depending on endogenous proteases of cereal flours. Later, the degradation process of proteins during sourdough fermentation, and accumulation of amino acids, peptides, and other metabolites in the dough are influenced by enzymes of sourdough microbiota, too. Certainly, the presence of antinutrient factors (such as protease inhibitors) affects protein bioconversion. Ganzle et al. (2008) reviewed the factors that influence the proteolysis process of proteins (i.e., glutenins and gliadins, and secalins) in wheat and rye sourdough, the microbiota of sourdough consisting of yeasts and lactic acid bacteria. They have concluded that pH and disulfide bonds' reduction by heterofermentative lactobacilli increases the accessibility of the protein substrate and activity of cereal proteases, amino acids accumulating by the action of intracellular peptidases of strain-specific lactobacilli. Substrate type and quality represent an important factor for the competitiveness of LAB in sourdough since specific strains have different abilities to adapt to a certain substrate. Consequently, scanty changes in substrate composition may influence the sourdough microbiota as a whole. In this regard, studies on LAB genomes and environmental conditions showed different competitive behavior between *Lactobacillus* species due to amino acids and peptides composition of the substrate (Vogelmann et al. 2009, Ganzle 2014). For example, Rizzello et al. (2014) found *Lactobacillus plantarum* as the dominant lactobacilli species in wheat sourdough with the addition of legume flour (chickpea, lentil, and bean). Vuyst et al. (2009) presented some metabolic pathways of carbohydrate and protein conversions by the exploitation of sourdough microbiota during the sourdough fermentation process, considering the microbiota diversity.

Lipids, phenolic compounds, and vitamins are minor components of cereals and pseudocereals, and their content varies with raw material type and flour extraction rate (higher in germ and bran). During sourdough fermentation, specific enzymes affect their metabolism. Similarly, there is a microbial enzymatic specificity for the substrate that contributes to LAB selection. Although present in trace quantities, some degradation products of these minor compounds are bioactive, and they can improve bread quality to a certain extent. For example, Adebo and Medina-Meza (2020) reviewed phenolic compound metabolism and observed some bioactive compounds (e.g., catechin, quercetin, etc.) were released in whole grain sourdough during co-fermentation with selected *Lactobacillus* strains by many metabolic pathways, depending on the type of cereal (i.e., wheat, rye malt, sorghum, millet, and rye) and combination of LAB species. Fernandez-Pelaez et al. (2020) mentioned relevant vitamins (e.g., vitamin E and folates) in cereal and pseudocereal products and noticed sourdough fermentation as a tool for the increment of vitamin content in bakery foodstuffs.

Mineral or ash content depends on the type of cereal grain and increases with flour extraction rate. Many metal ions, although in trace amounts, are enzyme cofactors. Mineral bioavailability is reduced by complexation of phytate with cations, but during fermentation at pH values below 5.5 under the endogenous grain, phytase generates the hydrolysis, and minerals sufficient for LAB growth are released (Hammes et al. 2005). Consequently, flour type or extraction rate is also related to differences in buffering capacity and acid production that influence cereal enzyme activity, microbial growth of sourdough, and dough properties (Lorenz and Bruemmer 2003, Clarke and Arendt 2005, Arendt et al. 2007).

Vogelmann et al. (2009) investigated the adaptability of yeasts and LAB to sourdough made from flours of different cereals, pseudocereals, and cassava, and they found some strains (e.g., *Lactobacillus helveticus*) were widely spread in many kinds of substrates, while others were restricted to a few substrates. Also, they asserted that LAB competitiveness in sourdough is generated either by the presence of substances with antimicrobial effects, such as tannins in amaranth, saponins in quinoa, and rutin in buckwheat, or by specific enzymatic equipment of lactic acid bacteria. Similarly, Ganzle (2014) commented on the conversion of phenolic compounds and the antimicrobial effect of some phenolic compounds in relationship with LAB selection during fermentation of sorghum sourdough.

In conclusion, grain type and flour composition are of utmost importance because this raw material represents an important source of nutrients (carbohydrates, proteins and amino acids, fatty acids, vitamins, minerals, and other growth factors) and energy for sourdough microbiota. The amount (ratio) and quality of the substrate, as well as the endogenous enzymes play a key role in sourdough fermentation and determine the biochemical and microbiological conversion of the main compounds in the dough with nutritional and sensorial effects on the overall quality of the sourdough bakery products. The knowledge on metabolic pathways and metabolite kinetics of the fermentation in sourdough allows optimization of the fermentation sourdough process to improve bread quality. Depending on its biochemical composition and own microbial load, flour substrate influences LAB and yeast dynamics and their specific fermentation associations in sourdough; and further, during one or multiple stages of sourdough bread-making, they may contribute to the quality improvement of the bread in terms of sensory, nutritional, and shelf-life attributes.

1.3.2 Acid Production

Sourdough bread is categorized as a high-acid bakery product based on its technological features. In general, sourdough bread has a pH between 4.4 and 4.92, depending on sourdough microbiota and amount used and flour extraction rate (Salovaara and Valjakka 1987, Martinez-Anaya et al. 1990). Acidification and pH are factors controlling the biological processes undergone by cereal enzymes, interdependent with microbial metabolism in sourdough. During sourdough fermentation, several types of acids are produced, mainly lactic acid (as a major component) and acetic acid; the acid profile yielded by different microbial strains is influenced by the wheat flour type (Salovaara and Valjakka 1987), LAB diversity (De Vuys et al. 2002), and other technological parameters. Galal et al. (1978) found that the total titratable acidity of studied sourdough consists of lactic acid, acetic acid, and six other minor acids (i.e., propionic, isobutyric, butyric, α-methyl n-butyric, isovaleric, and valeric acids). Lactobacilli produce more acids in sourdough than yeast, and collections of LAB species with or without yeasts have a synergistic effect on acid production (Martinez-Anaya et al. 1990), lactic/acetic acid ratio, and drop the sourdough pH during the fermentation process. Changes in environmental conditions, regarding the pH decrement of sourdough contribute to microbiota selection. Thus, the microbial population, which is adapted and strain-dependent for catabolism of carbohydrates and sourdough fermentation, has specific tolerance to the amount and type of acids (Ganzle et al. 2007, Casado et al. 2017).

Acid production causes the pH to decrease to 4.0 or less, at a level that changes the metabolic pathways in different flour substrates over sourdough fermentation. For instance, reducing the pH below 5.0 in wheat sourdough favors the action of certain endogenous cereal proteases and modifies protein composition and fractions of grain sourdough. Except for proteolysis, acidification supplies optimal conditions for other grain endogenous enzymes (as phytases) and controls many enzymatic activities

(Gobbetti et al. 1995, Clarke and Arendt 2005, Fernandez-Pelaez et al. 2020). Due to the synergic metabolism of LAB in sourdough at pH below 4.0, many microbial enzymes (e.g., phytase and aspartic proteases) become active, having final effects on the constituents' bioavailability and nutrition of sourdough bread (Huang et al. 2020, Fernandez-Pelaez et al. 2020).

The impact of pH on dough structure and bread properties has been studied for over a century. Assessment of dough structure and rheological properties revealed changes in protein association and aggregates, protein folding inwards or unfolding, interactions and bonds between protein fractions, non-protein constituents, etc. As Arendt et al. (2007) pointed out, the source of acidification influences the rheological properties of dough and alters the gluten matrix in wheat sourdough. As the effect of interactions between gluten proteins, starch and other compounds of wheat dough, by decreasing pH through sourdough fermentation, have registered changes in wheat dough properties as regards the swelling of gluten, extensibility and elasticity, firmness, viscosity, degree of softening, etc. (Clarke and Arendt 2005, Casado et al. 2017). But the technological effect of pH in sourdough made from other grain flours differs when compared to those of wheat due to differences in the flour substrate, mainly the protein type and quality. For instance, water-binding and gas-retaining properties in rye sourdough are not related to gluten proteins but to pentosan swelling and carbohydrate metabolism at an optimal pH of 4.9, etc. (Hammes and Ganzle 1998).

Through protein digestion at low pH sourdough, some amino acids are released in the dough, and during the baking process, either through Maillard reactions or through decarboxylation and transamination reactions, more flavor compounds are developed in the loaf's bread crumb and crust. In this respect, the sensorial quality of sourdough bakery products in terms of taste, flavor, texture, and bread crumb elasticity are explained over different mechanisms following various metabolic routes. However, analyzing the wheat sourdough bread, Galal et al. (1978) observed the total titratable acidity of bread was reduced during the baking process, while the pH of the end product was not significantly altered.

Hammes and Ganzle (1998) mentioned the contribution of acid formation to leavening and the delaying of bread staling as additional effects of pH dropping in sourdough bread-making. The acid production supplies a competitive selection of microbial strains with an antifungal effect that preserves bread against spoilage or undesirable fermentation and extends the shelf life of baked end products (Hammes and Ganzle 1998, Pepe et al. 2003).

1.3.3 Sensory Attributes of Sourdough Bread

Sourdough bread and yeasted bread differ from sensory and nutritional points of view. The prominent flavor and texture of sourdough bread influence its sensory properties. The intense flavor, acidic taste, and compactness of the crumb clearly distinguish sourdough bread from others (Catzeddu 2019). Ingredients used in bread-making, enzymatic reactions caused by yeasts and LAB, and the thermal reactions during baking affect the flavor of sourdough bread as well as its texture (Papadimitriou et al. 2019). The fermentation of maltose in flour by LAB by homofermentative and heterofermentative metabolism influences dough properties. Lactic acid produced by homofermentative metabolism, and lactic acid, CO_2, acetic acid, and ethanol produced by heterofermentative metabolism affects the flavor and texture attributes of sourdough bread. Production of acetic acid causes gluten to harden, whereas lactic acid causes more elastic gluten, influencing the texture of the bread. Acidity influences the gluten network causing softness, improving the extensibility of the dough and retention of CO_2. Long fermentation times and whole flours used in bread-making improves the flavor of sourdough bread, whereas, it influences the texture and volume of the bread negatively. Long fermentation time causes excessive acidity, and the result is a soft, less elastic dough with a reduced loaf volume and an increase in staling and bread firmness. The amino acid formation by fermentation affects the taste of the sourdough bread. Since yeasts use more amino acids for their metabolism, flavor of bakers' yeast bread is determined to be less than sourdough bread (Catzeddu 2019).

The high volatile content of sourdough bread improves sensory attributes compared to bakers' yeast bread. This indicates that acidification of the dough is not the only factor for the flavor of the bread. The formation of volatile compounds takes place in sourdough bread by the LAB fermentation process about 1–24 hours. As stated by Plessas et al. (2008), a complex aroma matrix was determined in sourdough breads obtained by *Kluyveromyces marxianus* and *L. delbrueckii subsp. Bulgaricus*. Plessas et al. (2008) also confirmed that the use of mixed starter culture (*Lactobacillus acidophilus* and *Lactobacillus sakei*) influenced the shelf-life of the bread positively, related to high lactic acid concentration with respect to other breads. The aroma, flavor, texture, and overall quality of this bread reported to be better than the other breads studied by Plessas et al. (2011). Acids, alcohols, aldehydes, esters, ketones, and pyrazines are the most cited chemical classes which are responsible for the flavor of bread, and these compound groups are in charge of the alcoholic, fruity, malty–acidic, and roasted odor of sourdough bread. In addition to volatile compounds, glutamate and glutathione generated during dough fermentation enhance the taste of sourdough bread, since they are active taste compounds giving umami taste (Papadimitriou et al. 2019, Petel 2017).

Sour characteristics of the sourdough bread were determined to be one of the important sensory attributes. The sour taste of the bread was not affected by the acetic acid content and fermentation phase, whereas pH of the bread correlates with sour taste. Savory or nonvolatile compounds, volatile compounds, and texture influence the sensory perception of bread. The savory compounds responsible for the sour taste of bread are usually lactic acid and acetic acid produced during dough fermentation. The addition of sodium chloride to the dough, and formation of glutamic acid during fermentation affect the taste and can change sour taste perception. The interaction of savory compounds and volatile compounds in the mouth together with their texture influenced the sourness, as well as the other sensory attributes of the sourdough bread (Rehman et al. 2006, Clement et al. 2020). LAB in sourdough bread improves the specific volume of bread as well as reducing the rate of staling. The metabolic products produced by sourdough LAB, such as organic acids, EPS, and enzymes were determined to be effective in the texture and staling of the bread. Production of the acids leading the pH to drop, increases the activity of amylases and proteases, which have a positive influence on staling of the bread (Arendt et al. 2007).

LAB also contributed to the sensory quality of bread by influencing the volatile formation during baking. The use of different flour or addition of different ingredients, such as fructose or citrate, to the dough affects LAB (Gobbetti 1995). The use of different flours in sourdough affects the flavor of dough which in turn affects the flavor and sensory properties of bread. Petel (2017) reported 68 and 82 volatile compounds in rye and wheat sourdough breads, respectively. Further studies should be conducted on sensory attributes of sourdough bread processed with different flours to consider the difference between sensory profiles. Callejo et al. (2015) stated the specific aroma attributes of sourdough bread as wheaty, dark beer, butter-like, and toasted grain in a study on the sensory attributes of bread.

Sourdough bread is made by at least two or more stages of processing to develop a certain microbiota. Sourdough microbiota is a collection of LAB with or without yeasts that may interact through various metabolic routes. Description of the sourdough microbiota and their usual applications are out of the scope herein, and these topics will be addressed in relevant chapters. However, the effect of the combination of LAB or LAB and yeast metabolites on the sensory attributes of bread should be emphasized. Different LAB starters affect the sensory profile of the sourdough bread. *Lactobacillus Brevis* or *L. Plantarum* were determined to intensify the roasted, fresh, pungent flavor, overall flavor, and aftertaste. The fermentation time, temperature, and flour quality also influence these attributes and the specific volume of bread (Katina et al. 2006). In a study by Rehman et al. (2006), wheat bread crumb produced with heterofermentative *L. Sanfranciscensis* informed to have a pleasant, mild, sour odor and taste, and sourdough bread produced by homofermentative *L. Plantarum* informed to have a metallic sour taste. The addition of *Saccharomyces Cerevisiae* enriched the aromatic flavor of sourdough bread (Rehman et al. 2006). The use of LAB and yeast in combination determined to affect the flavor of sourdough bread, and the desired and undesired sensory attributes of combination bread were determined to be less than the LAB fermented sourdough bread. LAB and yeast together influence carbon dioxide retention and affect the softness and volume of the bread positively. Xu (2020) indicated that the acidification of the bread by microbial metabolism improves the bread quality, and the use of mixed cultures affected the aroma and taste attributes of the bread in controlled conditions. Long time fermentation at low temperatures (10°C) contributed to more aroma and taste formation. Also, bread texture improved by the use of mixed cultures at low temperatures because of the LAB's forming polysaccharides at low temperature/long fermentation time. On the other hand, it has also been stated that uncontrolled LAB fermentation can cause unwanted flavor and texture attributes in bread. The process should be under control to avoid sourness and pungent flavor formed by excessive acidity (Katina et al., 2006, Catzeddu 2019).

Hansen (1989) reported that rye bread fermented with heterofermentative or homofermentative culture ended up with a crusty and spicy flavor. Homofermentative culture sourdough rye breads were determined to have sweet, mild sour, and flowery sensory attributes; whereas heterofermentative culture sourdough rye breads were determined to have mild sensory attributes. The addition of 5%, 10%, 15%, and 20% sourdough (*L. plantarum* + *S. Cerevisiae*) to yeast bread changed the flowery, sweet, and yeasty odor to mild sour, mild pricking, and fresh/aromatic odor, respectively. Bread, sweet, and dumpling taste of yeast bread was perceived as mild sour, aromatic, and spicy, respectively (Hansen 1996).

Lotong (2000) developed a lexicon for sourdough bread by evaluating 37 sourdough breads, and determined 26 and 28 attributes for bread crumb and crust, respectively. Sour aromas, such as dairy sour, lemon, malic acid, and vinegar were determined as sour characters for the sourdough bread lexicon. The sour characters for the crust were determined as too low, whereas sour aroma and taste for the crumb were high. The sour attributes dairy sour, lemon, malic acid, and vinegar were identified as sour characters in the crumb only.

However, as seen from the literature, studies on volatile compounds of sourdough bread together with sensory attributes are limited. In order to evaluate the impact of sourdough on bread quality, volatile compound analysis and sensory attributes of the bread should be studied together. Qualitative and quantitative analysis of the flavor of sourdough bread should be examined in further studies to determine the correlation between flavor compounds and sensory attributes.

REFERENCES

Adebo, O. A., and I. Gabriela Medina-Meza. 2020. "'Impact of fermentation on the phenolic compounds and antioxidant activity of whole cereal grains': A mini review." *Molecules* 25 (4): 927–945. https://doi.org/10.3390/molecules25040927.

ALjahdali, N., and F. Carbonero. 2019. "Impact of Maillard reaction products on nutrition and health: Current knowledge and need to understand their fate in the human digestive system." *Critical Reviews in Food Science and Nutrition* 59 (3): 474–487. https://doi.org/10.1080/10408398.2017.1378865.

Angelin, J, and M Kavitha. 2020. "Exopolysaccharides from probiotic bacteria and their health potential." *International Journal of Biological Macromolecules* 162: 853–865. https://doi.org/10.1016/j.ijbiomac.2020.06.190.

Arendt, E. K., Liam A. M. Ryan, and Fabio Dal Bello. 2007. "Impact of sourdough on the texture of bread." *Food Microbiology* 24 (2): 165–174. https://doi.org/10.1016/j.fm.2006.07.011.

Arora, K, H Ameur, A Polo, R Di Cagno, C G Rizzello, and M Gobbetti. 2021. "Thirty years of knowledge on sourdough fermentation: A systematic review." *Trends in Food Science and Technology* 108 (November 2020). 71–83. https://doi.org/10.1016/j.tifs.2020.12.008.

Bartkiene, Elena, Vadims Bartkevics, Vita Krungleviciute, Iveta Pugajeva, Daiva Zadeike, and Grazina Juodeikiene. 2017. "Lactic acid bacteria combinations for wheat sourdough preparation and their influence on wheat bread quality and acrylamide formation." *Journal of Food Science* 82 (10): 2371–2378. https://doi.org/10.1111/1750-3841.13858.

Bartkiene, Elena, Grazina Juodeikiene, Daiva Vidmantiene, Pranas Viskelis, and Dalia Urbonaviciene. 2011. "Nutritional and quality aspects of wheat sourdough bread using L. Luteus and L. Angustifolius flours fermented by Pedioccocus Acidilactici." *International Journal of Food Science and Technology* 46 (8): 1724–1733. https://doi.org/10.1111/j.1365-2621.2011.02668.x.

Baye, K., C. Mouquet-Rivier, C. Icard-Vernière, I. Rochette, and J. P. Guyot. 2013. "Influence of flour blend composition on fermentation kinetics and phytate hydrolysis of sourdough used to make injera." *Food Chemistry.* 138: 430–436. https://doi.org/10.1016/j.foodchem.2012.10.075.

Bender, D., V. Fraberger, P. Szepasvári et al. 2018. "Effects of selected lactobacilli on the functional properties and stability of gluten-free sourdough bread." *European Food Research and Technology* 244: 1037–1046. https://doi.org/10.1007/s00217-017-3020-1.

Bounaix, Marie Sophie, Valérie Gabriel, Sandrine Morel, Hervé Robert, Philippe Rabier, Magali Remaud-Siméon, Bruno Gabriel, and Catherine Fontagné-Faucher. 2009. "Biodiversity of exopolysaccharides produced from sucrose by sourdough lactic acid bacteria." *Journal of Agricultural and Food Chemistry* 57 (22): 10889–10897. https://doi.org/10.1021/jf902068t.

Callejo, María Jesús, María Eugenia Vargas-Kostiuk, and Marta Rodríguez-Quijano. 2015. "Selection, training and validation process of a sensory panel for bread analysis: Influence of cultivar on the quality of breads made from common wheat and spelt wheat." *Journal of Cereal Science* 61: 55–62. https://doi.org/10.1016/j.jcs.2014.09.008.

Canesin, Míriam Regina, and Cínthia Baú Betim Cazarin. 2021. "Nutritional quality and nutrient bioaccessibility in sourdough bread." *Current Opinion in Food Science* 40: 81–86. https://doi.org/10.1016/j.cofs.2021.02.007.

Casado, A., A. Álvarez, L. González, D. Fernández, J. L. Marcos, and M. E. Tornadijo. 2017. "Effect of fermentation on microbiological, physicochemical and physical characteristics of sourdough and impact of its use on bread quality." *Czech Journal Food Science* 35: 496–506. https://doi.org/10.17221/68/2017-CJFS.

Catzeddu, Pasquale. 2019. "Sourdough breads." *Flour and Breads and Their Fortification in Health and Disease Prevention* 177–188. https://doi.org/10.1016/B978-0-12-814639-2.00014-9.

Çetin-Babaoğlu, Hümeyra, Sultan Arslan-Tontul, and Nihat Akın. 2020. "Effect of immature wheat flour on nutritional and technological quality of sourdough bread." *Journal of Cereal Science* 94 (May). https://doi.org/10.1016/j.jcs.2020.103000.

Clarke, C. I., and E. Arendt. 2005. "A review of the application of sourdough technology to wheat breads." *Advances in Food & Nutrition Research* 49: 137–161. https://doi.org/10.1016/S1043-4526(05)49004-X.

Clement, Héliciane, Carole Prost, Cécile Rannou, Hubert Chiron, Maren Bonnand-Ducasse, Philippe Courcoux, and Bernard Onno. 2020. "Can instrumental characterization help predicting sour taste perception of wheat sourdough bread?" *Food Research International* 133 (August 2019): 109159. https://doi.org/10.1016/j.foodres.2020.109159.

Coda, R., R. Di Cagno, M. Gobbetti, and C. G. Rizzello. 2014. "Sourdough lactic acid bacteria: Exploration of non-wheat cereal-based fermentation." *Food Microbioogy* 37: 51–58. https://doi.org/10.1016/j.fm.2013.06.018

Cizeikiene, Dalia, Jolita Jagelaviciute, Mantas Stankevicius, and Audrius Maruska. 2020. "Thermophilic lactic acid bacteria affect the characteristics of sourdough and wholegrain wheat bread." *Food Bioscience* 38 (October). Elsevier Ltd: 100791. https://doi.org/10.1016/j.fbio.2020.100791.

Coda, R., C. G. Rizzello, D. Pinto, and M. Gobbetti. 2012. "Selected lactic acid bacteria synthesize antioxidant peptides during sourdough fermentation of cereal flours." *Applied and Environmental Microbiology* 78 (4): 1087–1096. https://doi.org/10.1128/AEM.06837-11.

Curiel, José Antonio, Rossana Coda, Isabella Centomani, Carmine Summo, Marco Gobbetti, and Carlo Giuseppe Rizzello. 2015. "Exploitation of the nutritional and functional characteristics of traditional italian legumes: The potential of sourdough fermentation." *International Journal of Food Microbiology* 196. Elsevier B.V.: 51–61. https://doi.org/10.1016/j.ijfoodmicro.2014.11.032.

Diana, Marina, Magdalena Rafecas, and Joan Quílez. 2014. "Free amino acids, acrylamide and biogenic amines in gamma-aminobutyric acid enriched sourdough and commercial breads." *Journal of Cereal Science* 60 (3): 639–644. https://doi.org/10.1016/j.jcs.2014.06.009.

Diowksz, Anna, Alicja Malik, Agnieszka Jásniewska, and Joanna Leszczyńska. 2020. "The inhibition of amylase and ACE enzyme and the reduction of immunoreactivity of sourdough bread." *Foods* 9 (5). https://doi.org/10.3390/foods9050656.

EFSA. 2011. "Scientific opinion on risk based control of biogenic amine formation in fermented foods." *EFSA Journal* 9 (10): 1–93. https://doi.org/10.2903/j.efsa.2011.2393.

Elkhalifa, Abd Elmoneim O., Rita Bernhardt, Francesco Bonomi, Stefania Iametti, Maria Ambrogina Pagani, and Marta Zardi. 2006. "Fermentation modifies protein/protein and protein/starch interactions in sorghum dough." *European Food Research and Technology* 222 (5–6): 559–564. https://doi.org/10.1007/s00217-005-0124-9.

Fekri, Arezoo, Mohammadali Torbati, Ahmad Yari Khosrowshahi, Hasan Bagherpour Shamloo, and Sodeif Azadmard-Damirchi. 2020. "Functional effects of phytate-degrading, probiotic lactic acid bacteria and yeast strains isolated from iranian traditional sourdough on the technological and nutritional properties of whole wheat bread." *Food Chemistry* 306 (May 2019). Elsevier: 125620. https://doi.org/10.1016/j.foodchem.2019.125620.

Fernández-Peláez, Juan, Candela Paesani, and Manuel Gómez. 2020. "Sourdough technology as a tool for the development of healthier grain-based products: An update." *Agronomy* 10 (12): 1962. https://doi.org/10.3390/agronomy10121962.

Fois, Simonetta, Piero Pasqualino Piu, Manuela Sanna, Tonina Roggio, and Pasquale Catzeddu. 2018. "Starch digestibility and properties of fresh pasta made with semolina-based liquid sourdough." *LWT: Food Science and Technology* 89 (November 2017). Elsevier: 496–502. https://doi.org/10.1016/j.lwt.2017.11.030.

Galal, A. M., Johnson, J. A. and E. Varriano-Marston. 1978. "Lactic and volatile (C2-C5) organic acids of San Francisco sourdough French bread." *Cereal Chemistry* 55: 461–468.

Galle, Sandra, and Elke K Arendt. 2014. "Exopolysaccharides from sourdough lactic acid bacteria." *Critical Reviews in Food Science and Nutrition* 54 (7): 891–901. https://doi.org/10.1080/10408398.2011.617474.

Galli, Viola, Lorenzo Mazzoli, Simone Luti, Manuel Venturi, Simona Guerrini, Paolo Paoli, Massimo Vincenzini, Lisa Granchi, and Luigia Pazzagli. 2018. "Effect of selected strains of lactobacilli on the antioxidant and anti-inflammatory properties of sourdough." *International Journal of Food Microbiology* 286 (July): 55–65. https://doi.org/10.1016/j.ijfoodmicro.2018.07.018.

Gänzle, M. G. 2014. "Enzymatic and bacterial conversions during sourdough fermentation." *Food Microbiology* 37: 2–10. https://doi.org/10.1016/j.fm.2013.04.007.

Gänzle, M. G., Vermeulen, N. and R. F. Vogel. 2007. "Carbohydrate, peptide and lipid metabolism of lactic acid bacteria in sourdough." *Food Microbiology* 24: 128–138. https://doi.org/10.1016/j.fm.2006.07.006.

Gänzle, Michael G., Jussi Loponen, and Marco Gobbetti. 2008. "Proteolysis in sourdough fermentations: Mechanisms and potential for improved bread quality." *Trends in Food Science & Technology* 19 (10): 513–521. https://doi.org/10.1016/j.tifs.2008.04.002.

Gobbetti, M., M. S. Simonetti, A. Corsetti, F. Santinelli, J. Rossi, and P. Damiani. 1995. "Volatile compound and organic acid productions by mixed wheat sour dough starters: Influence of fermentation parameters and dynamics during baking." *Food Microbiology* 12 (C): 497–507. https://doi.org/10.1016/S0740-0020(95)80134-0.

Gobbetti, Marco, Carlo G. Rizzello, Raffaella Di Cagno, and Maria De Angelis. 2014. "How the sourdough may affect the functional features of leavened baked goods." *Food Microbiology* 37 (February): 30–40. https://doi.org/10.1016/j.fm.2013.04.012.

Gobbetti, Marco, Maria De Angelis, Raffaella Di Cagno, Maria Calasso, Gabriele Archetti, and Carlo Giuseppe Rizzello. 2019. "Novel insights on the functional/nutritional features of the sourdough fermentation." *International Journal of Food Microbiology* 302 (April 2018): 103–113. https://doi.org/10.1016/j.ijfoodmicro.2018.05.018.

Gobbetti, M., M. De Angelis, R. Di Cagno, A. Polo, and C. G. Rizzello. 2020. "The sourdough fermentation is the powerful process to exploit the potential of legumes, pseudocereals and milling by-products in baking industry." *Critical Reviews in Food Science and Nutrition* 60: 2158–2173. https://doi.org/10.1080/10408398.2019.1631753.

Hammes, W. P., and M. G. Gänzle. 1998. "Sourdough breads and related products." In *Microbiology of Fermented Foods*, ed. B.J.B. Wood, 199–216. Boston: Springer.

Hammes, W. P., M. J. Brandt, K. L. Francis, J. Rosenheim, M. F. H.. Seitter, and S. A. Vogelmann. 2005. "Microbial ecology of cereal fermentations." *Trends in Food Science & Technology* 16: 4–11. https://doi.org/10.1016/j.tifs.2004.02.010

Hansen, Åse, and Birgit Hansen. 1996. "Flavour of sourdough wheat bread crumb." *European Food Research and Technology* 202 (3): 244–249. https://doi.org/10.1007/BF01263548.

Hansen, A., B. Lund, and M. J. Lewis. 1989. "Flavour of sourdough rye bread crumb." Lebensmittel-Wissenschaft und-Technologie 22 (4): 141–144.

Haros, Monica, Maria Bielecka, and Yolanda Sanz. 2005. "Phytase activity as a novel metabolic feature in bifidobacterium." *FEMS Microbiology Letters* 247 (2): 231–239. https://doi.org/10.1016/j.femsle.2005.05.008.

Hendek Ertop, Müge, and S. Beyza Öztürk Sarikaya. 2017. "The relations between hydroxymethylfurfural content, antioxidant activity and colorimetric properties of various bakery products." *Gıda/The Journal of Food* 42 (6): 834–843. https://doi.org/10.15237/gida.gd17033.

Huang, X., D. Schuppan, L. E. Rojas Tovar, V. F. Zevallos, J. Loponen and M. Gänzle. 2020. "Sourdough fermentation degrades wheat alpha-amylase/trypsin inhibitor (ATI) and reduces pro-inflammatory activity." *Foods*, 9 (7): 943. https://doi.org/10.3390/foods9070943.

Hüttner, E. K., F. Dal Bello and E. K. Arendt. 2010. "Identification of lactic acid bacteria isolated from oat sourdoughs and investigation into their potential for the improvement of oat bread quality." *European Food Research and Technology* 230: 849–857. https://doi.org/10.1007/s00217-010-1236-4.

Irakli, Maria, Aggeliki Mygdalia, Paschalina Chatzopoulou, and Dimitrios Katsantonis. 2019. "Impact of the combination of sourdough fermentation and hop extract addition on baking properties, antioxidant capacity and phenolics bioaccessibility of rice bran-enhanced bread." *Food Chemistry* 285 (January). Elsevier: 231–239. https://doi.org/10.1016/j.foodchem.2019.01.145.

Karaman, Kevser, Osman Sagdic, and M Zeki Durak. 2018. "Use of phytase active yeasts and lactic acid bacteria isolated from sourdough in the production of whole wheat bread." *LWT: Food Science and Technology* 91 (August 2017). Elsevier: 557–567. https://doi.org/10.1016/j.lwt.2018.01.055.

Katina, K., R. L. Heiniö, K. Autio, and K. Poutanen. 2006a. "Optimization of sourdough process for improved sensory profile and texture of wheat bread." *LWT: Food Science and Technology* 39 (10): 1189–1202. https://doi.org/10.1016/j.lwt.2005.08.001.

Katina, K., Salmenkallio-Marttila, M., Partanen, R., Forssell, P. and K. Autio. 2006b. "Effects of sourdough and enzymes on staling of high-fibre wheat bread." *LWT: Food Science and Technology* 39: 479–491. https://doi.org/10.1016/j.lwt.2005.03.013

Kopeć, A., M. Pysz, B. Borczak, E. Sikora, C. M. Rosell, C. Collar, and M. Sikora. 2011. "Effects of sourdough and dietary fibers on the nutritional quality of breads produced by bake-off technology." *Journal of Cereal Science* 54 (3): 499–505. https://doi.org/10.1016/j.jcs.2011.07.008.

Liljeberg, Helena G.M., and Inger M.E. Björk. 1996. "Delayed gastric emptying rate as a potential mechanism for lowered glycemia after eating sourdough bread: Studies in humans and rats using test products with added organic acids or an organic salt." *American Journal of Clinical Nutrition* 64 (6): 886–893. https://doi.org/10.1093/ajcn/64.6.886.

Lorenz, K. and J.-M. Bruemmer. 2003. "Preferments and sourdoughs in German breads." In *Handbook of Dough Fermentations*, ed. K. Kulp, and K. Lorenz, 247–267. New York: Marcel Dekker Inc.

Lotong, Varapha, Edgar Chambers IV, and Delores H. Chambers. 2000. "Determination of the sensory attributes of wheat sourdough bread." *Journal of Sensory Studies* 15 (3): 309–326. https://doi.org/10.1111/j.1745-459X.2000.tb00273.x.

Lukšič, Lea, Giovanni Bonafaccia, Maria Timoracka, Alena Vollmannova, Janja Trček, Tina Koželj Nyambe, Valentina Melini, Rita Acquistucci, Mateja Germ, and Ivan Kreft. 2016. "Rutin and Quercetin transformation during preparation of buckwheat sourdough bread." *Journal of Cereal Science* 69: 71–76. https://doi.org/10.1016/j.jcs.2016.02.011.

Luti, Simone, Lorenzo Mazzoli, Matteo Ramazzotti, Viola Galli, Manuel Venturi, Giada Marino, Martin Lehmann, et al. 2020. "Antioxidant and anti-inflammatory properties of sourdoughs containing selected lactobacilli strains are retained in breads." *Food Chemistry* 322 (April). Elsevier: 126710. https://doi.org/10.1016/j.foodchem.2020.126710.

Martinez-Anaya, M. A., B. Pitarch, P. Bayarri and C. Benedito de Barber. 1990. "Microflora of the sourdoughs of wheat flour bread×interactions between yeasts and lactic acid bacteria in wheat doughs and their effects on bread quality." *Cereal Chemistry* 67: 85–91.

Mastilović, J. and S. Popov. 2002. "Investigation of rheological properties of liquid sour dough." In Flour - Bread '01: Proceedings of International Congress, 3rd Croatian Congress of Cereal Technologists, Opatija, Croatia, 14–17 November 2001, ed. Ž. Ugarčić-Hardi, 53–62, ref. 6. Osijek: Faculty of Food Technology, University fo Josip Juraj Strossmayer.

Meroth, C. B., Hammes, W. P. and C. Hertel. 2004. "Characterisation of the microbiota of rice sourdoughs and description of *Lactobacillus spicheri* sp." *Systematic Applied Microbiology* 27: 151–159. https://doi.org/10.1078/072320204322881763.

Michalska, Anna, Miryam Amigo-Benavent, Henryk Zielinski, and Maria Dolores del Castillo. 2008. "Effect of bread making on formation of maillard reaction products contributing to the overall antioxidant activity of rye bread." *Journal of Cereal Science* 48 (1): 123–132. https://doi.org/10.1016/j.jcs.2007.08.012.

Mihhalevski, Anna, Ildar Nisamedtinov, Kristel Hälvin, Aleksandra Ošeka, and Toomas Paalme. 2013. "Stability of B-complex vitamins and dietary fiber during rye sourdough bread production." *Journal of Cereal Science* 57 (1): 30–38. https://doi.org/10.1016/j.jcs.2012.09.007.

Montemurro, Marco, Erica Pontonio, Marco Gobbetti, and Carlo Giuseppe Rizzello. 2019. "Investigation of the nutritional, functional and technological effects of the sourdough fermentation of sprouted flours." *International Journal of Food Microbiology* 302 (July 2018). Elsevier: 47–58. https://doi.org/10.1016/j.ijfoodmicro.2018.08.005.

Moroni, Alice V, Fabio Dal Bello, and Elke K Arendt. 2009. "Sourdough in gluten-free bread-making: An ancient technology to solve a novel issue?" *Food Microbiology* 26 (7). Elsevier Ltd: 676–684. https://doi.org/10.1016/j.fm.2009.07.001.

Nionelli, Luana, and Carlo Rizzello. 2016. "Sourdough-based biotechnologies for the production of gluten-free foods." *Foods* 5 (4): 65. https://doi.org/10.3390/foods5030065.

Nissen, Lorenzo, Seyedeh Parya Samaei, Elena Babini, and Andrea Gianotti. 2020. "Gluten free sourdough bread enriched with cricket flour for protein fortification: Antioxidant improvement and volatilome characterization." *Food Chemistry* 333 (May). Elsevier: 127410. https://doi.org/10.1016/j.foodchem.2020.127410.

Novotni, Dubravka, Nikolina Čukelj, Bojana Smerdel, Martina Bituh, Filip Dujmić, and Duška Ćurić. 2012. "Glycemic index and firming kinetics of partially baked frozen gluten-free bread with sourdough." *Journal of Cereal Science* 55 (2): 120–125. https://doi.org/10.1016/j.jcs.2011.10.008.

Olojede, A. O., A. I. Sanni, K. Banwo, and A. T. Adesulu-Dahunsi. 2020. "Sensory and antioxidant properties and in-vitro digestibility of gluten-free sourdough made with selected starter cultures." *LWT: Food Science and Technology* 129 (May). Elsevier: 109576. https://doi.org/10.1016/j.lwt.2020.109576.

Omedi, J. O. and W. Huang 2016. "Soy sourdough fermented by lactic acid bacteria starter (*L. plantarum*, and *L. sanfranciscensis*) concentration effect on dough fermentation, textural and shelf life properties of wheat bread." *MOJ Food Processing & Technology* 3: 327–334. https://doi.org/10.15406/mojfpt.2016.03.00075

Papadimitriou, Konstantinos, Georgia Zoumpopoulou, Marina Georgalaki, Voula Alexandraki, Maria Kazou, Rania Anastasiou, and Effie Tsakalidou. 2019. Chapter 6 - *Sourdough Bread*. *Innovations in Traditional Foods*. Elsevier Inc. https://doi.org/10.1016/B978-0-12-814887-7.00006-X.

Pejcz, Ewa, Anna Czaja, Agata Wojciechowicz-Budzisz, Zygmunt Gil, and Radosław Spychaj. 2017. "The potential of naked barley sourdough to improve the quality and dietary fibre content of barley enriched wheat bread." *Journal of Cereal Science* 77: 97–101. https://doi.org/10.1016/j.jcs.2017.08.007.

Pepe, O., Blaiotta, G., Moschetti, G., Greco, T. and F. Villani. 2003. "Rope-producing strains of *Bacillus spp.* from wheat bread and strategy for their control by lactic acid bacteria." *Applied and Environmental* 69 (4): 2321–2329. https://doi.org/10.1128/AEM.69.4.2321-2329.2003.

Pétel, Cécile, Bernard Onno, and Carole Prost. 2017. "Sourdough volatile compounds and their contribution to bread: A review." *Trends in Food Science and Technology* 59: 105–123. https://doi.org/10.1016/j.tifs.2016.10.015.

Plessas, S., A. Bekatorou, J. Gallanagh, P. Nigam, A. A. Koutinas, and C. Psarianos. 2008. "Evolution of aroma volatiles during storage of sourdough breads made by mixed cultures of Kluyveromyces Marxianus and Lactobacillus Delbrueckii Ssp. Bulgaricus or Lactobacillus Helveticus." *Food Chemistry* 107 (2): 883–889. https://doi.org/10.1016/j.foodchem.2007.09.010.

Plessas, S., A. Alexopoulos, I. Mantzourani, A. Koutinas, C. Voidarou, E. Stavropoulou, and E. Bezirtzoglou. 2011. "Application of novel starter cultures for sourdough bread production." *Anaerobe* 17 (6): 486–489. https://doi.org/10.1016/j.anaerobe.2011.03.022.

Poutanen, Kaisa, Laura Flander, and Kati Katina. 2009. "Sourdough and cereal fermentation in a nutritional perspective." *Food Microbiology* 26 (7): 693–699. https://doi.org/10.1016/j.fm.2009.07.011.

Riaz Rajoka, Muhammad Shahid, Yiguang Wu, Hafiza Mahreen Mehwish, Manisha Bansal, and Liqing Zhao. 2020. "Lactobacillus exopolysaccharides: New perspectives on engineering strategies, physiochemical functions, and immunomodulatory effects on host health." *Trends in Food Science and Technology* 103 (March): 36–48. https://doi.org/10.1016/j.tifs.2020.06.003.

Rieder, A., A. K. Holtekjølen, S. Sahlstrøm, and A. Moldestad. 2012. "Effect of barley and oat flour types and sourdoughs on dough rheology and bread quality of composite wheat bread." *Journal of Cereal Science* 55: 44–52. https://doi.org/10.1016/j.jcs.2011.10.003.

Rinaldi, M., M. Paciulli, A. Caligiani, et al. 2015. "Durum and soft wheat flours in sourdough and straight-dough bread-making." *Journal of Food Science and Technology*. 52: 6254–6265. https://doi.org/10.1007/s13197-015-1787-2.

Ripari, Valery, Yunpeng Bai, and Michael G. Gänzle. 2019. "Metabolism of phenolic acids in whole wheat and rye malt sourdoughs." *Food Microbiology* 77 (July 2018). Elsevier Ltd: 43–51. https://doi.org/10.1016/j.fm.2018.08.009.

Rizzello, Carlo G, Maria De Angelis, Raffaella Di Cagno, Alessandra Camarca, Marco Silano, Ilario Losito, Massimo De Vincenzi, et al. 2007. "Highly efficient gluten degradation by lactobacilli and fungal proteases during food processing: New perspectives for celiac disease." *Applied and Environmental Microbiology* 73 (14): 4499–4507. https://doi.org/10.1128/AEM.00260-07.

Rizzello, C. G., A. Cassone, R. Di Cagno, and M. Gobbetti. 2008. "Synthesis of angiotensin I-converting enzyme (ACE)-inhibitory peptides and γ-aminobutyric acid (GABA) during sourdough fermentation by selected lactic acid bacteria." *Journal of Agricultural and Food Chemistry* 56 (16): 6936–6943. https://doi.org/10.1021/jf800512u.

Rizzello, Carlo G., Luana Nionelli, Rossana Coda, and Marco Gobbetti. 2012. "Synthesis of the cancer preventive peptide lunasin by lactic acid bacteria during sourdough fermentation." *Nutrition and Cancer* 64 (1): 111–120. https://doi.org/10.1080/01635581.2012.630159.

Rizzello, Carlo Giuseppe, Maria Calasso, Daniela Campanella, Maria De Angelis, and Marco Gobbetti. 2014. "Use of sourdough fermentation and mixture of wheat, chickpea, lentil and bean flours for enhancing the nutritional, texture and sensory characteristics of white bread." *International Journal of Food Microbiology* 180. Elsevier B.V.: 78–87. https://doi.org/10.1016/j.ijfoodmicro.2014.04.005.

Rizzello, Carlo Giuseppe, Blanca Hernández-Ledesma, Samuel Fernández-Tomé, José Antonio Curiel, Daniela Pinto, Barbara Marzani, Rossana Coda, and Marco Gobbetti. 2015. "Italian legumes: Effect of sourdough fermentation on lunasin-like polypeptides." *Microbial Cell Factories* 14: 168. https://doi.org/10.1186/s12934-015-0358-6.

Rizzello, C. G., A. Lorusso, M. Montemurro, and M. Gobbetti. 2016. "Use of sourdough made with quinoa (Chenopodium quinoa) flour and autochthonous selected lactic acid bacteria for enhancing the nutritional, textural and sensory features of white bread." *Food Microbiology* 56: 1–13. https://doi.org/10.1016/j.fm.2015.11.018.

Rizzello, Carlo Giuseppe, Piero Portincasa, Marco Montemurro, Domenica Maria di Palo, Michele Pio Lorusso, Maria de Angelis, Leonilde Bonfrate, Bernard Genot, and Marco Gobbetti. 2019. "Sourdough fermented breads are more digestible than those started with baker's yeast alone: An in vivo challenge dissecting distinct gastrointestinal responses." *Nutrients* 11 (12) 2954. https://doi.org/10.3390/nu11122954.

Rodriguez-Ramiro, I., C. A. Brearley, S. F. A. Bruggraber, A. Perfecto, P. Shewry, and S. Fairweather-Tait. 2017. "Assessment of iron bioavailability from different bread making processes using an in vitro intestinal cell model." *Food Chemistry* 228: 91–98. https://doi.org/10.1016/j.foodchem.2017.01.130.

Rollán, G. C., C. L. Gerez and J. G. LeBlanc. 2019. "Lactic fermentation as a strategy to improve the nutritional and functional values of pseudocereals." *Frontiers in Nutrition* 6: 98. https://doi.org/10.3389/fnut.2019.00098.

Saa, Danielle Taneyo, Raffaella Di Silvestro, Giovanni Dinelli, and Andrea Gianotti. 2017. "Effect of sourdough fermentation and baking process severity on dietary fibre and phenolic compounds of immature wheat flour bread." *LWT* 83. Elsevier Ltd: 26–32. https://doi.org/10.1016/j.lwt.2017.04.071.

Saa, Danielle Taneyo, Raffaella Di Silvestro, Lorenzo Nissen, Giovanni Dinelli, and Andrea Gianotti. 2018. "Effect of sourdough fermentation and baking process severity on bioactive fiber compounds in immature and ripe wheat flour bread." *LWT: Food Science and Technology* 89 (October 2017): 322–328. https://doi.org/10.1016/j.lwt.2017.10.046.

Sahin, Aylin W, Emanuele Zannini, A. Coffey, and Elke K Arendt. 2019. "Sugar reduction in bakery products: Current strategies and sourdough technology as a potential novel approach." *Food Research International.* https://doi.org/10.1016/j.foodres.2019.108583.

Salim-ur-Rehman, Alistair Paterson, and John R. Piggott. 2006. "Flavour in sourdough breads: A review." *Trends in Food Science and Technology* 17 (10): 557–566. https://doi.org/10.1016/j.tifs.2006.03.006.

Salovaara, H. and T. Valjakka. 1987. "The effect of fermentation temperature, flour type, and starter on the properties of sour wheat bread." *International Journal Food Science & Technology* 22: 591–597. https://doi.org/10.1111/j.1365-2621.1987.tb00527.x

Scazzina, Francesca, Daniele Del Rio, Nicoletta Pellegrini, and Furio Brighenti. 2009. "Sourdough bread: Starch digestibility and postprandial glycemic response." *Journal of Cereal Science* 49 (3). Elsevier Ltd: 419–421. https://doi.org/10.1016/j.jcs.2008.12.008.

Schober, T., Schober, J., Bean, S. and D. Boyle. 2007. "Gluten-free sorghum bread improved by sourdough fermentation: Biochemical, rheological, and microstructural background." *Journal of Agriculture & Food Chemistry* 55: 5137–5146. https://doi.org/10.1021/jf0704155.

Sharma, Neha, Steffy Angural, Monika Rana, Neena Puri, Kanthi Kiran Kondepudi, and Naveen Gupta. 2020. "Phytase producing lactic acid bacteria: Cell factories for enhancing micronutrient bioavailability of phytate rich foods." *Trends in Food Science & Technology* 96 (February): 1–12. https://doi.org/10.1016/j.tifs.2019.12.001.

Shumoy, Habtu, Filip Van Bockstaele, Dilara Devecioglu, and Katleen Raes. 2018. "Effect of sourdough addition and storage time on in vitro starch digestibility and estimated glycemic index of tef bread." *Food Chemistry* 264 (April). Elsevier: 34–40. https://doi.org/10.1016/j.foodchem.2018.05.019.

Sîrbu, A. 2009. *Merceologie alimentară – Pâinea și alte produse de panificație (Science of Food Commodities: Bread and Other Bakery Products).* Bucharest: AGIR (in Romanian).

Teleky, B. E., Martău, A. G., Ranga, F., Chețan, F., and D.C. Vodnar. 2020. "Exploitation of lactic acid bacteria and baker's yeast as single or multiple starter cultures of wheat flour dough enriched with soy flour." *Biomolecules* 10: 778. https://doi.org/10.3390/biom10050778.

Tieking, Markus, and Michael G. Gänzle. 2005. "Exopolysaccharides from cereal-associated lactobacilli." *Trends in Food Science & Technology* 16 (1–3): 79–84. https://doi.org/10.1016/j.tifs.2004.02.015.

Torrieri, E., O. Pepe, V. Ventorino, P. Masi, and S. Cavella. 2014. "Effect of sourdough at different concentrations on quality and shelf life of bread." *LWT: Food Science and Technology* 56 (2): 508–516. https://doi.org/10.1016/j.lwt.2013.12.005.

Vogelmann, S. A., M. Seitter, U. Singer, M. J. Brandt, and C. Hertel. 2009. "Adaptability of lactic acid bacteria and yeasts to sourdoughs prepared from cereals, pseudocereals and cassava and use of competitive strains as starters." *International Journal of Food Microbiology* 130: 205–212. https://doi.org/10.1016/j.ijfoodmicro.2009.01.020.

Vuyst, L. De, S. Van Kerrebroeck, H. Harth, G. Huys, H. M. Daniel, and S. Weckx. 2014. "Microbial ecology of sourdough fermentations: Diverse or uniform?" *Food Microbiology* 37: 11–29. https://doi.org/10.1016/j.fm.2013.06.002.

Vuyst, L. De, V. Schrijvers, S. Paramithiotis, et al. 2002. "The biodiversity of lactic acid bacteria in Greek traditional wheat sourdoughs is reflected in both composition and metabolite formation." *Applied and Environmental Microbiology* 68: 6059–6069. https://doi.org/10.1128/aem.68.12.6059-6069.2002.

Vuyst, L. De, G. Vrancken, F. Ravyts, T. Rimaux, and S. Weckx. 2009. "Biodiversity, ecological determinants, and metabolic exploitation of sourdough microbiota." *Food Microbiology* 26: 666–675. https://doi.org/10.1016/j.fm.2009.07.012.

Zannini, E., C. Garofalo, L. Aquilanti, S. Santarelli, G. Silvestri, and F. Clementi. 2009. "Microbiological and technological characterization of sourdoughs destined for breadmaking with barley flour." *Food Microbiology* 26: 744–753. https://doi.org/10.1016/j.fm.2009.07.014.

Zhao, Cindy J., Ying Hu, Andreas Schieber, and Michael Gänzle. 2013. "Fate of ACE-inhibitory peptides during the bread-making process: Quantification of peptides in sourdough, bread crumb, steamed bread and soda crackers." *Journal of Cereal Science* 57 (3): 514–519. https://doi.org/10.1016/j.jcs.2013.02.009.

Xu, Dan, Huang Zhang, Jinzhong Xi, Yamei Jin, Yisheng Chen, Lunan Guo, Zhengyu Jin, and Xueming Xu. 2020. "Improving bread aroma using low-temperature sourdough fermentation." *Food Bioscience* 37 (July): 100704. https://doi.org/10.1016/j.fbio.2020.100704.

2 Cereals and Cereal Sourdoughs as a Source of Functional and Bioactive Compounds

*Monica Trif, Claudia Terezia Socol,
Sneh Punia Bangar, and Alexandru Vasile Rusu*

CONTENTS

2.1 Introduction 31
2.2 Structure of the Grain and Chemical Composition 33
    2.2.1 Dietary Fiber 37
    2.2.2 Proteins and Amino Acids 37
    2.2.3 Lipids 40
    2.2.4 Minerals and Vitamins 40
    2.2.5 Phytochemicals 40
2.3 Sourdough Potential for Improving Nutritional Properties 43
2.4 Health Benefits Associated with Cereals and Cereal Sourdough
    Consumption 46
2.5 Future Perspectives and Recommendations 52
References 53

2.1 INTRODUCTION

According to the European Commission report in 2021 related to cereals, the European Parliamentary Research Service (EPRS) in 2019 (Kelly, 2019), and Eurostat (2018), statistics have shown a high demand for wheat (*Triticum*) in industrial use (1.9% annually between 2016–2018), rye (*Secale cereale L.*), the second most commonly used grain for bread production after wheat (Tufail, et al., 2022), followed by maize (1.6%) and barley (*Hordeum vulgare L.*) ranking it fourth in the world (0.8%) (Figure 2.1) (Eurostat, 2011).

According to the Food and Agriculture Organization (FAO), around 11.3 million tons of rye were harvested worldwide in 2018. In comparison, the wheat was 735.2 million tons. Other cereals are grown in smaller quantities and include triticale (*x Triticosecale Wittmack*), a man-made cereal, developed by combining A & B

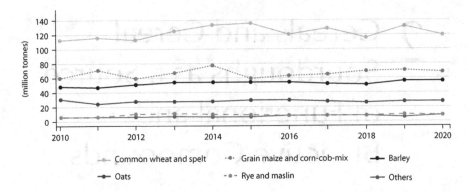

FIGURE 2.1 Most popular cereals production in Europe (EU-27). Note: 'Rye and maslin' include mixture of rye with other winter sown cereals. 'Others' includes rice, triticale, and sorghum. (Source: Eurostat, online data code: apro_cpnh1, data extracted in November 2021)

genome of durum wheat (*Triticum turgidum L.*) and R genome of rye (*Secale cereal L.*), oat, and spelt (European Commission, 2018).

The most popular cereals grown in Europe are wheat, by far the most popular in the EU, maize, and barley, each accounting for a third. In the EU wheat, rye, and triticale are typically winter crops, whereas maize and rice are summer crops (Eurostat, data extracted in November 2021). Barley is common in both winter and spring varieties (Bangar et al., 2022). The most important species of wheat are: common wheat (*Triticum vulgare*), which is available in numerous varieties, and hard wheat (*Triticum durum*) (Sharma et al., 2020).

The popularity of wheat, rye, more recently, barley, and oat has been noticed as a consumer preference for cereal foods based on the demonstrated health-promoting effects of their distinct biological active components with high nutritional value (Bangar et al., 2022; National Food Administration's Food Database, 2020). Wheat, rye, and barley have rather high fermentable oligo-, di-, monosaccharides, and polyols (FODMAP) levels. They are all carbohydrates, with oligosaccharides being particularly prevalent in cereals (Ispiryan et al., 2020; Schmidt and Sciurba, 2021).

Biologically active compounds are mostly located in the bran of cereal grains, while the germ is rich in unsaturated fatty acids, and its removal is necessary to prevent processes of lipid oxidation (Tufail et al., 2022). Products based on refined cereal flour, which is deprived of the bran and germ fractions, contain much less biologically active compounds, and therefore they are often supplemented with such substances (Bangar et al., 2022; Tufail et al., 2022).

Whole grain wheat in particular is a valuable part of a balanced diet, but products made from pure white flour are less healthy because they mainly consist of starch, which is converted into sugar in the body (Călinoiu and Vodnar, 2018; Demirel et al., 2021), and therefore in recent years, wheat has been criticized. Grain must have the husk, sprout, and endosperm in order to be called whole, regardless of whether it has been ground, milled, etc. People with gluten intolerance do not tolerate well the gluten contained in many types of grain. About 1% of the population suffers from celiac disease (an autoimmune disease), also known as gluten intolerance (Balakireva and Zamyatnin, 2016).

2.2 STRUCTURE OF THE GRAIN AND CHEMICAL COMPOSITION

Different cereal grains have a similar but complex structure that is characterized by different cell layers. The individual grain consists of the same three-part rough structure (Brouns et al., 2012; Călinoiu and Vodnar, 2018) (Figure 2.2):

- outer layer (7–20%) fruit peel, seed peel, (aleurone layer);
- endosperm (75–90%), aleurone layer;
- germ (3–15%).

The general composition of each individual part is shown in Table 2.1. The outer layer of the grain, the epidermis, is also its protective coat and holds the fruit together. The large endosperm represents the majority of the grain, and next to it is the small seedling, which has all the facilities for sprouting a new cereal plant (Laskowski et al., 2019). The content of the individual chemical composition and the bioactive compound distribution in the grain also vary within the cereal types (GMF, 2016).

Grain composition varies, and this influences how they may be used in further applications. Grains are a significant source of fiber, proteins, carbohydrates, minerals, vitamins, and phytochemicals, and regular inclusion of them into dietary habits has been proven to provide a number of health benefits (Zhu et al., 2017; Tieri et al., 2020; Garutti et al., 2022; Verstringe et al., 2023). A brief overview of the composition and average nutritional values of various cereal grains is presented in Table 2.2.

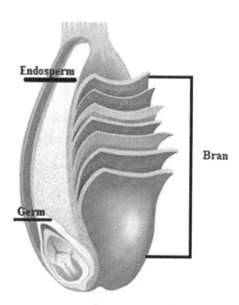

FIGURE 2.2 General grain structure. (Source: Adapted from Go for the Whole Grain Poster reference, Călinoiu and Vodnar, 2018)

TABLE 2.1
Structure of the Grain, Source of Bioactives, and Chemical Composition

Structure of the grain		Source of bioactives	Amount %	Dry matter %	Protein % (X6.25)d.m.	Lipids %	Ash %
Whole grain		Carbohydrates	100	100	9.9–17.7	2.2–3.1	1.42–2.24
Hull (husk, chaff)	Palea (dorsal)		16.0–18.0	—	—	—	—
	Larger lemma (ventral)						
Bran			3.0–30.0				
Seed coat (tegmen)	Testa/Seed coat/ Spermoderm	Alkylresorcinols, sterols, steryl ferulates	2.5 1.0	0.2–3.1	12.0–25.0	—	8.1–23.5
	Hyaline layer/Nucellar layer		—	—	—	—	—
Pericarp/Fruit coat	Outer pericarp	Insoluble dietary fibers (xylans, cellulose, lignin), Antioxidants bound to cell walls	3.0–5.0	3.5–9.5	5.0–8.0	0.6–1.0	6.9
	Epidermis/ Beeswing			3.2–5.8	3.6–9.8	0.7–1.1	1.8–4.0
	Hypodermis			2.6–4.4	2.1–5.7	0.8–1.2	1.1–1.7
	Inner pericarp Mesocarp (Tube cells) Endocarp (Cross cells)			0.5–1.5	7.8–12.3	0.4–0.6	6.2–15.9
Endosperm			83.0		7.0		

(*Continued*)

TABLE 2.1 (CONTINUED)
Structure of the Grain, Source of Bioactives, and Chemical Composition

Structure of the grain	Source of bioactives	Amount %	Chemical composition				
			Dry matter %	Protein % (X6.25)d.m.	Lipids %	Ash %
Aleurone cells/Aleurone layer Nucellar	Insoluble dietary fibers and very few soluble fibers (<5%) (xylans, β-glucans) Proteins, enzymes Phenolic compounds, ligans Vitamin E, B vitamins Minerals and phytic acid Lipids, plant sterols	6.0–9.0	4.6–10.4	23.4–41.3	7.0–11.5	4.9–20.0
Starchy endosperm/ Flour cells (Endorsperm cells with starch granules)	Protein, starch, very few fibers (2%)	80–85	80.1–86.5	8.8–15.9	1.0–2.2	0.5–0.74
Germ/Embryo	Lipids, plant sterols Antioxidants Vitamin E, B vitamins Minerals Enzymes	2.5–3.0	0.9–1.6	37.5–41.8	18.0	4–5
Empty layer Suction epithel Plumule Rudimentary leaves Primary root Root cap Micropyle Scutellum		—	1.3–2.0	30.3–36.3	35.0	6.8–9.5

TABLE 2.2
Composition of Various Cereal Grains

					Ingredient per 100g dry weight basis						
Cereal	Energy (kJ)	Energy (kcal)	Total dietary fibers (g)	Crude fiber (g)	Crude protein (g)	Carbohydrate (g)	Sugars (g)	Total lipid (g)	Water (g)	Ash (g)	Minerals (g)
Wheat	1372–1381	328–330	10.3	2.0	13.5	69.0–69.3	0.4	2.0–2.4	13–14	1.5	1.8
Rye	1360–1364	326	14.5	2.1	10.0–11.0	69.5–73.0	0.9	1.6–1.7	13–14	1.5	1.9
Triticale	1405	323	14.6		13.0	72.1	2.0	2.1	13–14	2.2	
Barley	1412–1415	338	9.9	1.6	10.0–12.0	70.0–71.2	0.8	1.6–2.1	13–14	1.2	2.3
Oats	1468–1542	351–368	10.0	1.6	13.0–16.0	62.5–62.8	0.9	6.5–7.0	13–14	1.6	2.9
Rice	1486–1497	326–358	3.5	0.7	7.0–7.4	73.5–75.0	0.6	2.2–3.1	13–14	1.2	1.2
Sorghum	1352	323	16.6		10.4	70.8	2.5		13–14	1.4	

2.2.1 Dietary Fiber

Cereals contain different amounts of fiber (Table 2.3). The differential carbohydrate content decreases the nutritional value of grains fed with hulls. The content of the individual types of fiber, the ratio of soluble to insoluble, and the distribution in the grain also vary within the cereal types. Dietary fiber is mainly found in the outer layers of the grain. There, the content of the insoluble dietary fibers – lignin, cellulose, and hemicellulose – predominates (Shewry et al., 2008; Dhingra et al., 2012; Prasadi and Joye, 2020).

Lignans were recently identified as an important group of bioactive substances that are widely found in plant foods. The cell walls of the plant are built from them, which is why they are biochemically part of the dietary fibers (Rodríguez-García et al., 2019). The lignan content in rye is particularly high, compared to other grains (Jonsson et al., 2018).

Soluble hemicelluloses, pectins, and ß-glucans are found in the endosperm (Williams et al., 2019; Prasadi and Joye, 2020). Therefore, almost only soluble fiber is found in the flour. As the degree of grinding increases, the total fiber content and the proportion of insoluble fiber also increase (Bangar et al., 2022). Rye and oats have a significantly higher proportion of soluble fiber than other types of grains (Frølich et al., 2013). Whole grain rye is particularly recommended as part of a healthy diet and in diabetes management, having a positive effect on blood sugar levels (Jonsson et al., 2018). Whole wheat flour contains all of the fiber from the grain. Wheat and rye mainly contain high-quality carbohydrates in the form of starch as well as soluble and insoluble fiber. Both provide a good 13 g of fiber per 100 g dry weight basis (Rakha et al., 2010). Contrary to popular belief, even light-colored type flours and the baked goods made from them contain considerable amounts of healthy fibers at 3 to 4 g per 100 g dry weight basis; often even more than many types of fruit and vegetables. Therefore, white flour products are also considered sources of dietary fiber according to EU criteria. Dark type flours provide 5 to 8 g of dietary fiber, whole grain products provide 10 to more than 13 g of dietary fiber and are considered to be high in fiber (Newman et al., 2019). When it comes to soluble fiber, the focus is on ß-glucans, which are characterized by their high viscosity and binding properties. The highest levels are found in oats and barley (El Khoury et al., 2012).

2.2.2 Proteins and Amino Acids

The crude protein content for cereal grains is between 8% and 14% on a dry weight basis. The crude protein content of wheat is greater than the average range. In terms of protein quality, the proteins, solubility, and amino acid content varies between cereal grains (Rani et al., 2021, Verstringe et al., 2023).

For people with celiac disease, gluten causes lasting damage to the intestinal mucosa. The bowel reacts with inflammation, which is often accompanied by abdominal pain or diarrhea (Di Sabatino et al., 2018). Digestion is impaired, and, ultimately, nutrients cannot be properly absorbed by the body. In particular, the paleo diet, a new diet trend, is based completely without grain (Bellini and Rossi,

TABLE 2.3
Dietary Fiber Composition of Various Cereal Grains (Andersson et al., 2009; Andersson et al., 2013; USDA, 2010; Verstringe et al., 2023; Andersson et al., 2013; Frølich et al., 2013)

						% of dry matter basis				
Cereal	Total DF	Polysaccharides	Arabinoxylan	Cellulose	β-glucan	Mannose	Galactose	Glucose	Uronic acids	Fructan	Klason Lignin
Wheat	12–13.5	2.4–9.5	4.4–6.9	2.5	0.5–1.0	1.0–4.0	3.0–7.0	31.0–33.0	3.0	1.3–2.3	0.8–2.0
Rye	15–20	6.8–12.5	8.0–12.1	1.0–3.0	1.3–2.2	3.0–5.0	3.0	28.0–31.0	4.0–5.0	4.1–6.4	0.3–3.0
Triticale	14.6		3.4–5.2		0.4–0.7					0.2–1.5	
Barley dehulled	8.2–18.8	7.9–15.1	4.2–5.4	1.9	2.4–8.3	3.0–4.0	1.0	51.0–69.0	5.0	1.6	0.7–3.5
Oat dehulled	10.2–10.6		2.0–14.5	1.3	2.2–5.0					0.1–0.2	1.4
Rice	0.7–19.2		1.1–13.0			1.0–5.0	2.0–5.0	57.0–72.0	9.0–10.0		3.9
Sorghum	2.5–9.0		3.0–18.0			3.0–9.0	2.0–9.0	51.0–61.0	11.0–13.0		0.4–2.5

2018). The consumption of wheat, as well as rye and barley, has very negative health effects for this population group. Gluten is basically made up of storage proteins and reserve proteins (Shewry, 2019). The different proteins in gluten explain why the flours have different baking properties. Rye contains much less gluten compared to wheat. Wheat has better baking properties due to its high gluten (protein complex) content, which means that doughs made from wheat flour can be made particularly easily and quickly (Pejcz et al., 2020).

According to the German Research Institute for Food Chemistry, the average gluten content of the most important cereals is as follows: wheat flour (type 405): 8.7 g gluten per 100 g, whole rye: 3.2 g gluten per 100 g, barley: 5.6 g gluten per 100 g (Scherf and Köhler, 2016). According to EU regulations, products may be designated as "very low gluten content" if they do not exceed the limit of 100 mg gluten/ kg food, and products labeled "gluten-free" can also contain gluten if a maximum content of 20 mg/kg is not exceeded (Miranda-Castro et al., 2016; Verma et al., 2017; Pernica et al., 2020).

In wheat, gliadin and glutenin ensure the formation of glue and thus the elasticity and firmness of the dough (Bangar et al., 2022). While rye has two types of gluten, secalin and secalinin, they do not have the same ability as wheat gluten (Deleu et al., 2020; Dziki et al., 2022). Therefore, the addition of acid (sourdough) is required so that the rye bread rises and forms a nice crumb. The adhesive protein gluten is particularly important for processing the flour into bread and baked goods, ensuring the dough "gelatinizes" (Khan et al., 2022).

Regarding amino acids, in general cereal grains may be low in lysine, tryptophan, threonine, and methionine (Table 2.4). Generally, the amino acid profile of cereal grains can be stated in the following order: oats > barley > wheat > corn > rice > rye > sorghum > millet (Anjum et al., 2005; Jiang et al., 2008; FAO, 2013; Siddiqi et al., 2020).

TABLE 2.4
Amino Acid Average Contents of Various Cereal Grains (Anjum et al., 2005; Jiang et al., 2008; FAO 2013; Brestenský et al., 2018; Siddiqi et al., 2020)

Variable (g/kg DM)	Wheat	Rye	Barley	Oats decorticated	Sorghum
Arginine	6.3	–	6.8	9.9	4.1
Histidine	2.8	1.29	2.9	4.8	2.4
Isoleucine	6.0	0.82	4.2	5.8	4.1
Leucine	11.0	8.25	8.5	11.0	14.1
Lysine	3.8	2.4	4.3	6.2	2.9
Methionine	2.3	1.6	2.1	3.7	2.0
Phenylalanine	6.0	–	5.9	7.5	5.6
Threonine	4.9	2.1	4.3	5.1	3.4
Thryptophan	1.4	0.7	1.7	1.8	0.9
Valine	5.8	5.59	6.0	14.7	6.4

2.2.3 LIPIDS

Lipids are present in wheat grain in small quantities, i.e. 1.7–2.0% dry weight, of which lipids in germ comprise 28.5% dry weight; aleurone layer 8% dry weight; endosperm 1.5%; and bran 1% dry weight (Surget and Barron 2005). Lipid content depends heavily on wheat variety. The lipid content ranges from less than 1% to greater than 6%. Unsaturated fatty acids are abundant in cereal grains. Most cereals contain linoleic (18:2), a major unsaturated fatty acid (wheat 55–60 (weight %), rye 57–65 (weight %), barley 51–60 (weight %), and palmitic (16:0), a major saturated fatty acid (wheat 17–24 (weight %), rye 12–19 (weight %), barley 19–22 (weight %) (Rosicka-Kaczmarek et al., 2015; Phillips et al., 2020; Slama et al., 2021; Snell et al., 2022).

2.2.4 MINERALS AND VITAMINS

In addition, cereals contain minerals such as potassium (K), magnesium (Mg) and iron (Fe) as well as B vitamins and secondary plant substances, in particular phytoestrogens, protease inhibitors, saponins and phytic acid (Mattila et al., 2005). Cereal grains are low in calcium (Ca) and high in phosphorus (P). A portion of the P may be bound with phytic acid in a phytate complex. For monogastrics, the availability of the bound P is low. In general, cereal grains contain low amounts of micronutrients (Table 2.5) (Ragaee et al., 2006; Danish Food Composition Databank; Afam et al., 2014; Frølich et al., 2013).

In terms of vitamins, cereal grains are fair sources of vitamin E but are low in A and D and most of the B complex vitamins (Table 2.6) (Matportalen, 2006; Danish Food Composition Databank, version 7.0. 2008; Frølich et al., 2013). The greatest part of the vitamin B1 content of the whole kernel is concentrated in the scutellum, i.e. 62% dry weight, whereas another 32% dry weight is contained in the aleurone cells (Shewry, 2014). The 62% dry weight is concentrated in the scutellum in only 1.5% dry weight of the kernel weight. The whole endosperm contains only 2.8% dry weight of the total vitamin B1 content, and the germ itself only 2% dry weight.

2.2.5 PHYTOCHEMICALS

Cereals reported with high antioxidant capacity include wheat and barley (Ragaee et al., 2006) (**Table 2.7** and **Table 2.8**). The largest concentrations of total hydroxycinnamic acids and their derivatives (such as p-coumaric, ferulic, and caffeic acids) were found in the wheat bran and rye bran fractions, according to research on the antioxidant capabilities of soluble extracts from wheat and rye (Gallardo et al., 2006; Devanand et al., 2015; Razgonova et al., 2021).

The outer layers of rye and wheat grains contain phytic acid, a strong chelator which promotes the development of the young grain and protects it from predators. Phytin stores important minerals and trace elements to make them available to the cereal seedling for growth (Raboy, 2020). But phytic acid is considered an

TABLE 2.5
Mineral Composition of Various Cereal Grains (Ragaee et al., 2006; Danish Food Composition Databank; Afam et al., 2014; Frølich et al., 2013)

	Ingredient mg/kg dry weight basis								
Cereal	Phosphorus (P)	Potassium (K)	Magnesium (Mg)	Calcium (Ca)	Sodium (Na)	Zinc (Zn)	Iron (Fe)	Manganese (Mn)	Copper (Cu)	Selenium (Se) (ug)
Wheat	1170	1550	250	170	20	8	12	5	1	5.8
Rye	3010	4380	930	330	50	28	28	22	3	2.8
Barley	2460	3290	670	270	40	13	24	13	1	2.1
Oats	4110	4000	1170	530	40	30	38	58	2	1.0
Rice	1030	1500	350	60	20	17	12	9	2	–
Sorghum	350	240	188	27	5	3	11	1	0.2	–

TABLE 2.6
Vitamins (Matportalen, 2006; Danish Food Composition Databank, version 7.0. 2008; Frølich et al., 2013)

	Ingredient mg/100g dry weight basis					
Cereal	Vitamin E (tocopherol)	Vitamin B1 (thiamine)	Vitamin B2 (riboflavin)	Vitamin B3 (niacin)	Vitamin B6 (pyridoxin)	Vitamin B9 (folate) (ug)
Wheat	1.2	0.4	0.1	4.1	0.3	35
Rye	1.0	0.4	0.2	1.2	0.3	48
Barley	0.4	0.2	0.1	4.5	0.3	25
Oats	0.8	0.5	0.1	1.6	0.2	45

TABLE 2.7
Total Antioxidant Activity of Some Cereal Grains (Zhu et al., 2015; Naji et al., 2016; Serafino et al., 2018; Kulichová et al., 2019)

Cereal	Phytochemicals
Wheat	800 ± 40 µmol gallic acid equivalents/100 g 76.70 ± 1.38 µmol of vitamin C equivalent/g of grain
Rye	3369.19 mg gallic acid equivalents/1000 g
Oats	650 ± 20 µmol gallic acid equivalents/100 g 74.67 ± 1.49 µmol of vitamin C equivalent/g of grain
Rice	560 ± 20 µmol gallic acid equivalents/100 g 55.77 ± 1.62 µmol of vitamin C equivalent/g of grain
Barley	166.21–713.25 gallic acid mg/100 g dry weight basis 1350 to 2290 mg of ferulic acid equivalents/100 g on the basis of lyophilized weight

TABLE 2.8
Phenolic Acids, Alkyl- and Alkenylresorcinols, Avenanthramides in Cereal Grains (Lampi et al., 2008; Nyström et al., 2008; Schlemmer et al., 2009; Frølich et al., 2013; Kulichová et al., 2019)

	Ingredient ug/g dry matter basis					
Cereal	Phytic acid	Tocols	Phenolic acids	Phytosterols	Alkylresorcinols	Avenanthramides
Wheat	390–1.350	28–80	326–1.171	670–960	220–650	Not present
Rye	540–1.460	44–67	491–1.082	1.089–1.420	797–1.231	Not present
Barley	380–1.116	46–69	254–675	899–1.153	32–103	Not present
Oats	420–1.160	16–36	351–873	618–682	Not present	42–91

anti-nutritive agent: it binds minerals (Ca, Mn, Mg, Fe), making absorption in the intestine difficult or even blocking the absorption of minerals (Kumar et al., 2010).

2.3 SOURDOUGH POTENTIAL FOR IMPROVING NUTRITIONAL PROPERTIES

Sourdough has great potential for improving the nutritional properties of cereal bread. Fermentation can increase nutritional value and enhance health-promoting effects (Verni et al., 2019). Whole grains have long been known for their high content of nutrients, bioactive substances, and their health-promoting properties as already described in the previous sections (Seal et al., 2021; Kaur et al., 2021). Sourdough fermentation stabilizes or increases the concentration of bioactive substances, such as vitamins and secondary plant compounds (Figure 2.3) (Schmidt and Sciurba, 2021).

Sourdough is a dough for the production of baked goods that is kept in fermentation temporarily or permanently by means of homo- ("only lactic acid-producing") and heterofermentative ("lactic as well as acetic acid-producing") lactic acid bacteria (LAB) and yeasts. In sourdough, "homofermentative" and "heterofermentative" LAB are essentially (Bangar et al., 2021; Fidan et al., 2022).

Sourdough is added as a leavening agent to loosen baked goods and makes rye dough more bakeable. Sourdoughs improve digestibility, aroma, flavor, shelf life, and cut of baked goods, and likewise, nutritional properties are improved (Pitsch et al., 2021).

Sourdough breads contain a variety of flavors and aromas. Only a small number of the more than 300 fragrance chemicals that are known actually determine flavor and aroma. Although many of the ingredients involved have a mild fragrance, flour already contains several of them (Pétel et al., 2017). These fragrance compounds have a strong development due to fermentation and the synthesis of esters (from ethanol and organic acids) in the sourdough (De Luca et al., 2021). While the carrying aroma compounds such as methylbutanol and diacetyl increase in concentration, the undesirable aromas are reduced; for instance, hexanol, a molecule with a grassy aroma, is degraded in sourdough. Despite the fact that whole grain breads have health benefits, it has several sensory qualities that may make consumers less likely to buy it. A good example is rye bread, which has a sour and a strong bitter flavor that is enhanced by the sourdough process (Koistinen et al., 2016).

Wheat doughs made from sourdough – especially whole wheat doughs – bind moisture and ensure better and easier processing due to good swelling. This is probably also due to the increased solubility of the pentosans. Gluten development is improved, with the dough becoming more extensible (Struyf et al., 2017).

In contrast to wheat doughs, rye doughs require acidity for easy baking. The causes are the nature of the gluten proteins, the gelatinization behavior of the starch and the characteristics of its enzymes (e.g. α-amylases) (Jekle et al., 2016). These α-amylases are highly active because rye cereals do not have any dormancy. Rye starch gelatinizes at a temperature of 49–56°C, resulting in an overlap with the optimum temperature of rye amylases, which is 45–50°C. Rye doughs do not form a

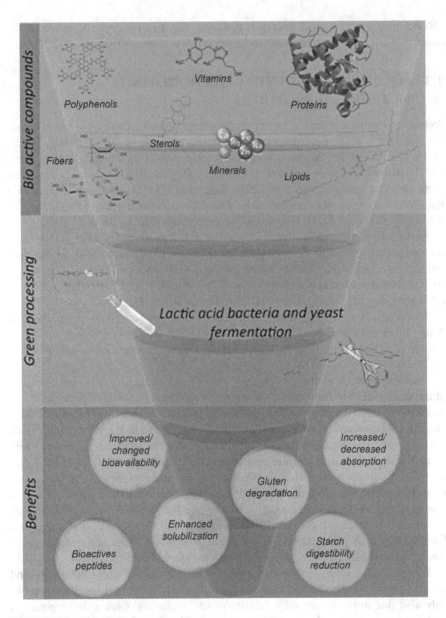

FIGURE 2.3 Potential of sourdough.

gluten "skeleton," the task of which (crumb formation) is mainly performed by water-binding pentosans during the baking process. The rye starch is deposited in these pentosans and forms a stable crumb after gelatinization (Oest et al., 2020). Without an acidic environment, the crumb would be broken down by the flour enzymes; the bread would remain flat and would be inedible. At a pH value of 4.1, the organic acids of sourdough cause the complete cessation of amylase activities (Stępniewska et al., 2021).

Rye is particularly characterized by its high proportion of pentosans. Pentosans belong to the group of hemicelluloses, also known as mucilage, and are an essential part of the cell wall (Skendi et al., 2020). They are important as dietary fiber in the human diet. The feeling of hunger decreases because during digestion pentosans bind with water and swell up (Hervik and Svihus, 2019).

Lactic acid bacteria can be found in flours, air, and water (Bangar et al., 2021). Therefore, sourdough can be made by mixing equal amounts of flour and water and leaving it covered at room temperature for about two days. If the dough then smells pleasantly sour, can be assumed that likely the LAB have taken hold, by producing organic acids and other metabolites that enhance flavor development, and have created an environment that prevents spoilage (Fidan et al., 2022). During the fermentation of bread sourdough, use of a combination of LAB (e.g. *Lactiplantibacillus plantarum*, *Lacticaseibacillus casei*, *Furfurilactobacillus rossiae*, etc.) and yeast produces compounds related to sour aromas of bread and contribute to bread flavor and aroma characteristics (Novotni et al., 2020). In addition to lactic acid, the heterolactic fermentation will produce other organic acids: for example, pyruvic acid can be degraded into acetic acid, formic acid, and ethanol under specific conditions, as well as acetic acid and carbon dioxide. Various conditions (fermentation time, mixing ratio, temperature) can influence the proportions of the individual fermentation products, and thus the properties and taste (effect) of the sourdough (Lau et al., 2021). During this process, enzymatic processes take place in which odor and flavor substances are created. In addition, precursors for flavor formation during the baking process are formed during dough ripening (Maillard reaction).

In principle, phytin in bread is insignificant for the human organism because it is not thermally stable; however, minerals remain bound in it. During digestion, phytin inhibits the release of minerals, trace elements and vitamins by binding these substances in a complex manner and also blocking digestive enzymes (Yu et al., 2018). Only the enzyme phytase can break down phytin and release the nutrients. In sourdough – due to the low pH value – the cereal's own enzymes break down phytate (Samtiya et al., 2020).

Although phytase is present in bread cereals, it can only be activated by long exposure to moisture in the grain or flour. Long swelling times and acidification of rye dough are therefore important, as LAB also participate in phytase release (Milanović et al., 2020). The darker flours (related to higher content of polyphenolic compounds as well) are associated with an increase in swelling and acidification (Ravisankar et al., 2020). Rye is particularly rich in phytase. Breads made from rye flour or rye meal generally contain less phytic acid than wheat bread (Hoppe, et al., 2019), especially if the breads have been professionally baked. When making rye bread, sufficient swelling time and acidification of the dough also play an important role with regard to the phytic acid content. The hydrolysis of phytic acid by an enzyme called phytase is known to increase the bioavailability of micronutrients (Frølich et al., 2013).

Because enzymes are concentrated in the embryo of the grain, at the periphery of the endosperm, and in the aleuronic layer, they are found in higher proportions in high-extraction flours and in lower proportions in white flours (Snell et al., 2022). Enzymes are biochemical catalysts produced by living cell organisms. Enzymes condition the volume, porosity, appearance of the kernel, and color and flavor of the shell, as well as the elasticity and consistency of the dough through their activity (Leitgeb et al., 2022). In conclusion, the quality and nutritional content of bakery goods prepared from cereal flour can be affected by the chemical makeup of the flour.

Soluble fiber content is also high in fermented wheat brans, and also in sourdough production from rye (Mihhalevski et. al., 2013). The prebiotic effect of soluble fibers (ß-glucans and arabinoxylans and their oligomeric products), galacto-oligosaccharides (GOSs), fructo-oligosaccharides (FOSs), resistant starch, phenolics, peptides, and other bioactive molecules contained in various cereal brans are the focus of optimization of fermentation and enzymatic pretreatment processes in order to use the value of cereal brans and to produce by-products. (Tsafrakidou et al., 2020).

2.4 HEALTH BENEFITS ASSOCIATED WITH CEREALS AND CEREAL SOURDOUGH CONSUMPTION

According to World Health Organization (WHO, 2013), the consumption of whole grains seems to mitigate the effects of non-communicable diseases (NCDs) such as heart disease, high blood pressure, type II diabetes, and obesity (Huang et al., 2022; Khan et al., 2022; Kwon et al., 2022).

The intestine has the largest immune system in the human body. The intestinal mucosa serves as a barrier between the bloodstream and the digestive contents (Vancamelbeke et al., 2017). The intestinal cells and the intestinal immune system are influenced by the microbiota of the intestinal mucosa (Afzaal et al., 2022). The entire immune system may be significantly impacted by alterations in the microflora. For optimal care of the intestinal flora, suitable foods, such as rye, are recommended to be consumed. Rye contains the largest percentage of dietary fiber in comparison to other cereals (Jonsson et al., 2018). This encourages the development of a healthy gut lactic acid and bifidobacteria flora. This causes specific food fibers to ferment, generating lactic and acetic acids that lower the pH of the colon. The foundation for a healthy and balanced cohabitation of gut bacteria is a pH that is slightly acidic. Rye's high dietary fiber content and wide range of more than 2000 bioactive components have positive impacts on preserving our bodies' overall health (Dziki, 2022). Many bioactive compounds, which among other things have an antioxidant effect, are transformed by intestinal bacteria and made more available to the body, thus achieving a greater effect.

In addition to obesity, diet is a lifestyle factor that is directly involved in altering the natural intestinal microbiota, with the resulting metabolites having the ability to modulate human health (Tözün and Vardareli, 2016; Rakha et al., 2022). Additionally, leptin genetic mutations associated with a high-fat diet and an obese condition affect microbiome profiles, which may lead to recommendations of new diet- and/or genetics-based therapies for obesity and other related comorbidities in people (Nagpal et al., 2020; Joung et al, 2021).

Obesity is a metabolic disorder and a major public health concern worldwide, being a critical risk factor and an etiological condition for many associated disorders. Obesity is also at the top of the list of disorders affecting humans all over the world and can be considered a global epidemic with a complex of interaction profiles related to environment, behavior, and genetics, which are still under debate – an essential issue in medicine, society, and economics (Shabana, et al., 2016; Socol et al., 2022).

Leptin is a key molecule in the hypothalamus which regulates food intake, body weight control, and energy balance. It is also a multifunctional hormone with different roles, not fully understood up to now (Fischer-Posovszky et al., 2010). The leptin molecule is involved in many physiological processes i.e. energy metabolism, endocrine and immune physiological mechanisms, playing major double roles as a hormone, on one side, and as a cytokine, on the other side. The leptin function is to regulate appetite and obesity development, as a minor increase of leptin levels leads to appetite and body weight decrement. Leptin hormone can be used in obesity treatment, showing positive effects on hunger, energy expenditure, and overall on behavior (Qadir and Ahmed, 2017; Socol et al., 2022; Rakha et al., 2022).

Cereal fiber-based diets (mainly wheat bran fiber) result in improved leptin resistance and sensitivity, as reported in high-fat/cholesterol diets of mice, by mechanisms consisting in increasing JAK2 and STAT3 protein expression which sensitizes leptin signaling and in decreasing SOCS3 protein expression, which blunts leptin signaling; wheat bran fiber shows a high content in insoluble fiber, which could influence more body weight management rather than bran fiber with a rich soluble fiber profile (Zhang et al. 2016; Socol et al., 2022).

Leptin mechanism is also related to its binding ability to leptin receptors and thus leads to downstream signaling initiation by tyrosine kinase JAK2 phosphorylation and transcription of factor signal transducer and STAT3 activators, followed by STAT phosphorylation and transcription modulation of key genes interfering in food intake and energy balance. SOCS3 is able to inhibit leptin functions by means of a negative leptin-induced STAT3 signaling feedback mechanism, by SOCS3 induction. SOCS3 inactivation was found in high-fat diet mice showing leptin resistance (Pedroso et al., 2014), but SOCS3 protein expression can be downregulated by a cereal fiber diet, thus improving leptin resistance and sensitivity related to the JAK2/STAT3 pathway, based on differences among cereal fiber types and its molecular structure with physiological roles that are not completely understood (Gemen et al., 2014).

A long-term cereal fiber diet has positive influence on the levels of cholesterol and triglycerides serum lipids, due to the soluble fiber potential to decrease plasma cholesterol. This is explained by the fact that soluble fiber is fermented by intestinal microflora, and thus fermentation may be able to change short-chain fatty acid production, thereby decreasing acetate synthesis and increasing propionate synthesis (Breneman and Tucker, 2013).

The gut microbiome is able to survive mostly due to dietary fiber, which is not digested in the upper part of the gastrointestinal segment; the fermentation of dietary fibers by the gut microbial community leads to short-chain fatty acid (SCFA) production (Afzaal et al., 2022). Mostly butyrate provides the energy supply for epithelia cells; 50–70% of acetate is taken up by the liver and used mostly as a substrate for cholesterol synthesis; and propionate is also taken up by the liver and used in various synthesis pathways, such as gluconeogenesis, lipogenesis, etc. Since GPR41 is a G protein-coupled receptor that is expressed by some gut epithelium cells and also serves as a sensor for SCFAs, SCFAs like propionate and acetate are ligands for this receptor. As a result, after binding to GPR41, leptin and peptide tyrosin–tyrosin are secreted, which modulates gut motility and lowers energy absorption from food (Priyadarshini et al., 2018; Hanhineva et al., 2012).

Serum leptin role as a cytokine is associated with markers of insulin resistance, and it reduces stored body fat. Diet-induced obesity may cause leptin resistance, described by high serum leptin levels and decreased leptin sensitivity (Socol et al., 2022).

A high-fat diet leads to high leptin levels, however, a high-fat diet supplemented with cereal fiber has been shown to inhibit serum leptin and to improve leptin resistance and sensitivity (Zhang et al. 2016).

It is well known that a fiber-rich diet is able to give protection against many diseases and obesity, whereas a red meat, animal fat, and low–fiber-based diet shows potential positive effect; high-fat diet and leptin profiles could lead to obesity, altering intestinal microbiome profiles and functionality (Nagpal et al., 2020).

Depending on the type of cereals, whole grains may also be a valuable source of bioactive phytochemicals that are useful for the prevention and treatment of a variety of disorders, including obesity (Xie et al., 2019). The colonic effect of cereal fiber is related to an increase in the fermentation SCFAs, a decrease in the production of free fatty acids and glucose by the liver, an increase in insulin sensitivity, and a decrease in insulin secretion (Mikušová et al. 2011; Socol et al., 2022).

Whole grain and dietary fiber can influence satiety by promoting and enhancing it, thus resulting in lower nutrient absorption, next to other effects consisting of glucose rate absorption reduction, insulin release delay, and interference of glycemic response, all together concurring to changes in body weight management (Mikušová et al. 2011). Satiety is also influenced by satiety-related hormone levels including glucagon-like peptide-1, peptide tyrosine–tyrosine (PYY), and leptin, cereal dietary fiber interfering responses and actions of those hormones (Mikušová et al. 2011).

The evidence that whole grain-based diets decrease the risk of obesity, next to the legumes' potential to concur to the effect of such diets, are still insufficient proofs of the protective influence of legumes on body weight (Mikušová et al. 2011), requiring further and complex approaches – maybe including leptin too.

Phytochemicals are non-essential nutrient bioactive compounds. The main classes of bioactive phytochemicals found in whole grains and rye are related to alkylresorcinols (wheat, rye) which are involved in obesity reduction; to lignans of the phenolics (e.g. in wheat and rye) acting in hormone modulation, α-carotene/β-carotene of the carotenoids (wheat) causing neutralization of free radicals (Suchowilska et a., 2020); and to sterols and stanols of the phytosterols (wheat), leading to low serum cholesterol levels, decreased lipid accumulation, and insoluble dietary fiber of the non-starchy polysaccharides (wheat). The last ones are also involved in the decrement of plasma cholesterol and lowering insulin resistance levels (Nurmi et al., 2008; Xie et al., 2019; Iftikhar et al., 2020). Alkylresorcinols (1,3-dihydroxy-5-n-alkylbenzenes), are a major group of phenolic lipids present in the whole grain of wheat and rye, which can be absorbed and detected in plasma and used as biomarkers for whole grain wheat/rye consumption (Xie et al., 2019).

Whole grain dietary fiber also shows potential to decrease serum estrogen levels by suppressing bacterial β-glucuronidase activity, increasing transient period and peristaltic activities at the gut level (Dong et al., 2011), influencing gut microbial structure and all related processes.

Sourdough fermentation by LAB can alter metabolites, may be involved in health-promoting mechanisms related to the whole grain of wheat and rye, and shows variation in different compound levels of branched-chain amino acids (BCAAs), BCAA derivate metabolites, small peptides with high BCAA content, microbial metabolites of phenolic acids, and other potentially bioactive molecules. These molecules might have a protective role in the development of many disorders associated with NCDs (Zanoaga et al., 2019; Koistinen et al, 2018; Jimenez-Pulido et al., 2022).

The pathophysiology of metabolic disorders, including obesity, involves particular classes of microbiota-derived metabolites like bile acids, SCFAs, BCAAs, trimethylamine N-oxide, tryptophan, and indole derivatives (Agus et al., 2021; Feng et al., 2022). BCAAs are essential amino acids that include valine, leucine, and isoleucine. BCAAs are important circulating nutrition signals, their related metabolites being positively associated with insulin resistance mechanisms. For example, BCAA levels were positively correlated with leptin ones, both in men and women (Katagiri et el., 2018); but there are also studies not reporting a significant correlation between BCAA and leptin levels – only an inverse association between BCAA and adiponectin levels (Nakamura, et al., 2014).

Metagenomics showed that bacterial communities usually consist of hundreds or thousands of taxa, mainly of two phyla: *Firmicutes* and *Bacteroidetes*. Gut microbial structure in obesity shows lower diversity, specifically a lower Bacteroidetes prevalence and a higher Firmicutes compared to the non-obesity state. Altered gut microbiota show potential to modify obesity and metabolic phenotype (Ridaura et al., 2013), based on transmissible and modifiable interactions between cereal-based diet and microbiome influence host lifestyle (Bifari et al., 2017).

In obese subjects, BCAA intake is associated with decreased body weight and body fat (Zemel et al., 2012). Cereal dietary fiber and phytochemicals i.e. phenolic acids and alkylresorcinols within matrix fiber, with proven *in vitro* antioxidative, antimicrobial, and anti-cancer effects, interfere with gastrointestinal microbiome resulting in alterations of metabolism mechanisms and generating effects, most all of which are unknown (Ross et al. 2004; Jimenez-Pulido et al., 2022; Li et al., 2008). To fully understand the altered metabolite profiles associated with sourdough fermentation's small-molecule metabolism, which primarily relates to the bioactive compounds' interference with and contribution to the health benefits of whole grain-derived products, as well as possibly including leptin-related mechanisms, more research is required (Koistinen et al., 2018).

After sourdough fermentation, various microbial metabolites of phenolic acid are present in high concentrations. These metabolites have different absorption and metabolic properties than their precursors, all these possibly being related to the bioactivity of these molecules (Sánchez-Maldonado et al., 2011; Koistinen et al., 2018; Pihlava et al., 2015).

Furthermore, there are studies demonstrating wheat availability as an independent predictor of obesity, demanding further research to differentiate the possible influence of whole grain and ultra-processed, refined, wheat-based meals (You and Henneberg, 2016).

Leptin plays a role in the intestinal epithelium which indicate an independence of food intake and influence on different antimicrobial peptide (AMP) gene expression, proving that leptin signaling can shape microbial composition (Rajala et al., 2014). Moreover, autoimmune infectious diseases or altered microbiota that lead to chronic inflammation processes show potential to influence leptin response, producing the resistance mechanism and, in this way, the obesity (Pérez-Pérez et al., 2020).

Fiber-rich foods and dietary products were proved to correlate with breast cancer prognosis (Andersen et al., 2020). But the response to the dietary products are much more complicated, therefore it should take into account the menopausal status, hormone receptor status (Estrogen receptor (ER) and Progesterone receptor (PR)), and the histological and molecular features of breast cancer. The intake of whole grain products was not associated with risk of breast cancer in a cohort of Danish postmenopausal women. No relationship has been noticed between the consumption of total or specific whole grain products and the risk of developing ER+, ER–, PR+, PR– and combined ER/PR status in breast cancer (Egeberg et al., 2009). Interestingly, high rye bread intake in adolescence and midlife correlated with an increased risk of late-life breast cancer (Haraldsdottir et al., 2018).

Alkylresorcinols (ARs) are considered valuable markers for consumption of rye and whole grain wheat. A Finish study, propose to correlate AR in breast cancer, considering the healthy dietary habits related to a reduction risk for breast cancer by its effect on enterohepatic circulation of estrogens (Aubertin-Leheudre et al., 2010).

The beneficial effect related to the consumption of the whole grain rye is the fact that it promoted a reduced concentration of inflammatory biomarkers, as an effect of high fiber content and a wide range of bioactive compounds (Jonsson et al., 2018). The beneficial protective effect is probably due to the fact that fiber diminishes the enterohepatic circulation of estrogens, leading to lower plasma estrogen concentrations. Regarding bioactive compounds, we mention the lignans and alkylresorcinols that have antioxidative and possibly anticarcinogenic effects (Adlercreutz et al., 2010). Phytic acid can protect tissues against oxidative reactions through its ability to sequester and inactivate pro-oxidative transition metals (Baublis et al., 2000).

Cereal grains are an excellent source of B vitamins: thiamine (B1), riboflavin (B2), niacin (B3), pyridoxine (B6), biotin (B7), and folic acid (B9). B vitamins are important for the nervous system, involved in cell division and tissue structure; important for metabolism (energy balance, protein, fat, and carbohydrate metabolism);

and function as coenzymes. Vitamins B1 and B2 also support various metabolic processes as cofactors. Folic acid is also important for detoxifying homocysteine. One hundred grams of rye a day could provide up to 50% of the recommended daily intake of several important vitamins and minerals (Garg et al., 2021).

German Institute for Nutritional Research examined the influence of dietary fiber on the risk of diabetes. While the long-term blood sugar level and, thus, the risk of diabetes rose continuously in the placebo group, the fiber group was able to maintain its long-term blood sugar level (The InterAct Consortium, 2015).

People with bowel problems often experience symptoms, such as diarrhea, constipation, gas, stomach pain, or nausea when ingesting FODMAPs. These carbohydrates are not inherently bad, but a low-FODMAP diet can be helpful for certain ailments such as irritable bowel syndrome (IBS). However, as randomized, double-blind, controlled crossover studies with patients have shown, the dough preparation until the dough is baked – plays a decisive role in terms of the tolerance of rye bread. The researchers concluded that the symptoms were less severe when the low-FODMAP rye bread was eaten (Laatikainen et al., 2016; Pirkola et al., 2018). Whole meal rye breads are unsuitable for IBS patients even though studies show a FODMAP reduction as a result of altered processing factors, such as longer proofing times or the use of sourdough (Schmidt and Sciurba, 2021).

Rye can be eaten by fructose intolerant consumers. The ratio of fructose and glucose in rye is also completely balanced, which further optimizes tolerance. However, if the intestine is weakened due to an intolerance such as fructose intolerance, symptoms from FODMAPs can also occur. In this sense, too, it is important to rely on low-FODMAP rye bread (Loponen and Gänzle, 2018).

Cereal grains are an excellent source of trace elements with high relevance for the human body (e.g. Fe, Zn, Mn). Fe is part of hemoglobin, the red blood pigment that stores and transports oxygen, and is also important for the immune system and a component of metabolic enzymes. Zn is also an important trace element for the immune system and important for the structure and function of cell walls (Prasadi and Joye, 2020). Together with Mn and Cu, it is part of the enzyme system. Mn is important for building bones, joints, and the nervous system (Horning et al., 2015).

In the human large intestine, lignans are converted into biologically effective hormone-like compounds, so-called phyto (i.e. plant) estrogens, with the help of bacteria from the intestinal flora. These phytoestrogens have a much lower bio-effectiveness than the body's own estrogens, but still seem to play a positive role in the metabolism of the hormones. A high concentration of the sex hormone estrogen is considered a risk factor for breast and uterine cancer. According to the results of cell tests, phytoestrogens block the receptors for human estrogens and thus reduce the risk of developing hormone-dependent cancers. The same applies to colon cancer, since colon cells also have estrogen receptors, thus phytoestrogens could possibly also block receptors in the colon and thus weaken the cancer-promoting effect of the estrogens (Bustamante-Barrientos et al., 2021). This may prevent or at least slow down tumor growth. Phytoestrogens from grain may strengthen the function of the blood vessels and have an antioxidant and immunomodulating effect.

The rye grain is an interesting source of natural compounds with potent biological activities. Increased consumption of foods containing rye, as a whole grain or as whole grain flour, can lead to the consumption of phytochemicals beneficial to the consumer's health (Jonsson et al., 2018; Kulichová et al., 2019).

Nutrition researchers have been able to detect high levels of phenolic acids, such as ferulic acid, in wheat in particular. These compounds primarily have an antioxidant effect, i.e. they render aggressive oxygen compounds harmless, which are formed in the body itself or through external influences. In this way they prevent these "free radicals" from damaging body cells and contributing to the development of cancer (Laddomada et al., 2015; Jimenez-Pulido et a., 2022).

2.5 FUTURE PERSPECTIVES AND RECOMMENDATIONS

Consumer preferences for cereal foods have been seen to be based on the popularity of wheat, rye, and, more recently, barley and oat, which are known to have health-promoting properties due to their distinctive biological active components and high nutritional content.

Fermentable oligo-, di-, monosaccharide and polyol levels in wheat, rye, and barley are considerably high. Sourdough is a dough used to make baked goods that is kept in fermentation either temporarily or permanently utilizing yeasts and bacteria that both produce homo- and heterofermentative lactic acid. Enhanced nutritional qualities also come with improved digestibility, aroma, flavor, and shelf life.

Dietary fiber, although mostly indigestible to humans, has a variety of effects in the human body. The best known is the satiating and digestive effect of fiber. Scientific studies have also shown that high-fiber meals lead to lower blood sugar levels in diabetics. The best sources of fiber are whole grain bread and baked goods. They contain all parts of the grain, including the outer layers in which the bioactive substances are particularly abundant. With a 42% share of the daily intake, cereal products are the most important suppliers of the feel-good factor "dietary fiber." Consuming enough fiber also improves the consumer's cholesterol levels, reduces inflammatory processes, and lowers their risk of obesity, heart attack, arteriosclerosis, and colon cancer (Eriksen, et al., 2020).

On one side, it is believed that almost all diseases of civilization can be cured simply by avoiding wheat in the daily diet. On the other side, in order to achieve a favorable concentration of recommended bioactive substances, scientists encourage consuming between 150 and 250 g (3–5 slices) of rye bread daily. The average dietary fiber intake in Germany is 20 g per day, but the German Nutrition Society recommends consuming at least 30 g of dietary fiber per day, half of which comes from grain products. With five slices of mixed bread and five servings of fruit and vegetables, this recommendation can be implemented easily and tastily, true to the motto "5 a day."

Mixed cereals can be referred as a valuable source of nutrients including dietary fiber, resistant starch, oligosaccharides, trace elements, vitamins, and other compounds related to disease prevention, mostly phytoestrogens and antioxidants.

REFERENCES

Adlercreutz H. Can rye intake decrease risk of human breast cancer? *Food Nutr Res.* 2010 Nov 10;54. doi: 10.3402/fnr.v54i0.5231

Agriculture, Forestry and Fishery Statistics: 2018 Edition, Eurostat, 2018 EU Agricultural Outlook for Markets and Income 2018-2030, European Commission, 2018.

Agus, A., Clément, K, Sokol, H. Gut microbiota-derived metabolites as central regulators in metabolic disorders. *Gut* 2021;70:1174–1182.

Andersen, J.L.M., Hansen, L., Thomsen, B.L.R., Christiansen, L.R., Dragsted, L.O., Olsen, A. Pre- and post-diagnostic intake of whole grain and dairy products and breast cancer prognosis: The Danish diet, cancer and health cohort. *Breast Cancer Res Treat.* 2020 Feb;179(3):743–753. doi: 10.1007/s10549-019-05497-1

Andersson, A.A.M., Andersson, R., Piironen, V., Lampi, A.M., Nyström, L., Boros, D., et al. Contents of dietary fibre components and their relation to associated bioactive components in whole grain wheat samples from the HEALTHGRAIN Diversity Screen. *Food Chem.* 2013;136:1243–1248.

Andersson, A.A.M., Kamal-Eldin, A., Fras, A., Boros, D., Aman, P. Alkylresorcinols in wheat varieties in the HEALTHGRAIN diversity screen. *J Agric Food Chem.* 2008a;56:9722–9725.

Andersson, A.A.M., Lampi, A.-M., Nyström, L., Piironen, V., Li, L., Ward, J.L., et al. Phytochemicals and dietary fiber components in barley varieties in the HEALTHGRAIN Diversity Screen. *J Agric Food Chem.* 2008b;56:9767–9776.

Andersson, R., Fransson, G., Tietjen, M., Åman, P. Content and molecular weight distribution of dietary fibre components in whole grain rye flour and bread. *J Agric Food Chem.* 2009;57:2004–2008.

Anjum, F.M., Ahmad, I., Butt, M.S., Sheikh, M.A., Pasha, I. Amino acid composition of spring wheats and losses of lysine during chapati baking. *J Food Compos Anal.* 2005;18:523–532. doi: 10.1016/j.jfca.2004.04.009

Aubertin-Leheudre, M., Koskela, A., Samaletdin, A., Adlercreutz, H. Plasma and urinary alkyl-resorcinol metabolites as potential biomarkers of breast cancer risk in Finnish women: A pilot study. *Nutr Cancer.* 2010;62(6):759–764. doi: 10.1080/01635581003693058

Afzaal, M., Saeed, F., Shah, Y.A., Hussain, M., Rabail, R., Socol, C.T., Hassoun, A., Pateiro, M., Lorenzo, J.M., Rusu, A.V., Aadil, R.M.. Human gut microbiota in health and disease: Unveiling the relationship. *Front Microbiol.* 2022;13:999001.

Balakireva, A.V., Zamyatnin, A.A.. Properties of gluten intolerance: Gluten structure, evolution, pathogenicity and detoxification capabilities. *Nutrients.* 2016;8:644.

Bangar, S.P., Sharma, N., Kumar, M., Ozogul, F., Purewal, S.S., Trif, M. Recent developments in applications of lactic acid bacteria against mycotoxin production and fungal contamination. *Food Biosci.* 2021;44:101444.

Bangar, S.P., Sandhu, K.S., Rusu, A., Trif, M., Purewal, S.S. Evaluating the effects of wheat cultivar and extrusion processing on nutritional, health-promoting, and antioxidant properties of flour. *Front Nutr.* 2022a;9:872589.

Bangar, S.P., Sandhu, K.S., Trif, M., Lorenzo, J.M. The effect of mild and strong heat treatments on in vitro antioxidant properties of barley (*Hordeum vulgare*) cultivars. *Food Anal Methods.* 2022b; 15, 2193–2201.

Bangar, S.P., Sandhu, K.S., Trif, M., Manjunatha, V., Lorenzo, J.M. Germinated barley cultivars: Effect on physicochemical and bioactive properties. *Food Anal Methods.* 2022c; 15: 2505–2512.

Bangar, S.P., Suri, S., Trif, M., Ozogul, F. Organic acids production from lactic acid bacteria: A preservation approach. *Food Bioscience*, 2022d;46:101615.

Baublis, A.J., Lu, C., Clydesdale, F.M., Decker, E.A. Potential of wheat-based breakfast cereals as a source of dietary antioxidant. *Journal of the American College of Nutrition* 2000;19(3) Supplement:308S–311S.

Bellini, M., Rossi, A. Is a low FODMAP diet dangerous? *Tech Coloproctol.* 2018;22:569–571.

Bifari, F., Ruocco, C., Decimo, I., Fumagalli, G., Valerio, A., Nisoli, E. Amino acid supplements and metabolic health: A potential interplay between intestinal microbiota and systems control. *Genes Nutr.* 2017 Oct 4;12:27. doi: 10.1186/s12263-017-0582-2

Breneman, C.B., Tucker, L. Dietary fibre consumption and insulin resistance the role of body fat and physical activity. *Br J Nutr.* 2013;110:375–383.

Brestenský, M., Nitrayová, S., Patráš, P. Comparison of two methods of protein quality evaluation in rice, rye and barley as food protein sources in human nutrition. *Potravinarstvo.* 2018; 8:1.

Brouns, F., Hemery, Y., Price, R., Anson, N.M. Wheat aleurone: Separation, composition, health aspects, and potential food use. *Crit Rev Food Sci Nutr.* 2012;52(6):553–568. doi: 10.1080/10408398.2011.589540

Bustamante-Barrientos, F.A., Méndez-Ruette, M., Ortloff, A., Luz-Crawford, P., Rivera, F.J., Figueroa, C.D., Molina, L. and Bátiz, L.F. The impact of estrogen and estrogen-like molecules in neurogenesis and neurodegeneration: Beneficial or harmful? *Front Cell Neurosci.* 2021;15:636176. doi: 10.3389/fncel.2021.636176

Călinoiu, L.F., Vodnar, D.C. Whole grains and phenolic acids: A review on bioactivity, functionality, health benefits and bioavailability. *Nutrients.* 2018;10(11):1615.

Danish Food Composition Databank, version 7.0. 2008. Available from: http://www.foodcomp.dk/

Danish Food Composition Databank. http://www.foodcomp.dk/fcdb_search.asp (accessed 1 August 2021).

De Luca, L.; Aiello, A.; Pizzolongo, F.; Blaiotta, G.; Aponte, M.; Romano, R. Volatile organic compounds in breads prepared with different sourdoughs. *Appl. Sci.* 2021;11:1330. doi: 10.3390/app11031330

Deleu, L.J.; Lemmens, E.; Redant, L.; Delcour, J.A. The major constituents of rye (Secale cereale L.) flour and their role in the production of rye bread, a food product to which a multitude of health aspects are ascribed. *Cereal Chem.* 2020;97:739–754.

Demirel, F.; Germec, M.; Turhan, I. Fermentable sugars production from wheat bran and rye bran: Response surface model optimization of dilute sulfuric acid hydrolysis. *Environ. Technol.* 2021;2:1–22.

Devanand, L.L.; Lu, Y.; John, K.M.M. Bioactive phytochemicals in wheat: Extraction, analysis, processing, and functional properties. *J Funct Foods* 2015;18:910–925.

Dhingra, D., Michael, M., Rajput, H., Patil, R.T. Dietary fibre in foods: A review. *J Food Sci Technol.* 2012;49(3):255–266. doi: 10.1007/s13197-011-0365-5

Di Sabatino, A., Lenti, M.V., Corazza, G.R., Gianfrani, C. Vaccine immunotherapy for celiac disease. *Front Med.* 2018;5:187.

Dong, J.Y.; He, K.; Wang, P.; Qin, L.Q. Dietary fiber intake and risk of breast cancer: A meta-analysis of prospective cohort studies. *Am J Clin Nutr.* 2011;94:900–905.

Dziki, D. Rye flour and rye bran: New perspectives for use. *Processes* 2022;10:293.

Egeberg R, Olsen A, Loft S, Christensen J, Johnsen NF, Overvad K, Tjønneland A. Intake of whole grain products and risk of breast cancer by hormone receptor status and histology among postmenopausal women. *Int J Cancer.* 2009 Feb 1;124(3):745–750. doi: 10.1002/ijc.23992

El Khoury, D., Cuda, C., Luhovyy, B.L., Anderson, G.H.. Beta glucan: Health benefits in obesity and metabolic syndrome. *J Nutr Metab.* 2012;2012:851362. doi: 10.1155/2012/851362

Eriksen, A.K.; Brunius, C.; Mazidi, M.; Hellström, P.M.; Risérus, U.; Iversen, K.N.; Fristedt, R.; Sun, L.; Huang, Y.; Nørskov, N.P.; et al. Effects of whole-grain wheat, rye, and lignan supplementation on cardiometabolic risk factors in men with metabolic syndrome: A randomized crossover trial. *Am J Clin Nutr.* 2020;111:864–876.

Eurostat, 2011; - Eurostat (online data code: apro_cpnh1) (https://ec.europa.eu/eurostat/statistics-explained/index.php?title=Agricultural_production_-_crops&oldid=583740)

FAO. *Dietary Protein Quality Evaluation in Human Nutrition: Report of an FAO Expert Consultation.* Rome: FAO, 2013.

Feng, W., Liu, J., Cheng, H., Zhang, D., Tan, Y., Peng, C. Dietary compounds in modulation of gut microbiota-derived metabolites. *Front Nutr.* 2022 Jul 19;9:939571. doi: 10.3389/fnut.2022.939571

Fidan, H.; Esatbeyoglu, T.; Šimat V; Trif, M.; Tabanelli, G.; Sensoy, I.; Ibrahim, S.A.; Özogul, F. Recent developments of lactic acid bacteria and their metabolites on foodborne pathogens and spoilage bacteria: Facts and gaps. *Food Biosci.* 2022; 47. https://www.sciencedirect.com/science/article/abs/pii/S2212429222002000

Fischer-Posovszky, P.; von Schnurbein, J.; Moepps, B.; Lahr, G.; Strauss, G.; Barth, T.F.; Kassubek, J.; Muhleder, H.; Moller, P.; Debatin, K.-M.; Gierschik, P. and Wabitsch, M. A new missense mutation in the Leptin gene causes mild obesity and hypogonadism without affecting T cell responsiveness. *J Clin Endocrinol Metab.* June 2010;95(6):2836–2840.

Frølich, W., Aman, P., & Tetens, I. Whole grain foods and health: A Scandinavian perspective. *Food Nutr. Res.* 2013;57. doi: 10.3402/fnr.v57i0.18503

Gallardo, C., Jimenez, L., Garcıa-Conesa, M.-T. Hydroxycinnamic acid composition and in vitro antioxidant activity of selected grain fractions. *Food Chem.* 2006;99:455–463.

Garg M, Sharma A, Vats S, Tiwari V, Kumari A, Mishra V, Krishania M. Vitamins in cereals: A critical review of content, health effects, processing losses, bioaccessibility, fortification, and biofortification strategies for their improvement. *Front Nutr.* 2021;16(8):586815.

Garutti M, Nevola G, Mazzeo R, Cucciniello L, Totaro F, Bertuzzi CA, Caccialanza R, Pedrazzoli P and Puglisi F (2022) The impact of cereal grain composition on the health and disease outcomes. *Front Nutr.* 9:888974. doi: 10.3389/fnut.2022.888974

Gemen, R., De Vries, J.F., Slavin, J.L. Relationship between molecular structure of cereal dietary fiber and health effects: Focus on glucose/insulin response and gut health. *Nutr Rev* 2011;69:22–33.

GMF. *2016 mit Nährstoffgehalten aus der Datenbank "Bundeslebensmittelschlüssel" (BLS, 3.02)*, des Max Rubner-Instituts, Karlsruhe, 2014.

Go for the Whole Grain Poster. Available online: https://nutritioneducationstore.com/products/go-for-the-whole-grain-poster (accessed on 25 April 2022).

Hanhineva, K.; Rogachev, I.; Aura, A.M.; Aharoni, A.; Poutanen, K.; Mykkänen, H. Identification of novel lignans in the whole grain rye bran by non-targeted LC-MS metabolite profiling. *Metabolomics* 2012;8:399–409.

Haraldsdottir, A., Torfadottir, J.E., Valdimarsdottir, U.A., Adami, H.O., Aspelund, T., Tryggvadottir, L., Thordardottir, M., Birgisdottir, B.E., Harris, T.B., Launer, L.J., Gudnason, V., Steingrimsdottir, L. Dietary habits in adolescence and midlife and risk of breast cancer in older women. *PLoS One.* 2018 May 30;13(5):e0198017. doi: 10.1371/journal.pone.0198017

Hervik, A.K., Svihus, B. The role of fiber in energy balance. *J Nutr Metab.* 2019 Jan 21;2019:4983657. doi: 10.1155/2019/4983657

Horning, K.J., Caito, S.W., Tipps, K.G., Bowman, A.B., Aschner, M. Manganese is essential for neuronal health. *Annu Rev Nutr.* 2015;35:71–108.

Hoppe, M., Ross, A.B., Svelander, C. et al. Low-phytate wholegrain bread instead of high-phytate wholegrain bread in a total diet context did not improve iron status of healthy Swedish females: A 12-week, randomized, parallel-design intervention study. *Eur J Nutr* 2019;58:853–864. doi: 10.1007/s00394-018-1722-1

Huang, Q.; Hao, L.; Wang, L.; Jiang, H.; Li, W.; Wang, S.; Jia, X.; Huang, F.; Wang, H.; Zhang, B.; Ding, G.; Wang, Z. Differential associations of intakes of whole grains and coarse grains with risks of cardiometabolic factors among adults in China. *Nutrients.* 2022;14:2109.

Iftikhar, M.; Zhang, H.; Iftikhar, A.; Raza, A.; Khan, M.; Sui, M.; Wang, J. Comparative assessment of functional properties, free and bound phenolic profile, antioxidant activity, and in vitro bioaccessibility of rye bran and its insoluble dietary fiber. *J Food Biochem.* 2020;44:e13388.

Ispiryan, L., Zannini, E., Arendt, E.K. Characterization of the FODMAP-profile in cereal-product ingredients. *J Cereal Sci.* 2020; 92. https://www.sciencedirect.com/science/article/abs/pii/S0733521019308392

Jekle, M., Mühlberger, K., & Becker, T.. Starch–gluten interactions during gelatinization and its functionality in dough like model systems. *Food Hydrocolloids.* 2016;54:196–201.

Jiang, X., Tian, J., Hao, Z., & Zhang, W. Protein content and amino acid composition in grains of wheat-related species. *Agricultural Sciences in China*, 2008;7:272–279.

Jideani, A.I.O., Silungwe, H., Takalani, T., Anyasi, T.A., Udeh, H., Omolola, A. *Antioxidant-Rich Natural Grain Products and Human Health, Antioxidant-Antidiabetic Agents and Human Health, Oluwafemi Oguntibeju*, IntechOpen, February 5th 2014. doi: 10.5772/57169

Jimenez-Pulido, I.J.; Daniel, R.; Perez, J.; Martínez-Villaluenga, C.; De Luis, D.; Martín Diana, A.B. Impact of protein content on the antioxidants, anti-inflammatory properties and glycemic index of wheat and wheat bran. *Foods* 2022;11:2049. doi: 10.3390/foods11142049

Jonsson, K., Andersson, R., Knudsen, K.E.B., Hallmans, G., Hanhineva, K., Katina, K., Kolehmainen, M., Kyro, C., Langton, M., Nordlund, E., Laerke, H.N., Olsen, A., Poutanen, K., Tjonneland, A., Landberg, R. Rye and Health: Where do we stand and where do we go? *Trends in Food Science & Technology* 2018; 79:78–87.

Joung, H., Chu, J., Kim, B.K., Choi, I.S., Kim, W., Park, T.S.. Probiotics ameliorate chronic low-grade inflammation and fat accumulation with gut microbiota composition change in diet-induced obese mice models. *Appl Microbiol Biotechnol.* 2021 Feb;105(3):1203–1213. doi: 10.1007/s00253-020-11060-6

Katagiri, R., Goto, A., Budhathoki, S., Yamaji, T., Yamamoto, H., Kato, Y., Iwasaki, M., Tsugane, S. Association between plasma concentrations of branched-chain amino acids and adipokines in Japanese adults without diabetes. *Sci. Rep.* 2018;8(1):1043.

Kaur, P.; Sandhu, K.S.; Bangar, S.P.; Purewal, S.S.; Kaur, M.; Ilyas, R.A.; Asyraf, M.R.M.; Razman, M.R. Unraveling the bioactive profile, antioxidant and dna damage protection potential of rye (Secale cereale) flour. *Antioxidants* 2021;10:1214.

Kelly, P.. *The EU Cereals Sector: Main Features, Challenges and Prospects,* 2019 EPRS (European Parliamentary Research Service. https://www.europarl.europa.eu/RegData/etudes/BRIE/2019/640143/EPRS_BRI(2019)640143_EN.pdf

Khan, J., Khan, M.Z., Ma, Y., Meng, Y., Mushtaq, A., Shen, Q., Xue, Y. Overview of the composition of whole grains' phenolic acids and dietary fibre and their effect on chronic non-communicable diseases. *Int J Environ Res Public Health.* 2022;19(5):3042. doi: 10.3390/ijerph19053042

Khan, J., Khurshid, S., Sarwar, A., Aziz, T., Naveed, M., Ali, U., Makhdoom, S.I., Nadeem, A.A., Khan, A.A., Sameeh, M.Y., Alharbi, A.A., Filimban, F.Z., Rusu, A.V., Göksen, G., Trif, M. Enhancing bread quality and shelf life via glucose oxidase immobilized on zinc oxide nanoparticles—A sustainable approach towards food safety. *Sustainability.* 2022;14:14255.

Koistinen, V.M.; Katina, K.; Nordlund, E.; Poutanen, K.; Hanhineva, K. Changes in the phytochemical profile of rye bran induced by enzymatic bioprocessing and sourdough fermentation. *Food Res Int.* 2016;89:1106–1115.

Koistinen, V.M., Mattila, O., Katina, K., Poutanen, K., Aura, A.M., Hanhineva, K. Metabolic profiling of sourdough fermented wheat and rye bread. *Sci Rep.* 2018 Apr 9;8(1):5684. doi: 10.1038/s41598-018-24149-w

Kulichová, K.; Sokol, J.; Nemeček, P.; Maliarová, M.; Maliar, T.; Havrlentová, M.; Kraic, J. Phenolic compounds and biological activities of rye (Secale cereale L.) grains. *Open Chem.* 2019;17:988–999.

Kumar, V., Sinha, A.K., Makkar, H.P., Becker, K. Dietary roles of phytate and phytase in human nutrition: A review. *Food Chem.* 2010;120:945–959.

Kwon Y-J, Lee HS, Park GE and Lee J.-W. Association between dietary fiber intake and all-cause and cardiovascular mortality in middle aged and elderly adults with chronic kidney disease. *Front Nutr.* 2022;9:863391. doi: 10.3389/fnut.2022.863391

Laatikainen R, Koskenpato J, Hongisto SM, Loponen J, Poussa T, Hillilä M, Korpela R. Randomised clinical trial: Low-FODMAP rye bread vs. regular rye bread to relieve the symptoms of irritable bowel syndrome. *Aliment Pharmacol Ther.* 2016; 44(5):460–470.

Laddomada, B., Caretto, S., Mita, G. Wheat bran phenolic acids: Bioavailability and stability in whole wheat-based foods. *Molecules* 2015;20:15666–15685. doi: 10.3390/molecules200915666

Lampi, A.-M., Nurmi, T., Ollilainen, V., Piironen, V. Tocopherols and tocotrienols in wheat genotypes in the HEALTHGRAIN diversity screen. *J Agric Food Chem.* 2008;56:9716–9721.

Laskowski, W., Górska-Warsewicz, H., Rejman, K., Czeczotko, M., Zwolińska, J. How important are cereals and cereal products in the average polish diet? *Nutrients* 2019;11:679.

Lau, S.W., Chong, A.Q., Chin, N.L., Talib, R.A., Basha, R.K. Sourdough microbiome comparison and benefits. *Microorganisms* 2021;9:1355.

Leitgeb, M., Knez, Ž., Hojnik Podrepšek, G. Enzyme activity and physiochemical properties of flour after supercritical carbon dioxide processing. *Foods* 2022;11:1826. doi: 10.3390/foods11131826

Li, L., Shewry, P.R., Ward, J.L.. Phenolic acids in wheat varieties in the HEALTHGRAIN diversity screen. *J Agric Food Chem.* 2008;56:9732–9739.

Loponen, J., Gänzle, M.G.. Use of sourdough in low FODMAP baking. *Foods.* 2018;7(7):96.

Matportalen. 2006. Available from: www.matportalen.no/matvaretabellen.

Mattila, P., Pihlava, J.M., Hellström, J. Contents of phenolic acids, alkyl- and alkenylresorcinols, and avenanthramides in commercial grain products. *J Agric Food Chem.* 2005;53(21):8290–8295. doi: 10.1021/jf051437z

Mihhalevski, A.; Nisamedtinov, I.; Hälvin, K.; Ošeka, A.; Paalme, T. Stability of B-complex vitamins anddietary fiber during rye sourdough bread production. *J. Cereal Sci.* 2013;57:30–38.

Mikušová, L., Šturdík, E., Holubková, A. Whole grain cereal food in prevention of obesity. *Acta Chemica Slovaca.* 2011;4(1):95–114.

Milanović, V., Osimani, A., Garofalo, C., Belleggia, L., Maoloni, A., Cardinali, F., Mozzon, M., Foligni, R., Aquilanti, L., Clementi, F. Selection of cereal-sourced lactic acid bacteria as candidate starters for the baking industry. *PLoS One.* 2020 Jul 23;15(7):e0236190. doi: 10.1371/journal.pone.0236190

Miranda-Castro, R., de-los-Santos-Álvarez, N., Miranda-Ordieres, A.J., Lobo-Castañón, M.J. Harnessing aptamers to overcome challenges in gluten detection. *Biosensors.* 2016;6(2):16.

Nagpal, R., Mishra, S.P., Yadav, H. Unique gut microbiome signatures depict diet-versus genetically induced obesity in mice. *Int J Mol Sci.* 2020 May 13;21(10):3434. doi: 10.3390/ijms21103434

Naji, E.A., Jasna, C.B., Gordana, C., Vesna, T.S., Jelena, V., Nebojsa, I. Powdered BARLEYSPROUTS: Composition, functionality and polyphenol digestibility. *Int J Food Sci Tech.* 2016;52(1):231–238.

Nakamura H, Jinzu H, Nagao K, Noguchi Y, Shimba N, Miyano H, Watanabe T, Iseki K. Plasma amino acid profiles are associated with insulin, C-peptide and adiponectin levels in type 2 diabetic patients. *Nutr Diabetes.* 2014 Sep 1;4:e133.

Newman, C.W., Newman, R.K., Fastnaught, C.E. Barley. In Johnson J., Wallace T., editors. *Whole Grains and Their Bioactives: Composition and Health.* John Wiley & Sons; Hoboken, NJ: 2019, pp. 135–167.

Novotni, D., Gänzle, M., Rocha, J.M. Chapter 5. Composition and activity of microbiota in sourdough and their effect on bread quality and safety. In C.M. Galanakis (Ed.). *Trends in Wheat and Bread Making*, Elsevier-Academic Press, 469 pp. Cambridge, MA, USA, 2020. Charis M. Galanakis (Editor), Galanaksis-TWBM-1632435, ISBN 978-0-12-821048-2.

Nurmi, T., Nyström, L., Edelmann, M., Lampi, A.-M., Piironen, V. Phytosterols in wheat genotypes in the HEALTHGRAIN diversity screen. *J Agric Food Chem.* 2008;56:9710–9715.

Nyström, L., Lampi, A.-M., Andersson, A.A.M., Kamal-Eldin, A., Gebruers, K., Courtin, C.M., et al. Phytochemicals and dietary fiber components in rye varieties in the HEALTHGRAIN diversity screen. *J Agric Food Chem.* 2008;56:9758–9766.

Oest, M.; Bindrich, U.; Voß, A.; Kaiser, H.; Rohn, S. Rye bread defects: Analysis of composition and further influence factors as determinants of dry-baking. *Foods* 2020;9:1900. doi: 10.3390/foods9121900

Prasadi, N.P.V., Joye, I.J. Dietary fibre from whole grains and their benefits on metabolic health. *Nutrients* 2020;12:3045.

Paras, S., Goudar, G., Longvah, T., Gour, V.S., Kothari, S.L., Wani, I.A.. Fate of polyphenols and antioxidant activity of barley during processing. *Food Rev. Int.* 2020. doi: 10.1080/87559129.2020.1725036

Pedroso, J.A., Buonfiglio, D.C., Cardinali, L.I., Furigo, I.C., Ramos-Lobo, A.M., Tirapegui, J., et al. Inactivation of SOCS3 in leptin receptor-expressing cells protects mice from diet-induced insulin resistance but does not prevent obesity. *Mol Metab.* 2014;3:608–618.

Pejcz, E., Spychaj, R., Gil, Z. Technological methods for reducing the content of fructan in rye bread. *Eur Food Res Technol* 2020;246(9):1839–1846.

Pérez-Pérez, A., Sánchez-Jiménez, F., Vilariño-García, T., Sánchez-Margalet, V. Role of Leptin in Inflammation and Vice Versa. *International Journal of Molecular Sciences.* 2020;21(16):5887

Pernica M., Boško R., Svoboda Z., Benešová K., Běláková S.. Monitoring of gluten in Czech commercial beers. *Czech J. Food Sci.* 2020;38:255–258.

Pétel, C., Onno, B., & Prost, C. (2017). Sourdough volatile compounds and their contribution to bread: A review. *Trends in Food Science and Technology*, 59, 105–123.

Phillips, H.N., Heins, B.J., Delate, K., Turnbull, R. Fatty acid composition dynamics of rye (*Secale cereale* L.) and wheat (*Triticum aestivum* L.) forages under cattle grazing. *Agronomy* 2020;10:813.

Pihlava, J.M.; Nordlund, E.; Heiniö, R.L.; Hietaniemi, V.; Lehtinen, P.; Poutanen, K. Phenolic compounds in wholegrain rye and its fractions. *J. Food Compos. Anal.* 2015;38:89–97.

Pirkola L, Laatikainen R, Loponen J, Hongisto SM, Hillilä M, Nuora A, Yang B, Linderborg KM, Freese R. Low-FODMAP vs regular rye bread in irritable bowel syndrome: Randomized SmartPill® study. *World J Gastroenterol.* 2018; 24(11):1259–1268.

Pitsch, J., Sandner, G., Huemer, J., Huemer, M., Huemer, S., Weghuber, J. FODMAP fingerprinting of bakery products and sourdoughs: Quantitative assessment and content reduction through fermentation. *Foods.* 2021;10(4):894.

Priyadarshini, M., Kotlo, K.U., Dudeja, P.K., Layden, B.T.. Role of short chain fatty acid receptors in intestinal physiology and pathophysiology. *Compr Physiol.* 2018 Jun 18;8(3):1091–1115. doi: 10.1002/cphy.c170050

Qadir, M.I., Ahmed, Z. lep expression and its role in obesity and type-2 diabetes. *Crit Rev Eukaryot Gene Expr.* 2017;27(1):47–51. doi: 10.1615/CritRevEukaryotGeneExpr.201 7019386

Raboy, V. *Low phytic acid* crops: Observations based on four decades of research. *Plants.* 2020;9:140.

Rajala, M.W., Patterson, C.M., Opp, J.S., Foltin, S.K., Young, V.B., Myers, M.G. Jr. Leptin acts independently of food intake to modulate gut microbial composition in male mice. *Endocrinology.* 2014;155(3):748–57.

Ragaee, S., Abdel-Aal, E.M., Noaman, M. Antioxidant activity and nutrient composition of selected cereals for food use. *Food Chem.* 2006;98, 32–38.

Rakha, A., Aman, P., Andersson, R. Characterisation of dietary fibre components in rye products. *Food Chem.* 2010;119:859–867.

Rakha, A., Mehak, F., Shabbir, M.A., Arslan, M., Ranjha, M.M.A.N., Ahmed, W., Socol, C.T., Rusu, A.V., Hassoun, A. and Aadil, R.M. Insights into the constellating drivers of satiety impacting dietary patterns and lifestyle. *Front Nutr.* 2022;9:1002619.

Rani, M., Singh, G., Siddiqi, R.A., Gill, B.S., Sogi, D.S. and Bhat, M.A. Comparative quality evaluation of physicochemical, technological, and protein profiling of wheat, rye, and barley cereals. *Front Nutr.* 2021;8:694679. doi: 10.3389/fnut.2021.694679

Ravisankar, S., Queiroz, V.A.V., Awika, J.M. Rye flavonoids: Structural profile of the flavones in diverse varieties and effect of fermentation and heat on their structure and antioxidant properties. *Food Chem.* 2020;324:126871.

Razgonova, M.P.; Zakharenko, A.M.; Gordeeva, E.I.; Shoeva, O.Y.; Antonova, E.V.; Pikula, K.S.; Koval, L.A.; Khlestkina, E.K.; Golokhvast, K.S. Phytochemical analysis of phenolics, sterols, and terpenes in colored wheat grains by liquid chromatography with tandem mass spectrometry. *Molecules.* 2021;26:5580. doi: 10.3390/molecules26185580

Ridaura, V.K., Faith, J.J., Rey, F.E., Cheng, J., Duncan, A.E., Kau, A.L., Griffin, N.W., Lombard, V., Henrissat, B., Bain, J.R., et al. Gut microbiota from twins discordant for obesity modulate metabolism in mice. *Science.* 2013;341(6150):1079–1089.

Rodríguez-García, C., Sánchez-Quesada, C., Toledo, E., Delgado-Rodríguez, M., Gaforio, J.J. Naturally Lignan-Rich foods: A dietary tool for health promotion? *Molecules.* 2019;24:917.

Rosicka-Kaczmarek, J., Miśkiewicz, K., Nebesny, E., & Makowski, B. *Composition and Functional Properties of Lipid Components from Selected Cereal Grains*, In *Plant Lipids Science, Technology, Nutritional Value and Benefits to Human Health*; Budryn, G., Żyżelewicz, D., Eds.; Transworld Research Network: Kerala, India, 2015; 119–145..

Ross, A.B., Kamal-Eldin, A., Aman, P. Dietary alkylresorcinols: Absorption, bioactivities, and possible use as biomarkers of whole-grain wheat- and rye-rich foods. *Nutr Rev.* 2004 Mar;62(3):81–95. doi: 10.1111/j.1753-4887.2004.tb00029.x

Samtiya, M., Aluko, R.E., Dhewa, T. Plant food anti-nutritional factors and their reduction strategies: an overview. *Food Prod Process and Nutr.* 2020;2:6

Sánchez-Maldonado, A.F., Schieber, A., Gänzle, M.G. Structure-function relationships of the antibacterial activity of phenolic acids and their metabolism by lactic acid bacteria. *J. Appl. Microbiol.* 2011;111:1176–1184.

Scherf, K.A., Köhler, P. Wheat and gluten: Technological and health aspects. *Ernahrungs Umschau* 2016;63(08):166–175.

Schlemmer, U., Frølich, W., Prieto, R., Grases, F. Phytates in foods: Bioavailability and significance for humans. *Mol Nutr Food Res.* 2009;53(Supplement 2):S330–S375.

Schmidt, M., Sciurba, E. Determination of FODMAP contents of common wheat and rye breads and the effects of processing on the final contents. *Eur Food Res Technol.* 2021;247:395–410.

Seal, C.J., Courtin, C.M., Venema, K., de Vries, J. Health benefits of whole grain: Effects on dietary carbohydrate quality, the gut microbiome and consequences of processing. *Compr Rev Food Sci Food Saf.* 2021;1–27. doi: 10.1111/1541-4337.12728

Serafino, S., Anna, I., Pasquale, C., Clara, F., Mario, R., Nicola, P., Ugo, M., Michele, S. Phenolic acids profile, nutritional and phytochemical compounds, antioxidant propertiesin colored barley grown in Southern Italy. *Food Res. Int.* 2018,113:221–233. doi: 10.1016/j.foodres.2018.06.072

Shabana, U., Shahid, S., Wah Li, K., Acharya, J., Cooper, J.A., Hasnain, S., Humphries, S.E. Effect of six type II diabetes susceptibility loci and an FTO variant on obesity in Pakistani subjects. *Eur J Human Genetics: EJHG*, 2016;24(6):903–910. doi: 10.1038/ejhg.2015.212

Sharma, N., Bhatia, S., Chunduri, V., Kaur, S., Sharma, S., Kapoor, P., Kumari, A., Garg, M. Pathogenesis of celiac disease and other gluten related disorders in wheat and strategies for mitigating them. *Front Nutr.* 2020;7:6.

Shewry, P.R. 2014. Minor components of the barley grain: Minerals, lipids, terpenoids, phenolics and vitamins. In: Shewry, P.R. and Ullrich, S.E. (ed.) *Barley: Chemistry and Technology*, 2nd ed. American Association of Cereal Chemists (AACC), St Paul, MN. pp. 169–192.

Shewry, P.R. What is gluten: Why is it special? *Front Nutr.* 2019;6:101.

Shewry, P.R., Piironen, V., Lampi, A.-M., Nyström, L., Li, L., Rakszegi, M., et al. Phytochemicals and dietary fiber components in oat varieties in the HEALTHGRAIN diversity screen. *J Agric Food Chem.* 2008;56:9777–9785.

Siddiqi, RA, Singh TP, Rani M, Sogi DS, Bhat MA.Diversity in grain, flour, amino acid composition, protein profiling, and proportion of total flour proteins of different wheat cultivars of North India. *Front. Nutr.* 2020;7:141. doi: 10.3389/fnut.2020.00141

Skendi, A., Zinoviadou, K.G., Papageorgiou, M., Rocha, J.M. Advances on the valorisation and functionalization of by-products and wastes from cereal-based processing industry. *Foods.* 2020;9:1243.

Slama, A., Cherif, A., Boukhchina, S. Importance of new edible oil extracted from seeds of seven cereals species. *J Food Qual.* 2021; 2021. https://doi.org/10.1155/2021/5531414

Snell, P., Wilkinson, M., Taylor, G.J., Hall, S., Sharma, S., Sirijovski, N., Hansson, M., Shewry, P.R., Hofvander, P., Grimberg, Å. Characterisation of grains and flour fractions from field grown transgenic oil-accumulating wheat expressing oat *WRI1*. *Plants.* 2022;11:889. doi: 10.3390/plants11070889

Socol, C.T., Chira, A., Martinez-Sanchez, M.A., Nuñez-Sanchez, M.A., Maerescu, C.M., Mierlita, D., Rusu, A.V., Ruiz-Alcaraz, A.J., Trif, M., Ramos-Molina, B. Leptin signaling in obesity and colorectal cancer. *Int J Mol Sci.* 2022;23:4713.

Stępniewska, S., Cacak-Pietrzak, G., Szafrańska, A., Ostrowska-Ligęza, E., Dziki, D. Assessment of the starch-amylolytic complex of rye flours by traditional methods and modern one. *Materials.* 2021;14:7603. doi: 10.3390/ma14247603

Struyf, N., Laurent, J., Verspreet, J., Verstrepen, K.J., Courtin, C.M. *Saccharomyces cerevisiae* and *Kluyveromyces marxianus* cocultures allow reduction of fermentable oligo-, di-, and monosaccharides and polyols levels in whole wheat bread. *J Agric Food Chem.* 2017;65(39):8704–8713.

Suchowilska, E., Bieńkowska, T., Stuper-Szablewska, K., Wiwart, M. Concentrations of phenolic acids, flavonoids and carotenoids and the antioxidant activity of the grain, flour and bran of triticum polonicum as compared with three cultivated wheat species. *Agriculture.* 2020;10:591.

Surget, A., Barron, C. Histologie du grain de blé. *Industrie des Céreales*. 2005;145:3–7.
The National Food Administration's Food Database, version 26/01/2020. Available from: http://www7.slv.se/Naringssok/Naringsamnen.aspx.
The InterAct Consortium. Dietary fibre and incidence of type 2 diabetes in eight European countries: The EPIC-interact study and a meta-analysis of prospective studies. *Diabetologia*. 2015;58:1394–1408.
Tieri M, Ghelfi F, Vitale M, Vetrani C, Marventano S, Lafranconi A, et al. Whole grain consumption and human health: An umbrella review of observational studies. *Int J Food Sci Nutr*. 2020;71:668–677. doi: 10.1080/09637486.2020.1715354
Tözün N, Vardareli E. Gut microbiome and gastrointestinal cancer: Les liaisons Dangereuses. *J Clin Gastroenterol*. 2016 Nov/Dec;50(Suppl 2) Proceedings from the 8th Probiotics, Prebiotics & New Foods for Microbiota and Human Health meeting held in Rome, Italy on September 13-15, 2015:S191–S196. doi: 10.1097/MCG.0000000000000714
Tsafrakidou, P., Michaelidou, A.M., Biliaderis, G.C. Fermented cereal-based products: Nutritional aspects, possible impact on gut microbiota and health implications. *Foods*. 2020 Jun 3;9(6):734. doi: 10.3390/foods9060734
Tufail, T., Ain, H.B.U., Saeed, F., Nasir, M., Basharat, S., Mahwish Rusu, A.V., Hussain, M., Rocha, J.M., Trif, M., Aadil, R.M. A Retrospective on the Innovative Sustainable Valorization of Cereal Bran in the Context of Circular Bioeconomy Innovations. *Sustainability* 2022 14: 14597.
U.S. Department of Agriculture, Agricultural Research Service, USDA Nutrient Data Laboratory. 23 (USDA National Nutrient Database for Standard Reference, 2010). https://www.ars.usda.gov/arsuserfiles/80400525/data/sr23/sr23_doc.pdf
Vancamelbeke, M., Vermeire, S. The intestinal barrier: A fundamental role in health and disease. *Expert Rev Gastroenterol Hepatol*. 2017 Sep;11(9):821–834. doi: 10.1080/17474124.2017
Verma, A.K., Gatti, S., Galeazzi, T., Monachesi, C., Padella, L., Baldo, G.D., Annibali, R., Lionetti, E., Catassi, C. Gluten contamination in naturally or labeled gluten-free products marketed in Italy. *Nutrients*. 2017;9:115.
Verni, M., Verardo, V., Rizzello, C.G. How fermentation affects the antioxidant properties of cereals and legumes. *Foods*. 2019;8:362. doi: 10.3390/foods8090362
Williams, B.A., Mikkelsen, D., Flanagan, B.M. et al. "Dietary fibre": Moving beyond the "soluble/insoluble" classification for monogastric nutrition, with an emphasis on humans and pigs. *J Animal Sci Biotechnol*. 2019;10:45.
Verstringe, S., Vandercruyssen, R., Carmans, H., Rusu, A.V., Bruggeman, G., Trif, M. Alternative proteins for food and feed. In: Galanakis, C.M. (eds) *Biodiversity, Functional Ecosystems and Sustainable Food Production*. Springer, Cham, 2023. doi: 10.1007/978-3-031-07434-9_10
World Health Organisation (WHO). *Prevention and Control of Non-communicable Diseases in the European Region: A Progress Report*. WHO Regional Office for Europe, Kopenhagen, 2013.
Xie, M., Liu, J., Tsao, R., Wang, Z., Sun, B., Wang, J. Whole grain consumption for the prevention and treatment of breast cancer. *Nutrients*. 2019;11(8):1769. doi: 10.3390/nu11081769
You, W., Henneberg, M. Cereal crops are not created equal: wheat consumption associated with obesity prevalence globally and regionally. *AIMS Public Health*. 2016;3:313–28.
Yu, X., Han, J., Li, H. et al. The effect of enzymes on release of trace elements in feedstuffs based on in vitro digestion model for monogastric livestock. *J Animal Sci Biotechnol*. 2018;9:73. doi: 10.1186/s40104-018-0289-2

Zanoaga, O., Braicu, C., Jurj, A., Rusu, A., Buiga, R., Berindan-Neagoe, I. Progress in research on the role of flavonoids in lung cancer. *Int J Mol Sci.* 2019;20:4291. doi: 10.3390/ijms20174291

Zemel, M.B., Bruckbauer, A. Effects of a leucine and pyridoxine-containing nutraceutical on fat oxidation, and oxidative and inflammatory stress in overweight and obese subjects. *Nutrients.* 2012;4(6):529–541.

Zhang, R., Jiao, J., Zhang, W., Zhang, Z., Zhang, W., Qin, L.Q., Han, S.F. Effects of cereal fiber on leptin resistance and sensitivity in C57BL/6J mice fed a high-fat/cholesterol diet. *Food Nutr Res.* 2016 Aug 16;60:31690. doi: 10.3402/fnr.v60.31690

Zhu, Y., Sang, S. Phytochemicals in whole grain wheat and their health-promoting effects. *Mol Nutr Food Res.* 2017;61:10347–10352. doi: 10.1002/MNFR.201600852

Zhu, Y., Li, T., Fu, X., Abbasi, A.M., Zheng, B., Liu, R.H. Phenolics content, antioxidantand antiproliferative activities of dehulled highland barley (Hordeum VulgareL.). *J Funct Foods.* 2015;19:439–450. doi: 10.1016/j.jff.2015.09.053

3 Non-cereal and Legume Based Sourdough Metabolites

Maria Aspri, Nikolina Čukelj Mustač, and Dimitrios Tsaltas

CONTENTS

3.1 Introduction 63
3.2 Fermentation of Pseudocereals and Legumes 66
3.3 Pseudocereal and Legume Sourdough Metabolites and Functional Properties 67
    3.3.1 Carbohydrate Components 67
    3.3.2 Anti-nutritional Compounds 69
    3.3.3 Vitamins 71
    3.3.4 γ-Aminobutyric Acid (GABA) 72
    3.3.5 Antimicrobial Compounds 74
    3.3.6 Phytochemicals and Antioxidant Activity 74
    3.3.7 Protein Digestibility and Bioactive Peptides 76
    3.3.8 Exopolysaccharides 77
3.4 Conclusion 79
Acknowledgments 79
References 79

3.1 INTRODUCTION

The nutritional importance of pseudocereals and legume consumption is increasingly recognized worldwide. Both provide energy, proteins, dietary fiber, micronutrients, and bioactive components, while the health benefits of consuming whole grains, including pseudocereals, and legumes are well established. In addition to improve dietary diversity, the consumption of legumes and pseudocereals contributes to the environmental and economic sustainability of agricultural production (Manners, Varela-Ortega, and van Etten 2020).

Grains of pseudocereals are edible seeds of dicotyledonous species that resemble true cereals (*Poaceae* family) in appearance, macronutrient content, processing routes, and utilization (Martinez-Villaluenga, Penas, and Hernandez-Ledesma 2020). Nevertheless, they are inferior to wheat in their technological properties, especially

DOI: 10.1201/9781003141143-4

from the perspective of the bakery industry. Pseudocereals are known as undemanding crops with high adaptability to different agronomic conditions. The most important pseudocereals are buckwheat (*Fagopyrum* sp.), quinoa (*Chenopodium quinoa* W.), and amaranth (*Amaranthus* sp.). While amaranth and quinoa are crops of the Southern Hemisphere, mainly the Andean region of South America (Bolivia, Ecuador, and Peru), the origin of buckwheat is linked to Central and West China, with the largest producers being Russian Federation, China, Ukraine, and the United States of America (Martinez-Villaluenga, Penas, and Hernandez-Ledesma 2020; FAO 2021). Chia (*Salvia hispanica* L.) is also a crop whose beneficial effects are widely recognized in the bakery industry. Chia seeds belong to the *Lamiaceae* family, and some authors consider it a pseudocereal (Petrova and Petrov 2020), although the chemical composition of these seeds is different from that of pseudocereals. Chia is native to southern Mexico and northern Guatemala, but the growing interest in this crop led to the spread of its cultivation in some other countries, such as Australia, Bolivia, Colombia, Peru, and Argentina. Chia has achieved "superfood" status, and its consumption has increased over the years, mainly due to its attributed health benefits. Its main macronutrients are lipids (~30%), followed by proteins (~19%), and carbohydrates (~4%), but it is best known for its high content of omega-3 fatty acids (~19.5%). The main benefits of consuming omega-3 fatty acids are reduced risk of coronary artery disease, hypertension, type 2 diabetes, rheumatoid arthritis, various cancers, and autoimmune diseases (Grancieri, Martino, and Gonzalez de Mejia 2019).

The legume family (*Leguminoseae, dicotyledons*) includes soybeans, peanuts, fresh peas, fresh beans, and pulses. Pulses are a group of leguminous crops that are harvested only for their seeds. On the other hand, legumes grown for oil (e.g. soybeans and peanuts) and legumes harvested green (e.g. green beans and green peas) are not considered pulses (Singh et al. 2017). There are several types of pulses, but the United Nations Food and Agriculture Organization recognizes 11 main types: beans, broad beans, peas, chickpeas, cowpeas, pigeon peas, lentils, bambara beans, vetches, lupins, and other "minor" pulses that do not have a major impact at the international level. Among the above mentioned, the most widely consumed pulses are dry pea (*Pisum sativum*), chickpea (*Cicer arietinum*), lentil (*Lens culinaris*), and dry beans, such as faba bean (*Vicia faba*), kidney bean (*Phaseolus vulgaris*), and lima bean (*Phaseolus lunatus*). Most of the pulses are produced in North America (Canada), Asia, and the Middle East (Roy, Boye, and Simpson 2010), although currently India is the leading producer of pulses in the world (Singh et al. 2017). Regular consumption of pulses has been associated with many health beneficial properties. Studies reported the lipid-lowering potential of pulse consumption and the indirect effect on cardiovascular disease risk factors as well as in the management of type 2 diabetes (Campos-Vega, Loarca-Piña, and Oomah 2010; Roy, Boye, and Simpson 2010; Ferreira et al. 2021).

Sourdough fermentation is an old biotechnological process, traditionally used in many countries. In modern times, it has been rediscovered as a way to improve the nutritional, technological, and organoleptic properties as well as the shelf life of cereal-based foods, especially leavened bakery products. Wheat and rye are the most explored cereals in sourdough fermentation. However, the increasing interest in pseudocereals and legumes as sourdough raw materials is reflected in recent research dealing with different aspects of lactic acid fermentation of non-wheat cereals and legumes. This interest is particularly related to the fact that pseudocereals and legumes do not contain gluten and are therefore suitable for people with celiac disease and non-celiac gluten sensitivity.

This chapter provides an overview of the findings on the sourdough microbiota of pseudocereals (including chia) and legumes (with a focus on pulses), as well as on nutritional and functional improvements of these raw materials by sourdough technology (see Figure 3.1).

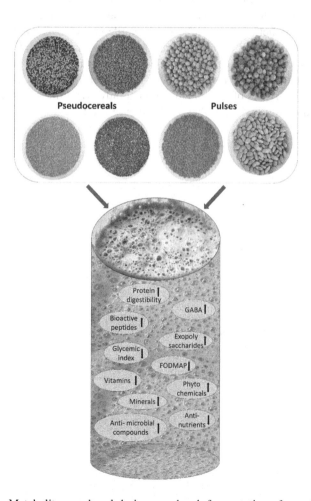

FIGURE 3.1 Metabolites produced during sourdough fermentation of pseudocereals and pulses.

3.2 FERMENTATION OF PSEUDOCEREALS AND LEGUMES

Sourdough fermentation is one of the oldest biotechnologies that is widely used in food production, as it converts grain flour into attractive, tasty, and digestible products. Sourdough is a unique food ecosystem: it is a mixture of flour and water fermented with lactic acid bacteria (LAB) and yeast microbiota. Based on the inocula used, 3 types of sourdough fermentation processes can be distinguished: backslopped (type 1), those initiated with starter cultures (type 2), and those initiated with a starter culture followed by backslopping (type 3) (De Vuyst, Van Kerrebroeck, and Leroy 2017) (see Figure 3.2).

Sourdough fermentation is a traditional process mainly used to improve volume, taste, nutritional value, and shelf life (Gobbetti et al. 2008). These beneficial effects on sourdough and sourdough-derived foods are due to the metabolic activities of the sourdough microflora, such as lactic acid fermentation, proteolysis, exopolysaccharide (EPS) production, and synthesis of volatile nutrients (Corsetti and Settanni 2007). Nevertheless, LAB are mainly responsible for all nutritional and functional benefits of sourdough fermentation, while yeasts are mainly responsible for CO_2 production and aroma formation. However, it has been found that the fermentative and proteolytic activities of sourdough LAB not only affect the sensory, technological, and baking properties, but they also increase the functional value (improving

Type 1- Spontaneous Sourdough
- Mother Dough- Backslopping
- Room temperature with a short to moderate fermentation time
- Firm Sourdough
- Traditional Process

Type 2- Starter Culture- Initiated Sourdough
- Addition of a starter culture
- Higher temperature (>30°C) than the backslopped and fermentation time from one to three days
- Liquid Sourdough
- Industrial

Type 3-Backslopped Starter Culture Initiated (Mixed) Sourdough
- Backslopped starter culture initiated or mixed sourdough fermentation processes can be used
- Appropriate temperature and time
- Dried sourdough
- Industrial

FIGURE 3.2 Characteristics of the 3 main types of sourdough fermentation.

bioactive compound content, reducing anti-nutritional factors, and increasing mineral absorption) of leavened bakery products in many studies (Gobbetti et al. 2014). Therefore, fermentation has been shown to be an effective option for processing legumes and pseudocereals to improve their nutritional and functional properties by removing anti-nutritional factors, such as tannins and phytic acid; synthesizing new compounds, such as group B vitamins; and releasing bioactive compounds, such as bioactive peptides, polysaccharides, more bioavailable isoflavones, γ-aminobutyric acid, and antioxidant compounds (Sáez et al. 2018).

The type of microorganisms involved in the fermentation process plays a key role in this process (Limón et al. 2015). Fermentation can occur spontaneously by the endogenous microbiota of legumes or can be controlled by inoculation of starter cultures (Sáez et al. 2018). Studies on the LAB microbiota of non-traditional grains and flours have increased in recent years, revealing a great biodiversity and opening a new perspective that can be exploited as a biotechnological tool through the selection of autochthonous LAB for sourdough fermentation (De Vuyst et al. 2014; Rodríguez et al. 2016a; Rodríguez et al. 2016b; Sáez et al. 2017; Sáez et al. 2018; Maidana et al. 2020; Galli et al. 2020).

3.3 PSEUDOCEREAL AND LEGUME SOURDOUGH METABOLITES AND FUNCTIONAL PROPERTIES

3.3.1 CARBOHYDRATE COMPONENTS

Pseudocereals, similar to cereals, are primarily sources of carbohydrates in the everyday diet, with starch being the main carbohydrate component. Reported starch contents for buckwheat, quinoa, and amaranth range between 63–82%, 58–64%, and 65–75%, respectively (Martinez-Villaluenga, Penas, and Hernandez-Ledesma 2020), while the main carbohydrates in chia seeds are dietary fibers (Grancieri, Martino, and Gonzalez de Mejia 2019). Legumes contain a slightly lower proportion of starch, with carbohydrates accounting for 50–65% of their weight (Ferreira et al. 2021).

The rate of starch digestion is related to the glycemic index (GI) of foods, a parameter defined as the area under the glucose response curve after food consumption (Montemurro, Coda, and Rizzello 2019). The importance of the GI of foods has been discussed in numerous studies, especially in the context of type 2 diabetes, and in the prevention and control of obesity and some metabolic risk factors, such as coronary heart disease. Foods are characterized by the "strength" they have to increase postprandial blood glucose levels, in reference to glucose or white bread (Taylor, Emmambux, and Kruger 2015). Glucose has a maximum GI value of 100, and other foods are characterized as low (<55), medium (55–69), or high (>70) GI, which is related to various intrinsic and extrinsic factors (Giuberti and Gallo 2017).

In general, cereal foods per se do not have low GI. GI value of cereal foods depends on the type of starch (rapidly digestible, slowly digestible, or resistant) (Taylor, Emmambux, and Kruger 2015), starch properties (amylose/amylopectin ratio, granule morphology, and physicochemical properties) (Zhu 2016), and the speed and type of processing (i.e. cooking, baking, and extrusion). Formulations generally have a high impact on GI due to the presence of lipids and proteins (Giuberti and Gallo 2017). Legumes are considered low glycemic foods, effective at reducing postprandial insulin response due to higher content of viscous soluble dietary fiber content and relatively higher amylose content compared to starch from cereal grains (Giuberti and Gallo 2017; Ferreira et al. 2021).

The effect of sourdough fermentation is often observed via starch digestibility and the associated GI of food products with added sourdough. It is known that sourdough technology generally reduces the GI of foods. The positive effect on postprandial glucose and insulin responses has been attributed to the production of lactic and acetic acids (Fernández-Peláez, Paesani, and Gómez 2020). Moreover, the decrease in amylase activity and trypsin inhibitors during sourdough fermentation could be another reason for the lower values of GI, as well as the increase in resistant starch content. As for pseudocereals, although they are often used strategically to reduce the GI of otherwise high-GI, gluten-free foods, studies have not always shown that sourdough fermentation of pseudocereals is beneficial for GI or that it reduces the rate of starch digestion (Giuberti and Gallo 2017). This may also be due to the specific properties of the starch present in pseudocereals. For example, buckwheat has a lower gelatinization temperature than teff and sorghum and a higher one than quinoa and white wheat. Granules are also larger than those of quinoa, hence it showed lower *in vitro* digestibility than quinoa and white wheat bread. In addition, the amylose content in buckwheat is higher than that of quinoa (Zhu 2016). Compared to sourdough fermentation of pseudocereals, sourdough fermentation of legume flours generally resulted in a decrease in GI but, like pseudocereals, further research on the different legume substrates is needed to better understand the effect of sourdough on postprandial blood glucose levels (Giuberti and Gallo 2017).

Another important factor in reducing type 2 diabetes and associated health conditions is the consumption of high-fiber foods. From this point of view, pseudocereals and legumes represent a valuable source of fiber in the diet (Martinez-Villaluenga, Penas, and Hernandez-Ledesma 2020; Ferreira et al. 2021). Dietary fiber includes non-starch polysaccharides, resistant starch, and non-digestible oligosaccharides. Dietary fibers can be classified based on different properties, but their water solubility is the one of the most frequently mentioned and used in the context of human nutrition. Zhu (2020) reviewed extensively the most important characteristics of dietary fiber polysaccharides of amaranth, buckwheat, and quinoa. The total dietary fiber content is up to 26% for quinoa and about 17% for amaranth and buckwheat (Martinez-Villaluenga, Penas, and Hernandez-Ledesma 2020), although the reported ranges varied depending on their genotype and the presence of bran/hull, etc. (Zhu 2020). Similar to true cereals, the ratio of soluble to insoluble dietary fiber content is 0.28 for quinoa, 0.34 for amaranth, and 0.22 for buckwheat (Martinez-Villaluenga, Penas, and Hernandez-Ledesma 2020). However, the difference from true cereals is that the dietary fiber composition of quinoa and amaranth could be more similar to vegetables and legumes than cereals, with higher levels of pectin, xyloglucans, and soluble dietary fiber compared to wheat and maize (Lamothe et al. 2015). Compared to other pseudocereals, chia seeds have higher dietary fiber content (35%), mainly in its insoluble form as in the case of pseudocereals. Lignin, cellulose, and hemicellulose make up the bulk of chia fiber composition and have a characteristic soluble type of fiber-mucilage. Chia mucilage has high water absorption capacity, allowing it to absorb 27 times its own weight, causing it to expand and take on a gel-like appearance (Grancieri, Martino, and Gonzalez de Mejia 2019). Pulses also provide a high amount of dietary fiber ranging between 15 and 30g/100g of dry weight. Insoluble dietary fiber is abundant in plants, so attempts are being made to increase the water solubility and functionality of these compounds through various processes, such as sourdough fermentation (Arora et al. 2021). However, there is limited information on changes in dietary fiber solubility during fermentation of legumes and pseudocereals.

Pseudocereals also contain small amounts of oligosaccharides and simple monosaccharides. Within these compounds, fermentable oligosaccharides (fructans, fructo-oligosaccharides and galacto-oligosaccharides), disaccharides (lactose), monosaccharides (fructose), and polyols (sorbitol, mannitol, maltitol, xylitol, polydextrose, and isomalt) (FODMAPs) are gaining increased interest. They are known to have beneficial physiological effects by increasing stool volume, improving calcium absorption and modulating immune function, as well as lowering serum cholesterol, triacylglycerol, and phospholipid levels. They also have a prebiotic effect, stimulating the growth of beneficial gut microbiota such as *bifidobacteria* (Catassi et al. 2017). Over the past 15 years, FODMAPs have received some negative attention due to various gastrointestinal problems in susceptible individuals and their link to irritable bowel syndrome (IBS) as well as to non-celiac wheat sensitivity (Catassi et al. 2017). IBS patients are advised to avoid foods high in this heterogeneous group of short-chain carbohydrates and polyols. The shortcoming of a reduced FODMAP diet is the potentially reduced intake of dietary fiber and associated minerals, vitamins, and antioxidant compounds. Indeed, foods that should be avoided in a low-FODMAP diet also include all legumes, i.e. pulses, as well as some cereals, particularly wheat and rye (Catassi et al. 2017; Ferreira et al. 2021). Among all crops, pulses have the highest amounts of oligosaccharides ranging from 71 to 145 mg/g dry weight (dw), the main compounds being α-galactoside, raffinose, stachyose, and verbascose (Singh et al. 2017). Pseudocereals, on the other hand, are considered low-FODMAP foods, although buckwheat contains fagopyritols that may have a similar effect (Ispiryan et al. 2021). The application of sourdough fermentation to legumes and pseudocereal grains and flours is a promising technique to reduce the FODMAP content of bread as sourdough microbiota degrades FODMAPs during fermentation, making these products suitable for consumption in low-FODMAP diets and safe for people with IBS symptoms. The partial or complete elimination of the raffinose family of oligosaccharides by sourdough fermentation has been demonstrated in many studies (Curiel et al. 2015; Coda et al. 2015; Montemurro et al. 2019; Galli et al. 2020). Curiel et al. (2015) reported that sourdough fermentation reduced raffinose content in traditional Italian legumes by up to 64%.

3.3.2 Anti-nutritional Compounds

Anti-nutritional compounds are those that interfere in some way with the digestive process. They are thought to play an important role in plant metabolism, stress, and pathogen resistance by helping the plant complete its life cycle under adverse conditions (Campos-Vega, Loarca-Piña, and Oomah 2010; Roy, Boye, and Simpson 2010).

The main antinutrient in pseudocereals is phytic acid, although significant amounts of saponins have also been detected in amaranth and quinoa, causing a bitter taste (Martinez-Villaluenga, Penas, and Hernandez-Ledesma 2020).

Phytic acid is the main storage form of phosphorus in plants. It is considered an antinutrient due to its strong ability to bind minerals, proteins, and starch, thereby decreasing their bioavailability. The importance of enzymatic degradation of phytic acid is widely recognized; for example, exogenous phytase is used in animal feed (Campos-Vega, Loarca-Piña, and Oomah 2010). In addition to exogenous enzyme application, sourdough technology has been explored to reduce phytic acid content. LAB reduces phytic acid content due to the activity of microbial phytases, improving mineral availability (Rizzello et al. 2010). Curiel et al. (2015) showed that sourdough fermentation of traditional Italian legumes with *L. plantarum* V48 and *L. brevis* AM increased phytase activity. Carrizo et al. (2017) isolated LAB from quinoa and amaranth (grains and sourdough) with high phytase activities that were subsequently used to make quinoa pasta. The results showed that the pasta fermented with the mixed culture containing *L. plantarum* CRL 2107 + *L. plantarum* CRL 1964 had higher concentrations of minerals in an *in vivo* mouse model (Carrizo et al. 2020). Rizzello et al. (2016) reported that the use of quinoa sourdough containing autochthonous LAB (*L. plantarum* T6B10 and *L. rossiae* T0A16) increased phytase activity during fermentation compared to non-fermented flour. In a recent study Castro-Alba et al. (2019) found that fermentation of pseudocereals (quinoa, canihua, and amaranth), grains, and flours, whether spontaneously or with *Lactiplantibacillus plantarum* 299v, was effective in degrading phytate.

Anti-nutritional compounds in pulses include compounds of a protein nature, such as lectins, trypsin inhibitors, chymotrypsin inhibitors, antifungal peptides, and ribosome-inactivating proteins; and non-protein compounds, such as alkaloids, phytic acid, phenolic compounds, and saponins (Roy, Boye, and Simpson 2010). Protein inhibitors inhibit the activity of proteases, amylases, lipases, glycosidases, and phosphatases. Among the best known protease inhibitors are the serine proteases trypsin and chymotrypsin, although tannins are also known to inhibit the activity of digestive enzymes and reduce the digestibility of most nutrients (Giuberti and Gallo 2017). Lectins or hemagglutinins are glycoproteins that agglutinate erythrocytes. Studies suggest that lectins may impair the immune response to ovalbumin and promote the development of food allergies. Saponins are undesirable compounds due to their hemolytic activity. Usually these compounds form insoluble complexes with 3-beta-hydroxysteroids, which then interact with bile acid and cholesterol. They may also occur in insoluble complexes with minerals, such as iron, zinc, and calcium. Oxalates are poorly soluble at intestinal pH, and oxalic acid is known to decrease calcium absorption in monogastric animals (Campos-Vega, Loarca-Piña, and Oomah 2010). Adequate processing can significantly reduce and inactivate these anti-nutritional factors.

The presence of anti-nutritional compounds is one of the most important limiting factors for dietary utilization of pulses. Some of these anti-nutritional compounds are heat-labile (e.g. protease inhibitors and lectins), so thermal treatments would eliminate the negative effects of their consumption (Samtiya, Aluko, and Dhewa 2020).

On the other hand, phytic acid, raffinose, tannins, and saponins are heat-stable, and various methods, such as dehulling, soaking, air classification, extrusion, or cooking have been used to reduce their negative effects on human consumption, but microbial fermentation offered the most promising prospects (Coda et al. 2015). There are several studies that have demonstrated the effect of sourdough fermentation of legumes and pseudocereals on anti-nutritional compounds. Spontaneous fermentation of kidney beans resulted in partial or complete removal of tannins, phytic acid, and trypsin inhibitory activity (Granito et al. 2002). The levels of α-galactosides, tannins, phytic acid, and trypsin inhibitory activity were reduced after lactic acid fermentation of grass pea with *Lactiplantibacillus plantarum* (Starzynska-Janiszewska and Stodolak 2011). Coda et al. (2015) reported that fermentation of faba bean with *Lactiplantibacillus plantarum* VTT E133328 resulted in 91% reduction in vicine and convicine content, a significant reduction in trypsin inhibitor activity and a significant increase in condensed tannins. The combination of legume sprouting and sourdough fermentation on wheat, barley, lentil, chickpea, and quinoa seeds reduced the levels of phytic acid, condensed tannins, trypsin inhibitory activity, and saponins (Montemurro et al. 2019).

3.3.3 Vitamins

Vitamins play an important role in many metabolic processes in the human body. However, vitamins cannot be synthesized in sufficient quantities (or at all) to meet the physiological needs of the body, so a daily intake through the diet via foods or synthetic sources is essential. Vitamins can be classified into 2 groups according to their solubility: a) the water-soluble vitamins (C and the group of B vitamins) and b) the fat-soluble vitamins (A, D, E, and K) (Tsafrakidou, Michaelidou, and Biliaderis 2020).

The most relevant vitamins in legumes and pseudocereals are vitamin E, group B vitamins, especially thiamine and folates. Quinoa seeds, for example, contain high levels of vitamin B_6 and folate, as well as riboflavin. Amaranth also showed the highest concentration of vitamin B_6. Vitamin E comprises isomers of tocopherols and tocotrienols, among which α-tocopherol has the highest vitamin E activity. Although the concentration of tocopherol in quinoa is similar to other cereal grains (37.5 and 77.7 mg/kg), the profile is somewhat different with the main components being γ-tocopherol followed by α-tocopherol. On the other hand, amaranth is reported to contain 3 times less total vitamin E compared to quinoa. Common buckwheat contains a similar vitamin E profile as quinoa but also the lowest content among pseudocereals (9.5–16.4 mg/kg). Carotenoids are known for their antioxidant activity, while some of them (α- and β-carotene) act as precursors of vitamin A. Among pseudocereals, quinoa was reported to be a better source of carotenoids (4.6–18.1 mg/kg dw) than amaranth (3.7–4.7 mg/kg dw) and buckwheat (3.6–3.7 mg/kg dw), while the dominant carotenoid in all pseudocereals is lutein (Martinez-Villaluenga, Penas, and Hernandez-Ledesma 2020). Chia seeds are a good source of vitamins B, C, and E, with reported levels (on dw) of 16 mg/kg vitamin C, 6.2 mg/kg thiamine, 1.7 mg/kg riboflavin, 8830 mg/kg niacin, and 490 mg/kg folate. Chia has higher niacin content than maize, soybean, and rice, while thiamine and riboflavin content is similar to that of rice and maize (Kaur and Bains 2019). Like pseudocereals, pulses contain vitamins such as thiamine, niacin, folate, riboflavin, pyridoxine, and vitamin E (Ferreira et al. 2021), with wide ranges of concentration reported depending on the type of pulses studied (Venkidasamy et al. 2019).

Despite the fact that pseudocereals and pulses contain certain vitamins, the vitamin content can be increased by sourdough fermentation with LAB or yeast bacteria. Many strains of LAB can synthesize vitamins and be used as starter or added cultures to fortify fermented products. Fortification of legumes and pseudocereals with vitamins is particularly important for populations that follow special diets by choice (e.g. vegans) or due to cultural habits, religious beliefs, or lack of other available food sources (Tsafrakidou, Michaelidou, and Biliaderis 2020). Therefore, the fortification of fermented legume and pseudocereal products with vitamins has been studied by many researchers.

In a study by Carrizo et al. (2016; 2017) LAB isolated from quinoa and amaranth (grains and sourdough) were able to produce riboflavin and folate. Although vitamin production was strain-specific and some strains were able to produce both vitamins, this would be important to develop novel fermented foods based on legumes or pseudocereals bio-fortified with folates and riboflavin. Carrizo et al. (2020) showed that pasta made from quinoa sourdough fermented with *Lactiplantibacillus plantarum* strains had higher vitamin content, especially vitamin B2 and B9, which were able to prevent and reverse vitamin deficiencies in different rodent models. In a recent study, Xie et al. (2021) investigated the *in situ* production of vitamin B12 in 11 cereal, pseudocereal, and legume materials by fermentation with *Propionibacterium freudenreichii* DSM 20271 and *Levilactobacillus brevis* (formerly Lactobacillus brevis) ATCC 14869. *P. freudenreichii* was used as the vitamin producer, and *Lev. brevis* was selected to improve the consistency and microbial safety of the process. The results showed that fermentation of cereals, pseudocereals, and legume materials with *P. freudenreichii* and *Lev. brevis* increased the concentration of vitamin B12.

3.3.4 γ-Aminobutyric Acid (GABA)

GABA is a non-protein amino acid with 4 carbon atoms and functions as an inhibitory neurotransmitter of the central nervous system (Sahab et al. 2020). Other physiological functions attributed to GABA include regulation of blood pressure, effect on neurological disorders, effect on psychiatric disorders (mood disorders, insomnia, antidepression, relaxing effect), boosting immunity, protection against cancer and cardio vascular disease (CVD), cellular regulation, and hormonal regulation (Diana, Quílez, and Rafecas 2014; Venturi et al. 2019). As a result, GABA is widely used in pharmaceuticals and functional foods, such as gammalone, dairy products, gabaron tea, and shochu (Coda, Rizzello, and Gobbetti 2010).

GABA is widely distributed in bacteria, plants, and vertebrates. The production of GABA in plants increases under biotic and abiotic stresses, such as drought, salt, wounding, hypoxia, infection, soaking, and germination (Briguglio et al. 2018). In animals, GABA is found in high concentrations in the brain and plays a fundamental role in neurotransmission through multiple pathways in the central nervous system and peripheral tissues. In plants and bacteria, it plays a metabolic role in the Krebs cycle, and in vertebrates it functions as a potent neuronal signal transmitter. GABA is formed from the irreversible α-decarboxylation of l-glutamic acid, which is catalyzed by glutamic acid decarboxylase (GAD), a specialized enzyme found in bacteria, plants, and animals (Galli et al. 2020). LAB is one of the most important GABA producers, as it has GAD activity. In this regard, there are many studies showing that fermentation of legumes and pseudocereals with LAB increases the amount of GABA in sourdough, which can lead to an increase in GABA content in the final product.

In a study by Coda, Rizzello, and Gobbetti (2010), LAB (*Lactiplantibacillus plantarum* C48 and *Lactococcus lactis subsp. lactis* PU1), previously selected for GABA biosynthesis, were used for sourdough fermentation of cereal, pseudocereal, and legume flours. The results showed that the use of a mixture of buckwheat, amaranth, chickpea, and quinoa flours subjected to sourdough fermentation by selected GABA-producing strains allowed the production of a GABA-enriched bread without affecting the textural and sensory properties of the product. Curiel et al. (2015) evaluated the composition of 19 traditional Italian legumes and the potential of sourdough fermentation with selected LAB to improve their nutritional and functional properties. *Lactiplantibacillus plantarum* C48 and *Levilactobacillus brevis* AM7 were used as selected starters. Compared to control doughs without bacterial inoculum, the concentrations of GABA increased and reached values up to 624 mg/kg. In another study Limón et al. (2015) investigated the effect of solid state fermentation (SSF) or liquid state fermentation (LSF) in kidney beans on the production of water-soluble extracts with potential antihypertensive effects (angiotensin I converting enzyme inhibitory (ACE-I) activity and GABA content). SSF was performed by *Bacillus subtilis*, while LSF was performed either by natural fermentation (NF) or by a *Lactiplantibacillus plantarum* strain (LPF). The results showed that LSF extracts exhibited potential antihypertensive activity due to their high GABA content (6.8–10.6 mg/g) and ACE-I activity (>90%) compared to SSF extracts, suggesting that the effects of fermentation on GABA content are due to the type of microorganism involved in the process. In addition, the higher GABA content found in NF compared to LPF extracts could be due to the higher number of LAB strains producing GAD in NF and also to the different pH of the 2 fermentations.

The pH seems to play a crucial role in GABA production by LAB. *Lactiplantibacillus plantarum* C48 synthesized the maximum concentration of GABA at pH 6. Rodríguez et al. (2016a) analyzed the diversity of LAB and the technological, functional, and safety properties of strains in spontaneously fermented quinoa sourdoughs. *Lactiplantibacillus plantarum* CRL1905 and *Leuconostoc mesenteroides* CRL1907, were selected as functional starter cultures for the production of gluten-free fermented quinoa products based on their ability to catalyze the conversion of L-glutamate into the bioactive metabolite GABA. In a recent study by Montemurro et al. (2019), the effect of germination and sourdough fermentation of wheat, barley, lentil, chickpea, and quinoa with *Lactobacillus rossiae* LB5, *Lactiplantibacillus plantarum* 1A7, and *Lactobacillus sanfranciscensis* DE9 on the nutritional and functional properties of the germinated flours was investigated. The results showed that the GABA concentration was significantly increased in the fermented flours compared to the non-fermented flours.

3.3.5 Antimicrobial Compounds

In addition to its functional properties, sourdough fermentation of legumes and pseudocereals is a natural way to extend the shelf life of the products by preventing microbial spoilage and delaying bread staling (Moroni, Dal Bello, and Arendt 2009). In the last decade, several studies have demonstrated the antibacterial and antifungal activity of LAB associated with sourdough fermentation. The inhibitory activity of sourdough LAB is mainly attributed to the rapid consumption of oxygen and fermentable carbohydrates by the bacteria, as well as the production of lactate and subsequent pH reduction. In addition, metabolites with antimicrobial activity such as organic acids, CO_2, ethanol, hydrogen peroxide, diacetyl and bacteriocins are also produced by sourdough-associated LAB (Moroni, Dal Bello, and Arendt 2009).

A study by Rodríguez et al. (2016a) investigated the diversity and technological, functional, and safety properties of LAB, present in spontaneously fermented quinoa sourdoughs. In this study, *Listeria monocytogenes* and *B. subtilis* were used as indicator organisms that represent contamination of the industrial environment; they can cause spoilage of bakery products and pose a health risk. The results showed that the antimicrobial activity against *Listeria* and *Bacillus* was due to the production of bacteriocins by LAB. In addition to antimicrobial activity, LAB, isolated from fermented quinoa sourdough, showed antifungal activity against *A. oryzae* and *P. roqueforti*. Mold growth is the most common microbial spoilage in bakery products, resulting in huge economic losses as well as reduced consumer safety due to the production of mycotoxins. Also, a study by Maidana et al. (2020) showed that LAB, isolated during the natural fermentation of chia dough, was unable to inhibit the growth of *L. monocytogenes*. However, the 2 strains *W. cibaria* CH5 and CH28 showed inhibitory activity against *Bacillus*. In addition, inhibition of *A. niger* and *P. roqueforti* by *E. casseliflavus, E. faecium, Lc. Lactis,* and *L. rhamnosus* was also observed, although the antifungal activity varied among the analyzed LAB strains.

3.3.6 Phytochemicals and Antioxidant Activity

The main significant phytochemicals found in pseudocereals are (poly)phenolic compounds, including flavonoids, phenolic acids, and tyrosol derivatives. They are distinguishable by their carbon backbone and ring saturation. Polyphenols play a role in specific biological functions by modulating the activity of enzymes and cell receptors. In addition, polyphenols have anti-allergic, antimicrobial, anti-cancer, anti-inflammatory effects, and estrogenic activity (Singh et al. 2017). These compounds are often associated with dietary fiber exerting synergistic biological effects. Depending on their binding to dietary fiber, they can occur in soluble free form, conjugated to sugars or other low molecular weight components, and in insoluble bound form. The majority of cereal phenolic acids occur in bound form, but in quinoa, free phenolic compounds contribute to the total phenolic content of 53–78%, with ferulic acid the most abundant compound. The second most abundant phenolic group found in quinoa are flavonoids, with a predominance of rutin, quercetin, and kaempferol. Anthocyanins can be found in significant amounts in colored versions of quinoa. Among the 3 pseudocereals, amaranth appears to have the lowest total phenolic content. Similar to quinoa and amaranth, ferulic acid was the main phenolic compound, followed by quercetin and other phenolic compounds, such as sesamin, tyrosol, and cardol. Among the pseudocereals, buckwheat is considered to be the richest source of phenolic compounds including tyrosol, alkylphenol, and phenolic acids. Tartary buckwheat has a higher content of soluble phenolic compounds and rutin compared to common buckwheat. Rutin, orientin, quercetin, quercitrin, homoorientin, vitexin, and isovitexin are the main flavonoids found in buckwheat (Martinez-Villaluenga, Penas, and Hernandez-Ledesma 2020). Chia seeds are known for their high concentration of antioxidant compounds, mainly phenolic acids (with dominance of rosmarinic acid, followed by protocatechuic, caffeic, and gallic acids), flavones, and flavanones (Grancieri, Martino, and Gonzalez de Mejia 2019). Phenolic acids, flavanols, flavan-3-ols, anthocyanins, and condensed tannins are the main polyphenolic groups found in legume seeds. For example, in lentils, peas and beans, the dominant phenolic compounds are flavonoids, phenolic acids, and procyanidins, and their concentration is directly related to their antioxidant activity. In fact, studies have correlated the health benefits of pulse consumption with the phenolic content of pulses (Singh et al. 2017), while the antioxidant activity of is one of the most studied food topics, as reactive oxygen species have been reported to be involved in many diseases such as cancer, diabetes, autoimmune diseases, various respiratory diseases, eye diseases, and schizophrenia (Campos-Vega, Loarca-Piña, and Oomah 2010). Consequently, efforts are being made to increase the antioxidant activity of foods.

Sourdough fermentation of legumes and pseudocereals can enhance the bioconversion of conjugated forms of phenolic compounds to their free forms, which improves their nutritional and functional properties, especially antioxidant activity (Kadiri 2017). A putative mechanism for the fermentation-induced enhancement of phenolic and antioxidant activities in legumes and pseudocereal grains is related to the enzymes present in the secretions of fermenting microbes, resulting in structural degradation of the cell wall matrix and increased accessibility of bound/conjugated phenolic compounds for enzymatic attack (Đorđević, Šiler-Marinković, and Dimitrijević-Branković 2010). Glycosyl hydrolases, esterases, tannin hydrolases, reductases, and decarboxylases are 5 enzyme groups that can facilitate enzymatic conversion of phenolic compounds in cereal grains (Gänzle 2014). Many studies showed the effects of LAB on the release of phenolic compounds in legumes and pseudocereals. Đorđević, Šiler-Marinković, and Dimitrijević-Branković (2010) showed that fermentation of legumes and pseudocereals increased total phenolic content and antioxidant activity compared to their unfermented forms. Carciochi et al. (2016) reported that fermentation of quinoa grains significantly increased the release of phenolic compounds and antioxidant activity compared to raw grains. In a study by Bustos et al. (2017), fermentation of chia dough with *Lactiplantibacillus plantarum* C8 improved the antioxidant activity and total phenolic content of the dough compared to those doughs made with unfermented flour. In a recent study by Rocchetti et al. (2019), fermentation of cooked quinoa and buckwheat by 2 autochthonous strains *Lactiplantibacillus paracasei* A1 2.6 and *Pediococcus pentosaceus* GS·B had a positive effect on total phenolic content and antioxidant activity of the seeds.

3.3.7 Protein Digestibility and Bioactive Peptides

Proteins are food macronutrients that contribute to normal growth and maintenance of the body as a source of energy and amino acids (Rizzello et al. 2012). The crude protein content of buckwheat is reported to be 6–14%, which is similar to that of true cereals. Quinoa and amaranth are slightly richer in protein content, with reported values of 9–17% for quinoa and 13–21% for amaranth. In addition, pseudocereals have a more balanced composition of essential amino acid and higher levels of sulfur-rich amino acids compared to cereals (Martinez-Villaluenga, Penas, and Hernandez-Ledesma 2020). Chia seeds contain about 19% protein, which is characterized by good digestibility (79%), similar to casein and beans – higher than maize, rice, and wheat proteins, but lower than amaranth (90%). The amino acid composition of chia seeds contains all the essential amino acids, the main amino acid being glutamine, with histidine present at low concentrations, and lysine is a limiting amino acid (Grancieri, Martino, and Gonzalez de Mejia 2019). Pulses are generally high in protein and are considered an environmentally sustainable dietary protein source. Chickpeas, lentils, and dry peas contain approximately 22–28% protein, depending on the species, variety, maturity, and agricultural conditions. Pulse proteins are low in sulfur-containing amino acids, such as methionine, cysteine, and tryptophan, but their lysine content is higher compared to cereals (Roy, Boye, and Simpson 2010; Ferreira et al. 2021).

However, protein content is not the indicator of protein digestibility. The hydrolysis of proteins into smaller peptides and amino acids is critical for their utilization by the human body. Several internal and external factors influence protein digestibility, which can be assessed by different quality scoring methods, such as the Protein Digestibility Corrected Amino Acid Score (PDCAAS) and the Digestible Indispensable Amino Acid Score (DIAAS). In general, plant proteins are inferior to animal proteins due to their limited digestibility and suboptimal amino acid profile, and this is also true for proteins from pseudocereals and legumes. For example, buckwheat proteins have low gastrointestinal absorption due to the high content of protease inhibitors and tannins and the low susceptibility of the proteins, especially the albumin fraction, to proteolytic activity. Food processing, including fermentation, has a significant influence on protein digestibility. During fermentation, a number of hydrolytic enzymes are released and hydrolyze proteins, releasing bioactive peptides.

Fermentation of legumes and pseudocereals under specific conditions and with specific bacterial strains can also lead to the production of peptides with specific bioactivities. Many of the physiological and functional properties of proteins are attributed to biologically active peptides, which are often encoded in the native sequence (Harvian et al. 2019). Bioactive peptides are released from their precursor proteins either by enzymes during gastrointestinal digestion or by proteolysis (e.g. microbial fermentation) during food processing. The formation of bioactive peptides during sourdough fermentation has been demonstrated in several studies. Bioactive peptides, usually composed of sequences of 3–20 amino acid units, can exert various physiological effects (Montemurro, Coda, and Rizzello 2019). In particular, legume hydrolysates and bioactive peptides had *in vitro* activities (against cancer and cardiovascular diseases) or their physiological manifestations, such as oxidative damage, inflammation, hypertension, and high cholesterol (Rizzello et al. 2015).

LAB has specific proteinase and peptidase activities, which are considered prerequisites for the release of bioactive peptides. Selected LAB had the ability to release antioxidant peptides from spelt and kamut under suitable sourdough fermentation conditions (Coda et al. 2012). Moreover, the *ex-vivo* antioxidant activity of the purified peptide fractions from spelt and kamut sourdough, tested on mouse fibroblasts artificially exposed to oxidative stress, was comparable to α-tocopherol (Coda et al. 2012). A study by Rizzello et al. (2012) showed that fermentation of cereal, pseudocereal, and legume flours with sourdough LAB (*Lb. curvatus* and *Lb. brevis*), selected for their proteolytic activity, were able to release lunasin, an anti-cancer peptide, suggesting new opportunities for the formulation of innovative functional foods. In another study by Rizzello et al. (2015), 19 traditional Italian legumes were subjected to fermentation with selected LAB strains, which showed the release of lunasin-like polypeptides as a result of proteolysis of native proteins. A significant inhibitory effect on the proliferation of human adenocarcinoma Caco-2 cells was observed with extracts from fermented legume doughs. In another study by Rizzello et al. (2017), the antioxidant potential of LAB used to ferment quinoa flour was investigated. As determined *in vitro*, the scavenging activity of water/salt soluble extracts (WSE) from fermented doughs was significantly higher than that of non-inoculated doughs. This study demonstrated the ability of autochthonous LAB to release peptides with antioxidant activity through proteolysis of native quinoa proteins.

3.3.8 Exopolysaccharides

Exopolysaccharides (EPS) of microbial origin are known to have potential applications as texturizers, emulsifiers, viscosity mediators, and gelling agents (Freitas, Alves, and Reis 2011). The absence of gluten in legume flours demonstrates poor technological properties for bread making that can be solved by *in situ* production of EPS during fermentation as an alternative to plant polysaccharides. Moreover, EPS may contribute to a variety of biologically active properties, such as antioxidant activity (Luo and Fang 2008), immune stimulation (Xu et al. 2009), antitumor, and antiviral activity (Tong et al. 2009; Wang, Ooi, and Ang 2007).

EPS from LAB have the potential to affect host health, and the consumption of these polymers can modulate the levels of beneficial bacteria, such as *bifidobacteria* and *lactobacilli*, in the gastrointestinal tract and stimulate immune function. This shows the promising potential of EPS beyond their physicochemical properties by also adding health benefits to consumers.

Based on their biosynthesis and composition, EPS can be classified as homopolysaccharides (HoPS) and heteropolysaccharides (HePS) (Valerio et al. 2020). While LAB can produce both types of polysaccharides, HoPS are the predominant EPS produced by *Weissella, Leuconostoc, Streptococcus, Lactobacillus* (Lynch, Coffey, and Arendt 2018), and HePS are generally produced in low amounts by mesophilic and thermophilic LAB (*Lactobacillus lactis, Lactobacillus casei, Lactobacillus plantarum, Lactobacillus rhamnosus, Lactobacillus helveticus*, etc.). HePS have irregular repeating units (e.g. galactose and rhamnose) (Galli et al. 2020), while HoPS consist of a monosaccharide unit (i.e. glucose or fructose) link to produce polysaccharides (i.e. glucans or fructans) (Lynch, Coffey, and Arendt 2018). Dextran, one of the most studied HoPS, is mainly composed of α-(1 → 6) glycosidic linkages along with some α-(1 → 2), α-(1 → 3), and α-(1 → 6) branching. The chemical structure of dextran (i.e. chain length, degree of branching, type of linkages of dextrans) is mainly influenced by the microorganism (Lacaze, Wick, and Cappelle 2007). Recently, *Weissela confusa* (Ck15) strain showed promising results for new legume (chickpea)-based sourdough products due to dextran production (Galli et al. 2020). HoPS have been shown to have prebiotic potential, while HePSs induce immunomodulatory effects as well as a positive role in rheological properties, texture, and mouth feel of fermented dairy products (Lynch, Coffey, and Arendt 2018; Valerio et al. 2020).

Studies on sourdough fermentation show great interest in EPS-producing bacteria. Combining pseudocereals and EPS-producing LABs may improve the overall quality (shelf life, flavor, dough structure) of the final products and their nutritional characteristics (Rühmkorf et al. 2012; Wolter et al. 2014). The same authors also demonstrated the suitability of pseudocereals as a substrate for EPS production using LAB strains. Bacterial EPS act as hydrocolloids in sourdoughs, increasing bread volume and decreasing crumb firmness (Zannini, Waters, and Arendt 2014; Rosell, Rojas, and De Barber 2001).

A study on the fermentation of quinoa flour by LAB identified *Lactobacillus, Enterococcus, Leuconostoc, Lactococcus, Pediococcus*, and *Weissella* capable of producing EPS. In particular, *Pediococcus pentosaceous* showed a higher yield of exopolysaccharide and rapid acidifying kinetics (dropping pH values below 4.0 in 24 hours) (Franco et al. 2020). Similarly, when quinoa flour was fermented with *Lactiplantibacillus plantarum* ATCC 8014, Chiş et al. (2020) observed increased carbohydrate content, higher amounts of organic acids and folic acid, and higher amounts of flavonoids and total phenols. These probably had a positive effect on the increased free radical scavenging activity. Lowering pH also leads to higher bioavailability of minerals, at least doubling their values. The same authors reported improved rheological properties of sourdough, probably due to the production of EPS.

Considering the increasing interest in unconventional flours (legumes, pseudocereals, etc.), sourdoughs made with alternative grains could be a good source of EPS-producing strains and new dough–LAB interactions could also emerge. LAB associated with food matrices represents an optimal source of potential starter cultures, thanks to their faster adaptation to the specific ecosystem. Moreover, the proportion of sourdough inoculum has been shown to influence not only the fermentation rate, but also the synthesis of EPS (Kaditzky and Vogel 2008) and thus sensory and rheological properties (Katina et al. 2006).

Great interest is shown in the combination of non-cereal flours with cereal flours (composite sourdoughs) and the incremental potential hidden in such sourdoughs due to the positive impact of both type of flours (Perri et al. 2021; Wang et al. 2018). EPS produced by LAB improves the texture, mouth feel, and stability of foods and are particularly useful in gluten-containing and gluten-free breads, where they contribute to volume and a softer texture (Riaz Rajoka et al. 2020).

3.4 CONCLUSION

Due to their high protein content, fiber, minerals, and other bioactive compounds, non-cereal grains, such as legumes and pseudocereals, can be a promising source of active compounds when incorporated into foods. This chapter highlights sourdough fermentation as an effective process improving the nutritional and functional properties of non-cereal grains, such as legumes and pseudocereals, in addition to the improvement of technological, rheological, sensory, and shelf-life features of the final baked goods. However, due to the great diversity of pseudocereals and legumes, more research is needed in order to fully uncover the beneficial effects of their sourdough fermentation, such as reduction of anti-nutritional factors, increase of protein digestibility, improvements in the bioavailability of minerals, as well as on the production of dietary components, such as vitamins and other bioactive compounds.

ACKNOWLEDGMENTS

This work is based upon work from COST Action 18101 SOURDOMICS – Sourdough biotechnology network towards novel, healthier, and sustainable food and bioprocesses (https://sourdomics.com/; https://www.cost.eu/actions/CA18101/#tabs|Name:overview).

REFERENCES

Arora, Kashika, Hana Ameur, Andrea Polo, Raffaella Di Cagno, Carlo Giuseppe Rizzello, and Marco Gobbetti. 2021. "Thirty years of knowledge on sourdough fermentation: A systematic review." *Trends in Food Science & Technology* 108:71–83. doi: 10.1016/j.tifs.2020.12.008.

Briguglio, M., B. Dell'Osso, G. Panzica, A. Malgaroli, G. Banfi, C. Zanaboni Dina, R. Galentino, and M. Porta. 2018. "Dietary neurotransmitters: A narrative review on current knowledge." *Nutrients* 10 (5):591. doi: 10.3390/nu10050591.

Bustos, Ana Yanina, Carla Luciana Gerez, Lina Goumana Mohtar Mohtar, Verónica Irene Paz Zanini, Mónica Azucena Nazareno, María Pía Taranto, and Laura Beatriz Iturriaga. 2017. "Lactic acid fermentation improved textural behaviour, phenolic compounds and antioxidant activity of chia (Salvia hispanica L.) dough." *Food Technology and Biotechnology* 55 (3):381–389.

Campos-Vega, Rocio, Guadalupe Loarca-Piña, and B. Dave Oomah. 2010. "Minor components of pulses and their potential impact on human health." *Food Research International* 43 (2):461–482. doi: 10.1016/j.foodres.2009.09.004.

Carciochi, Ramiro Ariel, Leandro Galván-D'Alessandro, Pierre Vandendriessche, and Sylvie Chollet. 2016. "Effect of germination and fermentation process on the antioxidant compounds of quinoa seeds." *Plant Foods for Human Nutrition* 71 (4):361–367.

Carrizo, Silvana L, Cecilia E Montes de Oca, Jonathan E Laiño, Nadia E Suarez, Graciela Vignolo, Jean Guy LeBlanc, and Graciela Rollán. 2016. "Ancestral Andean grain quinoa as source of lactic acid bacteria capable to degrade phytate and produce B-group vitamins." *Food Research International* 89:488–494.

Carrizo, Silvana L, Cecilia E Montes de Oca, María Elvira Hébert, Lucila Saavedra, Graciela Vignolo, Jean Guy LeBlanc, and Graciela Celestina Rollán. 2017. "Lactic acid bacteria from andean grain amaranth: A source of vitamins and functional value enzymes." *Journal of Molecular Microbiology and Biotechnology* 27 (5):289–298.

Carrizo, Silvana L, Alejandra de Moreno de LeBlanc, Jean Guy LeBlanc, and Graciela C Rollán. 2020. "Quinoa pasta fermented with lactic acid bacteria prevents nutritional deficiencies in mice." *Food Research International* 127:108735.

Castro-Alba, Vanesa, Claudia E Lazarte, Daysi Perez-Rea, Nils-Gunnar Carlsson, Annette Almgren, Björn Bergenståhl, and Yvonne Granfeldt. 2019. "Fermentation of pseudocereals quinoa, canihua, and amaranth to improve mineral accessibility through degradation of phytate." *Journal of the Science of Food and Agriculture* 99 (11):5239–5248.

Catassi, G., E. Lionetti, S. Gatti, and C. Catassi. 2017. "The low FODMAP diet: Many question marks for a catchy acronym." *Nutrients* 9 (3). doi: 10.3390/nu9030292.

Chiş, Maria Simona, Adriana Păucean, Simona Maria Man, Dan Cristian Vodnar, Bernadette-Emoke Teleky, Carmen Rodica Pop, Laura Stan, Orsolya Borsai, Csaba Balasz Kadar, and Adriana Cristina Urcan. 2020. "Quinoa sourdough fermented with Lactobacillus plantarum ATCC 8014 designed for gluten-free muffins: A powerful tool to enhance bioactive compounds." *Applied Sciences* 10 (20):7140.

Coda, Rossana, Carlo Giuseppe Rizzello, and Marco Gobbetti. 2010. "Use of sourdough fermentation and pseudo-cereals and leguminous flours for the making of a functional bread enriched of γ-aminobutyric acid (GABA)." *International Journal of Food Microbiology* 137 (2–3):236–245.

Coda, Rossana, Carlo Giuseppe Rizzello, Daniela Pinto, and Marco Gobbetti. 2012. "Selected lactic acid bacteria synthesize antioxidant peptides during sourdough fermentation of cereal flours." *Applied and Environmental Microbiology* 78 (4):1087–1096.

Coda, Rossana, Leena Melama, Carlo Giuseppe Rizzello, José Antonio Curiel, Juhani Sibakov, Ulla Holopainen, Marjo Pulkkinen, and Nesli Sozer. 2015. "Effect of air classification and fermentation by Lactobacillus plantarum VTT E-133328 on faba bean (Vicia faba L.) flour nutritional properties." *International Journal of Food Microbiology* 193:34–42.

Corsetti, Aldo, and Luca Settanni. 2007. "Lactobacilli in sourdough fermentation." *Food Research International* 40 (5):539–558.

Curiel, José Antonio, Rossana Coda, Isabella Centomani, Carmine Summo, Marco Gobbetti, and Carlo Giuseppe Rizzello. 2015. "Exploitation of the nutritional and functional characteristics of traditional Italian legumes: The potential of sourdough fermentation." *International Journal of Food Microbiology* 196:51–61.

De Vuyst, Luc, Simon Van Kerrebroeck, Henning Harth, Geert Huys, H-M Daniel, and Stefan Weckx. 2014. "Microbial ecology of sourdough fermentations: Diverse or uniform?" *Food Microbiology* 37:11–29.

De Vuyst, Luc, Simon Van Kerrebroeck, and Frédéric Leroy. 2017. "Microbial ecology and process technology of sourdough fermentation." *Advances in Applied Microbiology* 100:49–160.

Diana, Marina, Joan Quílez, and Magdalena Rafecas. 2014. "Gamma-aminobutyric acid as a bioactive compound in foods: A review." *Journal of Functional Foods* 10:407–420. doi: 10.1016/j.jff.2014.07.004.

Đorđević, Tijana M, Slavica S Šiler-Marinković, and Suzana I Dimitrijević-Branković. 2010. "Effect of fermentation on antioxidant properties of some cereals and pseudo cereals." *Food Chemistry* 119 (3):957–963.

FAO. 2021. "FAOSTAT." Food and Agriculture Organization of the United Nations. http://www.fao.org/faostat/en/#home.

Fernández-Peláez, Juan, Candela Paesani, and Manuel Gómez. 2020. "Sourdough technology as a tool for the development of healthier grain-based products: An update." *Agronomy* 10 (12). doi: 10.3390/agronomy10121962.

Ferreira, H., M. Vasconcelos, A.M. Gil, and E. Pinto. 2021. "Benefits of pulse consumption on metabolism and health: A systematic review of randomized controlled trials." *Critical Reviews in Food Science and Nutrition* 61 (1):85–96. doi: 10.1080/10408398.2020.1716680.

Franco, Wendy, Ilenys M Pérez-Díaz, Lauren Connelly, and Joscelin T Diaz. 2020. "Isolation of exopolysaccharide-producing yeast and lactic acid bacteria from quinoa (Chenopodium Quinoa) sourdough fermentation." *Foods* 9 (3):337.

Freitas, Filomena, Vitor D Alves, and Maria AM Reis. 2011. "Advances in bacterial exopolysaccharides: from production to biotechnological applications." *Trends in Biotechnology* 29 (8):388–398.

Galli, Viola, Manuel Venturi, Rossana Coda, Ndegwa Henry Maina, and Lisa Granchi. 2020. "Isolation and characterization of indigenous Weissella confusa for in situ bacterial exopolysaccharides (EPS) production in chickpea sourdough." *Food Research International* 138:109785.

Gänzle, Michael G. 2014. "Enzymatic and bacterial conversions during sourdough fermentation." *Food Microbiology* 37:2–10.

Giuberti, Gianluca, and Antonio Gallo. 2017. "Reducing the glycaemic index and increasing the slowly digestible starch content in gluten-free cereal-based foods: A review." *International Journal of Food Science & Technology* 53 (1):50–60. doi: 10.1111/ijfs.13552.

Gobbetti, Marco, Maria De Angelis, Raffaella Di Cagno, and Carlo Giuseppe Rizzello. 2008. "Sourdough/lactic acid bacteria." In *Gluten-free Cereal Products and Beverages*, 267–288. Elsevier.

Gobbetti, Marco, Carlo G Rizzello, Raffaella Di Cagno, and Maria De Angelis. 2014. "How the sourdough may affect the functional features of leavened baked goods." *Food Microbiology* 37:30–40.

Grancieri, M., H.S.D. Martino, and E. Gonzalez de Mejia. 2019. "Chia Seed (Salvia hispanica L.) as a Source of Proteins and Bioactive Peptides with Health Benefits: A Review." *Comprehensive Reviews in Food Science and Food Safety* 18 (2):480–499. doi: 10.1111/1541-4337.12423.

Granito, Marisela, Juana Frias, Rosa Doblado, Marisa Guerra, Martine Champ, and Concepción Vidal-Valverde. 2002. "Nutritional improvement of beans (Phaseolus vulgaris) by natural fermentation." *European Food Research and Technology* 214 (3):226–231.

Harvian, Zulvana Anggraeni, Andriati Ningrum, Sri Anggrahini, and Widiastuti Setyaningsih. 2019. "In silico approach in evaluation of jack bean (Canavalia ensiformis) Canavalin protein as precursors of bioactive peptides with dual antioxidant and angiotensin i-converting enzyme inhibitor." *Materials Science Forum* 984:85–94. doi: 10.4028/www.scientific.net/MSF.948.85

Ispiryan, L., R. Kuktaite, E. Zannini, and E.K. Arendt. 2021. "Fundamental study on changes in the FODMAP profile of cereals, pseudo-cereals, and pulses during the malting process." *Food Chemistry* 343:128549. doi: 10.1016/j.foodchem.2020.128549.

Kadiri, Oseni. 2017. "A review on the status of the phenolic compounds and antioxidant capacity of the flour: Effects of cereal processing." *International Journal of Food Properties* 20 (sup1):S798–S809.

Kaditzky, Susanne, and Rudi F Vogel. 2008. "Optimization of exopolysaccharide yields in sourdoughs fermented by lactobacilli." *European Food Research and Technology* 228 (2):291–299.

Katina, Kati, R-L Heiniö, Karin Autio, and Kaisa Poutanen. 2006. "Optimization of sourdough process for improved sensory profile and texture of wheat bread." *LWT-Food Science and Technology* 39 (10):1189–1202.

Kaur, Sukhdeep, and Kiran Bains. 2019. "Chia (Salvia hispanica L.): A rediscovered ancient grain, from Aztecs to food laboratories." *Nutrition & Food Science* 50 (3):463–479. doi: 10.1108/nfs-06-2019-0181.

Lacaze, G, M Wick, and S Cappelle. 2007. "Emerging fermentation technologies: Development of novel sourdoughs." *Food Microbiology* 24 (2):155–160.

Lamothe, L.M., S. Srichuwong, B.L. Reuhs, and B.R. Hamaker. 2015. "Quinoa (Chenopodium quinoa W.) and amaranth (Amaranthus caudatus L.) provide dietary fibres high in pectic substances and xyloglucans." *Food Chemistry* 167:490–496. doi: 10.1016/j.foodchem.2014.07.022.

Limón, Rocio I, Elena Peñas, M Inés Torino, Cristina Martínez-Villaluenga, Montserrat Dueñas, and Juana Frias. 2015. "Fermentation enhances the content of bioactive compounds in kidney bean extracts." *Food Chemistry* 172:343–352.

Luo, Dianhui, and Baishan Fang. 2008. "Structural identification of ginseng polysaccharides and testing of their antioxidant activities." *Carbohydrate Polymers* 72 (3):376–381.

Lynch, Kieran M, Aidan Coffey, and Elke K Arendt. 2018. "Exopolysaccharide producing lactic acid bacteria: Their techno-functional role and potential application in gluten-free bread products." *Food Research International* 110:52–61.

Maidana, Stefania Dentice, Cecilia Aristimuño Ficoseco, Daniela Bassi, Pier Sandro Cocconcelli, Edoardo Puglisi, Graciela Savoy, Graciela Vignolo, and Cecilia Fontana. 2020. "Biodiversity and technological-functional potential of lactic acid bacteria isolated from spontaneously fermented chia sourdough." *International Journal of Food Microbiology* 316:108425.

Manners, Rhys, Consuelo Varela-Ortega, and Jacob van Etten. 2020. "Protein-rich legume and pseudo-cereal crop suitability under present and future European climates." *European Journal of Agronomy* 113:125974. doi: 10.1016/j.eja.2019.125974.

Martinez-Villaluenga, C., E. Penas, and B. Hernandez-Ledesma. 2020. "Pseudocereal grains: Nutritional value, health benefits and current applications for the development of gluten-free foods." *Food and Chemical Toxicology* 137:111178. doi: 10.1016/j.fct.2020.111178.

Montemurro, M., R. Coda, and C.G. Rizzello. 2019a. "Recent advances in the use of sourdough biotechnology in pasta making." *Foods* 8 (4): 129. doi: 10.3390/foods8040129.

Montemurro, Marco, Erica Pontonio, Marco Gobbetti, and Carlo Giuseppe Rizzello. 2019b. "Investigation of the nutritional, functional and technological effects of the sourdough fermentation of sprouted flours." *International Journal of Food Microbiology* 302:47–58.

Moroni, Alice V, Fabio Dal Bello, and Elke K Arendt. 2009. "Sourdough in gluten-free bread-making: An ancient technology to solve a novel issue?" *Food Microbiology* 26 (7):676–684.

Perri, Giuseppe, Rossana Coda, Carlo Giuseppe Rizzello, Giuseppe Celano, Marco Ampollini, Marco Gobbetti, Maria De Angelis, and Maria Calasso. 2021. "Sourdough fermentation of whole and sprouted lentil flours: In situ formation of dextran and effects on the nutritional, texture and sensory characteristics of white bread." *Food Chemistry* 355:129638.

Petrova, P., and K. Petrov. 2020. "Lactic acid fermentation of cereals and pseudocereals: Ancient nutritional biotechnologies with modern applications." *Nutrients* 12 (4). doi: 10.3390/nu12041118.

Riaz Rajoka, Muhammad Shahid, Yiguang Wu, Hafiza Mahreen Mehwish, Manisha Bansal, and Liqing Zhao. 2020. "Lactobacillus exopolysaccharides: New perspectives on engineering strategies, physiochemical functions, and immunomodulatory effects on host health." *Trends in Food Science & Technology* 103:36–48. doi: 10.1016/j. tifs.2020.06.003.

Rizzello, Carlo Giuseppe, Luana Nionelli, Rossana Coda, Maria De Angelis, and Marco Gobbetti. 2010. "Effect of sourdough fermentation on stabilisation, and chemical and nutritional characteristics of wheat germ." *Food Chemistry* 119 (3):1079–1089.

Rizzello, Carlo Giuseppe, Luana Nionelli, Rossana Coda, and Marco Gobbetti. 2012. "Synthesis of the cancer preventive peptide lunasin by lactic acid bacteria during sourdough fermentation." *Nutrition and Cancer* 64 (1):111–120.

Rizzello, Carlo Giuseppe, Blanca Hernández-Ledesma, Samuel Fernández-Tomé, José Antonio Curiel, Daniela Pinto, Barbara Marzani, Rossana Coda, and Marco Gobbetti. 2015. "Italian legumes: Effect of sourdough fermentation on lunasin-like polypeptides." *Microbial Cell Factories* 14 (1):1–20.

Rizzello, Carlo Giuseppe, Anna Lorusso, Marco Montemurro, and Marco Gobbetti. 2016. "Use of sourdough made with quinoa (Chenopodium quinoa) flour and autochthonous selected lactic acid bacteria for enhancing the nutritional, textural and sensory features of white bread." *Food Microbiology* 56:1–13.

Rizzello, Carlo Giuseppe, Anna Lorusso, Vito Russo, Daniela Pinto, Barbara Marzani, and Marco Gobbetti. 2017. "Improving the antioxidant properties of quinoa flour through fermentation with selected autochthonous lactic acid bacteria." *International Journal of Food Microbiology* 241:252–261.

Rocchetti, Gabriele, Francesco Miragoli, Carla Zacconi, Luigi Lucini, and Annalisa Rebecchi. 2019. "Impact of cooking and fermentation by lactic acid bacteria on phenolic profile of quinoa and buckwheat seeds." *Food Research International* 119:886–894.

Rodríguez, R.L., Esteban Vera Pingitore, G Rollan, Pier Sandro Cocconcelli, C Fontana, L Saavedra, G Vignolo, and Elvira Maria Hebert. 2016a. "Biodiversity and technological-functional potential of lactic acid bacteria isolated from spontaneously fermented quinoa sourdoughs." *Journal of Applied Microbiology* 120 (5):1289–1301.

Rodríguez, R.L., Esteban Vera Pingitore, G Rollan, G Martos, L Saavedra, C Fontana, Elvira Maria Hebert, and G Vignolo. 2016b. "Biodiversity and technological potential of lactic acid bacteria isolated from spontaneously fermented amaranth sourdough." *Letters in Applied Microbiology* 63 (2):147–154.

Rosell, Cristima M, Jose A Rojas, and C Benedito De Barber. 2001. "Influence of hydrocolloids on dough rheology and bread quality." *Food Hydrocolloids* 15 (1):75–81.

Roy, F., J.I. Boye, and B.K. Simpson. 2010. "Bioactive proteins and peptides in pulse crops: Pea, chickpea and lentil." *Food Research International* 43 (2):432–442. doi: 10.1016/j. foodres.2009.09.002.

Rühmkorf, Christine, Heinrich Rübsam, Thomas Becker, Christian Bork, Kristin Voiges, Petra Mischnick, Markus J Brandt, and Rudi F Vogel. 2012. "Effect of structurally different microbial homoexopolysaccharides on the quality of gluten-free bread." *European Food Research and Technology* 235 (1):139–146.

Sáez, Gabriel D, Elvira M Hébert, Lucila Saavedra, and Gabriela Zárate. 2017. "Molecular identification and technological characterization of lactic acid bacteria isolated from fermented kidney beans flours (Phaseolus vulgaris L. and P. coccineus) in northwestern Argentina." *Food Research International* 102:605–615.

Sáez, Gabriel D, Lucila Saavedra, Elvira M Hebert, and Gabriela Zárate. 2018. "Identification and biotechnological characterization of lactic acid bacteria isolated from chickpea sourdough in northwestern Argentina." *LWT* 93:249–256.

Sahab, Novia RM, Edy Subroto, Roostita L Balia, and Gemilang L Utama. 2020. "γ-Aminobutyric acid found in fermented foods and beverages: Current trends." *Heliyon* 6 (11):e05526.

Samtiya, Mrinal, Rotimi E Aluko, and Tejpal Dhewa. 2020. "Plant food anti-nutritional factors and their reduction strategies: An overview." *Food Production, Processing and Nutrition* 2 (1):1–14.

Singh, B., J.P. Singh, K. Shevkani, N. Singh, and A. Kaur. 2017. "Bioactive constituents in pulses and their health benefits." *Journal of Food Science and Technology* 54 (4):858–870. doi: 10.1007/s13197-016-2391-9.

Starzynska-Janiszewska, Anna, and Bożena Stodolak. 2011. "Effect of inoculated lactic acid fermentation on antinutritional and antiradical properties of grass pea (Lathyrus sativus' Krab') flour." *Polish Journal of Food and Nutrition Sciences* 61 (4): 245–249.

Taylor, John R.N., M. Naushad Emmambux, and Johanita Kruger. 2015. "Developments in modulating glycaemic response in starchy cereal foods." *Starch - Stärke* 67 (1–2):79–89. doi: 10.1002/star.201400192.

Tong, Haibin, Fengguo Xia, Kai Feng, Guangren Sun, Xiaoxv Gao, Liwei Sun, Rui Jiang, Dan Tian, and Xin Sun. 2009. "Structural characterization and in vitro antitumor activity of a novel polysaccharide isolated from the fruiting bodies of Pleurotus ostreatus." *Bioresource Technology* 100 (4):1682–1686.

Tsafrakidou, Panagiota, Alexandra-Maria Michaelidou, and Costas G Biliaderis. 2020. "Fermented cereal-based products: Nutritional aspects, possible impact on gut microbiota and health implications." *Foods* 9 (6):734.

Valerio, Francesca, Anna Rita Bavaro, Mariaelena Di Biase, Stella Lisa Lonigro, Antonio Francesco Logrieco, and Paola Lavermicocca. 2020. "Effect of amaranth and quinoa flours on exopolysaccharide production and protein profile of liquid sourdough fermented by Weissella cibaria and Lactobacillus plantarum." *Frontiers in Microbiology* 11:967.

Venkidasamy, Baskar, Dhivya Selvaraj, Arti Shivraj Nile, Sathishkumar Ramalingam, Guoyin Kai, and Shivraj Hariram Nile. 2019. "Indian pulses: A review on nutritional, functional and biochemical properties with future perspectives." *Trends in Food Science & Technology* 88:228–242. doi: 10.1016/j.tifs.2019.03.012.

Venturi, M., V. Galli, N. Pini, S. Guerrini, and L. Granchi. 2019. "Use of selected lactobacilli to increase gamma-aminobutyric acid (GABA) content in sourdough bread enriched with Amaranth flour." *Foods* 8 (6): 218. doi: 10.3390/foods8060218 .

Wang, Hui, Engchoon Vincent Ooi, and Put O Ang. 2007. "Antiviral polysaccharides isolated from Hong Kong brown seaweed Hydroclathrus clathratus." *Science in China Series C: Life Sciences* 50 (5):611–618.

Wang, Yaqin, Päivi Sorvali, Arja Laitila, Ndegwa Henry Maina, Rossana Coda, and Kati Katina. 2018. "Dextran produced in situ as a tool to improve the quality of wheat-faba bean composite bread." *Food Hydrocolloids* 84:396–405.

Wolter, Anika, Anna-Sophie Hager, Emanuele Zannini, Michael Czerny, and Elke K Arendt. 2014. "Influence of dextran-producing Weissella cibaria on baking properties and sensory profile of gluten-free and wheat breads." *International Journal of Food Microbiology* 172:83–91.

Xie, Chong, Rossana Coda, Bhawani Chamlagain, Minnamari Edelmann, Pekka Varmanen, Vieno Piironen, and Kati Katina. 2021. "Fermentation of cereal, pseudo-cereal and legume materials with Propionibacterium freudenreichii and Levilactobacillus brevis for vitamin B12 fortification." *LWT* 137:110431.

Xu, Wentao, Fangfang Zhang, YunBo Luo, Liyan Ma, Xiaohong Kou, and Kunlun Huang. 2009. "Antioxidant activity of a water-soluble polysaccharide purified from Pteridium aquilinum." *Carbohydrate Research* 344 (2):217–222.

Zannini, Emanuele, Deborah M Waters, and Elke K Arendt. 2014. "The application of dextran compared to other hydrocolloids as a novel food ingredient to compensate for low protein in biscuit and wholemeal wheat flour." *European Food Research and Technology* 238 (5):763–771.

Zhu, Fan. 2016. "Buckwheat starch: Structures, properties, and applications." *Trends in Food Science & Technology* 49:121–135. doi: 10.1016/j.tifs.2015.12.002.

Zhu, Fan. 2020. "Dietary fiber polysaccharides of amaranth, buckwheat and quinoa grains: A review of chemical structure, biological functions and food uses." *Carbohydrate Polymers* 248:116819. doi: 10.1016/j.carbpol.2020.116819.

4 Innovative Technologies to Extract High-Value Compounds

*Anca C. Farcas, Sonia A. Socaci,
Dubravka Novotni, and Marco Garcia-Vaquero*

CONTENTS

4.1 Introduction 87
4.2 Compounds from Cereal and Pseudocereal By-products 89
    4.2.1 Composition of Cereal and Pseudocereal Waste at Harvest 89
    4.2.2 Milling By-products 89
    4.2.3 By-products of the Malting and Brewing Industries 93
4.3 Novel Approaches for the Utilization of Cereal By-products 96
    4.3.1 Sourdough Fermentation of Cereal By-products 97
    4.3.2 Novel Technologies for Extraction of Compounds 98
        4.3.2.1 Ultrasound-Assisted Extraction 98
        4.3.2.2 Microwave-Assisted Extraction 100
4.4 Future and Challenges of the Novel Technologies 102
Acknowledgments 105
References 105

4.1 INTRODUCTION

Meeting the increased demands of the World's growing population for food, while meeting the current and future shifts in dietary preferences of consumers and ensuring a sustainable use of resources, represents a huge challenge for the food and agricultural sectors in this century. The Food and Agriculture Organization (FAO) identified 5 main principles to support sustainable agriculture and food production which include (1) improvement of resource efficiency; (2) preservation, protection, and enhancement of natural resources; (3) protection of rural livelihoods, equity, and social well-being; (4) development of resilient people, communities, and ecosystems; and (5) responsible and effective governance (FAO, 2014).

    Cereals are essential and widely used as food, with about half of the World's cereal production destined for food consumption, and the rest being used as feed (36%) and other uses, such as ethanol production (Alexandratos and Bruinsma, 2012). Along with valuable grain, the agricultural production of cereals and pseudocereals generates

DOI: 10.1201/9781003141143-5

large amounts of lignocellulosic biomass, such as straw, stalk, stover, cob, and root residues. These are considered waste and, thus, offer an opportunity to improve the sustainability of cereal agricultural systems following the principles described by FAO. Detailed analysis of cereal processing reveals that the majority of cereal by-products are produced by the milling industries (bran and germ), malting and brewing industries (malt sprouts and brewers' spent grain), and bakery industries (waste bread) (Verni et al., 2019). With the exception of maize, wheat germ, and rice bran, which are commercially used for production of edible oil, only a small fraction (10%) of generated cereal by-products is used for food production (mostly as wholemeal flour or extruded breakfast cereals), while the rest is used as animal feed (Hossain et al., 2013; Steiner et al., 2015). Likewise, only 15% of cereal biomass is used as animal feed (and biofuels). Current practice is to bury straw in the soil or burn them, which is harmful for the environment and humans. By-products (30%) of maize processing into ethanol, mainly distillers' dried grains, are returned to the feed sector (Alexandratos and Bruinsma, 2012). The cereal sector contributes 22% of the estimated agricultural waste, co-products, and by-products generated in EU (Bedoić et al., 2019).

These identified cereal by-products differ in their chemical composition, depending on their botanical origin, cultivation practices, and processing strategies. High-value exploitable compounds from cereal by-products may include proteins and peptides, dietary fiber, lipids, polyphenols, and other functional compounds. The utilization of these valuable compounds from cereal and pseudocereal by-products has recently gained momentum as a sustainable strategy for the generation of new food ingredients. The extraction of high-value compounds has been traditionally performed by conventional protocols, such as steam distillation, maceration, and Soxhlet extraction (Garcia-Vaquero, Rajauria, and Tiwari, 2020). The use of these methods still endures in industry and scientific publications; however, their application is time-consuming, energy intensive, and requires the use of large amounts of solvents, raising both sustainability and safety concerns of these extraction processes (Garcia-Vaquero et al., 2020). The use of innovative technologies for the exploitation of compounds from different agri-food by-products has attracted the attention of researchers in recent years. Innovative technologies, such as ultrasound-assisted extraction (UAE) and microwave-assisted extraction (MAE) have been recently explored to recover multiple compounds from cereal by-products, achieving in general high yields of compounds, while reducing the time and solvents used during the process of extraction (Medina-Torres et al., 2017).

This chapter provides an overview of the chemical composition of cereal and pseudocereal by-products, including waste at harvest, milling by-products, and by-products of the malting and brewing industries. The principles of UAE and MAE have also been explained in this chapter together with the main extraction protocols developed using these technologies for the extraction of high-value compounds from cereal and pseudocereal by-products in the recent scientific literature. An overview of the current state of the art of the exploitation of cereal and pseudocereal by-products is also provided outlining current challenges, trends, and future research opportunities in the field using innovative technologies.

4.2 COMPOUNDS FROM CEREAL AND PSEUDOCEREAL BY-PRODUCTS

4.2.1 COMPOSITION OF CEREAL AND PSEUDOCEREAL WASTE AT HARVEST

Cereals are crops that produce large amounts of agricultural waste at harvest, most of which is straw, comprising dry stems and stalks (Bedoić et al., 2019). Together with legume straw, cereal straw such as wheat, maize, rice, oat, and barley represent the largest amount of lignocellulosic biomass worldwide (Espinosa et al., 2017). The amount of straw produced will depend on the crop and harvesting technique, ranging normally between 0.4 to 4 kg per 1 kg of harvested rice, wheat, barley, oat, rye, and triticale grain (Bedoić et al., 2019; Weiser et al., 2014).

The chemical composition of straw varies considerably depending on the cultivar, crop maturity, year and country, climatic conditions and agro-technical practices (Givens, Everington, and Adamson, 1989; Grove et al., 2003; Summerell and Burgess, 1989; Virk et al., 2019). Common cereal straws are mainly composed of cellulose, hemicellulose, and lignin and minerals. In addition, straw contains small amounts of starch, sugar, protein, phenolic compounds, and wax (Đorđević and Antov, 2018; Jørgensen et al., 2020).

Cereal and pseudocereal waste can be used as a raw material for the production of chemicals, enzymes, cellulose nanofibers, textiles, paper, rubber, bricks, and other construction and composite materials (Espinosa et al., 2017; Kuan et al., 2011; Singh and Arya, 2020; Smuga-Kogut et al., 2019; Weiser et al., 2014) as well as other valuable compounds for food production, such as poly- and oligosaccharides, sugars, proteins, minerals, and bioactive compounds (Álvarez et al., 2020; Gil-Ramirez et al., 2018; Jørgensen et al., 2020).

4.2.2 MILLING BY-PRODUCTS

For most grains, milling is the primary processing technique after cleaning them of all foreign matter, such as dust, stones, dirt, chaff, weed seeds, as well as broken and unsound grains. Impurities account for about 5-10% of grains destined for the food industry. Furthermore, some grains are subjected to dehulling (barley, oats, and rice), dehusking (buckwheat, rice), decortication (sorghum, millet), pearling (oats, barley) or polishing (rice) before milling. See examples of milling by-products in Figure 4.1.

Rice grains, for example, often undergo multiple processing steps, including dehusking, whitening, and polishing to produce polished white rice (Wang et al., 2017). The weight of the hull or husk is about 20% of the weight of the grain (Bora et al., 2018; Devisetti et al., 2014; Peanparkdee and Iwamoto, 2019). Grain hulls and husks are relatively low in protein or starch but rich in dietary fiber and bioactive compounds (Dziadek et al., 2016; Grove et al., 2003; Peanparkdee and Iwamoto, 2019). The main composition of milling by-products from several grains is summarized in Table 4.1.

FIGURE 4.1 Milling by-products from cereals (**a**) barley bran and hull, (**b**) maize germ, (**c**) rye bran and (**d**) wheat bran.

Grain milling can be done by a dry or wet process, depending on the type of grain and its intended purpose. Wet milling is mainly used to separate different fractions of the grain: starch, proteins, and lipids (Wronkowska, 2016). Dry milling is used to produce flour, semolina, grits, or meal from different grains. The process of dry milling aims to separate the parts of the grain, i.e. endosperm from germ and outer layers, as their presence in the flour reduces the product quality and shelf life. Numerous sifting operations are used between milling steps. Therefore, several streams of by-products are generated during the production of refined flour, mainly consisting of bran, germ, and a small amount of endosperm (in various proportions). In this process, 10–20% of the total grain weight is removed, depending on grain type, pre-processing and milling method (Hossain et al., 2013; Čukelj Mustač et al., 2020).

Bran is mainly composed of dietary fiber (>50% by dry weight), starch (10–35%), protein (8–19%), lipids (wheat 3–5%, rice 12–18%), minerals (4.8–6.3%), vitamins, and bioactive compounds (Habuš et al., 2021; Onipe, 2015, Prueckler et al., 2014; Sarfaraz et al., 2017; Wang et al., 2020). The germ is the most nutrient-rich part of the grain. Depending on its origin, the germ contains large amounts of carbohydrates

TABLE 4.1
The Proximate Composition of Milling By-products (% w/w)

By-product	Dry matter (% w/w)	Starch (% w/w)	Dietary fiber (% w/w)	Protein (% w/w)	Fat (% w/w)	Ash (% w/w)	Total phenolics (mg GAE/g)	References
Maize bran	–	–	35.8–41.9*	6.4–9.0*	2.7–3.5*	0.9–2.2*	–	Decimo et al. (2017)
Maize germ from wet milling	–	6.9–11.6*	–	13.1–18.4*	36.4–40.9*	1.4–2.3*	–	Johnston et al. (2005)
Maize germ from dry milling	–	19.8–21.2	–	15.4–17.5	18.1–23.0	–	–	Johnston et al. (2005)
Wheat germ	89.8	7.2	–	–	9.8	18.0	4	De Vasconcelos et al. (2013)
Wheat bran	86.1–97.7	10.4–20.7 (8.8–25.6)[a]	36.1–60.2	12.6–17.8	2.6–4.3	3.6–6.6	2.4	Habuš et al. (2021); Kamal-Eldin et al. (2009); Sarfaraz et al. (2017)
Wheat shorts	87.1–87.3	20.5–31.8*	48.6–53.4*	15.8–19.0*	3.8–4.2*	5.7–6.3*	2.5	Sarfaraz et al. (2017)
Wheat red dog	88.2	34.9[a]	38.1	19.3	5.4	4.8	2.7	Sarfaraz et al. (2017)
Rice bran	93.5	26.6[a]	20.1	15.5	–	9.1	–	Casas et al. (2019)
Defatted rice bran	88.7	26.7–32.5[a]	21.5–25.9	14.7–18.6	5.3	14.0	–	Casas et al. (2019); Liu et al. (2020)
Rye bran	90.6–93.1	13.2–28.3*	41.1–47.5*	5.7–18.6*	2.7*–6.6	2.8–7.2	5.2*[c]	Agil and Hosseinian (2014); Kamal–Eldin et al. (2009); Nordlund et al. (2013)
Triticale bran	91.4	31.3–43.0[a]	34.1[b]	6.3	6.4	6.2	–	Agil and Hosseinian (2014)

(Continued)

TABLE 4.1 (CONTINUED)
The Proximate Composition of Milling By-products (% w/w)

By-product	Dry matter (% w/w)	Starch (% w/w)	Dietary fiber (% w/w)	Protein (% w/w)	Fat (% w/w)	Ash (% w/w)	Total phenolics (mg GAE/g)	References
Oat bran	90.3	27–62[a]	9.7	13.8–23.1	6–12.8	1.7–7.0	–	Liu and Wise (2021); Liu et al. (2020)
Barley bran	92.3	51.2	17.1	13.6	4.5	2.1	–	Karimi et al. (2018)
Red sorghum mill-feed	88.5	43.8	20.0	13.4	5.0	2.0	–	Alvarenga et al. (2018)
Proso millet bran	–	1.8–26.3*	36.2–71.8*	4.9–13.9*	1.9–9.0*	5.8–7.9*	0.3*	Čukelj Mustač et al., 2020
Foxtail millet defatted bran	91.5–93.9	10.2–18.6	42.6–60.7	10.2–12.9	6.6–9.6	6.2–7.8	0.2–0.3	Amadou et al., (2011); Zhu et al. (2018)
Buckwheat hulls	–	0.2–2.2	76.5–80.7	5.1–5.7	0.5–0.8	1.9–2.1	0.4–0.5	Dziadek et al. (2016)
Quinoa dry milling-fiber fraction	88.0	18.7	68.2	14.2	3.7	2.7	0.8*	Ballester-Sanchez et al. (2020)
Quinoa wet milling-fiber fraction	83.1	22.9	59	15.3	6.0	3.1	0.9*	Ballester-Sanchez et al. (2020)
Quinoa germ	7.1	27.2[a]	–	36.5	31.7	4.3	–	Mufari et al. (2018)

* Dry weight basis, [a] carbohydrates, [b] cellulose+lignin, [c] total phenolic acids.

(about 50%), proteins (15–37%), lipids (10–41%), dietary fiber, minerals, vitamins E and B, and bioactive compounds (Table 4.2) (Boukid et al., 2018; De Vasconcelos et al., 2013; Mufari et al., 2018). The maize kernel contains the largest germ, which accounts for about 11% of the kernel weight, while wheat germ accounts for only 2–3% of the kernel (Albuquerque et al., 2014). Because of its higher nutritional and economic value, but also a shorter shelf life, the germ is generally separated from the bran. Germ can be used for the production of culinary or pharmaceutical oil, so it is often considered a co-product (Johnston et al., 2005). The resulting defatted germ meal is rich in protein and dietary fiber (Albuquerque et al., 2014; Lakshmi et al., 2017).

4.2.3 BY-PRODUCTS OF THE MALTING AND BREWING INDUSTRIES

Malt is an important ingredient in the manufacture of beer and other beverages, baked goods, and breakfast cereals. The grain most commonly processed into malt is barley, but other grains such as wheat, maize, sorghum, or rice are also used. Malting involves three steps: soaking, germination, and drying (or kilning) (Mussatto et al., 2006). The main by-products of the malting industry are malt sprouts consisting of roots, sprouts, and hulls (Almendinger et al., 2020). Further by-products are generated during beer production. The main by-product (85%) of the beer (and spirits) industries is the brewer's spent grain (BSG) (Stojceska, 2011) that is obtained from the malted grain after milling, mashing with water, and filtering (Mussatto et al., 2006). BSG constitutes up to 30% (weight per weight, w/w) of the initial malt grain (Del Rio et al., 2013) and contains a mixture of its husk, pericarp, seed coat, and fragments of the endosperm (Steiner et al., 2015; Mussatto et al., 2006). Another by-product of the brewing process is spent brewer's yeast (SBY), accounting for about 1.5–2.5% of the total beer production (Puligundla et al., 2020). See images of the main by-products from the malting and brewing industries in Figure 4.2.

These by-products are susceptible to microbiological spoilage if not stabilized, e.g., by drying. Brewery by-products (spent grain and spent yeast) may contain high levels of mycotoxins, but generally within recommended regulatory limits (Mastanjević et al., 2018).

The chemical composition of malt sprouts and BSG varies, depending on the grain variety, growing and harvesting conditions, characteristics of the added hops, malting and mashing conditions, and composition of brewing adjuvants (Steiner et al., 2015; Mussatto et al., 2006). The typical components of malt sprouts and BSG (on a dry weight (dw) basis) are dietary fiber (45–59%), protein (15–30%), lipids (4–24%) and ash (2–5%) (Del Río et al., 2013; Lynch et al., 2016; Mussatto et al., 2006; Santos et al., 2003; Severini et al., 2015; Steiner et al., 2015). The BSG polysaccharides are mainly hemicellulose (20–25%), cellulose (12–25%), lignin (12–28%), and trace amounts of starch (Lynch et al., 2016; Mussatto et al., 2006). Malt sprouts and BSG contain free reducing sugars (maltose, glucose, fructose, xylose, and arabinose) and free amino acids derived from the hydrolysis of starch or proteins (Almendinger et al., 2020; Waters et al., 2012). The predominant lipids of the BSG are triglycerides (67% of total fat), followed by free fatty acids (18%) and diglycerides (8%), along

TABLE 4.2
Summary of UAE for the Extraction of Bioactive Compounds from Cereals and Cereal By-products

Cereals and by-products	Targeted compounds	UAE processing conditions	Solvent	References
Barley	β-glucans	UAE improved the extraction efficiency of β-glucans and their molecular weight compared to that obtained using conventional solvent extraction techniques. The main advantage of UAE is related with the reduction of process time and energy consumption (3 min versus 3 h and 170 kJ/L versus 1460 kJ/L). The best results were obtained in the range 250–425 kJ/L that achieved high yields (>40.5%) of high molecular weight β-glucans (>260 kDa).	Water	Benito-Román et al. (2013)
Defatted oat	Phenolic compounds and β-glucans	UAE generated extracts with higher yields of phenolics compounds (1.5 times higher) and β-glucans (5.73%) compared to conventional solvent extraction.	Ethanol	Chen et al. (2018)
Purple glutinous rice bran	Polysaccharides	Optimum UAE (solid–liquid ratio 1:20 w/v, 70°C, 20min and static power 150 W) achieved yields of polysaccharides of 4%, significantly higher than those achieved by hot water extraction (0.8%). Extracts from UAE had better antioxidant properties compared to those extracted by conventional techniques.	Water	Surin et al. (2020)
Brewers spent grain (BSG) by-products	Arabinoxylans	UAE reduced the time of extraction (25min) and energy consumption compared to alkaline treatment (7 h), to recover similar amounts of arabinoxylans (60%) from BSG, leading to the production of starch-free arabinoxylans-rich extracts.	Water	Reis et al. (2015)
Oat hulls	Hemicellulose	UAE and alkali pretreatment combined (10 min UAE in water, followed by incubation in 5 M NaOH at 80°C for 9 h) led to the solubilization of 72% of hulls' hemicellulose.	Water	Schmitz et al. (2021)

(Continued)

TABLE 4.2 (CONTINUED)
Summary of UAE for the Extraction of Bioactive Compounds from Cereals and Cereal By-products

Cereals and by-products	Targeted compounds	UAE processing conditions	Solvent	References
BSG by-products	Proanthocyanins	UAE (80% acetone/water (v/v), 55 min, and 400 W) achieved 2 times higher extraction yields of proanthocyanins compared to conventional solid–liquid extraction method.	80% acetone/water (v/v)	Martín-García et al. (2019)
Rice bran by-product	Proteins	UAE (solid: liquid ratio of 0.43, power of 48.25% amplitude, and extraction time 30 min) achieved the highest yields of protein (39.85%) from rice bran.	Water	Iscimen and Hayta (2018)
BSG by-products	Proteins	UAE (81.4 min, ultrasonic power of 88.2 W, 2:100 solid: solvent ratio) achieved the highest yield of proteins (104.2 mg/g BSG) from BSG by-products.	Sodium carbonate buffer, pH 10	Tang et al. (2010)
Barley and corn grain	Lipids	UAE (liquid: solid ratio 4:1 for different extraction times: 2, 10, 20 min) had a positive effect on the yields of oil extracted from the grain. Compared with the static method, the average oil yield increased from 0.9% to 1.3% for barley and 1.7% to 3.2% for the corn when using UAE.	n-hexane	Gordon et al. (2018)

FIGURE 4.2 By-products from the malting and brewing industries (a) malt sprout and (b) brewer's spent grain (BSG).

with steroid compounds (approximately 5%), such as free and conjugated sterols and small amounts of n-alkanes and alkylresorcinols (Del Río et al., 2013). In addition, malt sprouts and BSG contain many minerals and phenolic compounds (phenolic acids, flavonoids, lignans, alkylphenols, and procyanidins) with antioxidant activity (Almendinger et al., 2020; Birsan et al., 2019; Lynch et al., 2016; Waters et al., 2012).

SBY is a rich source of proteins (45–64% dw), carbohydrates (35–45% dw) (mainly dietary fiber, but also oligo- and monosaccharides), and minerals (6–14% dw) including Na, K, Ca, Mg, Zn, Fe, Mn, and B vitamins, while it contains little fat (1.3–4.9%) (Amorim et al., 2016; Liu et al., 2008; Mathias et al., 2015; Podpora et al., 2016; Puligundla et al., 2020; Vieira et al., 2016). However, due to the high content (6–15%) of nucleic acids, the application of SBY in human nutrition is limited (Podpora et al., 2016; Viera et al., 2016).

4.3 NOVEL APPROACHES FOR THE UTILIZATION OF CEREAL BY-PRODUCTS

As seen from the previous review of the literature, there are multiple by-products at different stages during the processing of cereals. These by-products contain high amount of nutrients including proteins and peptides (Amorim et al., 2016), poly- and oligosaccharides (Liu et al., 2008; Severini et al., 2015; Steiner et al., 2015), lipids (Del Río et al., 2013; Patel et al., 2018), and phenolic antioxidants (Almendinger et al., 2020; Birsan et al., 2019; Farcas et al., 2015; Meneses et al., 2013). These compounds are currently underutilized and could represent an excellent source of pharmaceutical ingredients (Lordan et al., 2019) or functional foods (Podpora et al., 2016; Waters et al., 2012). Significant efforts have been explored aiming to elucidate novel processing strategies to utilize or extract valuable compounds from these by-products, including sourdough fermentation and the use of novel technologies, such as ultrasound-assisted extraction (UAE) or microwave-assisted extraction (MAE) to achieve high yields of these valuable compounds.

4.3.1 SOURDOUGH FERMENTATION OF CEREAL BY-PRODUCTS

Sourdough-like fermentation processing of cereal by-products or waste such as bran, germ, wasted bread, or BSG can be performed spontaneously or by using selected starter cultures to enhance their nutritional features. The selection of appropriate fermentation conditions (starter culture, aeration, temperature, time, addition of sugars or enzymes) is crucial to achieve a tailored quality of by-products. Favored changes concern fiber content and composition, improved digestibility and bioavailability of macro- and micronutrients, and synthesis of bioactive compounds and some vitamins (Verni et al., 2019).

Sourdough fermentation of maize, wheat, or rye bran leads to considerable solubilization of dietary fiber such as arabinoxylans with activation of cereal hemicellulase enzymes and glycolytic activity of lactic acid bacteria (LAB) (Decimo et al., 2017; Katina et al., 2012; Manini et al., 2016; Nikinmaa et al., 2020). Nevertheless, solubilization of arabinoxylans and endogenous xylanase activity of bran are influenced by the type of bran, fermentation type and conditions (Katina et al., 2012).

Some sourdough LAB strains have the ability to synthesize exopolysaccharides such as glucan or fructan *in situ* during fermentation of cereal by-products (Abedfar et al., 2018; 2020; Kajala et al., 2016; Koirala et al., 2021; Manini et al., 2016; Nikinmaa et al., 2020). Exopolysaccharides have beneficial effects on human health acting as prebiotic fiber and antioxidants. Nevertheless, some people with irritable bowel syndrome, Crohn's disease, or non-celiac wheat sensitivity (NCWS) are sensitive to fermentable oligosaccharides, disaccharides, monosaccharides, and polyols (FODMAPs) including fructan. Among wheat milling fractions, fructan content is highest in bran (3.4–4.0%) and shorts (3.2–4.1%) (Haskå et al., 2008; Ispiryan et al., 2020). Similarly, rye bran contains 7.5 % dw of fructan (Nordlung et al., 2013). Wheat germ contains 1.2% of fructan and 3.3% of total FODMAPs (Tuck et al., 2018). FODMAPs level can be substantially reduced during sourdough fermentation using a consortium of LAB, expressing extracellular fructanases or β-fructosidase (Fang et al., 2021; Loponen and Gänzle, 2018), and/or using yeast with extracellular inulinase or invertase activity (Nilsson et al., 1987; Struyf et al., 2017). Moreover, wheat germ agglutinin associated with NCWS can be reduced by sourdough fermentation with *Lactobacillus sakei* TMW1.22 due to thiol exchange reactions (Tovar and Gänzle, 2021).

Furthermore, sourdough fermentation of cereal by-products affects the solubilization of protein from the matrix (and thus digestibility) as well as the formation of bioactive peptides and free amino acids (Arte et al., 2019; Babini et al., 2020; Coda et al. 2014; Liu et al., 2017; Nikinmaa et al., 2020; Nordlund et al., 2013; Rizzello et al., 2010; Verni et al., 2020; Zalán et al., 2015). Acidification promotes the activity of endogenous proteases and LAB proteolytic activity (Gänzle, 2014). Thus, microbial fermentation is an alternative, more affordable method for producing bioactive peptides from protein substrates than enzymatic bioprocessing (Babini et al., 2020). The mixture of peptides and organic acids synthesized during sourdough fermentation of wheat germ shows antifungal activity (Rizzello et al., 2011). During the fermentation of bran or wheat germ, the content of an important neurotransmitter – aminobutyric acid (metabolite of glutamate) can be increased (Jin et al., 2013; Kim et al., 2019; Rizzello et al., 2012).

Fermentation of wheat, oat, and barley bran with certain LAB and yeast increases the content of folates (Katina et al., 2012; Kariluoto et al., 2014; Korhola et al., 2014). Wheat bran is a promising matrix for *in situ* production of vitamin B_{12} that is not present in plants, by fermentation with *Propionibacterium freudenreichii* (Xie et al., 2018).

In addition, fermentation contributes to the bioconversion and release of phenolic acids (Decimo et al., 2017; Gupta et al., 2013; Katina et al. 2007; 2012; Nordlung et al., 2013; Savolainen et al., 2014; Rizzello et al., 2010; Verni et al., 2020). The synthesis or release of various bioactive compounds can occur under the action of ferulic acid esterase activity of LAB and xylanase activity of bran and depends on the fermentation conditions, especially temperature, pH, and time (Gänzle, 2014; Katina et al., 2007). Nevertheless, sourdough fermentation of by-products decreases the content of alkylresorcinols (Bartkiene et al., 2018). Sourdough fermentation is known for successful reduction of the phytic acid level in cereal by-products (Decimo et al., 2017; Rizzello et al., 2010). Phytic acid is considered an antinutrient because it chelates minerals, reducing their bioaccessibility and bioavailability. Phytate reduction is related to moderate acidification and activation of endogenous cereal phytases associated with the phytate-degrading ability of some LAB and yeast strains (Gänzle, 2014; Manini et al., 2016; Nuobariene et al., 2012).

4.3.2 NOVEL TECHNOLOGIES FOR EXTRACTION OF COMPOUNDS

4.3.2.1 Ultrasound-Assisted Extraction

Ultrasound-assisted extraction (UAE) is based on the application of acoustic energy for the extraction of different bioactive molecules from various plant matrices. The technique is based on the cavitation process induced by compression and rarefaction cycles associated with the passage of ultrasounds (20 kHz–100 MHz frequency) through the sample (Figure 4.3). The implosion of the cavitation bubbles induces interparticle collisions which result, among others, in particle disruption and enhanced diffusion of the extractable biomolecules into the solvent (Panzella et al., 2020).

In general, UAE enhances the extraction efficiency and yields of compounds while decreasing the extraction times and reducing the energy consumption and solvent used during these extraction processes. Furthermore, it is economically feasible, reproducible, and can be easily integrated along with other classical or nonconventional techniques in order to improve the efficiency of treatments (Medina-Torres et al., 2017).

Cereal by-products continue to represent an unexploited source of various compounds and fractions with high nutritional value that could serve as sustainable material for the extraction of bioactive compounds for multiple purposes (Skendi et al., 2020). Some representative examples that highlight the effectiveness of UAE treatment in releasing the bioactive fractions from different cereals and by-products are summarized in Table 4.2.

Innovative Technologies to Extract High-Value Compounds

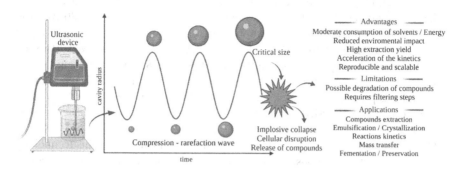

FIGURE 4.3 Mechanism, advantages, and applications of ultrasound-assisted extraction (UAE). Image created with BioRender.com.

As previously mentioned, cereal by-products have high hemicellulose content that needs to be processed in order to exploit these compounds as a source of dietary fiber with prebiotic activities. Different processing strategies, including autoclaving, ultrasonication, microwave, deep eutectic solvents, and alkaline treatments have been applied in order to solubilize hemicellulose from oat hulls. UAE followed by alkaline solvents has been explored as a suitable pretreatment. High solubilization yields of hemicellulose, of approximately 74%, were achieved using a UAE treatment of 10 minutes followed by 9 hours of incubation of the biomass with 5 M NaOH at 80°C (Schmitz et al., 2021). UAE of rice straw for 30 min increased yields of lignin separation from 72.8% to 84.7%, and UAE using alkaline solvents led to a more thermostable and high-purity lignin (Vu et al., 2017).

UAE has also been applied to decompose the lignocellulosic structure of BSG in order to increase the recovery yield or bioavailability of bioactive compounds, such as prebiotic sugars (Reis et al., 2015), proteins (Tang et al., 2010), and polyphenolic compounds (Socaci et al., 2018; Martin-García et al., 2019; Alonso-Riaño et al., 2020). Arabinoxylans are one of the most abundant polysaccharides of cereal grain cell walls; therefore the generated by-product is considered a sustainable biomass for prebiotic extraction. Compared with conventional enzymatic treatment, the extraction yield of arabinoxylans from wheat bran noticeably increased with intensification of UAE power from 50 to 200 W (Wang et al., 2014). A comparison with the conventional procedure for retrieval of horseweed functional fiber revealed that the optimized UAE protocol was more sustainable, resulting in a 38% lower cumulative energy demand, 6.6 times reduced water consumption, and 1.2 times less total processing time (Balicki et al., 2020).

Overall, UAE has reported huge advantages by reducing the extraction time, solvent consumption, and extraction temperature, while achieving high yields of compounds from several biological matrices. However, in order to achieve these beneficial results, UAE protocol have to be optimized according to the matrix of choice, as well as the chemical structure and stability of the desired compounds. For example, high-intensity UAE can break down polymers once extracted, affecting the biological and other physical properties of these molecules (Reis et al., 2015). Tang et al. (2010) optimized UAE (extraction time, ultrasonic power, and solid–liquid ratio) to extract proteins from BSG. Optimum UAE conditions achieved 104.3 mg protein/g of BSG using an extraction time of 81.4 min, 88.2 W of ultrasonic power and 2:100 BSG: solvent ratio. Additionally, Rahman and Lamsal (2021) reported the effect of UAE parameters (energy density, time of sonication, substrate, slurry ratio, temperature) on the physicochemical, structural, and functional characteristics of the protein.

One of the main advantages of UAE is the high compatibility of this technology with other extraction methods. The use of ultrasound-microwave-assisted extraction (UMAE) hybrid combining ultrasound and microwave extraction forces simultaneously have shown promising effects in terms of energy consumption and extraction efficiency (Barrera et al., 2014). The efficiency of UMAE has been demonstrated when extracting phenolic compounds from cereal husks. UMAE using 70% ethanol solutions as solvent of extraction achieved 3.6 times higher yields of phenolic dyes from sorghum husk compared to conventional solvent extraction methods (Wizi et al., 2018). Moreover, the compounds extracted by the UMAE method had high contents of 3-deoxyanthocyanins and higher thermal stabilities (Wizi et al., 2018). UMAE using alkali solvents (0.3 M sodium hydroxide) was also used to extract non-water-extractable arabinoxylans from corn bran (Jiang et al., 2019). Maximum yield of 27% arabinoxylans was achieved from corn bran when using UMAE at 500 W of ultrasonic power, 25 min of microwave synergetic time and a 1:30 biomass: solvent ratio (Jiang et al., 2019).

4.3.2.2 Microwave-Assisted Extraction

Classical extraction techniques, used for the recovery of bioactive compounds from vegetable matrices include Soxhlet, solid–liquid extraction, and liquid–liquid extraction. These techniques have several drawbacks, the main ones being the high volumes of solvents used and the long extraction times needed, but also the decreased extraction efficiency of these methods (Angiolillo et al., 2015; Socaci et al., 2018). To overcome these limitations, novel techniques were developed and optimized in order to increase the extraction yield and to reduce the extraction time and the volume of solvents used, these being considered more environmentally friendly procedures compared to conventional extraction processes. Microwave-assisted extraction (MAE) is one of the most promising novel technologies used for the extraction of bioactive compounds from a wide range of matrices. MAE uses electromagnetic waves with a frequency ranging from 300 MHz to 300 GHz that induces changes in the cell structure resulting in the release of the bioactive compounds into the extraction solvent (Angiolillo et al., 2015). MAE targets the moisture content of the samples, and as the water evaporates inside the cells this creates an increased intracellular pressure leading to the breakdown of cell walls and the release of intracellular compounds (Gomez et al. 2020). One of the main differences between MAE and other conventional heating methods that facilitate the extraction processes relates to

Innovative Technologies to Extract High-Value Compounds 101

the mechanisms of heating-release of compounds. During MAE, both the heat and mass transfer of the extracted compounds follow the same direction from within the cells to the media, while during conventional heating, the heat transfer occurs from the media to the inside of the cells and the release of compounds follows an opposite direction (see Figure 4.4).

Moreover, MAE efficiency depends on several factors, including solid–liquid ratio, solvent used, extraction time, and temperature. Chemometric approaches, such as response surface methodology, can be successfully used to optimize these extraction parameters, assessing those that significantly affect the efficiency of MAE procedures (Setyaningsih et al., 2015).

In recent years there has been a shift in the way that agricultural waste is addressed. This waste is no longer regarded exclusively as residues contributing to environmental pollution, but instead agricultural wastes are being perceived as unconventional sources of compounds that can be recovered and further exploited

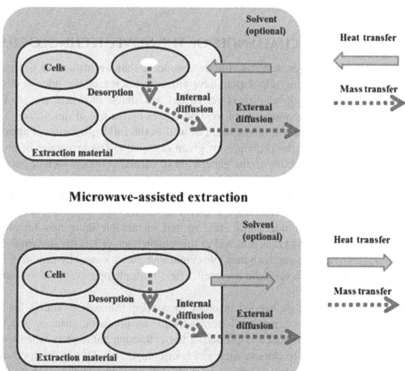

FIGURE 4.4 Comparison of heat and mass transfer between MAE and conventional heating during extraction processes. Image reproduced from Gomez et al. (2020) with permission from Elsevier.

in the development of value-added products and bioactive ingredients. The use of agricultural waste as a source of bioactive compounds has the advantage that waste is a low-cost and renewable source compared to the original biomass (Socaci et al., 2016). MAE is a suitable and efficient technique that can be used for the extraction of bioactive compounds from complex matrices, such as cereal waste or by-products. Several representative examples that highlight the effectiveness of MAE treatment in releasing the bioactive fractions from different cereals and by-products are summarized in Table 4.3.

Other studies also used microwave technology as a pretreatment in order to optimize the extraction of different bioactive compounds from complex vegetable matrices. For example, in a recent study performed by Norazlina et al., 2021, the microwave-assisted–acidic hydrolysis method was used to extract xylose from rice straw. The two treatments were combined in order to promote the saccharification of rice straw by removing lignin and hemicelluloses and increase the xylose accessibility. As previously mentioned, MAE can be also combined with other novel technologies, such as UAE in UMAE. This technique is very promising in terms of extraction time and extraction efficiency, the advantages of UAE and MAE being simultaneously exploited (Chemat et al., 2017).

4.4 FUTURE AND CHALLENGES OF THE NOVEL TECHNOLOGIES

Current global challenges, such as climate change, global warming, and depletion of resources for a growing population, have currently increased the interest of the scientific community in developing effective and sustainable food practices. In this sense, increasing the sustainability of food processes by the use of innovative technologies has currently gained momentum, as well as the full exploitation of agricultural biomass following the principles of green chemistry. The principles of green chemistry applied specifically to the generation of natural products include aspects related to the sustainability of the original biomass for extraction, ensuring its full utilization using a biorefinery concept, as well as the use of water, energy, and safety of the extraction procedures and final products (Chemat, Vian, and Cravotto, 2012). The extraction of compounds using efficient and sustainable innovative technologies, such as UAE and MAE as outlined in this chapter, as well as the sustainability of the biomass of extraction, with multiple cereal and pseudocereal by-products currently under-utilized and yet to be explored, offers multiple opportunities for future research in this field.

Future research will be needed in order to develop efficient protocols for the extraction of high-value compounds from multiple by-products, ensuring the full utilization of each by-product by optimizing the extraction parameters when using innovative technologies to achieve high yields of compounds. Of equal importance are the reduction of the use of organic solvents, the time and energy of these processes, thereby increasing the sustainability of the future extraction processes developed. Moreover, the exploration of the biological properties of these compounds can also increase the commercial value of the final products that could be exploited as nutraceuticals or functional foods. Functional foods have been defined as processed

TABLE 4.3
Summary of MAE for the Extraction of Bioactive Compounds from Cereals and Cereal by-products

Cereals and by-products	Targeted compounds	MAE processing conditions	References
Wheat and rye stillage	Sugar	The MAE treatment improves decomposition of lignocellulose present in the biomass of wheat and rye stillage for the production of cellulosic ethanol. The highest glucose concentration after pretreatment (>156 mg/g^{-1} of DW) and the highest yield of cellulose hydrolysis after 24 h of the process (over 75%) were obtained using microwave power of 300 W, 15 min, 54 PSI.	Mikulskia and Kłosowski (2020)
Brewer's spent grains (BSG)	Fermentable sugars from brewing waste	MAE pretreatment is effective for fractionating BSG without using acid or alkali catalysts. Pretreatment optimal conditions were 192.7°C and 5.4 min achieving recovery yields of 82% (43 g fermentable sugars/100 g BSG).	López-Linares et al. (2019)
Sorghum	3-deoxyanthocyanin (pigment)	Compared to control, MAE (up to 1200 W/100°C) significantly improved both the extraction yield (3100 vs 1520 mg/g) and the structural stability of the of 3-deoxyanthocyanin pigment.	Herrman et al. (2020)
Purple corn cob	Phenolic compounds	The highest anthocyanin content (185.1 mg/100 g) was obtained using MAE (555 W) for 19 min and 1.5 M HCl–95% ethanol as extraction solvent.	Yang and Zhai (2010)
BSG	Phenolic compounds (ferulic acid)	Optimum MAE (100°C, 15 min and solvent: sample ratio 20:1) achieved yields of ferulic acid of 1.31 (w/w) that was 5-fold higher than those achieved by conventional solid–liquid techniques. The extraction time, energy, and solvent consumption were considerably reduced using MAE.	Moreira et al. (2012)
Wheat, rice, oat brans	Phenolic compounds	Compared with conventional method (1.24–2.87 mg GAE/g), MAE (2450 MHZ, 3.5 min) significantly increased the total phenolic compound content extracted (2.20–4.09 mg GAE/g).	Dar and Sharma (2011)
Sesame bran	Protein and phenolic compounds	The highest protein yield (90.6%) and total phenolic content (8.2 mg GAE/g) were extracted using MAE (49°C, 98 min) combined with enzymatic extraction (1.94 AU/100 g enzyme concentration).	Gorguc et al. (2019)

(Continued)

TABLE 4.3 (CONTINUED)
Summary of MAE for the Extraction of Bioactive Compounds from Cereals and Cereal by-products

Cereals and by-products	Targeted compounds	MAE processing conditions	References
Rice bran	Protein	Protein yields extracted by MAE were 1.54-fold higher compared to those achieved when using alkaline extraction, while the protein digestibility was similar. Optimal MAE parameters were 1000 W of microwave power, 90 seconds of extraction time, and a solid: liquid ratio of 0.89 g rice bran/10 mL of distilled water.	Phongthai et al. (2016)

or natural products with potential health benefits when included in the diet beyond those of basic nutrition (Granato et al., 2020). The global market value of functional foods was valued at USD 64.75 billion in 2017, and this market is projected to reach USD 94.21 billion by 2023 (Marketsandmarkets, 2018). Thus, the extraction of compounds from underutilized agricultural by-products and the exploration of the biological activities of these compounds will continue to attract industry and researchers guided by a strong market and consumers' demand for functional foods.

Moreover, the sequential use of several innovative extraction technologies, as well as the simultaneous application of one or several extraction forces seems to be a new promising field of research to explore further. In this sense, the aforementioned simultaneous application of ultrasounds and microwaves in UMAE is currently being explored for the extraction of multiple high-value compounds such as carbohydrates and phenolic compounds from seaweeds (Garcia-Vaquero, Ummat, Tiwari, and Rajauria, 2020) and pigments from tomatoes (Lianfu and Zelong, 2008). More research will be needed for the exploration of combined innovative technologies to extract high-value compounds from cereal and pseudocereal by-products.

Overall, the exploration of multiple innovative technologies to develop sustainable and green extraction processes, together with the high number of under-exploited cereal and pseudocereal by-products, as well as the exploration of the biological properties of these compounds for the functional food market, offer an unparalleled field of research and opportunities in the agri-food sector.

ACKNOWLEDGMENTS

The authors thank the COST Action 18101 SOURDOMICS "Sourdough biotechnology network towards novel, healthier and sustainable food and bioprocesses" (https://www.cost.eu/actions/CA18101/#tabs|Name:overview). Anca C. Farcas and Sonia A. Socaci's work was supported by grants of the Romanian Ministry of Education and Research, CNCS – UEFISCDI, project number PN-III-P4-ID-PCE-2020-2306 and project number PN-III-P4-ID-PCE-2020-1847, within PNCDI III.

REFERENCES

Abedfar, A., Hosseininezhad, M., Sadeghi, A., Raeisi, M., Feizy, J. 2018. Investigation on "spontaneous fermentation" and the productivity of microbial exopolysaccharides by Lactobacillus plantarum and Pediococcus pentosaceus isolated from wheat bran sourdough. *LWT-Food Science and Technology*, 96:686–693.

Abedfar, A., Abbaszadeh, S., Hosseininezhad, M., Taghdir, M. 2020. Physicochemical and biological characterization of the EPS produced by L. acidophilus isolated from rice bran sourdough. *LWT- Food Science and Technology*, 127:109373.

Agil, R., Hosseinian, F. 2014. Determination of water-extractable polysaccharides in triticale bran. *Journal of Food Composition and Analysis*, 34:12–17.

Albuquerque, C.S., Rabello, C.B.V., Santos, M.J.B., Lima, M.B.D., Silva, E.P.D., Lima, T.S., Ventura, D.P., Dutra Jr, W.M., 2014. Chemical composition and metabolizable energy values of corn germ meal obtained by wet milling for layers. *Brazilian Journal of Poultry Science*, 16:107–112.

Alexandratos, N., Bruinsma, J. 2012. *World Agriculture Towards 2030/2050: The 2012 Revision*. ESA Working paper No. 12-03. FAO.

Almendinger, M., Rohn, S., Pleissner, D. 2020. Malt and beer-related by-products as potential antioxidant skin-lightening agents for cosmetics. *Sustainable Chemistry and Pharmacy*, 17:100282.

Alonso-Riaño, P., Sanz Diez M.T., Blanco, B., Beltrán, S., Trigueros, E., Benito-Román, O. 2020. Water ultrasound-assisted extraction of polyphenol compounds from Brewer's spent grain: Kinetic study, extract characterization, and concentration. *Antioxidants*, 9(3):265.

Alvarenga, I.C., Ou, Z., Thiele, S., Alavi, S., Aldrich, C.G. 2018. Effects of milling sorghum into fractions on yield, nutrient composition, and their performance in extrusion of dog food. *Journal of Cereal Science*, 82:121–128.

Álvarez, C., González, A., Alonso, J.L., Sáez, F., Negro, M.J., Gullón, B. 2020. Xylooligosaccharides from steam-exploded barley straw: Structural features and assessment of bifidogenic properties. *Food and Bioproducts Processing*, 124:131–142.

Amadou, I., Amza, T., Shi, Y.H., Le, G.W. 2011. Chemical analysis and antioxidant properties of foxtail millet bran extracts. *Songklanakarin Journal of Science & Technology*, 33:509–515.

Amorim, M.M., Pereira, J.O., Monteiro, K.M., Ruiz, A.L., Carvalho, J.E., Pinheiro, H., Pintado, M. 2016. Antiulcer and antiproliferative properties of spent brewer's yeast peptide extracts for incorporation into foods. *Food & Function*, 7:2331–2337.

Angiolillo, L.A., Del Nobile, M., Conte, A. 2015. The extraction of bioactive compounds from food residues using microwaves. *Current Opinion in Food Science*, 5:93–98. https://doi.org/10.1016/j.cofs.2015.10.001.

Arte, E., Huang, X., Nordlund, E., Katina, K. 2019. Biochemical characterization and technofunctional properties of bioprocessed wheat bran protein isolates. *Food Chemistry*, 289:103–111.

Babini, E., Taneyo-Saa, D.L., Tassoni, A., Ferri, M., Kraft, A., Grän-Heedfeld, J., Bretz K., Roda A., Michelini E., Calabretta M.M., Guillon F., Tagliazucchi D., Martini S., Nissen L., Gianotti, A. 2020. Microbial fermentation of industrial rice-starch byproduct as valuable source of peptide fractions with health-related activity. *Microorganisms*, 8:986.

Balicki, S., Pawlaczyk-Graja, I., Gancarz, R., Capek, P., Wilk, K.A. 2020. Optimization of Ultrasound-Assisted Extraction of Functional Food Fiber from Canadian Horseweed (*Erigeron canadensis* L. *ACS Omega*, 5 (33):20854–20862. https://doi.org/10.1021/acsomega.0c02181.

Ballester-Sánchez, J., Fernández-Espinar, M.T., Haros, C.M. 2020. Isolation of red quinoa fibre by wet and dry milling and application as a potential functional bakery ingredient. *Food Hydrocolloids*, 101:105513.

Barrera Vázquez, M.F., Comini, L.R., Martini, R.E., Núñez Montoya, S.C., Bottini, S., Cabrera, J.L., 2014. Comparisons between conventional, ultrasound-assisted andmicrowave-assisted methods for extraction of anthraquinones from Heterophyllaeapustulata Hook f. (*Rubiaceae*). *Ultrasonics Sonochemistry*, 21:478–484.

Bartkiene, E., Bartkevics, V., Krungleviciute, V., Juodeikiene, G., Zadeike, D., Baliukoniene, V. et al. 2018. Application of hydrolases and probiotic Pediococcus acidilactici BaltBio01 strain for cereal by-products conversion to bioproduct for food/feed. *International Journal of Food Sciences and Nutrition*, 69:165–175.

Bedoić, R., Ćosić, B., Duić, N. 2019. Technical potential and geographic distribution of agricultural residues, co-products and by-products in the European Union. *Science of the Total Environment*, 686:568–579.

Benito-Román, Ó., Alonso, E., Cocero, M.J. 2013. Ultrasound-assisted extraction of β-glucans from barley. *LWT: Food Science and Technology*, 50:57–63.

Birsan, R.I., Wilde, P., Waldron, K.W., Rai, D.K., 2019. Recovery of polyphenols from brewer's spent grains. *Antioxidants*, 8:380.
Bora, P., Ragaee, S., Marcone, M., 2018. Effect of parboiling on decortication yield of millet grains and phenolic acids and in vitro digestibility of selected millet products. *Food Chemistry*, 274: 718–725.
Boukid, F., Folloni, S., Ranieri, R., Vittadini, E. 2018. A compendium of wheat germ: Separation, stabilization and food applications. *Trends in Food Science & Technology*, 78:120–133.
Casas G.A., Helle, N.L., Knud, E.B.K., Hans, H.S. 2019. Arabinoxylan is the main polysaccharide in fiber from rice coproducts, and increased concentration of fiber decreases in vitro digestibility of dry matter. *Animal Feed Science and Technology*, 247:255–261.
Chemat, F., Rombaut, N., Sicaire, A.G., Meullemiestre, A., Fabiano-Tixier, A.S., Abert-Vian, M. 2017. Ultrasound assisted extraction of food and natural products. Mechanisms, techniques, combinations, protocols and applications. A review. *Ultrasonics Sonochemistry*, 34:540–560.
Chemat, F., Vian, M.A., Cravotto, G. 2012. Green extraction of natural products: Concept and principles. *International Journal of Molecular Sciences*, 13(7), 8615–8627.
Chen, C., Wang, L., Wang, R. et al. 2018. Ultrasound-assisted extraction from defatted oat (Avena sativa L.) bran to simultaneously enhance phenolic compounds and β-glucan contents: Compositional and kinetic studies. *Journal of Food Engineering*, 222:1–10.
Coda, R., Kärki, I., Nordlund, E., Heiniö, R.L., Poutanen, K., Katina, K. 2014. Influence of particle size on bioprocess induced changes on technological functionality of wheat bran. *Food Microbiology*, 37:69–77.
Čukelj Mustač, N., Novotni, D., Habuš, M., Drakula, S., Nanjara, L., Voučko, B., et al. 2020. Storage stability, micronisation, and application of nutrient-dense fraction of proso millet bran in gluten-free bread. *Journal of Cereal Science*, 91:102864.
Dar, B.N., Sharma, S. 2011. Total phenolic content of cereal brans using conventional and microwave assisted extraction. *American Journal of Food Technology*, 6: 1045–1053.
De Vasconcelos, M.C.B.M., Bennett, R., Castro, C.A.B.B., Cardoso, P., Saavedra, M.J., Rosa, E.A. 2013. Study of composition, stabilization and processing of wheat germ and maize industrial by-products. *Industrial Crops and Products*, 42:292–298.
Decimo, M., Quattrini, M., Ricci, G., Fortina, M.G., Brasca, M., Silvetti, T., et al. 2017. Evaluation of microbial consortia and chemical changes in spontaneous maize bran fermentation. *AMB Express*, 7:1–13.
Del Río, J.C., Prinsen, P., Gutiérrez, A. 2013. Chemical composition of lipids in brewer's spent grain: A promising source of valuable phytochemicals. *Journal of Cereal Science*, 58:248–254.
Devisetti, R., Yadahally, S.N., Bhattacharya, S. 2014. Nutrients and antinutrients in foxtail and proso millet milled fractions: Evaluation of their flour functionality. *LWT-Food Science and Technology*, 59:889–895.
Đorđević, T., Antov, M. 2018. The influence of hydrothermal extraction conditions on recovery and properties of hemicellulose from wheat chaff: A modeling approach. *Biomass and Bioenergy*, 119:246–252.
Dziadek, K., Kopeć, A., Pastucha, E., Piątkowska, E., Leszczyńska, T., Pisulewska, E., et al. 2016. Basic chemical composition and bioactive compounds content in selected cultivars of buckwheat whole seeds, dehulled seeds and hulls. *Journal of Cereal Science*, 69:1–8.
Espinosa, E., Sánchez, R., Otero, R., Domínguez-Robles, J., Rodríguez, A. 2017. A comparative study of the suitability of different cereal straws for lignocellulose nanofibers isolation. *International Journal of Biological Macromolecules*, 103:990–999.

Fang, S., Yan, B., Tian, F., Lian, H., Zhao, J., Zhang, H., Chen, W., Fan, D. 2021. β-fructosidase FosE activity in Lactobacillus paracasei regulates fructan degradation during sourdough fermentation and total FODMAP levels in steamed bread. *LWT- Food Science and Technology*, 145:111294.

FAO. 2014. *Building a Common Vision for Sustainable Food and Agriculture: Principles and Approaches*. FAO.

Farcas, A.C., Socaci, S.A., Dulf, F.V., Tofana, M., Mudura, E., Diaconeasa, Z. 2015. Volatile profile, fatty acids composition and total phenolics content of brewers' spent grain by-product with potential use in the development of new functional foods. *Journal of Cereal Science*, 64:34–42.

Gänzle, M.G. 2014. Enzymatic and bacterial conversions during sourdough fermentation. *Food Microbiology*, 37:2–10.

Garcia-Vaquero, M., Rajauria, G., Tiwari, B. 2020a. Conventional extraction techniques: Solvent extraction. In M.D. Torres, S. Kraan, H. Dominguez (Eds.), *Sustainable Seaweed Technologies* (pp. 171–189). Cambridge, MA, United States. Elsevier.

Garcia-Vaquero, M., Ummat, V., Tiwari, B., Rajauria, G. 2020b. Exploring ultrasound, microwave and ultrasound–microwave assisted extraction technologies to increase the extraction of bioactive compounds and antioxidants from brown macroalgae. *Marine Drugs*, 18(3), 172.

Gil-Ramirez, A., Salas-Veizaga, D.M., Grey, C., Karlsson, E.N., Rodriguez-Meizoso, I., Linares-Paste´n, J.A. 2018. Integrated process for sequential extraction of saponins, xylan and cellulose T from quinoa stalks (*Chenopodium quinoa Willd.*). *Industrial Crops and Products*, 121:54–65.

Givens, D.I., Everington, J.M., Adamson, A.H. 1989. Chemical composition, digestibility in vitro, and digestibility and energy value in vivo of untreated cereal straws produced on farms throughout England. *Animal Feed Science and Technology*, 26:323–335.

Gomez, L., Tiwari, B., Garcia-Vaquero, M. 2020. Emerging extraction techniques: Microwave-assisted extraction. In M.D. Torres, S. Kraan, H. Dominguez (Eds.), *Sustainable Seaweed Technologies* (pp. 207–224): Elsevier.

Gordon, R., Chapman, J., Power, A., Chandra S., Roberts J, Cozzolino D. 2018. Comparison of ultrasound-assisted extraction with static extraction as pre-processing method before gas chromatography analysis of cereal lipids. *Food Analytical Methods*, 11:3276–3281.

Gorguc, A., Ozer, P., Yılmaz, F.M. 2019. Microwave-assisted enzymatic extraction of plant protein with antioxidant compounds from the food waste sesame bran: Comparative optimization study and identification of metabolomics using LC/Q-TOF/MS. *Journal of Food Processing and Preservation*, 44, e14304.

Granato, D., Barba, F.J., Kovačević, D.B., Lorenzo, J.M., Cruz, A.G., Putnik, P. 2020. Functional foods: Product development, technological trends, efficacy testing, and safety. *Annual Review of Food Science and Technology*, 11(1), 93–118. https://doi.org/10.1146/annurev-food-032519-051708.

Grove, A.V., Hepton, J., Hunt, C.W. 2003. Chemical composition and ruminal fermentability of barley grain, hulls, and straw as affected by planting date, irrigation level, and variety. *The Professional Animal Scientist*, 19:273–280.

Gupta, S., Jaiswal, A.K., Abu-Ghannam, N. 2013. Optimization of fermentation conditions for the utilization of brewing waste to develop a nutraceutical rich liquid product. *Industrial Crops and Products*, 44:272–282.

Habuš, M., Novotni, D., Gregov, M., Štifter, S., Čukelj Mustač, N., Voučko, B., Ćurić, D. 2021. Influence of particle size reduction and high-intensity ultrasound on polyphenol oxidase, phenolics, and technological properties of wheat bran. *Journal of Food Processing and Preservation*, 45:e15204.

Haskå, L., Nyman, M., Andersson, R. 2008. Distribution and characterisation of fructan in wheat milling fractions. *Journal of Cereal Science*, 48:768–774.

Herrman, D.A., Brantsen, J.F., Ravisankar, S., Lee, K.M., Awika, J.M. 2020. Stability of 3-deoxyanthocyanin pigment structure relative to anthocyanins from grains under microwave assisted extraction. *Food Chemistry*, 333:127494.

Hossain, K., Ulven, C., Glover, K. et al. 2013. Interdependence of cultivar and environment on fibre composition in wheat bran. *Australian Journal of Crop Science*, 7:525–531.

İşçimen, E.M., Hayta, M. 2018. Optimisation of ultrasound assisted extraction of rice bran proteins: Effects on antioxidant and antiproliferative properties. *Quality Assurance and Safety of Crops & Foods*, 10(2):165–174.

Ispiryan, L., Zannini, E., Arendt, E.K. 2020. Characterization of the FODMAP-profile in cereal-product Ingredients. *Journal of Cereal Science*, 92:102916.

Jiang, Y., Bai, X., Lang, S., Zhao, Y., Liu, C., Yu, L. 2019. Optimization of ultrasonic-microwave assisted alkali extraction of arabinoxylan from the corn bran using response surface methodology. *International Journal of Biological Macromolecules*, 128, 452–458.

Jin, W., Kim, M., Kim, K., 2013. Utilization of barley or wheat bran to bioconvert glutamate to γ-aminobutyric acid (GABA). *Journal of Food Science*, 78:C1376–C1382.

Johnston, D.B., McAloon, A.J., Moreau, R.A., Hicks, K.B., Singh, V. 2005. Composition and economic comparison of germ fractions from modified corn processing technologies. *Journal of the American Oil Chemists' Society*, 82:603–608.

Jørgensen, H., Thomsen, S.T., Schjoerring, J.K. 2020. The potential for biorefining of triticale to protein and sugar depends on nitrogen supply and harvest time. *Industrial Crops and Products*, 149:112333.

Kajala, I., Mäkelä, J., Coda, R., Shukla, S., Shi, Q., Maina, N.H., Juvonen, R., Ekholm, P., Goyal A., Tenkanen M., Katina, K. 2016. Rye bran as fermentation matrix boosts in situ dextran production by Weissella confusa compared to wheat bran. *Applied Microbiology and Biotechnology*, 100:3499–3510.

Kamal-Eldin, A., Lærke, H.N., Knudsen, K.E.B., Lampi, A.M., Piironen, V., Adlercreutz, H., et al. 2009. Physical, microscopic and chemical characterisation of industrial rye and wheat brans from the Nordic countries. *Food & Nutrition Research*, 53:1912.

Kariluoto, S., Edelmann, M., Nyström, L., Sontag-Strohm, T., Salovaara, H., Kivelä, R., Herranen M., Korhola M., Piironen, V. 2014. In situ enrichment of folate by microorganisms in beta-glucan rich oat and barley matrices. *International Journal of Food Microbiology*, 176:38–48.

Karimi, R., Azizi, M.H., Xu, Q., Naghizadeh-Raeisi, S. 2018. Enzymatic removal of starch and protein during extraction of dietary fiber from barley bran. *Journal of Cereal Science*, 83:259–265.

Katina, K., Laitila, A., Juvonen, R., Liukkonen, K.H., Kariluoto, S., Piironen, V., Landberg, R., Åman P. Poutanen, K. 2007. Bran fermentation as a means to enhance technological properties and bioactivity of rye. *Food Microbiology*, 24:175–186.

Katina, K., Juvonen, R., Laitila, A., Flander, L., Nordlund, E., Kariluoto, S., et al. 2012. Fermented wheat bran as a functional ingredient in baking. *Cereal Chemistry*, 89:126–134.

Kim, S.Y., Kim, K.J., Chung, H.C., Han, G.D. 2019. Use of rice bran for preparation of GABA (γ-aminobutyric acid)-rich sourdough. *Food Science and Technology Research*, 25:399–404.

Koirala, P., Maina, N.H., Nihtilä, H., Katina, K., Coda, R. 2021. Brewers' spent grain as substrate for dextran biosynthesis by Leuconostoc pseudomesenteroides DSM20193 and Weissella confusa A16. *Microbial Cell Factories*, 20:1–13.

Kuan, C.Y., Yuen, K.H., Bhat, R., Liong, M.T. 2011. Physicochemical characterization of alkali treated fractions from corncob and wheat straw and the production of nanofibres. *Food Research International*, 44:2822–2829.

Korhola, M., Hakonen, R., Juuti, K., Edelmann, M., Kariluoto, S., Nystrom, L. Sontag, -Strohm, T., Piironen, V. 2014. Production of folate in oat bran fermentation by yeasts isolated from barley and diverse foods. *Journal of Applied Microbiology*, 117:679–689.

Lakshmi, R.K., Kumari, K., Reddy, P. 2017. Corn germ meal (CGM)-Potential feed ingredient for livestock and poultry in India-A review. *International Journal of Livestock Research*, 7:39–50.

Lianfu, Z., Zelong, L. 2008. Optimization and comparison of ultrasound/microwave assisted extraction (UMAE) and ultrasonic assisted extraction (UAE) of lycopene from tomatoes. *Ultrasonics Sonochemistry*, 15(5), 731–737.

Liu, F., Chen, Z., Shao, J., Wang, C., Zhan, C. 2017. Effect of fermentation on the peptide content, phenolics and antioxidant activity of defatted wheat germ. *Food Bioscience*, 20:141–148.

Liu, J., Jin, S., Song, H., Huang, K., Li, S., Guan, X., Wang, Y. 2020. Effect of extrusion pretreatment on extraction, quality and antioxidant capacity of oat (Avena Sativa L.) bran oil. *Journal of Cereal Science*, 95:102972.

Liu, K., Wise, M.L. 2021. Distributions of nutrients and avenanthramides within oat grain and effects on pearled kernel composition. *Food Chemistry*, 336:127668.

Liu, X., Wang, Q., Cui, S., Liu, H. 2008. A new isolation method of β-d-glucans from spent yeast Saccharomyces cerevisiae. *Food Hydrocolloids*, 22:239–247.

López-Linares, J.C., García-Cubero, M.T., Lucas, S., González-Benito, G., Coca, M. 2019. Microwave assisted hydrothermal as greener pretreatment of brewer's spent grains for biobutanol production. *Chemical Engineering Journal*, 368:1045–1055.

Loponen, J., Gänzle, M. 2018. Use of sourdough in low FODMAP baking. *Foods*, 7:96.

Lordan, R., O'Keeffe, E., Tsoupras, A., Zabetakis, I. 2019. Total, neutral, and polar lipids of brewing ingredients, by-products and beer: Evaluation of antithrombotic activities. *Foods*, 8:171.

Lynch, K.M., Steffen, E.J., Arendt, E.K. 2016. Brewers' spent grain: A review with an emphasis on food and health. *Journal of the Institute of Brewing*, 122:553–568.

Manini, F., Casiraghi, M.C., Poutanen, K., Brasca, M., Erba, D., Plumed-Ferrer, C. 2016. Characterization of lactic acid bacteria isolated from wheat bran sourdough. *LWT-Food Science and Technology*, 66:275–283.

Mathias, T.R.D.S., Alexandre, V.M.F., Cammarota, M.C., de Mello, P.P.M., Sérvulo, E.F.C. 2015. Characterization and determination of brewer's solid wastes composition. *Journal of the Institute of Brewing*, 121:400–404.

Meneses, N.G., Martins, S., Teixeira, J.A., Mussatto, S.I. 2013. Influence of extraction solvents on the recovery of antioxidant phenolic compounds from brewer's spent grains. *Separation and Purification Technology*, 108:152–158.

Marketsandmarkets. 2018. https://www.marketsandmarkets.com/Market-Reports/functional-food-ingredients-market-9242020.html. Retrieved 20th May 2021.

Martin-Garcia, B., Pasini, F., Verardo, V. et al. 2019. Optimization of sonotrode ultrasonic-assisted extraction of proanthocyanidins from brewers' spent grains. *Antioxidants*, 8(8):282.

Mastanjević, K., Šarkanj, B., Warth, B., Krska, R., Sulyok, M., Mastanjević, K., Šantek, B., Krstanović, V. 2018. Fusarium culmorum multi-toxin screening in malting and brewing by-products. *LWT-Food Science and Technology*, 98:642–645.

Medina-Torres, N., Ayora-Talavera, T., Espinosa-Andrews, H., Sánchez-Contreras, A., Pacheco, N. 2017. Ultrasound assisted extraction for the recovery of phenolic compounds from vegetable sources. *Agronomy*, 7(3):47. https://doi.org/10.3390/agronomy7030047.

Mikulski, D., Kłosowski, G. 2020. Microwave-assisted dilute acid pretreatment in bioethanol production from wheat and rye stillages. *Biomass and Bioenergy*, 136:105528. https://doi.org/10.1016/j.biombioe.2020.105528.

Moreira, M.M., Morais, S., Barros, A.A., Delerue-Matos, C., Guido, L.F. 2012. A novel application of microwave-assisted extraction of polyphenols from brewer's spent grain with HPLC-DAD-MS analysis. *Analytical and Bioanalytical Chemistry*, 403(4):1019–1029. https://doi.org/10.1007/s00216-011-5703-y.

Mufari, J.R., Miranda-Villa, P.P., Calandri, E.L. 2018. Quinoa germ and starch separation by wet milling, performance and characterization of the fractions. *LWT-Food Science and Technology*, 96:527–534.

Mussatto, S.I., Dragone, G., Roberto, I.C., 2006. Brewers' spent grain: Generation, characteristics and potential applications. *Journal of Cereal Science*, 43:1–14.

Nikinmaa, M., Kajala, I., Liu, X., Nordlund, E., Sozer, N. 2020. The role of rye bran acidification and in situ dextran formation on structure and texture of high fibre extrudates. *Food Research International*, 137:109438.

Nilsson, U., Öste, R., Jägerstad, M. 1987. Cereal fructans: Hydrolysis by yeast invertase, in vitro and during fermentation. *Journal of Cereal Science*, 6:53–60.

Norazlina, I., Dhinashini, R.S., Nurhafizah, I., Norakma, M.N., Noor Fazreen, D. 2021. Extraction of xylose from rice straw and lemongrass via microwave assisted. *Materials Today: Proceedings*. https://doi.org/10.1016/j.matpr.2021.02.307.

Nordlund, E., Katina, K., Aura, A.-M., Poutanen, K. 2013. Changes in bran structure by bioprocessing with enzymes and yeast modifies the in vitro digestibility and fermentability of bran protein and dietary fibre complex. *Journal of Cereal Science*, 58:200–208.

Nuobariene, L., Hansen, Å.S., Arneborg, N. 2012. Isolation and identification of phytase-active yeasts from sourdoughs. *LWT-Food Science and Technology*, 48:190–196.

Onipe, O.O., Jideani, A.I., Beswa, D. 2015. Composition and functionality of wheat bran and its application in some cereal food products. *International Journal of Food Science & Technology*, 50:2509–2518.

Panzella, L., Moccia, F., Nasti, R., Marzorati, S., Verotta, L., Napolitano, A. 2020. Bioactive phenolic compounds from agri-food wastes: An update on green and sustainable extraction methodologies. *Frontiers in Nutrition*, 7:60.

Patel, A., Mikes, F., Bühler, S., Matsakas, L. 2018. Valorization of brewers' spent grain for the production of lipids by oleaginous yeast. *Molecules*, 23:3052.

Peanparkdee, M., Iwamoto, S. 2019. Bioactive compounds from by-products of rice cultivation and rice processing: Extraction and application in the food and pharmaceutical industries. *Trends in Food Science & Technology*, 86:109–117.

Phongthai, S., Lim, S.T., Rawdkuen, S. 2016. Optimization of microwave-assisted extraction of rice bran protein and its hydrolysates properties. *Journal of Cereal Science*, 70:146–154.

Podpora, B., Świderski, F., Sadowska, A., Rakowska, R., Wasiak-Zys, G. 2016. Spent brewer's yeast extracts as a new component of functional food. *Czech Journal of Food Sciences*, 34:554–563.

Prueckler, M., Siebenhandl-Ehn, S., Apprich, S., Hoeltinger, S., Haas, C., Schmid, E., Kneifel, W. 2014. Wheat bran-based biorefinery 1: Composition of wheat bran and strategies of functionalization. *LWT-Food Science and Technology*, 56(2):211–221.

Puligundla, P., Mok, C., Park, S. 2020. Advances in the valorization of spent brewer's yeast. *Innovative Food Science & Emerging Technologies*, 62:102350.

Rahman, M., Lamsal, B. 2021. Ultrasound-assisted extraction and modification of plant-based proteins: Impact on physicochemical, functional, and nutritional properties. *Comprehensive Reviews in Food Science and Food Safety*, 20(2):1457–1480.

Reis, S.F., Coelho, E., Coimbra, M.A., Abu-Ghannam, N. 2015. Improved efficiency of brewer's spent grain arabinoxylans by ultrasound-assisted extraction. *Ultrasonics Sonochemistry*, 24:155–164.

Rizzello, C.G., Nionelli, L., Coda, R., De Angelis, M., Gobbetti, M. 2010. Effect of sourdough fermentation on stabilisation, and chemical and nutritional characteristics of wheat germ. *Food Chemistry*, 119:1079–1089.

Rizzello, C.G., Cassone, A., Coda, R., Gobbetti, M. 2011. Antifungal activity of sourdough fermented wheat germ used as an ingredient for bread making. *Food Chemistry*, 127:952–959.

Rizzello, C.G., Coda, R., Mazzacane, F., Minervini, D., Gobbetti, M. 2012. Micronized by-products from debranned durum wheat and sourdough fermentation enhanced the nutritional, textural and sensory features of bread. *Food Research International*, 46:304–313.

Santos, M., Jiménez, J., Bartolomé, B., Gómez-Cordovés, C., del Nozal, M. 2003. Variability of brewer's spent grain within a brewery. *Food Chemistry*, 80:17–21.

Sarfaraz, A., Azizi, M.H., Gavlighi, H.A., Barzegar, M. 2017. Physicochemical and functional characterization of wheat milling co-products: Fine grinding to achieve high fiber antioxidant-rich fractions. *Journal of Cereal Science*, 77:228–234.

Savolainen, O.I., Coda, R., Suomi, K., Katina, K., Juvonen, R., Hanhineva, K., Poutanen, K. 2014. The role of oxygen in the liquid fermentation of wheat bran. *Food Chemistry*, 153:424–431.

Schmitz, E., Karlsson, E.N., Adlercreutz, P. 2021. Ultrasound assisted alkaline pre-treatment efficiently solubilises hemicellulose from oat hulls. *Waste Biomass Valorization*. https://doi.org/10.1007/s12649-021-01406-0.

Setyaningsih, W., Saputro, I.E., Palma, M., Barroso, C.G. 2015. Optimisation and validation of the microwave-assisted extraction of phenolic compounds from rice grains. *Food Chemistry*, 169:141–149.

Severini, C., Azzollini, D., Jouppila, K., Jussi, L., Derossi, A., De Pilli, T. 2015. Effect of enzymatic and technological treatments on solubilisation of arabinoxylans from brewer's spent grain. *Journal of Cereal Science*, 65:162–166.

Singh, G., and Arya, S.K. 2020. A review on management of rice straw by use of cleaner technologies: Abundant opportunities and expectations for Indian farming. *Journal of Cleaner Production*, 291:125278.

Skendi, A., Zinoviadou, K.G., Papageorgiou, M., Rocha, J.M. 2020. Advances on the valorisation and functionalization of by-products and wastes from cereal-based processing industry. *Foods*, 9, 1243. https://doi.org/10.3390/foods9091243.

Smuga-Kogut, M., Walendzik, B., Szymanowska-Powalowska, D., Kobus-Cisowska, J., Wojdalski, J., Wieczorek, M., Cielecka-Piontek, J. 2019. Comparison of bioethanol preparation from triticale straw using the ionic liquid and sulfate methods. *Energies*, 12:1155.

Socaci, S.A., Farcas, A.C., Vodnar, D., Tofana, M. 2016. Food wastes as valuable sources of bioactive compounds. In: *Superfood and Functional Food-Development of Superfood and its Role in Medicine*, Intech, pp. 75–93.

Socaci, S.A. Farcas, A.C, Galanakis, C. 2018a. Introduction in functional components for membrane separations. In: *Separation of Functional Molecules in Food by Membrane Technology*, Academic Press is an imprint of Elsevier, pp. 31–77.

Socaci, S.A., Farcas, A.C., Diaconeasa, Z.M., Vodnar, D.C., Rusu, B., Tofana, M. 2018b. Influence of the extraction solvent on phenolic content, antioxidant, antimicrobial and antimutagenic activities of brewers' spent grain. *Journal of Cereal Science*, 80:180–187.

Steiner, J., Procopio, S., Becker, T. 2015. Brewer's spent grain: Source of value-added polysaccharides for the food industry in reference to the health claims. *European Food and Research Technology*, 241: 303–315.

Stojceska, V. 2011. Dietary fiber from brewer's spent grain as a functional ingredient in bread making technology. In *Flour and Breads and their Fortification in Health and Disease Prevention*, edited by Victor R. Preedy, Ronald Ross Watson and Vinood B. Patel, (pp. 171–181). Academic Press.

Struyf, N., Laurent, J., Verspreet, J., Verstrepen, K.J., Courtin, C.M. 2017. Saccharomyces cerevisiae and Kluyveromyces marxianus cocultures allow reduction of fermentable oligo-, di-, and monosaccharides and polyols levels in whole wheat bread. *Journal of Agricultural and Food Chemistry*, 65:8704–8713.

Summerell, B.A., Burgess, L.W. 1989. Decomposition and chemical composition of cereal straw. *Soil Biology and Biochemistry*, 21:551–559.
Surin, S., You, S., Seesuriyachan, P. et al. 2020. Optimization of ultrasonic-assisted extraction of polysaccharides from purple glutinous rice bran (Oryza sativa L.) and their antioxidant activities. *Scientific Reports*, 10:10410. https://doi.org/10.1038/s41598-020-67266-1.
Tang, D.S., Tian, Y.J., He, Y.Z., Li, L., Hu, S.Q., Li, B. 2010. Optimisation of ultrasonic-assisted protein extraction from brewer's spent grain. *Czech Journal of Food Sciences*, 28:9–17.
Tovar, L.E.R., and Gänzle, M.G. 2021. Degradation of wheat germ agglutinin during sourdough fermentation. *Foods*, 10:340.
Tuck, C., Ly, E., Bogatyrev, A., Costetsou, I., Gibson, P., Barrett, J., Muir, J. 2018. Fermentable short chain carbohydrate (FODMAP) content of common plant-based foods and processed foods suitable for vegetarian- and vegan-based eating patterns. *Journal of Human Nutrition and Dietetics*, 31:422–435.
Verni, M., Rizzello, C.G., Coda, R. 2019. Fermentation biotechnology applied to cereal industry by-products: Nutritional and functional insights. *Frontiers in Nutrition*, 6:42.
Verni, M., Pontonio, E., Krona, A., Jacob, S., Pinto, D., Rinaldi, F., et al. 2020. Bioprocessing of brewers' spent grain enhances its antioxidant activity: Characterization of phenolic compounds and bioactive peptides. *Frontiers in Microbiology*, 11:1831.
Vieira, E.F., Carvalho, J., Pinto, E., Cunha, S., Almeida, A.A., Ferreira, I.M.P.L.V.O. 2016. Nutritive value, antioxidant activity and phenolic compounds profile of brewer's spent yeast extract. *Journal of Food Composition and Analysis*, 52:44–51.
Virk, P., Xianglin, L., Blümmel, M. 2019. A note on variation in grain and straw fodder quality traits in 437 cultivars of rice from the varietal groups of aromatic, hybrids, Indica, new planting types and released varieties in the Philippines. *Field Crops Research*, 233:96–100.
Vu, N., Tran, H.T., Bui, N., Vu, C.D., Nguyen, H.V. 2017. Lignin and cellulose extraction from Vietnam's rice straw using ultrasound-assisted alkaline treatment method. *International Journal of Polymer Science*, 1–8. ID 1063695. https://doi.org/10.1155/2017/1063695.
Wang, H., Geng, H., Chen, J., Wang, X., Li, D., Wang, T., et al. 2020. Three phase partitioning for simultaneous extraction of oil, protein and polysaccharide from rice bran. *Innovative Food Science & Emerging Technologies*, 65:102447.
Wang, J., Sun, B., Liu, Y., Zhang, H. 2014. Optimisation of ultrasound-assisted enzymatic extraction of arabinoxylan from wheat bran. *Food Chemistry* 150:482–488.
Wang, X., Chen, H., Fu, X., Li, S., Wei, J. 2017. A novel antioxidant and ACE inhibitory peptide from rice bran protein: Biochemical characterization and molecular docking study. *LWT-Food Science and Technology*, 75:93–99.
Waters, D.M., Jacob, F., Titze, J., Arendt, E.K., Zannini, E. 2012. Fibre, protein and mineral fortification of wheat bread through milled and fermented brewer's spent grain enrichment. *European Food Research and Technology*, 235:767–778.
Weiser, C., Zeller, V., Reinicke, F., Wagner, B., Majer, S., Vetter, A., Thraen, D. 2014. Integrated assessment of sustainable cereal straw potential and different straw-based energy applications in Germany. *Applied Energy*, 114:749–762.
Wizi, J., Wang, L., Hou, X., Tao, Y., Ma, B., Yang, Y. 2018. Ultrasound-microwave assisted extraction of natural colorants from sorghum husk with different solvents. *Industrial Crops and Products*, 120:203–213.
Wronkowska, M. 2016. Wet-milling of cereals. *Journal of Food Processing and Preservation*, 40(3):572–580.

Xie, C., Coda, R., Chamlagain, B., Edelmann, M., Deptula, P., Varmanen, P., Piironen, V., Katina, K. 2018. In situ fortification of vitamin B12 in wheat flour and wheat bran by fermentation with Propionibacterium freudenreichii. *Journal of Cereal Science*, 81:133–139.

Yang, Z., Zhai, W. 2010. Optimization of microwave-assisted extraction of anthocyanins from purple corn (Zea mays L.) cob and identification with HPLC–MS. *Innovative Food Science & Emerging Technologies*, 11(3):470–476.

Zalán, Z., Hegyi, F., Szabó, E.E., Maczó, A., Baka, E., Du, M., Liao, Y., Jianquan, K. 2015. Bran fermentation with lactobacillus strains to develop a functional ingredient for sourdough production. *International Journal of Nutrition and Food Sciences*, 4:409–419.

Zhu, Y., Chu, J., Lu, Z., Lv, F., Bie, X., Zhang, C., Zhao, H. 2018. Physicochemical and functional properties of dietary fiber from foxtail millet (Setaria italic) bran. *Journal of Cereal Science*, 79:456–461.

Section II

Enzymes from Sourdough Cultures

5 Introduction to Sourdough Enzymology

Bogdan Păcularu-Burada, Marina Pihurov, Mihaela Cotârleţ, Elena Enachi, and Gabriela-Elena Bahrim

CONTENTS

5.1 Introduction 117
5.2 Microorganisms with Implications in Sourdough Biotechnology 120
    5.2.1 Diversity of Sourdough Lactic Acid Bacteria Strains 121
    5.2.2 Diversity of Sourdough Yeasts 124
    5.2.3 Unconventional Probiotic Starter Cultures Used in Sourdough Fermentation 125
5.3 Microbial Enzymes Involved in Sourdough Bioconversions 127
    5.3.1 Carbon-Based Compounds 127
        5.3.1.1 Bioconversion of Carbohydrates 127
        5.3.1.2 Production of Exopolysaccharides 129
    5.3.2 Biotransformation of Lipids 130
    5.3.3 Nitrogen-Based Compounds 131
    5.3.4 Bioconversion of Phenolic Compounds 132
    5.3.5 Minimization of Anti-nutritional Factors 133
5.4 Concluding Remarks and Prospects 134
Acknowledgments 134
References 135

5.1 INTRODUCTION

Cereal-based products are important for the daily intake of nutrients, providing carbohydrates, proteins, dietary fibers, vitamins, and other health-promoting compounds. The nutritional, sensorial, and functional features of cereals and derived food products can be enhanced by fermentation (Galimberti et al. 2021; Weckx, Van Kerrebroeck, and De Vuyst 2019). This biotechnological process was successfully used for whole grain flours, brans, gluten-free raw materials, and sprouts to increase the beneficial effects on the structural characteristics and shelf life extension of the resulting baked goods (Gobbetti et al. 2014; Zotta et al. 2006).

The annual consumption of bread in Europe was estimated between 46 kg and 100 kg per capita, this amount being different among countries (Palla et al. 2020).

Massive amounts of baked products are wasted, due to consumers' practices and storage conditions of the products. These days enzymes play a crucial role in the quality and shelf life of various food products, including baked goods (Teigiserova, Bourgine, and Thomsen 2021; Alpers et al. 2021). Furthermore, value-added products or by-products from other food industry sectors are currently being explored to reduce waste generation following the circular economy concept. Following this sustainability trend, different enzymes have been produced from wheat, rice, and corn residues. Another important factor for reducing the level of additives is the exploitation of beneficial micoorganisms (LAB and yeasts) for their enzymatic properties according to their metabolic characteristics for multiple technological and functional traits (Dahiya et al. 2020; Novotni, Gänzle, and Rocha 2021).

Nowadays, the modern approaches offered by the fermentation processes (complex substrates, metabolic diversity in starter cultures, optimization of fermentation parameters) open novel perspectives for the production and utilization of the metabiotics (prebiotics, probiotics, postbiotics, and paraprobiotics) as advanced tools for the technological and health-related properties of sourdoughs (Cuevas-González, Liceaga, and Aguilar-Toalá 2020; Aguilar-Toalá et al. 2018; Torres et al. 2020; Vladimirovich 2016).

The beneficial effects of probiotic microorganisms were intensively studied mainly by *in vivo* tests, but it is difficult to maintain the cells' viability, according to the accepted definition and characteristics of probiotics. Thus, the new approach of biotics production by fermentation boosts the concepts of postbiotics and paraprobiotics with valuable *in vitro* and health effects in the field of functional foods. Paraprobiotics, also known as "inactivated/dead/non-viable microbial cells of probiotics" (intact or ruptured and containing probiotic cell components upon lysis) or crude cell extracts (i.e. with complex mixtures of metabolites) when administered in sufficient amounts, can confer several benefits to consumers. Postbiotics are defined as soluble high- and low-molecular weight, biologically active metabolites released by food-grade microorganisms during their growth and fermentation in complex culture mediums, foods, or the gut. They have impact on the nutritional value, quality, and safety of food products and ingredients as well as *in vivo* beneficial effects (i.e. secreted proteins, peptides, enzymes, amino acids, organic acids, short-chain fatty acids, and vitamins). Some cellular components released in the environmental substrate after the cell lysis are also considered postbiotics. Specific metabolites, such as reuterin, biosynthetized by *Lactobacillus reuteri* are considered probioceuticals/probiotaceuticals (Lin et al. 2019; Liang et al. 2021). These concepts could also be applied for functional sourdough production as a healthy bioingredient for food or feed formulation.

Usually, the mixture of flours, water, salt, and sometimes other ingredients that are spontaneously fermented is defined as a sourdough (De Vuyst, Van Kerrebroeck, and Leroy 2017; Weckx, Van Kerrebroeck, and De Vuyst 2019), a natural bioingredient used as a bread improver or leavening agent (Zamaratskaia, Gerhardt, and Wendin 2021). The consortia of yeast and LAB strains are involved in the flavor biotransformation and allergenicity minimization due to proteolysis (Suo, Chen, and Wang 2021; Fujimoto et al. 2019; Dan Xu et al. 2019). The postbiotics (i.e. organic

Introduction to Sourdough Enzymology

acids, ethanol, CO_2) from sourdough contribute to the enhanced volume, texture, and flavor of bread, as well as exerting antimicrobial properties against bread spoilage microorganisms (Filipčev and Olivera Bodroža-Solarov 2007; De Vuyst and Neysens 2005). The functional characteristics of the fermented products are also improved by the generation of bioactive peptides, short-chain fatty acids (SCFAs), vitamins, and phenolic derivatives (Torres et al. 2020). Therefore, the technological and biotic properties of sourdough are determined by multiple parameters, influenced by the composition and characteristics of the fermentation substrates, properties of starter cultures, association in co-cultures, and their concentration as well as the fermentative conditions.

Sourdough's fermentation and its overall quality is correlated to the individual and complex contribution of the factors involved in the bioprocess. Microorganisms can biosynthetize a large type and quantity of enzymes located extra- and intracellularly. Thus, microbiota (autochthonous microorganisms and exogen starter cultures) is able to produce enzymes with implications in many bioprocesses, such as fermentation and bioconversion, which transform the substrate in valuable bioactives. The biocatalysts' functionality (cells and enzymes) depends on some intercorrelated factors (intrinsic, extrinsic, and biological) which alter the properties of the fermented product mainly due to the bioactives produced (Figure 5.1).

Consequently, various raw materials are used for sourdough production in Europe, such as wheat (*Triticum* spp.), spelt (*Triticum aestivum* subsp. *spelta* L.), rye (*Secale cereale* L.), corn (*Zea mays* L.), oat (*Avena sativa* L.), and barley (*Hordeum vulgare* L.) (De Vuyst, Van Kerrebroeck, and Leroy 2017). The latest trends in sourdough technology are focused on exploiting underutilized pseudocereals like amaranth (*Amaranthus* spp.), buckwheat (*Fagopyrum esculentum*), quinoa (*Chenopodium quinoa* Willd.), and chia (*Salvia hispanica* L.) with promising results (Papadimitriou et al. 2019). As starters, lactic acid bacteria (LAB) and yeasts are usually involved in co-cultures with symbiotic functionality (Furukawa et al. 2013). Wild strains or

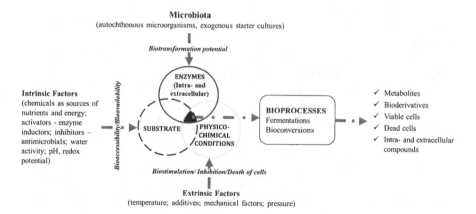

FIGURE 5.1 The correlation between biocatalysts (microorganisms and enzymes) functionality and their potential on substrate biotransformation.

selected starter cultures are currently used in sourdough bread making to improve the pre- and postbiotic characteristics of the fermented products. Furthermore, LAB concentration ranging between 10^7 and 10^9 CFU/g and yeast concentration between 10^5 and 10^7 CFU/g will determine a reduction of the anti-nutritional factors by phytases and a unique flavor due to alcohols, esters, and peptides (Luti et al. 2020; Gänzle 2014; Dan Xu et al. 2020). The optimization of the biotechnological conditions for sourdough fermentation should be focused on: (1) the substrate compositional formulation and its water activity, (2) the inoculum type and concentration, and (3) the adapted physicochemical conditions for fermentation to ensure an optimal metabolic biotransformation of the substrate into biotics with multiple purposes (Păcularu-Burada et al. 2020).

The preservation and stabilization techniques for sourdough have been intensively studied to offer the best solutions for the baking industry. Sourdoughs were subjected to different drying protocols to conclude that, in some cases, freeze-drying was more efficient than spray-drying regarding the cells' viability. Hence, the quality, properties, and survivability rate of LAB from the resulting powders are strongly influenced by the cryoprotectants and the microencapsulation protocols that are used (Caglar, Ermis, and Durak 2021; Farahmandi et al. 2021).

5.2 MICROORGANISMS WITH IMPLICATIONS IN SOURDOUGH BIOTECHNOLOGY

Bacteria and yeasts from sourdough microbiota should be able to convert the major chemical complex compounds from the substrates (carbohydrates, proteins, lipids, and phenols) into simpler compounds that can be used as nutrients in the fermentation process and as postbiotics with technological and functional relevance. Sourdough represents a stressful acidic environment, dominated by LAB strains (~10^9 CFU/g) which work in symbiosis with yeasts. Under these specific conditions, only some specialized and well-adapted strains will survive and produce valuable postbiotics. The starter cultures used in the sourdough can be wild strains from artisanal cultures or selected ones as co-cultures of bacteria and yeasts (Fang et al. 2021; Comasio et al. 2020).

Usually, the microbial strains involved in the spontaneous fermentation of sourdough must be able to develop synergistic relationships as LAB and yeast consortia are frequently found in sourdoughs (De Vuyst and Neysens 2005; De Vuyst, Van Kerrebroeck, and Leroy 2017). It is well known that yeast strains support the LAB strains during fermentation offering an enhanced aroma profile after glycerol metabolism and ethanol production. Microbiologically stable sourdoughs with an extended shelf life were obtained by the mixed fermentation of LAB and yeast strain consortia in symbiosis (Vogelmann and Hertel 2011). The microbial diversity of sourdoughs represents an opportunity for innovation in the field of metagenomics, metaproteomics, and metabolomics (Suo, Chen, and Wang 2021). Moreover, several scientific studies have highlighted that sourdough microbiota is influenced by the origin of the fermented products, especially in China (Fu et al. 2020).

The artisanal probiotic consortia of bacteria and yeasts from kefir grains (milk and water) and SCOBY (symbiotic culture used to obtain the probiotic beverage named kombucha) were recently used for sourdough production with very good results, obtaining metabiotic products with *in vitro* and *in vivo* health benefits (Tu et al. 2019; Soares, de Lima, and Reolon Schmidt 2021). The advantages of these cultures are their high adaptability, protection, and metabolic diversity, which results in fermented products with multiple beneficial properties.

The selection and utilization of some performant starter cultures offer the possibility to identify strains with particular metabolic properties – based on the extracellular enzyme biosynthesis with impact on the substrate's bioconversion – that enhance the technological, nutritional, and sensorial characteristics of the fermented products, their microbiological stability, and safety (Chen et al. 2021; Velasco et al. 2021; Păcularu-Burada et al. 2020).

5.2.1 Diversity of Sourdough Lactic Acid Bacteria Strains

The bacterial populations from the sourdough microbiota have a complex metabolic activity that ensures its superior techno-functional properties. More than 10 species of LAB were found in this type of micro-ecosystem, besides another 40 species of bacterial strains (Arora et al. 2021). Sourdoughs made from cereals and pseudocereals are mainly characterized by the presence of *Lactobacillus* spp. species (*L. plantarum, L. sanfranciscensis, L. brevis,* and *L. paralimentarius*). The development of other LAB strains in the sourdough occurs after successive backsloppings (De Vuyst and Neysens 2005). At the beginning of the fermentation process, only *Enterococcus durans* was identified, while after several backslopping steps, *Pediococcus pentosaceus, E. durans, L. garviae, Weissella cibaria,* and *W. paramesenteroides* strains increased their relevance in the fermented product (Sáez et al. 2018). The metabolic properties of certain strains, such as the extracellular enzymes biosynthesized by the LAB strains, can be a decisive selection criterion for future utilization of these microorganisms. Thereby, the members of *Lactobacillus* spp. genus were able to produce only lactic acid from carbohydrates (homofermentative strains) that were isolated from sourdough. Furthermore, heterofermentative strains, such as *Weissella* spp. (*W. cibaria* and *W. confusa*), *Pediococcus* spp. (*Pc. acidilactici* and *Pc. pentosaceus*), and *Leuconostoc* spp. (*Leu. mesenteroides* and *Leu. citreum*), contribute to an enriched aroma profile of sourdough (De Vuyst, Van Kerrebroeck, and Leroy 2017; Ripari, Bai, and Gänzle 2019). In the last decades, sourdough's diverse microbiota has been exploited for the isolation and identification of novel strains of homofermentative LAB strains, namely *L. crustorum, L. mindensis, L. nantensis, L. nodensis,* and *L. songhuajiangensis,* or facultatively heterofermentative strains of *L. frumenti, L. namurensis, L. secaliphilus, L. hammesii, L. siliginis,* and *L. acidifarinae* (De Vuyst, Van Kerrebroeck, and Leroy 2017).

Some *Streptococcus* spp., *Lactococcus* spp., and *Enterococcus* spp. homo- or heterofermentative strains were isolated from traditionally fermented products and thereafter used as starter cultures for sourdough fermentation (García Vilanova et al. 2015; Papadimitriou et al. 2019).

During the formation of cereal kernels, different species of LAB strains belonging to *Lactobacillus* spp., *Lactococcus* spp., *Pediococcus* spp., and *Weissella* spp. were located on different layers of the seeds, the most performant strains surviving the storage and milling processes and becoming an important part of the specific microbiota of the sourdoughs (Minervini et al. 2018). *L. vaccinostercus*, *L. sakei*, *L. pentosus*, and *L. alimentarius* strains were identified in some sourdoughs made in Japanese bakeries (Fujimoto et al. 2019), whereas in rye-based sourdoughs from Sweden, Belgium, and Greece the strains of *L. sanfranciscensis* were the most abundant ones (Palla et al. 2020). On the same substrate, *L. brevis* was the dominant strain from some Iranian, Italian, or Russian sourdoughs. Conclusively, the geographical area of the sourdough and the used flours represent decisive factors on the diversity or dominance of the beneficial LAB strains (Xing et al. 2020; Zhao, Mu, and Sun 2019). In-depth studies were conducted in order to support this behavior of the LAB strains. In fact, gluten-free flours spontaneously fermented were used to isolate novel strains of *Leu. holzapfelii*, *L. vaginalis*, *L. gallinarum*, *L. pontis*, and *L. graminis* using buckwheat, sorghum, rice, or teff flours (Moroni, Arendt, and Bello 2011). Protein- or starch-based matrices made of maize, cowpea, and peanut, containing a cassava hydrolysate supplemented with provitamin A, represent innovative solutions toward ready-to-use starter cultures consisting of competitive LAB strains with applications in sourdough technology (Kyereh and Sathivel 2021; Oguntoye, Ezekiel, and Oridupa 2021).

Besides the isolation of LAB strains, their genetic identification is essential for a better discrimination and classification of the isolates. *Lactobacillus* spp., *Lactococcus* spp., *Leuconostoc* spp., *Pediococcus* spp., and *Streptococcus* spp. are the most common genera of LAB (Peng et al. 2020) belonging to *Lactobacillae* family, and as such, *Bifidobacterium* spp. strains from *Actinobacteria* family are included among the LAB strains as well (Petrova and Petrov 2020). Accordingly, for genetic classification, DNA-based methods were applied in the literature. *L. brevis*, *L. parabuchneri*, *Pediococcus* spp., and *W. viridescens* strains that originated from Turkish sourdoughs were analyzed by Fourier-transform infrared spectroscopy (FTIR) and included in the above-mentioned classes of LAB (Karaman, Sagdic, and Durak 2018; Çakır, Arıcı, and Durak 2020). The analysis of the extracted DNA sequences using bioinformatics and imagistics (i.e. Illumina sequencing databases, multi fragment melting technique) were successfully applied for the identification of the *Lactobacillus pentosus* strain, isolated from fermented vegetables (Yi et al. 2020), or for the identification of *L. kefiri* and *L. kefiranofaciens* originated from milk kefir grains (H. Wang, Wang, and Guo 2020; Kesmen et al. 2020). Due to the fact that *Lactobacillus* spp. taxonomic evolution started from the former *Lactobacillus* species, the new genera were reorganized, some of the most common novel taxa being nominated as *Companilactobacillus* spp., *Fructilactobacillus* spp., *Paucilactobacillus* spp., *Lacticaseibacillus* spp., *Levilactobacillus* spp., *Lactiplatibacillus* spp., *Ligilactobacillus* spp., *Limosilactobacillus* spp.,

Introduction to Sourdough Enzymology

Loigolactobacillus spp., *Schleiferilactobacillus* spp., and *Lentilactobacillus* spp. (Oshiro, Zendo, and Nakayama 2021; Esen and Çetin 2021; Zheng et al. 2021). The LAB's former and new genera and strains' abundance in sourdough microbiota is presented in Figure 5.2, considering their metabolic characteristics during the fermentation process.

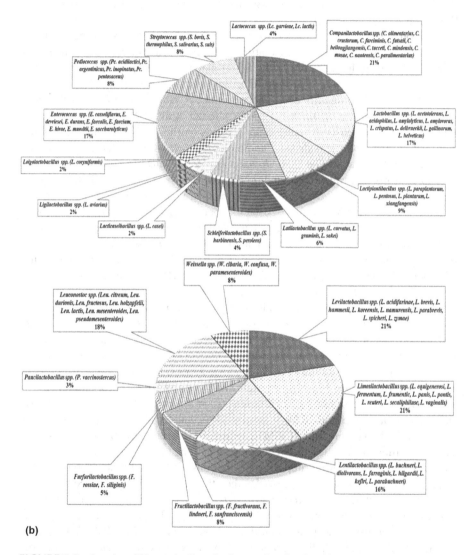

FIGURE 5.2 Lactic acid bacteria diversity in sourdough microbiota according to their metabolic properties: (a) homofermentative strains and (b) heterofermentative strains. Adapted from Oshiro, Zendo, and Nakayama (2021).

5.2.2 Diversity of Sourdough Yeasts

Spontaneously fermented sourdoughs were suitable sources for the isolation of performant yeast strains, the first isolated strain being the *Kazachstania exigua* (later reclassified as *Saccharomyces exiguus*) in the 1970s. Further in-depth studies highlighted that this yeast strain can develop mutual relationships with LAB strains in sourdoughs. *Saccharomyces exiguus* cannot metabolize maltose, being aided in this process by *L. sanfranciscensis* strains (De Vuyst et al. 2016; De Vuyst, Van Kerrebroeck, and Leroy 2017).

Sourdoughs that were subjected to successive backsloppings developed a stable microbiota and were used for the isolation of different yeast strains, the same strain being identified in the literature by its initial name or by its reclassified taxonomy (i.e. *K. humilis* or *Candida humilis*; *Torulaspora delbrueckii* or *C. colliculosa*; *Wickerhamomyces anomalus* or *Pichia anomala*; *P. kudriavzevii* or *Issatchenkia orientalis*; *C. glabrata* or *Yarrowia* spp.) (Alfonzo et al. 2021; Arora et al. 2021). The dominance of yeasts in sourdoughs depends on the yeast's adaptation systems that relate to the intrinsic, extrinsic, and biological factors of the fermentation ecosystems. While some strains of *C. humilis* and *S. exiguus* are sourdough-specific microorganisms that can survive and produce diverse beneficial metabolites within sourdough, *S. cerevisiae* and *W. anomalus* strains are found in various environments or food products (De Vuyst et al. 2016). In spontaneous sourdoughs, *S. cerevisiae* was found in association with other yeast strains and their presence is influenced by the geographical area. *S. cerevisiae* along with *C. humilis* were found in Italian sourdoughs, and *W. anomalus* was found in consortium with *S. cerevisiae* in sourdoughs originating from Belgium, China, and Japan (Huys, Daniel, and De Vuyst 2013; Palla et al. 2019; Fujimoto et al. 2019; Zhao, Mu, and Sun 2019; T. Liu et al. 2018). The associations between maltose-negative *K. barnettii* or maltose-positive *S. cerevisiae* strains and glucose/fructose consuming *L. plantarum* or *L. sanfranciscensis* were identified in sourdoughs. Furthermore, LAB consortia comprising *L. sanfranciscensis*, *L. plantarum*, *L. paralimentarius*, *L. curvatus*, and *L. germinis* strains were part of the stabile sourdough's microbiota (De Vuyst, Van Kerrebroeck, and Leroy 2017; Weckx, Van Kerrebroeck, and De Vuyst 2019). The metabolic interactions between a co-culture of LAB strains (*L. plantarum* and *L. sanfranciscensis*) and *S. cerevisiae* were studied by Sieuwerts, Bron, and Smid (2018) by cultivation in a chemically defined medium (CDM) containing different carbon sources, as well as in a sourdough-like medium. The above-mentioned research concluded that *L. plantarum* and *S. cerevisiae* support each other only when the fermentation medium contains glucose, fructose, or lactose. On the other hand, the CO_2 produced by yeast stimulates the activity of *L. sanfranciscensis*; this process is also enhanced by an unknown growth factor biosynthesized by *S. cerevisiae*. These yeast–LAB consortia contribute to the diversification of sourdough-baked goods without baker's yeast and characterized by a pleasant flavor. Sourdough bread with a softer crumb, better color, and lower gluten content was obtained with a yeast-containing sourdough. It seemed that the alkaline medium resulting after the yeast's fermentation was a factor that maximized the proteolysis (Xi et al. 2020). The Chinese traditional fermented beverage made of wheat and pea flours was the source for the isolation of competitive non-*Saccharomyces* spp. yeasts (*P. kudriavzevii* and *Meyerozyma guilliermondii*), which increased various sensorial and technological properties of a wheat-based sourdough bread in association with *L. sanfranciscensis*. Various volatile compounds and organic acids were determined in this product. The action of yeast on the properties of the final baking product is influenced mainly by its concentration and fermentation parameters. It was demonstrated that an increased yeast concentration would have an influence on the volatile compounds found in the sourdough breads, whereas an extended fermentation time would be the cause of unwanted off-flavor volatiles derived from the oxidative processes of lipids (Xi et al. 2021; Dan Xu et al. 2020).

Compared to the protocols applied for the LAB identification based on 16 sequences, yeast genotyping requires ribosomal DNA/ RNA and the analysis of 26 sequences that must be further subjected to polymerase chain reaction (PCR), matrix assisted laser desorption/ionization – time of flight–mass spectrometry (MALDI-TOF-MS) or FTIR methods. These assays were successfully applied for the genetic identification of *Saccharomyces* spp., *Pichia* spp., *Torulaspora* spp., and *Kazachstania* spp. (Karaman, Sagdic, and Durak 2018; Jin et al. 2021; Fraberger et al. 2020; Johansson et al. 2021).

Some exogenous yeast starter cultures with valuable enzymatic potential, like *Y. lipolytica*, *Candida* spp., *Kluyveromyces* spp., and *Endomycopsis fibuligera* are considered important candidates for sourdough fermentation as producers of lipases, proteases, and amylases with impact on the substrate's bioconversion and metabolite production with functional properties (antioxidant bioactive peptides, antimicrobial fatty acids) (Palla et al. 2020; Liu, Ji, and Huang 2015).

5.2.3 Unconventional Probiotic Starter Cultures Used in Sourdough Fermentation

Water and milk kefir grains are characterized by a wild consortia of microorganisms (LAB, acetic acid bacteria, and yeasts) living together in stable matrices made of kefiran or other polysaccharides with metabiotic properties (Prado et al. 2015; Garrote, Abraham, and De Antoni 2010). These artisanal cultures are frequently employed in fermented beverage production (Plessas et al. 2011). Therefore, kefir grains, or the resulting fermented products, were used for the isolation of some homo- and heterofermentative LAB strains belonging to *Lactobacillus*, *Lactococcus*, *Leuconostoc*, and *Streptococcus* genera. The yeast strains belonging to *Candida* spp., *Kluyveromyces* spp., *Saccharomyces* spp., and *Torulaspora* spp. originated as well, from kefir grains and kefir-based products (Filipčev and Olivera Bodroža-Solarov 2007; Hermann et al. 2016; Seo, Jeong, and Kim 2020). Generally, the isolation and cultivation under controlled conditions of these microorganisms are difficult tasks. These microorganisms work very well in symbiosis or consortium developed during successive fermentations. Moreover, these consortia are characterized by microorganisms that develop positive and supportive relationships that reduce other microbiological contaminants (Păcularu-Burada, Georgescu, and Bahrim 2020). *L. hordei* was isolated from water kefir, and in-depth studies highlighted that the processes leading to dextran, fructose, mannitol, peptides, citrate, vitamins, and unsaturated fatty acids were supported by specific enzymes. Enzymatic reactions and their resulting metabolites seemed to be boosted by other microorganisms from the water kefir grain consortium (Xu et al. 2019). Furthermore, a newly isolated strain of *Bifidobacterium aquikefiri* originated from water kefir grains, along with other new yeast strains identified as *Dekkera bruxellensis* and *Oenococcus aquikefiri*, which can convert fructose into mannitol (Verce, De Vuyst, and Weckx 2019); whereas the probiotic potential of *Kl. lactis* and *S. unisporus* strains isolated from milk kefir grains was reported by Gut et al. (2019).

The level of bioactive compounds, such as phytoestrogens (isoflavones), was enhanced after *in vitro* and *in vivo* digestion studies that involved a fermented kefir-based product enriched with soybean sprouts (de Melo et al. 2020). Microorganisms from a water kefir grains consortium were able to synthesize proteases, thus increasing the level of soybeans bioactive peptides along with a minimization of saponins (Azi et al. 2020). Bacteriocins, exopolysaccharides (EPS), and other postbiotic compounds produced during the water or milk kefir grains' fermentation depend on the diversity of the microorganisms from the consortium and other biotechnological factors (i.e. type of milk and its composition, fermentation time and temperature, and geographical area) (Bengoa et al. 2019; H. Wang, Wang, and Guo 2020).

The utilization of milk kefir grains was extended toward sourdough fermentation due to the incrementation of some functional properties (i.e. enhanced sensorial characteristics, retarded staling and spoilage, extended shelf life, antibacterial, antiinflammatory, and antioxidant activities) (Mantzourani et al. 2014; Yilmaz-Ersan et al. 2018; Oliveira and Castro 2020; Sindi et al. 2020). The heterofermentative strains of *Lentilactobacillus hilgardii* and *Liquorilactobacillus nagelii* isolated from water kefir were used for the fermentation of a wheat-based liquid sourdough to conclude that the LAB strains' activity and their metabolites were strongly influenced by the availability of the carbon sources from the flour-based substrate, mainly glucose and fructose being selectively used as nutrients by these above-mentioned strains (Comasio et al. 2020). The water kefir grains that were subjected to sterilization and freeze-drying were used as a hydrocolloid replacer in bread making, proving the diverse applications of kefir grains as quality enhancers of food products (Hermann et al. 2016).

Furthermore, the beverage made of green or dark sugared tea, kombucha, is fermented by a symbiotic culture of bacteria and yeasts and has various health-promoting effects, such as antilipidemic, antidiabetic, antioxidant, antimicrobial, and anticarcinogenic. Kombucha utilization in sourdough bread making determined enhanced properties in terms of texture, volume, flavor and shelf life (Mohd Roby et al. 2020; Coelho et al. 2020; Jafari et al. 2020).

5.3 MICROBIAL ENZYMES INVOLVED IN SOURDOUGH BIOCONVERSIONS

5.3.1 CARBON-BASED COMPOUNDS

5.3.1.1 Bioconversion of Carbohydrates

LAB strains' enzymatic activities improve the overall characteristics of the fermented products due to the enhanced bioavailability of micronutrients and peptides or to the release of prebiotic fibers from the hydrolysis of starch. Moreover, similar outcomes are also achieved through the activity of endogenous enzymes from the fermentation substrates, especially when spontaneously fermented sourdoughs are involved (Cizeikiene et al. 2020).

Starch is the main carbohydrate used by plants in metabolic processes. Consequently, starch is stored in seeds in cereals and pseudocereals, and it represents the principal source of carbon and energy for the microorganisms during sourdough fermentation (Dhaka, Muthamilarasan, and Prasad 2021). The metabolic processes that occur in the microbial or plant cells are biocatalized by enzymes with specific properties. Hence, for starch metabolism, several classes of enzymes were identified. These enzymes, including starch synthases (SS) and phosphorylases (PHO), are responsible for the elongation of linear glucan chains – amylose units – respectively for the phosphorylation by ADP-glucose phosphorylase (AGPase); starch branching enzymes (SBE) that ensure branch formation – amylopectin units – through the insertion of α-1,6 glycosidic bonds; debranching enzymes (DBE) comprising two major classes, namely isoamylases (ISA) – debranching enzymes of amylopectin – and pullulanases (PUL), that use amylopectin as a substrate and generate dextrins and pullulans; disproportionating enzymes (DPE), using maltotriose as a substrate that is further hydrolysed into glucose and other maltose-oligosaccharides (Dhaka, Muthamilarasan, and Prasad 2021; Xu et al. 2021; O'Neill et al. 2015).

Therefore, some authors reported that LAB strains originating from Italian or Chinese sourdoughs were able to metabolize starch. Newly identified *Levilactobacillus brevis*, *Lactoplantibacillus paraplantarum*, *L. plantarum*, *Leuconostoc* spp., *F. rossiae*, and *F. sanfranciscensis* strains were tested for their enzymatic activities alongside some specific bakery-stress conditions (i.e. probiotic potential and resistance to acid, ethanol, sucrose, and NaCl) to be further used as selected sourdough starters with industrial applications (Padmavathi et al. 2018; Reale et al. 2020; Y. Xu et al. 2020). Hence, the yeast strains isolated from a sourdough and identified as *Kl. lactis* and *Kl. aestuarii* were studied for their probiotic and starch hydrolysis capacity. It was concluded that these strains needed at least 60 min to reach a starch hydrolysis rate of 80% (Fekri et al. 2019).

Cereal flours commonly used for sourdough making (i.e. rye and wheat) are not rich in fermentable sugars, thus their utilization by microorganisms during the fermentation process is possible when enzymes transform starch and generate fractions that are further converted into basic carbohydrate-based nutrients. During spontaneous sourdough fermentation, the activity of endogenous enzymes from flours can be enhanced by LAB-synthesized enzymes that lead to maltose, glucose, and other maltodextrins (Van Kerrebroeck, Maes, and De Vuyst 2017). Maltose is produced at the beginning of the sourdough fermentation when α- and/or β-amylases cleave the starch. Amylases have the highest yield at pH values of 4.5–5.0 (Weckx, Van Kerrebroeck, and De Vuyst 2019). Taking these facts into consideration, the sourdough starters' selection procedures must be focused on the metabolic pathways of carbohydrates and on the specificity of the analyzed strains and fermentation substrates (Păcularu-Burada et al. 2020). Some *F. sanfranciscensis* strains proved a maltase-producing ability in the sourdough that increased the amount of glucose released after hydrolysis; another *F. sanfranciscensis* strain produced an homoexopolysaccharide made of glucose monomers using maltose as the nutrient (Zhang et al. 2019). *L. plantarum* was able to cleave maltose or sucrose, sugars that were involved afterwards in the metabolic processes of this LAB strain, the excess being metabolized by a *S. cerevisiae* strain (Sieuwerts, Bron, and Smid 2018). Mature sourdoughs that are characterized by a stable microbiota comprising beneficial and well-adapted microorganisms were capable of glucose, fructose, and sucrose metabolization – thus resulting in a fermented product and a sourdough bread with reduced glycemic index, that facilitates consumption by people suffering from metabolic disorders. Conclusively, an extended sourdough fermentation could have positive impacts on the starch content and its hydrolysis rate, on the one side, and on the bioavailability's maximization of the fermentable oligosaccharides, disaccharides, monosaccharides and polyols (FODMAPS), frequently used as prebiotics, on the other side (Canesin and Cazarin 2021).

As mentioned earlier, LAB strains use the available carbohydrates to produce various compounds with technological and functional applications. Hence, the metabolic features of LAB strains are clearly classified and differentiated, even though the same enzymatic biocatalysts (i.e. hydrolases, isomerases, kinases, dehydrogenases, and phosphorylases) contribute to postbiotic metabolite synthesis. Specifically, the LAB strains with homofermentative metabolism use the Embden-Meyerhof-Parnas (EMP) pathway to convert hexoses (glucose) into pyruvate as the major metabolite. In addition, pentoses can be involved in the homofermentative metabolism of LAB. For this process, hexoses (when pentoses are not available) need to be firstly converted into pentoses (phosphorylated forms of ribulose or xylulose) that are further transformed into pyruvate through the pentose phosphate pathway (PPP). The resulting pyruvate from the above-mentioned pathway is converted into lactate as the unique end product. On the other hand, some LAB strains possess a heterofermentative metabolism that can be briefly characterized by: (i) lactate production from hexoses, following the previously described EMP pathway and (ii) ethanol and acetate biosynthesis from pentoses when phosphoketolase is the main catalyst (Salmerón et al. 2020; Păcularu-Burada, Georgescu, and Bahrim 2020). Likewise, the heterofermentative strains were employed in the fermentative processes for their ability to convert pyruvate into propionic acid or flavor compounds (i.e. diacetyl and acetoin), such compounds resulting after cultivation under stressful conditions represented by aerobiosis or glycerol (Sauer et al. 2017).

The heterofermentative strains of *L. sanfranciscensis*, *L. hilgardii*, and *L. pontis* were competitive strains that produce glucose, organic acids, and volatile compounds after the maltose breakdown due to the maltose phosphorylase activity. Fructose was enzymatically produced from mannitol by *L. fermentum* strain with an encoded gene responsible for the mannitol dehydrogenase's synthesis in a simulated sourdough-like medium. Specific genes were identified in *L. rossiae* and *Weissella* spp. genomes, their properties being related to the dehydrogenases and oxydases' action on lipids with pyruvate as an important metabolite (Weckx, Van Kerrebroeck, and De Vuyst 2019; Michael G. Gänzle, Vermeulen, and Vogel 2007).

Enhanced flavor profiles, described as caramel or cheese like, and buttery or fruity aroma are manufactured when sourdoughs fermented by newly classified heterofermentative strains belonging to *Paucilactobacillus* spp., *Limosilactobacillus* spp., *Furfurilactobacillus* spp., *Fructilactobacillus* spp., *Leuconostoc* spp., and *Levilactobacillus* spp. are employed (Oshiro, Zendo, and Nakayama 2021; Zamaratskaia, Gerhardt, and Wendin 2021).

5.3.1.2 Production of Exopolysaccharides

EPS are synthesized by a wide range of microorganisms. Four different pathways for EPS production were determined, namely Wxz/Wzy dependent pathway, ATP-binding cassette (ABC) transporter dependent pathway, synthase dependent pathway, and sucrase mediated pathway (Rana and Upadhyay 2020). EPS are released extracellularly after synthesis, and thus the production yield of EPS is strongly influenced by various factors, the most important being the ratio between carbon and nitrogen (C: N); others are the minerals available in the fermentation medium, the strain used, and finally, the fermentation time and temperature (Chavan and Chavan 2011). The enzyme responsible for the depletion of sugars in the first stages of hydrolysis will be decisive on the structure of the produced EPS. For example, when glucansucrases and glucosyltransferases act on sucrose, homopolysaccharides containing glucose monomers will be produced. Likewise, homopolysaccharides with fructans would be produced when fructansucrases and fructosyltransferases catalyse such reactions. Important representatives of glucans are alternans, dextrans, mutans, reuterans, β-glucans, inulin, and levans that are included among fructans. In contrast, heteropolysaccharides are composed of at least 2 different monomeric units (Oshiro, Zendo, and Nakayama 2021; Daba, Elnahas, and Elkhateeb 2021).

Fructan- and glucan-based EPS were produced by *L. sanfranciscensis*, *L. reuteri*, or *L. acidophilus* strains in wheat or sorghum sourdoughs when a sucrose supplementation was ensured (Gobbetti et al. 2014). EPS positively impact the quality and properties of sourdough-baked goods, as several studies demonstrated that these biopolymers possess antioxidant properties, being healthy alternatives for the chemically synthetized hydrocolloids that contribute to the products' shelf life. The synbiotic effect determined by the simultaneous activity of probiotics and prebiotics represents a key feature toward a healthy lifestyle. Fructans produced by lactobacilli seemed to support the colonization of *Bifidobacterium* spp. in the colon, thus demonstrating their positive impact on human health (Chavan and Chavan 2011; Weckx, Van Kerrebroeck, and De Vuyst 2019). The anticarcinogenic effect was determined after *in vitro* tests using colonic cancer cells and EPS. The authors of this study demonstrated that *L. kefiri* originating from milk kefir grains were able to produce a heteropolysaccharide containing glucose and galactose (Riaz Rajoka et al. 2019). Another heteropolysaccharide comprising glucose, galactose, arabinose, mannose, and rhamnose was produced by an *L. kefiranofaciens* strain, isolated from kefir grains, and possessed antibacterial effects against *Listeria monocytogenes* and *Salmonella enteritidis* (Jeong et al. 2017). Other important EPS-producing strains belonging to *Weissella* spp., *Lactococcus* spp., and *Leuconostoc* spp. were studied to conclude that the strains originating from cereals produced inulin from sucrose (Sahin et al. 2019), due to the inulosucrase encoded gene that was identified as well in the genome of *L. gasseri* strain (Ni et al. 2018).

The current perspectives in food industries are focused on salt and sugar reduction, such objectives being reachable using sourdough fermentations with polyol-producing LAB and/or yeasts. *L. sanfranciscensis*, *Leu. mesenteroides*, *Leu. citreum*, *Leu. oenos*, and *Candida* spp. strains produced mannitol, erythritol, or xylitol from sucrose, xylose, or fructose, the obtained food products showing reduced glycemic index suitable for low-calorie diets (Behare et al. 2020; Lipinska-Zubrycka et al. 2020; Jeske et al. 2018).

5.3.2 Biotransformation of Lipids

Lipidomics focuses on studying and characterizing lipids and their derived compounds (Balkir, Kemahlioglu, and Yucel 2021). Lipids are minor constituents of flours. Sourdough mixing allows the incorporation of oxygen into the matrix. Thus, mixing is an important mechanical factor in the oxidative process of lipids. Furthermore, lipases, lipoxygenases, hydratases, and alcohol dehydrogenases are specialized enzymes that catalyze the bioconversion of lipids through the hydrolysis of fatty acids into hydroxyperoxy acids that are finally converted into hydroxy-fatty acids and aldehydes, compounds responsible for the bread or sourdough aroma. Specific amino acids, like cysteine, contribute to the lipids' hydrolyzation when microbial enzymes are produced during the sourdough fermentation process (Gänzle 2014; De Vuyst, Van Kerrebroeck, and Leroy 2017). Despite the pleasant odor determined by the protein or lipid hydrolytic reactions, it is also worth mentioning that sourdough fermentation using *Lactobacillus* spp. strains contributes to the minimization of some off-flavor compounds described as metallic, green, or fatty (Gänzle, Vermeulen, and Vogel 2007). Linoleic, linolenic, and oleic fatty acids were metabolized frequently by some lactobacilli that can synthesize alcohol dehydrogenases. As a result, the LAB lipidomics is intensively studied for a better understanding of the metabolic processes (Weckx, Van Kerrebroeck, and De Vuyst 2019). Recently, published papers have highlighted that probiotic strains, especially the ones belonging to *Lactobacillus* genus, have a cholesterol-lowering capacity determined by various mechanisms, such as intracellular transport, extracellular adhesion, or enzymatic hydrolysis by bile salt hydrolase (Khare and Gaur 2020; Prete et al. 2020; Wang et al. 2019).

Furthermore, monounsaturated and polyunsaturated fatty acids were produced from palmitic (16:0) or stearic (18:0) acid by selected strains of *Y. lipolytica*, *T.*

delbrueckii, *Saccharomyces* spp., *Kluyveromyces* spp., *Pichia* spp., or *Candida* spp. after catalytic reactions assisted by carboxylases, synthases, reductases, and desaturases, thus increasing their applications in various food industry sectors (Mbuyane, Bauer, and Divol 2021; Horincar et al. 2017; Parfene et al. 2013). Additional studies using a *P. pastoris* strain, conducted by Nurul Furqan and Akhmaloka (2020), proved that the lipase production is possible after a targeted insertion of the responsible gene into the chromosome, this enzyme yield being strongly influenced by the fermentation parameters, namely pH and temperature.

5.3.3 Nitrogen-Based Compounds

Besides carbon, the microorganisms from sourdough need nitrogen sources to fulfill their nutritional requirements. During sourdough fermentation, a two-step proteolysis occurs: the primary proteolysis, carried out by endogenous flour proteases; in a second stage, the peptidases produced by LAB strains are responsible for the secondary proteolysis that liberates the protein fractions (peptides) or amino acids with antimicrobial, antioxidant, antiproliferative, or flavor-enhancing properties (Galli et al. 2018; Oshiro, Zendo, and Nakayama 2021). It should be mentioned that the proteolysis mediated by LAB or yeasts is not a key feature for all the strains as the cleavage of proteins varies depending on several biotechnological factors. Therefore, some *Lactobacillus* spp. strains were able to reduce gluten, a process facilitated by the acidic pH of the sourdough (Gobbetti et al. 2014; Zotta et al. 2006). Other interesting approaches use flours rich in proteases (i.e. rye, wheat, amaranth, soybeans, or legumes) that are fermented by non-protease-producing LAB, thus their acidification potential would activate the enzymes from the flours that will result in a sourdough with high amounts of free amino acids and bioactive peptides, like lunasins, with diverse biological activities (Gänzle, Vermeulen, and Vogel 2007; Galli et al. 2018).

Studies involving *L. sanfranciscensis*, *L. plantarum*, and *Lactococcus lactis* strains were subjected to comparative analysis. *L. sanfranciscensis* possesses extracellular proteinases, intracellular transporters for complex oligo- and dipeptides, probably resulting after the primary hydrolysis of gluten or yeast extract, containing amino acids that this strain was not able to synthetize (i.e. lysine, glutamic, and aspartic acids). The encoded genes responsible for the transportation of peptides and their intracellular cleavage seemed to be more active during the exponential growth of *L. sanfranciscensis* strains (Oshiro, Zendo, and Nakayama 2021; Gänzle, Vermeulen, and Vogel 2007; Chavan and Chavan 2011).

Two major pathways characterize the amino acids metabolism of LAB: (i) arginine deiminase pathway (ADI) and (ii) glutamate decarboxylase pathway (GAD). The bacterial strains can use yeast extract to provide available arginine that is the starting substrate for the ADI pathway that uses deiminases, transferases, and kinases to produce ornithine as a major flavor compound. The increased concentration of ornithine can be obtained under acidic conditions when a specific co-factor, such as citrulline, was added (Oshiro, Zendo, and Nakayama 2021; Weckx, Van Kerrebroeck, and De Vuyst 2019). GAD pathway is characterized by glutaminases and glutamate decarboxylases/dehydrogenases that use glutamate or glutamine, major components of the wheat gluten proteins, to generate α-ketoglutarate as an intermediate compound and finally γ-aminobutyric acid (GABA) with various health-promoting effects (i.e. anticarcinogenic, antidiabetic, and antidepressant) that can be exploited in the food and feed industry (Diez-Gutiérrez et al. 2020). Citrate, α-ketoglutarate, or monosodium glutamate displayed a co-factor potential in the enzymatic catalysis (Park et al. 2021). Hence, the above-mentioned co-factors proved to facilitate the conversion of other amino acids (i.e. histidine, phenylalanine, leucine, isoleucine, and valine) into GABA (Yang, He, and Wu 2021). GABA biosynthesis was determined for *L. mindensis*, *L. nantendis*, *L. paraplantarum*, *L. helveticus*, *L. buchneri*, and *L. paracasei* strains or when a co-culture comprising *L. plantarum* and *Lc. lactis* strains was involved in a proteinaceous sourdough fermentation, made of chickpea, amaranth, quinoa, and buckwheat, carried out under optimized conditions and taking into consideration the extended fermentation time that seemed to facilitate GABA accumulation (Gobbetti et al. 2014; Weckx, Van Kerrebroeck, and De Vuyst 2019). Other LAB strains, except lactobacilli, originating from sourdoughs or other fermented products were used for the assessment of proteolytic activity, thus highlighting that *Weissella* spp., *Lactococcus* spp., *Enterococcus* spp., and *Pediococcus* spp. strains had this feature that could be used in the sourdough technology to increase the overall quality of the baked goods (Galli et al. 2018; Sáez et al. 2018; Heredia-Sandoval et al. 2016).

Ehrlich pathway is commonly used by yeast strains to produce alcohols, esters, and CO_2 as the main metabolites that contribute to the aroma profile after the yeast fermentation. *Saccharomyces* spp. and *Y. lipolytica* strains were subjected to metabolomics that highlighted that methionine and aromatic or branched amino acids were enzymatically converted primarily into phenolics, and flavor compounds were produced thereafter (i.e. vanillin, phenylethanol, and butanol derivatives) (Liu et al. 2020; Generoso et al. 2015).

5.3.4 Bioconversion of Phenolic Compounds

Phenolic compounds from cereals, pseudocereals, and legumes contribute majorly to the management of oxidative stress and its side-effects, therefore, other important benefits could be attributed to these compounds. Various types of colored wheat were analyzed regarding their bioactive composition to identify specific compounds. More specifically, benzoic acids (p-hydroxybenzoic, vanillic, protochatechuic, and caffeic), cinnamic acids (ferulic, p-coumaric, and sinapic), flavonols, chalcones, anthocyanins (delphinidin and cyanidin derivatives), flavones (apigenin and luteolin), isoflavones (genistein and daidzein), and flavonols (quercetin and kaempferol) were responsible for the color and functional properties of the seeds (Gupta, Meghwal, and Prabhakar 2021). Moreover, legumes have high content of phenolic acids, flavonoids, stilbenes, or tannins – condensed tannins are considered anti-nutritional factors that, after the enzymatic hydrolysis by tannases, have enhanced bioavailability (Conti et al. 2021). Phenolic compounds possess positive and controversial benefits because, in some cases, they can cause protein precipitation or enzymatic inhibition; thus it was observed that foods rich in phenolic compounds are characterized by a lower glycemic index that contributes to weight loss (Ripari, Bai, and Gänzle 2019).

Hydroxycinnamic and hydroxybenzoic acids are the major classes of phenolic acids identified in wheat and rye. The level of these bioactive compounds seemed to be higher in the whole grain flours of sorghum, wheat, and millet when refined flours were analyzed by comparison (Weckx, Van Kerrebroeck, and De Vuyst 2019). The sinapic, caffeic, p-coumaric, and ferulic acids were identified as well in the rye and wheat sourdoughs, their bounded forms being reduced after sourdough fermentation or sourdough bread baking (Konopka et al. 2014). This outcome could possibly be the result of enzymatic processes determined by LAB strains that demonstrated the ability to release the bounded/ esterified phenolics to their free forms or derivatives with flavoring potential (De Vuyst, Van Kerrebroeck, and Leroy 2017; Koistinen et al. 2018). Specifically, the conversion of phenolic compounds is catalyzed by decarboxylases and reductases that generate phenolic acid derivatives. Lactobacilli are the most performant group of LAB involved in the hydrolysis of bioactive compounds, as these strains produce esterases (i.e. tannin acyl hydrolases or feruloyl esterases) that selectively act upon the targeted phenolic compounds (Gänzle 2014; Novotni, Gänzle, and Rocha 2021). Fritsch et al. (2017) determined that *L. fermentum*, *L. platarum*, *L. acidophilus*, and *L. gasseri* strains can synthesize cinnamoyl esterases responsible for the hydroxycinnamic acids cleavage. The LAB strain *L. gasseri* had its maximum enzymatic activity at 30°C at neutral or slightly alkaline pH values. The phenolic acid decarboxylases encoded in the *Pediococcus* spp. strain genome were analyzed in order to maximize their activity in the wheat and rye malted sourdoughs. The concluding remarks highlighted that the best results would be determined by mixed cultures of pediococci and lactobacilli due to different metabolisms of the strains, with respect to the bioconversion of phenolic compounds (Ripari, Bai, and Gänzle 2019).

5.3.5 Minimization of Anti-nutritional Factors

The most studied anti-nutritional factors are phytic acid, trypsin inhibitors, condensed tannins, and saponins, their concentrations varying among different food matrices. The fermentation of these compounds with selected microorganisms, especially LAB strains and yeasts, proved to be one of the most efficient for the reduction of anti-nutritional factors, even though other processes (i.e. extrusion, thermal treatments, soaking, and germination) were studied and seemed to contribute to the minimization of these compounds. Considering its versatility and cost-efficiency, fermentation is applied frequently, the mechanisms behind the reduction of the anti-nutritional factors could be explained by the exogenous enzymes (originating from the flours) that are activated at low pH values or by enzymes produced by LAB and yeast strains (Weckx, Van Kerrebroeck, and De Vuyst 2019; Gänzle 2014; De Vuyst, Van Kerrebroeck, and Leroy 2017). Co-cultures containing *S. cerevisie* and *Pc. pentosaceus*, or *P. membranifaciens* and *L. brevis* strains, contributed to the degradation of phytic acid in the whole wheat sourdough bread (Karaman, Sagdic, and Durak 2018). Tanases, β-glucosidases, and α-galactosidases synthesized by *L. plantarum* strains determined the depletion of some faba bean-specific anti-nutritional factors, namely vicine and convicine after 48 hours of fermentation (Rizzello et al.

2016). Furthermore, the level of anti-nutritional factors decreased (20–80%) after the spontaneous fermentation of hemp, chia, wheat, lentils, chickpea, quinoa, and pea flours. The different minimization ratios are mainly related to the plant-based matrix subjected to fermentation and to the microorganisms involved (Arora et al. 2021; Jagelaviciute and Cizeikiene 2020).

5.4 CONCLUDING REMARKS AND PROSPECTS

LAB and yeasts have been used since ancient times to obtain fermented products with valuable technological and functional properties. The interest in these beneficial microorganisms increased over time as they can positively impact consumers' lifestyle and health. Nowadays, bread making with sourdoughs has become a common and frequent practice in bakeries worldwide, this field being intensively studied by researchers, firstly for a comprehensive understanding of the processes and, secondly, to offer suitable industrial-scale solutions for its production. The traditional fermentation process can be improved to facilitate the transition toward a modern approach by varying the types of fermentation substrate, starter cultures, and fermentation processes (systems and parameters) in order to enhance the *in vitro* and *in vivo* functionality of sourdoughs. Legumes or gluten-free pseudocereals are underutilized, thus offering interesting perspectives for their valorization in sourdough biotechnology due to their diverse nutritional and phytochemical composition. In the future, further research will be needed on the identification of performant and competitive LAB and/or yeast strains originating from sourdoughs or adapted to live in such ecosystems, involving here the in-depth genomic and metabolomic studies. The selected microorganisms need to possess multiple properties (i.e. probiotic properties, prebiotics, postbiotics, and paraprobiotics), the attention being drawn on the enzymatic activities of these biocatalysts that support the actual challenges as well as on sustainable research, development, and innovation. The characterization, standardization, and stabilization of the postbiotics and paraprobiotics in order to obtain functional ingredients used for food products, feed, and nutraceuticals are challenging goals in the new era of metabiotics that promote a healthy life for both humans and animals. These effects positively impact the quality of life through an intelligent and efficient valorization of the agri-food bioresources that support the principles of a circular economy.

ACKNOWLEDGMENTS

The work of Bogdan Păcularu-Burada was supported by the project ANTREPRENORDOC in the framework of Human Resources Development Operational Programme 2014–2020 financed from the European Social Fund under Contract number 36355/23.05.2019 HRD OP/380/6/13 – SMIS Code: 123847. This chapter is also based upon the research activities of COST Action 18101 SOURDOMICS – Sourdough biotechnology network towards novel, healthier, and sustainable food and bioprocesses, supported by COST (European Cooperation in Science and Technology) and grant of the Romanian Ministry of Research,

Innovation and Digitization, CNCS/CCCDI – UEFISCDI, project number PCE 159/2021, within PNCDI III.

REFERENCES

Aguilar-Toalá, J.E., R. Garcia-Varela, H.S. Garcia, V. Mata-Haro, A.F. González-Córdova, B. Vallejo-Cordoba, and A. Hernández-Mendoza. 2018. "Postbiotics: An Evolving Term within the Functional Foods Field." *Trends in Food Science & Technology* 75. Elsevier Ltd: 105–14. doi:10.1016/j.tifs.2018.03.009.

Alfonzo, Antonio, Delphine Sicard, Giuseppe Di Miceli, Stéphane Guezenec, and Luca Settanni. 2021. "Ecology of Yeasts Associated with Kernels of Several Durum Wheat Genotypes and Their Role in Co-Culture with Saccharomyces Cerevisiae during Dough Leavening." *Food Microbiology* 94 (4). Academic Press: 103666. doi:10.1016/j.fm.2020.103666.

Alpers, Thekla, Roland Kerpes, Mariana Frioli, Arndt Nobis, Ka Ian Hoi, Axel Bach, Mario Jekle, and Thomas Becker. 2021. "Impact of Storing Condition on Staling and Microbial Spoilage Behavior of Bread and Their Contribution to Prevent Food Waste." *Foods* 10 (1). MDPI AG: 76. doi:10.3390/foods10010076.

Arora, Kashika, Hana Ameur, Andrea Polo, Raffaella Di Cagno, Carlo Giuseppe Rizzello, Marco Gobbetti. 2021. "Thirty Years of Knowledge on Sourdough Fermentation: A Systematic Review." *Trends in Food Science & Technology* 108 (December). Elsevier Ltd: 71–83. doi:10.1016/j.tifs.2020.12.008.

Azi, Fidelis, Chuanhai Tu, Ling Meng, Li Zhiyu, Mekonen Tekliye Cherinet, Zahir Ahmadullah, Mingsheng Dong, Mekonen Tekliye Cherinet, Zahir Ahmadullah, and Mingsheng Dong. 2020. "Metabolite Dynamics and Phytochemistry of a Soy Whey-Based Beverage Bio-Transformed by Water Kefir Consortium." *Food Chemistry* 342: 128225. doi:10.1016/j.foodchem.2020.128225.

Balkir, Pinar, Kemal Kemahlioglu, and Ufuk Yucel. 2021. "Foodomics: A New Approach in Food Quality and Safety." *Trends in Food Science & Technology* 108. Elsevier Ltd: 49–57. doi:10.1016/j.tifs.2020.11.028.

Behare, Pradip V., Shahneela Mazhar, Vincenzo Pennone, and Olivia McAuliffe. 2020. "Evaluation of Lactic Acid Bacteria Strains Isolated from Fructose-Rich Environments for Their Mannitol-Production and Milk-Gelation Abilities." *Journal of Dairy Science* 103 (12). Elsevier: 11138–51. doi:10.3168/jds.2020-19120.

Bengoa, A.A., C. Iraporda, G.L. Garrote, and A.G. Abraham. 2019. "Kefir Micro-Organisms: Their Role in Grain Assembly and Health Properties of Fermented Milk." *Journal of Applied Microbiology* 126 (3). Blackwell Publishing Ltd: 686–700. doi:10.1111/jam.14107.

Caglar, Nagihan, Ertan Ermis, and Muhammed Zeki Durak. 2021. "Spray-Dried and Freeze-Dried Sourdough Powders: Properties and Evaluation of Their Use in Breadmaking." *Journal of Food Engineering* 292. Elsevier BV: 110355. doi:10.1016/j.jfoodeng.2020.110355.

Çakır, Elif, Muhammet Arıcı, and Muhammed Zeki Durak. 2020. "Biodiversity and Techno-Functional Properties of Lactic Acid Bacteria in Fermented Hull-Less Barley Sourdough." *Journal of Bioscience and Bioengineering* 130 (5): 450–56. doi:10.1016/j.jbiosc.2020.05.002.

Canesin, Míriam Regina, and Cínthia Baú Betim Cazarin. 2021. "Nutritional Quality and Nutrient Bioaccessibility in Sourdough Bread." *Current Opinion in Food Science* 37. Elsevier Ltd: 81–86. doi:10.1016/j.cofs.2021.02.007.

Chavan, Rupesh S., and Shraddha R. Chavan. 2011. "Sourdough Technology: A Traditional Way for Wholesome Foods: A Review." *Comprehensive Reviews in Food Science and Food Safety* 10 (3): 169–82. doi:10.1111/j.1541-4337.2011.00148.x.

Chen, Xi, Ruifang Mi, Biao Qi, Suyue Xiong, Jiapeng Li, Chao Qu, Xiaoling Qiao, Wenhua Chen, and Shouwei Wang. 2021. "Effect of Proteolytic Starter Culture Isolated from Chinese Dong Fermented Pork (Nanx Wudl) on Microbiological, Biochemical and Organoleptic Attributes in Dry Fermented Sausages." *Food Science and Human Wellness* 10 (1). Elsevier B.V.: 13–22. doi:10.1016/j.fshw.2020.05.012.

Cizeikiene, Dalia, Jolita Jagelaviciute, Mantas Stankevicius, and Audrius Maruska. 2020. "Thermophilic Lactic Acid Bacteria Affect the Characteristics of Sourdough and Whole-Grain Wheat Bread." *Food Bioscience* 38: 100791. doi:10.1016/j.fbio.2020.100791.

Coelho, Raquel Macedo Dantas, Aryelle Leite de Almeida, Rafael Queiroz Gurgel do Amaral, Robson Nascimento da Mota, and Paulo Henrique M. de Sousa. 2020. "Kombucha: Review." *International Journal of Gastronomy and Food Science* 22. AZTI-Tecnalia: 100272. doi:10.1016/j.ijgfs.2020.100272.

Comasio, Andrea, Simon Van Kerrebroeck, Henning Harth, Fabienne Verté, and Luc De Vuyst. 2020. "Potential of Bacteria from Alternative Fermented Foods as Starter Cultures for the Production of Wheat Sourdoughs." *Microorganisms* 8 (10). MDPI AG: 1534. doi:10.3390/microorganisms8101534.

Conti, Maria Vittoria, Lorenzo Guzzetti, Davide Panzeri, Rachele De Giuseppe, Paola Coccetti, Massimo Labra, and Hellas Cena. 2021. "Bioactive Compounds in Legumes: Implications for Sustainable Nutrition and Health in the Elderly Population." *Trends in Food Science & Technology*. Elsevier. doi:10.1016/j.tifs.2021.02.072.

Cuevas-González, P.F., A.M. Liceaga, and J.E. Aguilar-Toalá. 2020. "Postbiotics and Paraprobiotics: From Concepts to Applications." *Food Research International* 136 (October): 109502. doi:10.1016/j.foodres.2020.109502.

Daba, Ghoson M., Marwa O. Elnahas, and Waill A. Elkhateeb. 2021. "Contributions of Exopolysaccharides from Lactic Acid Bacteria as Biotechnological Tools in Food, Pharmaceutical, and Medical Applications." *International Journal of Biological Macromolecules* 173: 79–89. doi:10.1016/j.ijbiomac.2021.01.110.

Dahiya, Seema, Bijender Kumar Bajaj, Anil Kumar, Santosh Kumar Tiwari, and Bijender Singh. 2020. "A Review on Biotechnological Potential of Multifarious Enzymes in Bread Making." In *Process Biochemistry*. Elsevier Ltd. doi:10.1016/j.procbio.2020.09.002.

Dhaka, Annvi, Mehanathan Muthamilarasan, and Manoj Prasad. 2021. "A Comprehensive Study on Core Enzymes Involved in Starch Metabolism in the Model Nutricereal, Foxtail Millet (Setaria Italica L.)." *Journal of Cereal Science* 97. Academic Press: 103153. doi:10.1016/j.jcs.2020.103153.

Diez-Gutiérrez, Lucía, Leire San Vicente, Luis Javier Luis, María del Carmen Villarán, María Chávarri, Luis Javier R. Barrón, María del Carmen Villarán, and María Chávarri. 2020. "Gamma-Aminobutyric Acid and Probiotics: Multiple Health Benefits and Their Future in the Global Functional Food and Nutraceuticals Market." *Journal of Functional Foods* 64. Elsevier Ltd: 103669. doi:10.1016/j.jff.2019.103669.

Esen, Yusuf, and Bülent Çetin. 2021. "Bacterial and Yeast Microbial Diversity of the Ripened Traditional Middle East Surk Cheese." *International Dairy Journal* 117. Elsevier Ltd: 105004 doi:10.1016/j.idairyj.2021.105004.

Fang, Siyi, Bowen Yan, Fengwei Tian, Huizhang Lian, Jianxin Zhao, Hao Zhang, Wei Chen, and Daming Fan. 2021. "β-Fructosidase FosE Activity in Lactobacillus Paracasei Regulates Fructan Degradation during Sourdough Fermentation and Total FODMAP Levels in Steamed Bread." *LWT - Food Science and Technology* 145. Academic Press: 111294. doi:10.1016/j.lwt.2021.111294.

Farahmandi, Kajal, Shadi Rajab, Fatemeh Tabandeh, Mahvash Khodabandeh Shahraky, Amir Maghsoudi, and Morahem Ashengroph. 2021. "Efficient Spray-Drying of Lactobacillus Rhamnosus PTCC 1637 Using Total CFU Yield as the Decision Factor." *Food Bioscience* 40. Elsevier Ltd: 100816. doi:10.1016/j.fbio.2020.100816.

Fekri, Arezoo, Mohammadali Torbati, Ahmad Yari Khosrowshahi, Hasan Bagherpour Shamloo, Sodeif Azadmard-Damirchi, Ahmad Yari Khosrowshahi, Hasan Bagherpour Shamloo, and Sodeif Azadmard-Damirchi. 2019. "Functional Effects of Phytate-Degrading, Probiotic Lactic Acid Bacteria and Yeast Strains Isolated from Iranian Traditional Sourdough on the Technological and Nutritional Properties of Whole Wheat Bread." *Food Chemistry* 306. Elsevier BV: 125620. doi:10.1016/j.foodchem.2019.125620.

Filipčev, Bojana Šimurina, and Marija Olivera Bodroža-Solarov. 2007. "Effect of Native and Lyophilized Kefir Grains on Sensory and Physical Attributes of Wheat Bread." *Journal of Food Processing and Preservation* 31 (3): 367–77. doi:10.1111/j.1745-4549.2007.00134.x.

Fraberger, Vera, Christine Unger, Christian Kummer, and Konrad J. Domig. 2020. "Insights into Microbial Diversity of Traditional Austrian Sourdough." *LWT: Food Science and Technology* 127. Academic Press: 109358. doi:10.1016/j.lwt.2020.109358.

Fritsch, Caroline, André Jänsch, Matthias A. Ehrmann, Simone Toelstede, and Rudi F. Vogel. 2017. "Characterization of Cinnamoyl Esterases from Different Lactobacilli and Bifidobacteria." *Current Microbiology* 74 (2). Springer New York LLC: 247–56. doi:10.1007/s00284-016-1182-x.

Fu, Wenhui, Huan Rao, Yang Tian, and Wentong Xue. 2020. "Bacterial Composition in Sourdoughs from Different Regions in China and the Microbial Potential to Reduce Wheat Allergens." *LWT: Food Science and Technology* 117: 108669. doi:10.1016/j.lwt.2019.108669.

Fujimoto, Akihito, Keisuke Ito, Noriko Narushima, and Takahisa Miyamoto. 2019. "Identification of Lactic Acid Bacteria and Yeasts, and Characterization of Food Components of Sourdoughs Used in Japanese Bakeries." *Journal of Bioscience and Bioengineering* 127 (5). Elsevier B.V.: 575–81. doi:10.1016/j.jbiosc.2018.10.014.

Furukawa, Soichi, Taisuke Watanabe, Hirohide Toyama, and Yasushi Morinaga. 2013. "Significance of Microbial Symbiotic Coexistence in Traditional Fermentation." *Journal of Bioscience and Bioengineering* 116 (5). Elsevier: 533–39. doi:10.1016/j.jbiosc.2013.05.017.

Galimberti, Andrea, Antonia Bruno, Giulia Agostinetto, Maurizio Casiraghi, Lorenzo Guzzetti, and Massimo Labra. 2021. "Fermented Food Products in the Era of Globalization: Tradition Meets Biotechnology Innovations." *Current Opinion in Biotechnology* 70: 36–41. doi:10.1016/j.copbio.2020.10.006.

Galli, Viola, Lorenzo Mazzoli, Simone Luti, Manuel Venturi, Simona Guerrini, Paolo Paoli, Massimo Vincenzini, Lisa Granchi, and Luigia Pazzagli. 2018. "Effect of Selected Strains of Lactobacilli on the Antioxidant and Anti-Inflammatory Properties of Sourdough." *International Journal of Food Microbiology* 286. Elsevier B.V.: 55–65. doi:10.1016/j.ijfoodmicro.2018.07.018.

Gänzle, Michael G. 2014a. "BREAD | Sourdough Bread." In *Encyclopedia of Food Microbiology*, 309–15. Elsevier. doi:10.1016/B978-0-12-384730-0.00045-8.

Gänzle, Michael G. 2014b. "Enzymatic and Bacterial Conversions during Sourdough Fermentation." *Food Microbiology* 37. Academic Press: 2–10. doi:10.1016/j.fm.2013.04.007.

Gänzle, Michael G., Nicoline Vermeulen, and Rudi F. Vogel. 2007. "Carbohydrate, Peptide and Lipid Metabolism of Lactic Acid Bacteria in Sourdough." *Food Microbiology* 24 (2): 128–38. doi:10.1016/j.fm.2006.07.006.

García Vilanova, María, Carmen Díez, Brando Quirino, and J. Iñaki Álava. 2015. "Microbiota Distribution in Sourdough: Influence of High Sucrose Resistant Strains." *International Journal of Gastronomy and Food Science* 2 (2): 98–102. doi:10.1016/j.ijgfs.2015.01.002.

Garrote, Graciela L., Anala G. Abraham, and Graciela L. De Antoni. 2010. "Microbial Interactions in Kefir: A Natural Probiotic Drink." In *Biotechnology of Lactic Acid Bacteria*, edited by Fernanda Mozzi, Raúl R. Raya, and Graciela M. Vignolo, 327–40. Wiley-Blackwell. doi:10.1002/9780813820866.ch18.

Generoso, Wesley Cardoso, Virginia Schadeweg, Mislav Oreb, and Eckhard Boles. 2015. "Metabolic Engineering of Saccharomyces Cerevisiae for Production of Butanol Isomers." In *Current Opinion in Biotechnology*. Elsevier Ltd. doi:10.1016/j.copbio.2014.09.004.

Gobbetti, Marco, Carlo G. Rizzello, Raffaella Di Cagno, and Maria De Angelis. 2014. "How the Sourdough May Affect the Functional Features of Leavened Baked Goods." *Food Microbiology* 37: 30–40. doi:10.1016/j.fm.2013.04.012.

Gupta, Rachna, Murlidhar Meghwal, and Pramod K. Prabhakar. 2021. "Bioactive Compounds of Pigmented Wheat (Triticum Aestivum): Potential Benefits in Human Health." *Trends in Food Science & Technology* 110. Elsevier BV: 240–52. doi:10.1016/j.tifs.2021.02.003.

Gut, Abraham Majak, Todor Vasiljevic, Thomas Yeager, and Osaana N. Donkor. 2019. "Characterization of Yeasts Isolated from Traditional Kefir Grains for Potential Probiotic Properties." *Journal of Functional Foods* 58. Elsevier Ltd: 56–66. doi:10.1016/j.jff.2019.04.046.

Heredia-Sandoval, Nina, Maribel Valencia-Tapia, Ana Calderón de la Barca, and Alma Islas-Rubio. 2016. "Microbial Proteases in Baked Goods: Modification of Gluten and Effects on Immunogenicity and Product Quality." *Foods* 5(3):59. doi:10.3390/foods5030059.

Hermann, Maria, Kerstin Kronseder, Jennifer Sorgend, Tharalinee Ua-Arak, and Rudi F. Vogel. 2016. "Functional Properties of Water Kefiran and Its Use as a Hydrocolloid in Baking." *European Food Research and Technology* 242 (3): 337–44. doi:10.1007/s00217-015-2543-6.

Horincar, Georgiana, Vicentiu Bogdan Horincar, Davide Gottardi, and Gabriela Bahrim. 2017. "Tailoring the Potential of Yarrowia Lipolytica for Bioconversion of Raw Palm Fat for Antimicrobials Production." *LWT: Food Science and Technology* 80. Academic Press: 335–40. doi:10.1016/j.lwt.2017.02.026.

Huys, Geert, Heide-Marie Daniel, and Luc De Vuyst. 2013. "Taxonomy and Biodiversity of Sourdough Yeasts and Lactic Acid Bacteria." In *Handbook on Sourdough Biotechnology*, 105–54. Springer US. doi:10.1007/978-1-4614-5425-0_5.

Jafari, Reyhaneh, Nafiseh Sadat Naghavi, Kianoush Khosravi-Darani, Monir Doudi, and Kahin Shahanipour. 2020. "Kombucha Microbial Starter with Enhanced Production of Antioxidant Compounds and Invertase." *Biocatalysis and Agricultural Biotechnology* 29. Elsevier Ltd: 101789. doi:10.1016/j.bcab.2020.101789.

Jagelaviciute, Jolita, and Dalia Cizeikiene. 2020. "The Influence of Non-Traditional Sourdough Made with Quinoa, Hemp and Chia Flour on the Characteristics of Gluten-Free Maize/Rice Bread." *LWT: Food Science and Technology* 137. Academic Press: 110457. doi:10.1016/j.lwt.2020.110457.

Jeong, Dana, Dong-Hyeon Kim, Il-Byeong Kang, Hyunsook Kim, Kwang-Young Song, Hong-Seok Kim, and Kun-Ho Seo. 2017. "Characterization and Antibacterial Activity of a Novel Exopolysaccharide Produced by Lactobacillus Kefiranofaciens DN1 Isolated from Kefir." *Food Control* 78 (August). Elsevier Ltd: 436–42. doi:10.1016/j.foodcont.2017.02.033.

Jeske, Stephanie, Emanuele Zannini, Kieran M. Lynch, Aidan Coffey, and Elke K. Arendt. 2018. "Polyol-Producing Lactic Acid Bacteria Isolated from Sourdough and Their Application to Reduce Sugar in a Quinoa-Based Milk Substitute." *International Journal of Food Microbiology* 286. Elsevier B.V.: 31–36. doi:10.1016/j.ijfoodmicro.2018.07.013.

Jin, Juhui, Thi Thanh Hanh Nguyen, Sanjida Humayun, Sung Hoon Park, Hyewon Oh, Sangyong Lim, Il Kyoon Mok, Yan Li, Kunal Pal, and Doman Kim. 2021. "Characteristics of Sourdough Bread Fermented with Pediococcus Pentosaceus and Saccharomyces Cerevisiae and Its Bio-Preservative Effect against Aspergillus Flavus." *Food Chemistry*. doi:10.1016/j.foodchem.2020.128787.

Johansson, Linnea, Jarkko Nikulin, Riikka Juvonen, Kristoffer Krogerus, Frederico Magalhães, Atte Mikkelson, Maija Nuppunen-Puputti, et al. 2021. "Sourdough Cultures as Reservoirs of Maltose-Negative Yeasts for Low-Alcohol Beer Brewing." *Food Microbiology* 94 (April). Academic Press: 103629. doi:10.1016/j.fm.2020.103629.

Karaman, Kevser, Osman Sagdic, and M. Zeki Durak. 2018. "Use of Phytase Active Yeasts and Lactic Acid Bacteria Isolated from Sourdough in the Production of Whole Wheat Bread." *LWT: Food Science and Technology* 91: 557–67. doi:10.1016/j.lwt.2018.01.055.

Kerrebroeck, Simon Van, Dominique Maes, and Luc De Vuyst. 2017. "Sourdoughs as a Function of Their Species Diversity and Process Conditions, a Meta-Analysis." *Trends in Food Science & Technology* 68 (October). Elsevier Ltd: 152–59. doi:10.1016/j.tifs.2017.08.016.

Kesmen, Zülal, Özge Kılıç, Yasin Gormez, Mete Çelik, and Burcu Bakir-Gungor. 2020. "Multi Fragment Melting Analysis System (MFMAS) for One-Step Identification of Lactobacilli." *Journal of Microbiological Methods* 177. Elsevier: 106045. doi:10.1016/j.mimet.2020.106045.

Khare, Aditi, and Smriti Gaur. 2020. "Cholesterol-Lowering Effects of Lactobacillus Species." *Current Microbiology* 77(4): 638–44. Springer. doi:10.1007/s00284-020-01903-w.

Koistinen, Ville M., Outi Mattila, Kati Katina, Kaisa Poutanen, Anna-Marja Aura, and Kati Hanhineva. 2018. "Metabolic Profiling of Sourdough Fermented Wheat and Rye Bread." *Scientific Reports* 8 (1): 5684. doi:10.1038/s41598-018-24149-w.

Konopka, Iwona, Małgorzata Tańska, Alicja Faron, and Sylwester Czaplicki. 2014. "Release of Free Ferulic Acid and Changes in Antioxidant Properties during the Wheat and Rye Bread Making Process." *Food Science and Biotechnology* 23 (3). Kluwer Academic Publishers: 831–40. doi:10.1007/s10068-014-0112-6.

Kyereh, Emmanuel, and Subramaniam Sathivel. 2021. "Viability of Lactobacillus Plantarum NCIMB 8826 Immobilized in a Cereal-Legume Complementary Food "Weanimix" with Simulated Gastrointestinal Conditions." *Food Bioscience* 40. Elsevier Ltd: 100848. doi:10.1016/j.fbio.2020.100848.

Liang, Nuanyi, Věra Neužil-Bunešová, Václav Tejnecký, Michael Gänzle, and Clarissa Schwab. 2021. "3-Hydroxypropionic Acid Contributes to the Antibacterial Activity of Glycerol Metabolism by the Food Microbe Limosilactobacillus Reuteri." *Food Microbiology* 98. Elsevier BV: 103720. doi:10.1016/j.fm.2020.103720.

Lin, Tzu-Lung, Ching-Chung Shu, Wei-Fan Lai, Chi-Meng Tzeng, Hsin-Chih Lai, and Chia-Chen Lu. 2019. "Investiture of next Generation Probiotics on Amelioration of Diseases – Strains Do Matter." *Medicine in Microecology* 1–2. Elsevier BV: 100002. doi:10.1016/j.medmic.2019.100002.

Lipinska-Zubrycka, Lidia, Robert Klewicki, Michal Sojka, Radoslaw Bonikowski, Agnieszka Milczarek, and Elzbieta Klewicka. 2020. "Anticandidal Activity of Lactobacillus Spp. in the Presence of Galactosyl Polyols." *Microbiological Research* 240. Elsevier GmbH: 126540. doi:10.1016/j.micres.2020.126540.

Liu, Hu-Hu, Xiao-Jun Ji, and He Huang. 2015. "Biotechnological Applications of Yarrowia Lipolytica: Past, Present and Future." *Biotechnology Advances* 33 (8). Elsevier Inc.: 1522–46. doi:10.1016/j.biotechadv.2015.07.010.

Liu, Quanli, Yi Liu, Yun Chen, and Jens Nielsen. 2020. "Current State of Aromatics Production Using Yeast: Achievements and Challenges." *Current Opinion in Biotechnology*. 65: 65–74. Elsevier Ltd. doi:10.1016/j.copbio.2020.01.008.

Liu, Tongjie, Yang Li, Faizan A. Sadiq, Huanyi Yang, Jingsi Gu, Lei Yuan, Yuan Kun Lee, and Guoqing He. 2018. "Predominant Yeasts in Chinese Traditional Sourdough and Their Influence on Aroma Formation in Chinese Steamed Bread." *Food Chemistry* 242: 404–11. doi:10.1016/j.foodchem.2017.09.081.

Luti, Simone, Lorenzo Mazzoli, Matteo Ramazzotti, Viola Galli, Manuel Venturi, Giada Marino, Martin Lehmann, et al. 2020. "Antioxidant and Anti-Inflammatory Properties of Sourdoughs Containing Selected Lactobacilli Strains Are Retained in Breads." *Food Chemistry* 322: 126710. doi:10.1016/j.foodchem.2020.126710.

Mantzourani, I., S. Plessas, G. Saxami, A. Alexopoulos, A. Galanis, and E. Bezirtzoglou. 2014. "Study of Kefir Grains Application in Sourdough Bread Regarding Rope Spoilage Caused by Bacillus Spp." *Food Chemistry* 143 (January). Elsevier Ltd: 17–21. doi:10.1016/j.foodchem.2013.07.098.

Mbuyane, Lethiwe Lynett, Florian Franz Bauer, and Benoit Divol. 2021. "The Metabolism of Lipids in Yeasts and Applications in Oenology." *Food Research International* 110: 110142. Elsevier Ltd. doi:10.1016/j.foodres.2021.110142.

Melo, Ester Lopes de, Aline Moreira Pinto, Camila Lins Bilby Baima, Heitor Ribeiro da Silva, Iracirema da Silva Sena, Brenda Lorena Sanchez-Ortiz, Abrahão Victor Tavares de Lima Teixeira, et al. 2020. "Evaluation of the in Vitro Release of Isoflavones from Soybean Germ Associated with Kefir Culture in the Gastrointestinal Tract and Anxiolytic and Antidepressant Actions in Zebrafish (Danio Rerio)." *Journal of Functional Foods* 70. Elsevier Ltd: 103986. doi:10.1016/j.jff.2020.103986.

Minervini, Fabio, Anna Lattanzi, Francesca Rita Dinardo, Maria De Angelis, and Marco Gobbetti. 2018. "Wheat Endophytic Lactobacilli Drive the Microbial and Biochemical Features of Sourdoughs." *Food Microbiology* 70: 162–71. doi:10.1016/j.fm.2017.09.006.

Mohd Roby, Bizura Hasida, Belal J. Muhialdin, Muna Mahmood Taleb Abadl, Nor Arifah Mat Nor, Anis Asyila Marzlan, Sarina Abdul Halim Lim, Nor Afizah Mustapha, and Anis Shobirin Meor Hussin. 2020. "Physical Properties, Storage Stability, and Consumer Acceptability for Sourdough Bread Produced Using Encapsulated Kombucha Sourdough Starter Culture." *Journal of Food Science* 85 (8). Blackwell Publishing Inc.: 2286–95. doi:10.1111/1750-3841.15302.

Moroni, Alice V., Elke K. Arendt, and Fabio Dal Bello. 2011. "Biodiversity of Lactic Acid Bacteria and Yeasts in Spontaneously-Fermented Buckwheat and Teff Sourdoughs." *Food Microbiology* 28 (3): 497–502. doi:10.1016/j.fm.2010.10.016.

Ni, Dawei, Yingying Zhu, Wei Xu, Yuxiang Bai, Tao Zhang, and Wanmeng Mu. 2018. "Biosynthesis of Inulin from Sucrose Using Inulosucrase from Lactobacillus Gasseri DSM 20604." *International Journal of Biological Macromolecules* 109. Elsevier B.V.: 1209–18. doi:10.1016/j.ijbiomac.2017.11.120.

Novotni, Dubravka, Michael Gänzle, and João Miguel Rocha. 2021. "Composition and Activity of Microbiota in Sourdough and Their Effect on Bread Quality and Safety." In *Trends in Wheat and Bread Making*, edited by Charis M. Galanakis, 129–72. Elsevier. doi:10.1016/B978-0-12-821048-2.00005-2.

Nurul Furqan, Baiq Repika, and Akhmaloka. 2020. "Heterologous Expression and Characterization of Thermostable Lipase (Lk1) in Pichia Pastoris GS115." *Biocatalysis and Agricultural Biotechnology* 23. Elsevier Ltd: 101448. doi:10.1016/j.bcab.2019.101448.

O'Neill, Ellis C., Clare E.M. Stevenson, Krit Tantanarat, Dimitrios Latousakis, Matthew I. Donaldson, Martin Rejzek, Sergey A. Nepogodiev, Tipaporn Limpaseni, Robert A. Field, and David M. Lawson. 2015. "Structural Dissection of the Maltodextrin Disproportionation Cycle of the Arabidopsis Plastidial Disproportionating Enzyme 1 (DPE1)." *Journal of Biological Chemistry* 290 (50): 29834–53. doi:10.1074/jbc.M115.682245.

Oguntoye, Modupeola A., Olufunke O. Ezekiel, and Olayinka A. Oridupa. 2021. "Viability of Lactobacillus Rhamnosus GG in Provitamin A Cassava Hydrolysate during Fermentation, Storage, in Vitro and in Vivo Gastrointestinal Conditions." *Food Bioscience* 40. Elsevier: 100845. doi:10.1016/j.fbio.2020.100845.

Oliveira, Iuri Magalhães de Alencar, and Ruann Janser Soares de Castro. 2020. "Kefir Fermentation as a Bioprocess to Improve Lentils Antioxidant Properties: Is It Worthwhile?" *Brazilian Journal of Food Technology* 23: e2019120. FapUNIFESP (SciELO). Doi:10.1590/1981-6723.12019.

Oshiro, Mugihito, Takeshi Zendo, and Jiro Nakayama. 2021. "Diversity and Dynamics of Sourdough Lactic Acid Bacteriota Created by a Slow Food Fermentation System." *Journal of Bioscience and Bioengineering* 131(4): 333–40. Elsevier. doi:10.1016/j.jbiosc.2020.11.007.

Păcularu-Burada, Bogdan, Luminița Anca Georgescu, and Gabriela-Elena Bahrim. 2020a. "Current Approaches in Sourdough Production with Valuable Characteristics for Technological and Functional Applications." *The Annals of the University Dunarea de Jos of Galati Fascicle VI: Food Technology* 44 (1): 132–48. doi:10.35219/foodtechnology.2020.1.08.

Păcularu-Burada, Bogdan, Luminița Anca Georgescu, Mihaela Aida Vasile, João Miguel Rocha, and Gabriela-Elena Bahrim. 2020b. "Selection of Wild Lactic Acid Bacteria Strains as Promoters of Postbiotics in Gluten-Free Sourdoughs." *Microorganisms* 8 (5). MDPI AG: 643. doi:10.3390/microorganisms8050643.

Padmavathi, Tallapragada, Rayavarapu Bhargavi, Purushothama Rao Priyanka, Naige Ranganath Niranjan, and Pogakul Veerabhadrappa Pavitra. 2018. "Screening of Potential Probiotic Lactic Acid Bacteria and Production of Amylase and Its Partial Purification." *Journal of Genetic Engineering and Biotechnology* 16 (2). Academy of Scientific Research and Technology: 357–62. doi:10.1016/j.jgeb.2018.03.005.

Palla, Michela, Monica Agnolucci, Antonella Calzone, Manuela Giovannetti, Raffaella Di Cagno, Marco Gobbetti, Carlo Giuseppe Rizzello, and Erica Pontonio. 2019. "Exploitation of Autochthonous Tuscan Sourdough Yeasts as Potential Starters." *International Journal of Food Microbiology* 302: 59–68. doi:10.1016/j.ijfoodmicro.2018.08.004.

Palla, Michela, Massimo Blandino, Arianna Grassi, Debora Giordano, Cristina Sgherri, Mike Frank Quartacci, Amedeo Reyneri, Monica Agnolucci, and Manuela Giovannetti. 2020. "Characterization and Selection of Functional Yeast Strains during Sourdough Fermentation of Different Cereal Wholegrain Flours." *Scientific Reports* 10 (1): 12856. doi:10.1038/s41598-020-69774-6.

Papadimitriou, Konstantinos, Georgia Zoumpopoulou, Marina Georgalaki, Voula Alexandraki, Maria Kazou, Rania Anastasiou, and Effie Tsakalidou. 2019. "Sourdough Bread." In *Innovations in Traditional Foods*, 127–58. Elsevier. doi:10.1016/B978-0-12-814887-7.00006-X.

Parfene, Georgiana, Vicentiu Horincar, Amit Kumar Tyagi, Anushree Malik, and Gabriela Bahrim. 2013. "Production of Medium Chain Saturated Fatty Acids with Enhanced Antimicrobial Activity from Crude Coconut Fat by Solid State Cultivation of Yarrowia Lipolytica." *Food Chemistry* 136 (3–4). Food Chem: 1345–49. doi:10.1016/j.foodchem.2012.09.057.

Park, Su Jeong, Dong Hyun Kim, Hye Jee Kang, Minhye Shin, Soo-Yeon Yang, Jungwoo Yang, and Young Hoon Jung. 2021. "Enhanced Production of γ-Aminobutyric Acid (GABA) Using Lactobacillus Plantarum EJ2014 with Simple Medium Composition." *LWT: Food Science and Technology* 137: 110443. doi:10.1016/j.lwt.2020.110443.

Peng, Kaidi, Mohamed Koubaa, Olivier Bals, and Eugène Vorobiev. 2020. "Recent Insights in the Impact of Emerging Technologies on Lactic Acid Bacteria: A Review." *Food Research International* 137: 109544. doi:10.1016/j.foodres.2020.109544.

Petrova, Penka, and Kaloyan Petrov. 2020. "Lactic Acid Fermentation of Cereals and Pseudocereals: Ancient Nutritional Biotechnologies with Modern Applications." *Nutrients* 12 (4): 1118. doi:10.3390/nu12041118.

Plessas, Stavros, Athanasios Alexopoulos, Argyro Bekatorou, Ioanna Mantzourani, Athanasios A. Koutinas, and Eugenia Bezirtzoglou. 2011. "Examination of Freshness Degradation of Sourdough Bread Made with Kefir through Monitoring the Aroma Volatile Composition during Storage." *Food Chemistry* 124 (2): 627–33. doi:10.1016/j.foodchem.2010.06.086.

Prado, Maria R., Lina Marcela Blandón, Luciana P.S. Vandenberghe, Cristine Rodrigues, Guillermo R. Castro, Vanete Thomaz-Soccol, and Carlos R. Soccol. 2015. "Milk Kefir: Composition, Microbial Cultures, Biological Activities, and Related Products." *Frontiers in Microbiology* 6. Frontiers Media S.A.: 1177. doi:10.3389/fmicb.2015.01177.

Prete, Roberta, Sarah Louise Long, Alvaro Lopez Gallardo, Cormac G. Gahan, Aldo Corsetti, and Susan A. Joyce. 2020. "Beneficial Bile Acid Metabolism from Lactobacillus Plantarum of Food Origin." *Scientific Reports* 10 (1). Nature Research: 1165. doi:10.1038/s41598-020-58069-5.

Rana, Sonali, and Lata Sheo Bachan Upadhyay. 2020. "Microbial Exopolysaccharides: Synthesis Pathways, Types and Their Commercial Applications." *International Journal of Biological Macromolecules* 157. Elsevier B.V.: 577–83. doi:10.1016/j.ijbiomac.2020.04.084.

Reale, Anna, Teresa Zotta, Rocco G. Ianniello, Gianfranco Mamone, and Tiziana Di Renzo. 2020. "Selection Criteria of Lactic Acid Bacteria to Be Used as Starter for Sweet and Salty Leavened Baked Products." *LWT - Food Science and Technology* 133. Academic Press: 110092. doi:10.1016/j.lwt.2020.110092.

Riaz Rajoka, Muhammad Shahid, Hafiza Mahreen Mehwish, Huiyan Fang, Arshad Ahmed Padhiar, Xierong Zeng, Mohsin Khurshid, Zhendan He, and Liqing Zhao. 2019. "Characterization and Anti-Tumor Activity of Exopolysaccharide Produced by Lactobacillus Kefiri Isolated from Chinese Kefir Grains." *Journal of Functional Foods* 63 (6). Elsevier: 103588. doi:10.1016/j.jff.2019.103588.

Ripari, Valery, Yunpeng Bai, and Michael G. Gänzle. 2019. "Metabolism of Phenolic Acids in Whole Wheat and Rye Malt Sourdoughs." *Food Microbiology* 77. Academic Press: 43–51. doi:10.1016/j.fm.2018.08.009.

Rizzello, Carlo Giuseppe, Ilario Losito, Laura Facchini, Kati Katina, Francesco Palmisano, Marco Gobbetti, and Rossana Coda. 2016. "Degradation of Vicine, Convicine and Their Aglycones during Fermentation of Faba Bean Flour." *Scientific Reports* 6 (1). Nature Publishing Group: 32452. doi:10.1038/srep32452.

Sáez, Gabriel D., Lucila Saavedra, Elvira M. Hebert, and Gabriela Zárate. 2018. "Identification and Biotechnological Characterization of Lactic Acid Bacteria Isolated from Chickpea Sourdough in Northwestern Argentina." *LWT: Food Science and Technology* 93. Academic Press: 249–56. doi:10.1016/j.lwt.2018.03.040.

Sahin, Aylin W., Emanuele Zannini, Aidan Coffey, and Elke K. Arendt. 2019. "Sugar Reduction in Bakery Products: Current Strategies and Sourdough Technology as a Potential Novel Approach." *Food Research International* 126. Elsevier Ltd: 108583. doi:10.1016/j.foodres.2019.108583.

Salmerón, Ivan, Samuel B. Pérez-Vega, Néstor Gutiérrez-Méndez, and Ilderbando Pérez-Reyes. 2020. "Lactic Acid Bacteria in Preservation and Functional Foods." In *Food Microbiology and Biotechnology : Safe and Sustainable Food Production*, edited by Guadalupe Virginia, Nevárez-Moorillón, Arely, Prado-Barragán, José Luis, Martínez-Hernández, Aguilar, Cristóbal Noé, 137–61. Apple Academic Press, Inc. doi:10.31142/ijtsrd23951.

Sauer, Michael, Hannes Russmayer, Reingard Grabherr, Clemens K. Peterbauer, and Hans Marx. 2017. "The Efficient Clade: Lactic Acid Bacteria for Industrial Chemical Production." *Trends in Biotechnology* 35 (8). Elsevier Ltd: 756–69. doi:10.1016/j.tibtech.2017.05.002.

Seo, Kun-Ho, Jaewoon Jeong, and Hyunsook Kim. 2020. "Synergistic Effects of Heat-Killed Kefir Paraprobiotics and Flavonoid-Rich Prebiotics on Western Diet-Induced Obesity." *Nutrients* 12 (8). Multidisciplinary Digital Publishing Institute: 2465. doi:10.3390/nu12082465.

Sieuwerts, Sander, Peter A. Bron, and Eddy J. Smid. 2018. "Mutually Stimulating Interactions between Lactic Acid Bacteria and Saccharomyces Cerevisiae in Sourdough Fermentation." *LWT: Food Science and Technology* 90: 201–6. doi:10.1016/j.lwt.2017.12.022.

Sindi, Abrar, Md. Bahadur Badsha, Barbara Nielsen, and Gülhan Ünlü. 2020. "Antimicrobial Activity of Six International Artisanal Kefirs against Bacillus Cereus, Listeria Monocytogenes, Salmonella Enterica Serovar Enteritidis, and Staphylococcus Aureus." *Microorganisms* 8 (6). MDPI AG: 849. doi:10.3390/microorganisms8060849.

Soares, Marcelo Gomes, Marieli de Lima, and Vivian Consuelo Reolon Schmidt. 2021. "Technological Aspects of Kombucha, Its Applications and the Symbiotic Culture (SCOBY), and Extraction of Compounds of Interest: A Literature Review." *Trends in Food Science & Technology* 110. Elsevier: 539–50. doi:10.1016/j.tifs.2021.02.017.

Suo, Biao, Xinyi Chen, and Yuexia Wang. 2021. "Recent Research Advances of Lactic Acid Bacteria in Sourdough: Origin, Diversity, and Function." *Current Opinion in Food Science* 37. Elsevier: 66–75. doi:10.1016/j.cofs.2020.09.007.

Teigiserova, Dominika Alexa, Joseph Bourgine, and Marianne Thomsen. 2021. "Closing the Loop of Cereal Waste and Residues with Sustainable Technologies: An Overview of Enzyme Production via Fungal Solid-State Fermentation." *Sustainable Production and Consumption* 27. Elsevier BV: 845–57. doi:10.1016/j.spc.2021.02.010.

Torres, Sebastian, Hernán Verón, Luciana Contreras, and Maria I. Isla. 2020. "An Overview of Plant-Autochthonous Microorganisms and Fermented Vegetable Foods." *Food Science and Human Wellness* 9 (2): 112–23. doi:10.1016/j.fshw.2020.02.006.

Tu, Chuanhai, Fidelis Azi, Jin Huang, Xiao Xu, Guangliang Xing, and Mingsheng Dong. 2019. "Quality and Metagenomic Evaluation of a Novel Functional Beverage Produced from Soy Whey Using Water Kefir Grains." *LWT: Food Science and Technology* 113. Academic Press: 108258. doi:10.1016/j.lwt.2019.108258.

Velasco, Lina, Myriam Loeffler, Isabel Torres, and Jochen Weiss. 2021. "Influence of Fermentation Temperature on in Situ Heteropolysaccharide Formation (Lactobacillus Plantarum TMW 1.1478) and Texture Properties of Raw Sausages." *Food Science & Nutrition* 9 (3). Wiley-Blackwell: 1312–22. doi:10.1002/fsn3.2054.

Verce, Marko, Luc De Vuyst, and Stefan Weckx. 2019. "Shotgun Metagenomics of a Water Kefir Fermentation Ecosystem Reveals a Novel Oenococcus Species." *Frontiers in Microbiology* 10 (3). Frontiers Media S.A.: 479. doi:10.3389/fmicb.2019.00479.

Vladimirovich, Sinitsa Aleksandr. 2016. *Metabiotic Composition to Ensure Colonisation Resistance of Human Intestinal Microbiocenosis*. 2015113356/15, issued 2016. https://patents.google.com/patent/RU2589818C1/en.

Vogelmann, Stephanie A., and Christian Hertel. 2011. "Impact of Ecological Factors on the Stability of Microbial Associations in Sourdough Fermentation." *Food Microbiology* 28 (3): 583–89. doi:10.1016/j.fm.2010.11.010.

Vuyst, Luc De, and Patricia Neysens. 2005. "The Sourdough Microflora: Biodiversity and Metabolic Interactions." *Trends in Food Science & Technology* 16:43–56. doi:10.1016/j.tifs.2004.02.012.

Vuyst, Luc De, Henning Harth, Simon Van Kerrebroeck, and Frédéric Leroy. 2016. "Yeast Diversity of Sourdoughs and Associated Metabolic Properties and Functionalities." *International Journal of Food Microbiology* 239: 26–34. doi:10.1016/j.ijfoodmicro.2016.07.018.

Vuyst, Luc De, Simon Van Kerrebroeck, and Frédéric Leroy. 2017. "Microbial Ecology and Process Technology of Sourdough Fermentation." *Advances in Applied Microbiology* 100: 49–160. doi:10.1016/bs.aambs.2017.02.003.

Wang, Guangqiang, Wenli Huang, Yongjun Xia, Zhiqiang Xiong, and Lianzhong Ai. 2019. "Cholesterol-Lowering Potentials of Lactobacillus Strain Overexpression of Bile Salt Hydrolase on High Cholesterol Diet-Induced Hypercholesterolemic Mice." *Food and Function* 10 (3). Royal Society of Chemistry: 1684–95. doi:10.1039/c8fo02181c.

Wang, Hao, Cuina Wang, and Mingruo Guo. 2020. "Autogenic Successions of Bacteria and Fungi in Kefir Grains from Different Origins When Sub-Cultured in Goat Milk." *Food Research International* 138. Elsevier Ltd: 109784. doi:10.1016/j.foodres.2020.109784.

Weckx, Stefan, Simon Van Kerrebroeck, and Luc De Vuyst. 2019. "Omics Approaches to Understand Sourdough Fermentation Processes." *International Journal of Food Microbiology* 302: 90–102. doi:10.1016/j.ijfoodmicro.2018.05.029.

Xi, Jinzhong, Dan Xu, Fengfeng Wu, Zhengyu Jin, and Xueming Xu. 2020. "Effect of Na2CO3 on Quality and Volatile Compounds of Steamed Bread Fermented with Yeast or Sourdough." *Food Chemistry* 324: 126786. doi:10.1016/j.foodchem.2020.126786.

Xi, Jinzhong, Qiyan Zhao, Dan Xu, Yamei Jin, Fengfeng Wu, Zhengyu Jin, and Xueming Xu. 2021. "Volatile Compounds in Chinese Steamed Bread Influenced by Fermentation Time, Yeast Level and Steaming Time." *LWT: Food Science and Technology* 141: 110861. doi:10.1016/j.lwt.2021.110861.

Xing, Qinhui, Susanne Dekker, Konstantina Kyriakopoulou, Remko M. Boom, Eddy J. Smid, and Maarten A.I. Schutyser. 2020. "Enhanced Nutritional Value of Chickpea Protein Concentrate by Dry Separation and Solid State Fermentation." *Innovative Food Science & Emerging Technologies* 59. Elsevier Ltd: 102269. doi:10.1016/j.ifset.2019.102269.

Xu, Dan, Yao Zhang, Kaixing Tang, Ying Hu, Xueming Xu, and Michael G. Gänzle. 2019. "Effect of Mixed Cultures of Yeast and Lactobacilli on the Quality of Wheat Sourdough Bread." *Frontiers in Microbiology* 10: 2113. doi:10.3389/fmicb.2019.02113.

Xu, Dan, Huang Zhang, Jinzhong Xi, Yamei Jin, Yisheng Chen, Lunan Guo, Zhengyu Jin, and Xueming Xu. 2020. "Improving Bread Aroma Using Low-Temperature Sourdough Fermentation." *Food Bioscience* 37. Elsevier Ltd: 100704. doi:10.1016/j.fbio.2020.100704.

Xu, Di, Julia Bechtner, Jürgen Behr, Lara Eisenbach, Andreas J. Geißler, and Rudi F. Vogel. 2019. "Lifestyle of Lactobacillus Hordei Isolated from Water Kefir Based on Genomic, Proteomic and Physiological Characterization." *International Journal of Food Microbiology* 290. Elsevier B.V.: 141–149. doi:10.1016/j.ijfoodmicro.2018.10.004.

Xu, Pei, Shi-Yu Zhang, Zhi-Gang Luo, Min-Hua Zong, Xiao-Xi Li, and Wen-Yong Lou. 2021. "Biotechnology and Bioengineering of Pullulanase: State of the Art and Perspectives." *World Journal of Microbiology and Biotechnology* 37 (3). Springer Science and Business Media B.V.: 43. doi:10.1007/s11274-021-03010-9.

Xu, Yihan, Tao Zhou, Huiqin Tang, Xiqiang Li, Yujing Chen, Limin Zhang, and Jianhua Zhang. 2020. "Probiotic Potential and Amylolytic Properties of Lactic Acid Bacteria Isolated from Chinese Fermented Cereal Foods." *Food Control* 111. Elsevier Ltd: 107057. doi:10.1016/j.foodcont.2019.107057.

Yang, Huan, Muwen He, and Chongde Wu. 2021. "Cross Protection of Lactic Acid Bacteria during Environmental Stresses: Stress Responses and Underlying Mechanisms." *LWT: Food Science and Technology* 144 (June). Academic Press: 111203. doi:10.1016/j.lwt.2021.111203.

Yi, Lanhua, Teng Qi, Yang Hong, Lili Deng, and Kaifang Zeng. 2020. "Screening of Bacteriocin-Producing Lactic Acid Bacteria in Chinese Homemade Pickle and Dry-Cured Meat, and Bacteriocin Identification by Genome Sequencing." *LWT: Food Science and Technology* 125. Academic Press: 109177. doi:10.1016/j.lwt.2020.109177.

Yilmaz-Ersan, Lutfiye, Tulay Ozcan, Arzu Akpinar-Bayizit, and Saliha Sahin. 2018. "Comparison of Antioxidant Capacity of Cow and Ewe Milk Kefirs." *Journal of Dairy Science* 101 (5). Elsevier Inc.: 3788–98. doi:10.3168/jds.2017-13871.

Zamaratskaia, Galia, Karin Gerhardt, and Karin Wendin. 2021. "Biochemical Characteristics and Potential Applications of Ancient Cereals - An Underexploited Opportunity for Sustainable Production and Consumption." *Trends in Food Science & Technology* 107. Elsevier: 114–23. doi:10.1016/j.tifs.2020.12.006.

Zhang, Guohua, Weizhen Zhang, Lijun Sun, Faizan A. Sadiq, Yukun Yang, Jie Gao, and Yaxin Sang. 2019. "Preparation Screening, Production Optimization and Characterization of Exopolysaccharides Produced by Lactobacillus Sanfranciscensis Ls-1001 Isolated from Chinese Traditional Sourdough." *International Journal of Biological Macromolecules* 139 (October). Elsevier B.V.: 1295–1303. doi:10.1016/j.ijbiomac.2019.08.077.

Zhao, Zheng, Taihua Mu, and Hongnan Sun. 2019. "Microbial Characterization of Five Chinese Traditional Sourdoughs by High-Throughput Sequencing and Their Impact on the Quality of Potato Steamed Bread." *Food Chemistry* 274: 710–17. doi:10.1016/j.foodchem.2018.08.143.

Zheng, Yin, Li, Zekun Jin, Peipei An, Shang-Tian Yang, Yongtao Fei, and Gongliang Liu. 2021. "Characterization of Fermented Soymilk by Schleiferilactobacillus Harbinensis M1, Based on the Whole-Genome Sequence and Corresponding Phenotypes." *LWT: Food Science and Technology* 144: 111237. Academic Press. doi:10.1016/j.lwt.2021.111237.

Zotta, Teresa, Paolo Piraino, Annamaria Ricciardi, Paul L.H.H. McSweeney, and Eugenio Parente. 2006. "Proteolysis in Model Sourdough Fermentations." *Journal of Agricultural and Food Chemistry* 54 (7): 2567–74. doi:10.1021/jf052504s.

6 Major Classes of Sourdough Enzymes

Maria Aspri and Dimitrios Tsaltas

CONTENTS

6.1 Introduction 147
6.2 General Role of Enzymes in Bread Making 148
6.3 Endogenous Enzymes Present in Cereal Grains and Flours 149
6.4 Sourdough Microbiota 149
6.5 Carbohydrate Degrading Enzymes 150
6.6 Proteolytic Enzymes 151
6.7 Fiber Degrading Enzymes 154
    6.7.1 Exopolysaccharides 154
        6.7.1.1 Phytate and Phytic Acid Activities 155
    6.7.2 Lipoxygenase 155
    6.7.3 Lipases 156
    6.7.4 Phenolic Compounds 156
6.8 Conclusions 157
Acknowledgments 157
References 158

6.1 INTRODUCTION

Sourdough fermentation is one of the oldest food biotechnological applications, fermenting cereal grains by the action of microorganisms present in or added to the preparation. Therefore, sourdough can be defined as a stable ecosystem composed of lactic acid bacteria (LAB) and yeasts used in the production of bakery goods (Fernández-Peláez et al., 2020). Sourdough fermentation has been studied for its effect on the organoleptic, structural, nutritional, and shelf-life properties of leavened baked goods.

Modifications of cereal grain nutrients by microorganisms contained in sourdough to improve the characteristics of baked goods have attracted great research interest in recent years. Sourdough could be considered an indispensable tool for unleashing the potential of wheat, rye, and whole grain flours. Acidification, proteolysis, and activation of several endogenous cereal enzymes, as well as the synthesis of microbial active metabolites, cause several changes during sourdough fermentation. These changes affect mainly the dough and baked good matrix and also influence the nutritional and functional quality of the final product. Sourdough fermentation has been shown to decrease the glycemic index of bread, improve the properties and bioavailability of the dietary fiber complex, and promote mineral, vitamin, and phytochemical intake.

This chapter discusses the main classes of enzymes present in cereals, sourdough, and final sourdough bread products. In the cereal and bakery industry, various classes of enzymes such as amylases, proteases, and oxidoreductases are being studied to improve the handling properties of dough and quality of baked goods.

6.2 GENERAL ROLE OF ENZYMES IN BREAD MAKING

In bread making, there are three types of enzymes to consider: endogenous enzymes found in flours, enzymes related to yeast and LAB metabolic activity, and exogenous enzymes, which are added to the dough (Figure 6.1). The role of enzymes in the production of bread flavor is crucial. Enzymatic activities vary in intensity depending on the stage of bread making, but they all begin when the flour is hydrated during mixing and proceed in a steady way until the temperature during baking degrades its protein structure.

The metabolism of sourdough microbiota and the action of endogenous cereal enzymes are interdependent (Gänzle, 2014). LAB acidification is a key factor in regulating the activity of cereal enzymes as well as the solubility of substrates, especially gluten proteins and phytate. Sourdough fermentation is generally dominated by obligate heterofermentative LAB (De Vuyst and Neysens, 2005).

Based on the different metabolic requirements, carbohydrate metabolism and heterofermentative lactobacilli utilize a variety of dough ingredients as electron

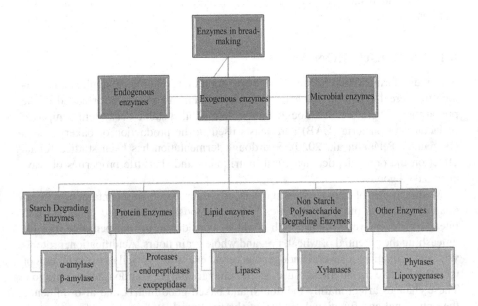

FIGURE 6.1 Role of enzymes in bread making.

acceptors for reduced co-factor regeneration (Gänzle et al., 2007). Thus, heterolactic metabolism affects enzymatic activity by reducing the redox potential of sourdoughs and the accumulation of low-molecular weight thiol compounds. On the other hand, cereal enzymes provide substrates for bacterial growth.

6.3 ENDOGENOUS ENZYMES PRESENT IN CEREAL GRAINS AND FLOURS

Cereal grains contain a significant number of specific enzymes, and the variation in their activity levels affects the quality of cereal raw materials. Differences in cultivars utilized and climatic conditions during growing and harvesting account for this diversity. Amylases, proteases, lipoxygenases, polyphenol oxidases, and peroxidases are among the technologically relevant enzymes found in wheat flour. The distribution of enzymes is not uniform throughout the seed: several enzymes are located in the outer, aleurone, and bran layers of the kernel, as well as in the germ (Oghbaei and Prakash, 2016). This means that when milling, enzymes are unevenly distributed in different fractions, the endosperm fractions generally having the lowest enzyme activity. On the other hand, those fractions that are rich in bran and dietary fiber may also be rich in endogenous enzymes. For example, alpha-amylase is found mainly in the pericarp of wheat grains, with small quantities present in the aleurone layer and seed coat (Rani et al., 2001). Protease is concentrated in the endosperm, germ, and aleurone layers. Lipoxygenase is abundant in the scutellum and embryo. Polyphenol oxidase and peroxidase are predominant in the bran layer. These enzymes are inactive during storage of grains and flour, but they become active when water is added and play an important role in determining the functional properties of flours (Rani et al., 2001).

6.4 SOURDOUGH MICROBIOTA

Sourdough microbiota is composed of yeasts and LAB (De Vuyst et al., 2014). Sourdough microorganisms usually originate from flours and other ingredients, the environment, or may be exogenously added as starters in commercially available products or in artisanally produced in-house preparations. Metabolism of sourdough microbiota in combination with the activity of cereal enzymes determines the quality of the products (Gänzle, 2014). Therefore, in order to achieve consistent product quality, it is necessary to control the composition and activity of the fermented microbiota. Sourdoughs are dominated by LAB occurring at concentrations above 10^8 CFU/g, which may be in coexistence or possibly in symbiosis with yeasts whose numbers are usually one or two logarithmic magnitudes lower than LAB (Chavan and Chavan, 2011).

Sourdough LAB generally belong to the genera *Lactobacillus*, *Pediococcus*, *Leuconostoc*, and *Weissella* (Gänzle et al., 2007). The majority of the species (>23 species) belong to the genus *Lactobacillus* which has been recently reclassified (Zheng et al., 2020). The species *Fructilactobacillus sanfranciscensis* (previously *L. sanfranciscensis*), *Lactiplantibacillus plantarum* (previously *Lactobacillus plantarum*),

Limosilactobacillus pontis (previously *Lactobacillus pontis*), and *Lactobacillus rossiae* are recognized as key organisms in sourdoughs (Yazar and Tavman, 2012). Sourdough yeast belong to more than 20 species (Chavan and Chavan, 2011; Zheng et al., 2020). The most prevalent yeast species are *Saccharomyces cerevisiae*, *Saccharomyces exiguus*, *Candida milleri*, *Pichia norvegensis*, *Hansenula anomala*, and *Candida krusei* (Yazar and Tavman, 2012).

6.5 CARBOHYDRATE DEGRADING ENZYMES

The hydrolysis of starch is the primary source of fermentable carbohydrates. Fermentation activates endogenous flour enzymes, such as amylases, which degrade starch into maltodextrin, maltose, and glucose, depending on the type of flour (Gänzle, 2014). Maltose followed by sucrose, glucose, and fructose, along with some trisaccharides, such as maltotriose and raffinose, are the main carbohydrates available for fermentation (Chavan and Chavan, 2011). The pH of the cereal matrix is eventually shifted to levels too low for most endogenous cereal enzymes during sourdough fermentation, which results in enzyme activity changes (De Vuyst et al., 2017).

Sourdough yeasts use the Embden-Meyerhof-Parnas (EMP) pathway to convert flour carbohydrates (maltose, sucrose, glucose, and fructose) into pyruvate, generating both adenosine triphosphate (ATP) and reducing power (NADH + H$^+$), and then convert pyruvate to ethanol and CO_2, regenerating the co-factor NAD$^+$ used in the upper part of the EMP pathway (De Vuyst et al., 2017; De Vuyst et al., 2021) (Figure 6.2). Redox balance (NAD$^+$ regeneration) is further aided by the production of glycerol from dihydroxy acetone phosphate produced by the EMP route, as well as the formation of succinic acid via the reductive tricarboxylic acid (TCA) cycle (De Vuyst et al., 2021). Production of CO_2 is important during fermentation, as it causes dough to rise during proofing and its expansion during baking. Ethanol also affects the properties of the dough (strengthens the gluten network), dissolving fat-soluble compounds (lipids) and dispersing them in the matrix, while most of it evaporates during the baking process. Both glycerol (produced as osmoprotectant) and succinate (causing a pH decrease) affect the rheology of the dough by improving gas retention and gluten formation, respectively (De Vuyst et al., 2017).

The prevalence of obligate heterofermentative lactobacilli in sourdough is due to their adapted carbohydrate metabolism (Figure 6.3). Maltose phosphorylase activity and the phosphogluconate pathway are the main reasons for the dominance of *Limosilactobacillus fermentum*, *Lim. Reuteri*, and *Fructilactobacillus sanfranciscensis*, since sucrose fermentation ability by sucrose phosphorylase in *Lim. reuteri* and a glucan- and fructansucrase or levansucrase activities in *Lim. reuteri* and *Fru. Sanfranciscensis*, are significant factors affecting microbial growth in these environments (De Vuyst et al., 2017). Also, the use of fructose and other substrates (e.g., oxygen, lipid oxidation-derived aldehydes, such as hexanal, 2-nonenal, and 2,4-decadienal; oxidized glutathione of the gluten network) act as alternative external electron acceptors and may achieve co-factor regeneration and activation of the acetate kinase pathway. The fermentation of hexoses and pentoses, by *Companilactobacillus*

Major Classes of Sourdough Enzymes

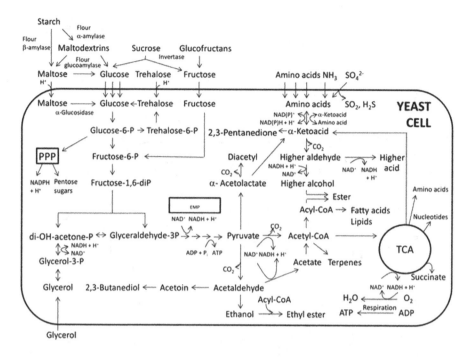

FIGURE 6.2 Overview of the metabolism of yeasts during sourdough fermentation. Only the uptake of appropriate substrates and the most relevant pathways are represented. EMP, Embden-Meyerhof-Parnas pathway; PPP, pentose phosphate pathway; TCA, tricarboxylic acid cycle (De Vuyst et al., 2017).

paralimentarius and *Lactiplantibacillus plantarum*, is another example of heterofermentative lactobacilli fermenting a wide range of carbohydrates (Gänzle and Gobbetti, 2013; Bintsis, 2018).

These strictly homofermentative lactobacilli usually transport carbohydrates by facilitated diffusion, and maltose and sucrose metabolism is not exposed to carbon catabolite repression (Ganzle, 2015). Homofermentative lactobacilli transport most carbohydrates via the phosphoenolpyruvate-dependent phosphotransferase system; and maltose and sucrose metabolism is generally affected by carbon catabolite repression (Gänzle and Follador, 2012). In citrate-positive LAB species such as *Lc. lactis*, conversion of citrate to pyruvate results in the synthesis of acetoin/diacetyl, acetate, and/or lactate, hence contributing to redox balance and/or ATP production.

6.6 PROTEOLYTIC ENZYMES

Proteinases and peptidases are two types of proteolytic enzymes (proteases). Proteinases catalyze protein degradation into smaller peptide fractions; peptidases hydrolyze specific peptide bonds or completely break down peptides into amino acids. The degradation of proteins during sourdough fermentation is dependent on

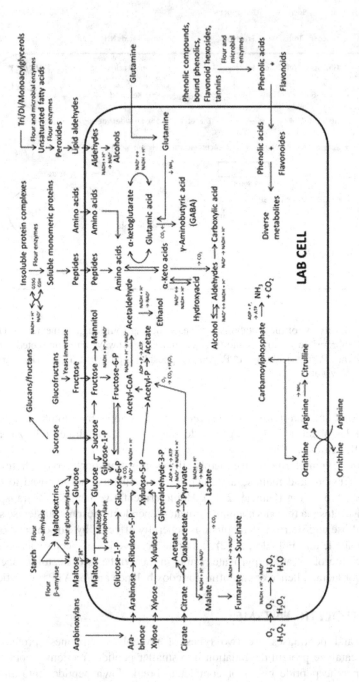

FIGURE 6.3 Overview of the metabolism of LAB during sourdough fermentation. Only the uptake of appropriate substrates and the most relevant pathways are represented (De Vuyst et al., 2017).

Major Classes of Sourdough Enzymes 153

bacterial metabolic activity and cereal enzymes (Gänzle et al., 2008) (Figure 6.4). This markedly affects the overall quality of baked goods (flavor, volume, and texture). Gluten proteins contribute to dough hydration and gas retention, while amino acid conversions contribute to flavor formation (Gänzle, 2014).

The solubility of gluten proteins increases with the acidification and accumulation of low-molecular weight thiols, making them more susceptible to enzymatic degradation (Gänzle, 2014). The primary proteolytic activity of endogenous cereal proteases is promoted by acidification and the reduction of disulfide bonds of gluten by heterofermentative lactobacilli, resulting in the release of polypeptides of various sizes (Di Cagno et al., 2014).

Sourdough LAB's intracellular peptidases complete proteolysis and liberate free amino acids, which are then subjected to diverse catabolic processes by the same microorganisms (Gobbetti et al., 2014). LAB, in general, have multiple amino acid auxotrophies and rely on their proteolytic system to meet their dietary amino acid requirements (Gänzle et al., 2008). The proteolytic system of LAB consists of the

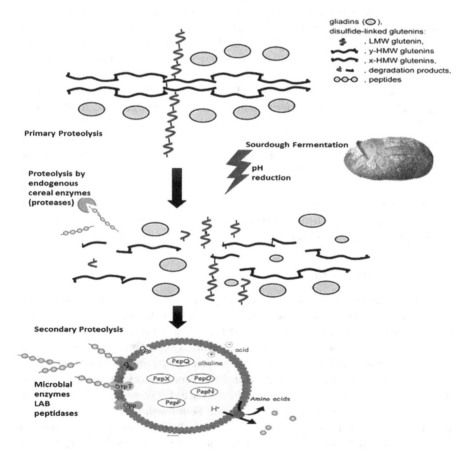

FIGURE 6.4 Overview of proteolysis during sourdough fermentation.

cell-envelope-associated serine proteinase (CEP), amino acid and peptide transport systems, and a range of intracellular peptidases (Gänzle et al., 2008).

At first, proteinase degrade proteins to smaller peptides which are considered as the main source of organic nitrogen for the cells. Proteinase activity, although not widely expressed by most lactobacilli, supports the growth of the non-proteolytic ones. Endopeptidases, aminopeptidases, a tripeptidases, and a dipeptidase have been reported in LAB (Christensen et al., 1999). Specific proline hydrolyzing peptidases are also needed (Savijoki et al., 2006). When the substrates are rich in peptides, the endogenous proteinase activity is downregulated.

6.7 FIBER DEGRADING ENZYMES

Dietary fiber is defined as non-digestible carbohydrates of ≥ 3 monomeric units that are naturally present in foods, also including isolated or synthetic fibers with demonstrated physiological benefits, such as resistant starches (Zhang et al., 2018). Dietary fiber consumption has been shown to reduce the risk of cardiovascular diseases, diabetes, hypertension, obesity, and gastrointestinal problems (Gobbetti et al., 2014). Dietary fiber can be classified according to its solubility in water as: soluble dietary fiber, including inulin, β-glucans, and other non-starch polysaccharides; and insoluble dietary fiber, including lignin, cellulose, and some hemicelluloses (Zhang et al., 2018).

In the sourdough process, the fibers change their chemical and physical properties depending on the degree of fermentation (Fernández-Peláez et al., 2020). During the baking process, enzymes can alter the ratio of insoluble to soluble fibers. In sourdough breads, dietary fiber can undergo two types of enzymatic hydrolysis. Hydration of flour activates several hydrolytic enzymes intrinsic to cereal grains, such as hemicellulase, that degrade hemicelluloses (Hansen et al., 2002). On the other hand, LAB produce enzymes with glycolytic activity, which are also able to act on the fiber present in the dough. Fermentation also decreases the molecular weight of β-glucans, which can adversely affect the physiological activities of barley β-glucan, particularly in relation to glucose and lipid metabolism.

6.7.1 EXOPOLYSACCHARIDES

Exopolysaccharides (EPS) are extracellularly released microbial polysaccharides. The amount and structure of these polysaccharides are determined by the particular microorganisms and available carbon substrate (Chavan and Chavan, 2011). EPS from LAB have the potential to affect host health. Consumption of such polymers can modulate the levels of beneficial bacteria such as *bifidobacteria* and *lactobacilli* in the gastrointestinal tract and stimulate immune functions (Lynch et al., 2018). This shows potential beyond physicochemical functions, as EPS could be components that add functionality and benefit consumer health. Sourdough LAB synthesize a wide structural variety of EPS through the activity of glycosyltransferases (Gobbetti et al., 2014). The viscoelastic properties of the dough are influenced by EPS generated during sourdough fermentation, which has a positive impact on bread rheology, texture, and shelf life (particularly starch retrogradation). EPS are found in two forms – cell bound and released. The former is found as a tightly bound capsule of microorganisms, while the latter is a loosely adherent slime layer that microbial cells secrete into the surrounding environment. Based on their biosynthesis and composition, they are classified into homopolysaccharides (HoPS) and heteropolysaccharides (HePS) (Valerio et al., 2020). LAB can produce both types of polysaccharides, and generally it can be said that HoPS are produced by the genera of *Weissella, Leuconostoc, Streptococcus,* and *Lactobacillus* (Lynch et al., 2018), while HePS are generally produced in lower amounts by mesophilic and thermophilic LAB (*Lacticaseibacillus casei, Lactiplantibacillus plantarum, Lacticaseibacillus rhamnosus, Lactobacillus helveticus,* etc.) (Valerio et al., 2020). HoPS consist of one monosaccharide (mostly fructose or glucose) (Galle and Arendt, 2014). In contrast to HoPS, HePS are composed of irregular repeating units (e.g. galactose and rhamnose) that are synthesized from sugar nucleotides by the activity of intracellular glycosyltransferases (Galle and Arendt, 2014).

6.7.1.1 Phytate and Phytic Acid Activities

Whole grains and cereal products are good sources of minerals, mainly calcium, potassium, magnesium, iron, zinc, and phosphorus (Gobbetti et al., 2014). Phytate (phytic acid) can act as an anti-nutritional factor by reducing the bioavailability and absorption of minerals such as calcium, magnesium, and iron (De Vuyst et al., 2017). During sourdough fermentation, endogenous cereal enzymes are able to degrade the phytic acid. While yeasts may contribute to phytate hydrolysis, the role of LAB in phytate degradation is probably limited to acidification-stimulating endogenous flour phytase activity (De Vuyst et al., 2017). Acidic conditions (pH 3.5–5) produced in sourdough fermentation stimulate the enzymatic hydrolysis of phytate (Gänzle, 2014). Complexes of phytates with divalent cations are insoluble above pH 5.0 and, therefore, are resistant to enzymatic hydrolysis. Below pH 3.5, wheat and rye phytases are inhibited (Leenhardt et al., 2005).

6.7.2 Lipoxygenase

Wheat flour contains very little lipoxygenase activity which is found in germ and bran, but it is abundant in soy, faba beans, and peas (Martínez-Anaya, 1996). Lipoxygenases have structural specificity for their substrate and catalyze the oxygenation of unsaturated fatty acids containing *cis, cis*-1,4-pentadiene groups (linoleic, linolenic, and araquidonic acids) in the presence of molecular oxygen (Martínez-Anaya, 1996). The oxidative reaction leads to the production of free radicals, which induce the formation of peroxides and hydroperoxides that destroy carotenoid pigments, oxidize sulfhydryl groups of proteins to disulfide groups, and decompose them into carbonyl compounds. Lipoxygenases have many functions in the bakery industry, including: (1) bleaching of flour and dough, (2) strengthening the structure of gluten to increase its mixing tolerance, which ultimately improves the loaf volume and texture, (3) production of carbonyl compounds that influence bread flavor, and (4) destruction of liposoluble vitamins (provitamin A) and essential fatty acids.

6.7.3 LIPASES

Lipids are present in small amounts in sourdough due to the low lipid content of the flours compared to other constituents like starch or protein, but they have a significant impact on bread quality. Wheat grain contains 3–4% lipids, depending on genetics and growth conditions, while wheat flour contains 1–2.5% lipids, depending on the raw materials and milling conditions (Schaffarczyk et al., 2014). Lipids are subdivided into starch lipids (40%), which, as their name implies, are located within the starch granule, and non-starch lipids (60%) (Gerits et al., 2014). Non-starch lipids include both free and bound lipids, and they are located outside the starch granule. They have significant structural roles in the gluten backbone and, hence, have an impact on rheological characteristics. On the other hand, starch lipids do not have a significant impact on dough features, so they don't have an impact on the quality of the final bread (Rocha et al., 2012).

Whole wheat flour is highly susceptible to enzymatically formed fatty acid (FA) rancidity due to endogenous lipase activity. High storage temperatures and long storage times promotes enzymatic rancidity. Enzymatic rancidity can affect the final product due to loss of flour functionality, nutritional quality, and organoleptic acceptability. Lipase activity in white flour is usually low enough to avoid endogenous lipid hydrolysis; thus, whole wheat grain is more susceptible to such deteriorations as most lipase activity is located in the bran and germ fractions (Gerits et al., 2014). The low pH achieved by sourdough fermentation reduces fat rancidity by lowering the lipase activity. This reduction in lipid rancidity in the sourdough positively improves the aroma and flavor of the final product by minimizing unpleasant aromas (Pétel et al., 2017). The action of the enzymes present in the dough modifies the lipid profile of the flours by partial hydrolysis of the triglycerides and diglycerides (Fernández-Peláez et al., 2020). This increases the percentage of monoglycerides and maintains a stable sterol ester content, but the effect on the nutritional characteristics of the bread is limited due to the low level of lipids in the product.

6.7.4 PHENOLIC COMPOUNDS

Phenolic compounds, such as phenolic acids and alkylresorcinols, are considered anti-nutritional factors that impart bitter taste and inhibit the digestion of starch and proteins (Gänzle, 2014). On the other hand, phenolic compounds have beneficial health effects as antioxidants and are precursor compounds for flavor formation in bread making. LAB harbor a diverse set of enzymes for conversion of phenolic compounds.

Phenolic acid metabolism by lactobacilli is mediated by esterases or tannases, phenolic acid reductases, and phenolic acid decarboxylases (Pswarayi et al., 2022) (Figure 6.5). Bioconversion of phenolic acids is strain-specific.

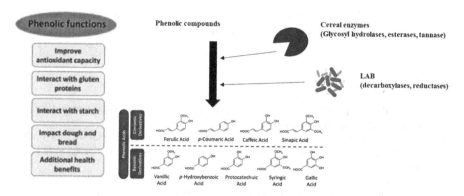

FIGURE 6.5 Conversion of phenolic compounds during sourdough fermentation by cereal enzymes and LAB.

6.8 CONCLUSIONS

In conclusion, the effect of sourdough fermentation on bread quality is dependent on enzymatic and microbial conversions at the dough stage, and the activity of cereal enzymes is an important determinant of sourdough microbial ecology. This chapter highlights the main classes of enzymes available in cereals, sourdough, and, in general, in the baking process. An interdependent relationship between microbiota metabolism and cereal enzymes is pivotal for sourdough development. A key element to consider is the fact that activities are strain-dependent, and as a result, the various cereal species select active consortia that are well adapted to the substrate. Research in this field and identification of such strains is expected to bring forward innovation in sourdough with improved starter cultures. Very recent studies using proteomics and peptidomics are demonstrating promising tools that are able to evaluate the degree of enzymatic proteolysis and characterize the product of hydrolysis. Through the identification of intermediate- and small-sized peptides generated during LAB sourdough fermentation, these tools provide very accurate data of the final products. Such technologies are very much in need since digestibility problems are a major health concern of our day. Also, in recent years, application of sourdough into other cereal-based food products such as noodles, desserts, and snacks seems to be the next phase of exploration in order for sourdough benefits to impact a wider population and gain stronger market share.

ACKNOWLEDGMENTS

This work is based upon the work from COST Action 18101 SOURDOMICS – Sourdough biotechnology network towards novel, healthier and sustainable food and bioprocesses (https://sourdomics.com/; https://www.cost.eu/actions/CA18101/#tabs|Name:overview), where the authors Maria Aspri and Dimitris Tsaltas were chosen. SOURDOMICS is supported by COST (European Cooperation in Science and Technology). COST is a funding agency for research and innovation networks.

COST Actions help connect research initiatives across Europe and enable scientists to grow their ideas by sharing them with their peers – thus boosting their research, career, and innovation.

REFERENCES

Bintsis, T. (2018). Lactic acid bacteria as starter cultures: An update in their metabolism and genetics. *AIMS Microbiology* 4(4), 665.

Chavan, R.S., and Chavan, S.R. (2011). Sourdough technology: A traditional way for wholesome foods: A review. *Comprehensive Reviews in Food Science and Food Safety* 10(3), 169–182.

Christensen, J.E., Dudley, E.G., Pederson, J.A., and Steele, J.L. (1999). Peptidases and amino acid catabolism in lactic acid bacteria. *Antonie van Leeuwenhoek* 76(1), 217–246.

De Vuyst, L., and Neysens, P. (2005). The sourdough microflora: Biodiversity and metabolic interactions. *Trends in Food Science & Technology* 16(1–3), 43–56.

De Vuyst, L., Van Kerrebroeck, S., Harth, H., Huys, G., Daniel, H.-M., and Weckx, S. (2014). Microbial ecology of sourdough fermentations: Diverse or uniform? *Food Microbiology* 37, 11–29.

De Vuyst, L., Van Kerrebroeck, S., and Leroy, F. (2017). Microbial ecology and process technology of sourdough fermentation. *Advances in Applied Microbiology* 100, 49–160.

De Vuyst, L., Comasio, A., and Kerrebroeck, S.V. (2021). Sourdough production: Fermentation strategies, microbial ecology, and use of non-flour ingredients. *Critical Reviews in Food Science and Nutrition*, 1–33.

Di Cagno, R., Rizzello, C.G., and Gobbetti, M. (2014). Adverse reactions to gluten: Exploitation of sourdough fermentation. In *Wheat and Rice in Disease Prevention and Health*. Elsevier, 171–177.

Fernández-Peláez, J., Paesani, C., and Gómez, M. (2020). Sourdough technology as a tool for the development of healthier grain-based products: An update. *Agronomy* 10(12), 1962.

Galle, S., and Arendt, E.K. (2014). Exopolysaccharides from sourdough lactic acid bacteria. *Critical Reviews in Food Science and Nutrition* 54(7), 891–901.

Gänzle, M., and Follador, R. (2012). Metabolism of oligosaccharides and starch in lactobacilli: A review. *Frontiers in Microbiology* 3, 340.

Gänzle, M., and Gobbetti, M. (2013). Physiology and biochemistry of lactic acid bacteria. In *Handbook on Sourdough Biotechnology*. Springer, 183–216.

Gänzle, M.G. (2014). Enzymatic and bacterial conversions during sourdough fermentation. *Food Microbiology* 37, 2–10.

Ganzle, M.G. (2015). Lactic metabolism revisited: Metabolism of lactic acid bacteria in food fermentations and food spoilage. *Current Opinion in Food Science* 2, 106–117.

Gänzle, M.G., Vermeulen, N., and Vogel, R.F. (2007). Carbohydrate, peptide and lipid metabolism of lactic acid bacteria in sourdough. *Food Microbiology* 24(2), 128–138.

Gänzle, M.G., Loponen, J., and Gobbetti, M. (2008). Proteolysis in sourdough fermentations: Mechanisms and potential for improved bread quality. *Trends in Food Science & Technology* 19(10), 513–521.

Gerits, L.R., Pareyt, B., Decamps, K., and Delcour, J.A. (2014). Lipases and their functionality in the production of wheat-based food systems. *Comprehensive Reviews in Food Science and Food Safety* 13(5), 978–989.

Gobbetti, M., Rizzello, C.G., Di Cagno, R., and De Angelis, M. (2014). How the sourdough may affect the functional features of leavened baked goods. *Food Microbiology* 37, 30–40.

Hansen, H.B., Andreasen, M., Nielsen, M., Larsen, L., Knudsen, B.K., Meyer, A., et al. (2002). Changes in dietary fibre, phenolic acids and activity of endogenous enzymes during rye bread-making. *European Food Research and Technology* 214(1), 33–42.

Leenhardt, F., Levrat-Verny, M.-A., Chanliaud, E., and Rémésy, C. (2005). Moderate decrease of pH by sourdough fermentation is sufficient to reduce phytate content of whole wheat flour through endogenous phytase activity. *Journal of Agricultural and Food Chemistry* 53(1), 98–102.

Lynch, K.M., Coffey, A., and Arendt, E.K. (2018). Exopolysaccharide producing lactic acid bacteria: Their techno-functional role and potential application in gluten-free bread products. *Food Research International* 110, 52–61.

Martínez-Anaya, M.A. (1996). Enzymes and bread flavor. *Journal of Agricultural and Food Chemistry* 44(9), 2469–2480.

Oghbaei, M., and Prakash, J. (2016). Effect of primary processing of cereals and legumes on its nutritional quality: A comprehensive review. *Cogent Food & Agriculture* 2(1), 1136015.

Pétel, C., Onno, B., and Prost, C. (2017). Sourdough volatile compounds and their contribution to bread: A review. *Trends in Food Science & Technology* 59, 105–123.

Pswarayi, F., Qiao, N., Gaur, G., and Gänzle, M. (2022). Antimicrobial plant secondary metabolites, MDR transporters and antimicrobial resistance in cereal-associated lactobacilli: Is there a connection? *Food Microbiology* 102, 103917.

Rani, K., Rao, U.P., Leelavathi, K., and Rao, P.H. (2001). Distribution of enzymes in wheat flour mill streams. *Journal of Cereal Science* 34(3), 233–242.

Rocha, J.M., Kalo, P.J., and Malcata, F.X. (2012). Composition of neutral lipid classes and content of fatty acids throughout sourdough breadmaking. *European Journal of Lipid Science and Technology* 114(3), 294–305. doi: 10.1002/ejlt.201100208.

Savijoki, K., Ingmer, H., and Varmanen, P. (2006). Proteolytic systems of lactic acid bacteria. *Applied Microbiology and Biotechnology* 71(4), 394–406.

Schaffarczyk, M., Østdal, H., and Koehler, P. (2014). Lipases in wheat bread making: Analysis and functional effects of lipid reaction products. *Journal of Agricultural and Food Chemistry* 62(32), 8229–8237.

Valerio, F., Bavaro, A.R., Di Biase, M., Lonigro, S.L., Logrieco, A.F., and Lavermicocca, P. (2020). Effect of amaranth and quinoa flours on exopolysaccharide production and protein profile of liquid sourdough fermented by Weissella cibaria and Lactobacillus plantarum. *Frontiers in microbiology* 11, 967.

Yazar, G., and Tavman, Ş. (2012). Functional and technological aspects of sourdough fermentation with Lactobacillus sanfranciscensis. *Food Engineering Reviews* 4, 171–190.

Zhang, H., Wang, H., Cao, X., and Wang, J. (2018). Preparation and modification of high dietary fiber flour: A review. *Food Research International* 113, 24–35.

Zheng, J., Wittouck, S., Salvetti, E., Franz, C.M., Harris, H.M., Mattarelli, P., O'toole, P.W., Pot, B., Vandamme, P., Walter, J. and Watanabe, K., and Lebeer, S. (2020). A taxonomic note on the genus Lactobacillus: Description of 23 novel genera, emended description of the genus Lactobacillus Beijerinck 1901, and union of Lactobacillaceae and Leuconostocaceae. *International journal of systematic and evolutionary microbiology* 70(4), 2782–2858.

7 Discovery, Characterization, and Databases of Enzymes from Sourdough

Zeynep Agirbasli and Sebnem Harsa

CONTENTS

7.1 Introduction 161
7.2 Enzymes in Carbohydrate Metabolism 162
    7.2.1 General Enzymes in Carbohydrate Metabolism............................162
    7.2.2 EPS Producer Enzymes 169
7.3 Enzymes in Protein Metabolism 173
    7.3.1 Proteinase and Peptidases..173
    7.3.2 ACE Inhibitor Activity 175
    7.3.3 Other Bioactive Peptide-Forming Activities 175
7.4 Enzymes in Phenolic Metabolism 176
7.5 Phytase...177
7.6 Enzymes in Stress Metabolism..180
    7.6.1 Glutamate Decarboxylase..180
    7.6.2 Glutaminase and ADI Pathway Enzymes.....................................181
    7.6.3 Glutathione Reductase 183
7.7 Other Enzymatic Formation of Bioactive Compounds in Sourdough..........184
7.8 Conclusion 185
References..185

7.1 INTRODUCTION

Before the discovery of baker's yeast, sourdough in baking applications provided bread structure by acidifying and CO_2 production for leavening bread dough. Today, the diverse bioactivities in sourdough have helped to popularize these traditional breads (Brandt 2007, Gänzle, Vermeulen, and Vogel 2007). This accelerating awareness in bioactive compounds from microorganisms led to the discovery and characterization of contributing enzymes. Most bioactivity-associated sourdough studies focused on exopolysaccharides (EPS) (İspirli et al. 2020), gamma amino butyric acid (GABA) (Villegas et al. 2016), phytic acid content (Yildirim and Arici 2019), and bioactive peptide formation (Rizzello et al. 2015). The associated enzymes forming these compounds constitute a well investigated enzyme portfolio of sourdough. During sourdough fermentation, microbiota and the outcomes from their activity play a crucial role in the formation of bioactive compounds due to the metabolic conversion of components, such as phenolic compounds, proteins, and carbohydrates. Excellent reviews on carbohydrate, lipid, protein, and phytic acid metabolism of the microbial population of sourdough were reviewed by Hammes and Gänzle (1998); Gobbetti et al. (2005); Gänzle, Loponen, and Gobbetti (2008); Gänzle (2014); (Sharma et al. 2020); and Maske et al. (2021). The metabolism of rye phenolics and endogenous proteases of cereal flours are explained in detail by Koistinen et al. (2018) and Guzmán-Ortiz et al. (2019). Since the microcosmos in the sourdough influences the conversion of bioactive metabolites excessively, microbial enzymes play a crucial role in this review. Besides providing general information on microbial enzymatic conversions in sourdough, this chapter emphasizes biochemical and molecular techniques for the identification of several key microbial enzymes in sourdough and their characteristic properties.

7.2 Enzymes in Carbohydrate Metabolism

7.2.1 General Enzymes in Carbohydrate Metabolism

The organoleptic and textural characteristics of bread are affected by organic acids due to the carbohydrate metabolism of lactic acid bacteria (LAB). Availability of oxygen and other electron acceptors, such as fructose, play a key role in the formation of metabolites like mannitol or acetic acid (Stolz, Vogel, and Hammes 1995). *Fructilactobacillus sanfranciscensis* is one of the most studied sourdough species, and its carbohydrate metabolism has been explored by the characterization of enzymes in genetic and/or biochemical level (Table 7.1). Gänzle, Vermeulen, and Vogel (2007) highlighted carbohydrate metabolism pathways by summarizing the heterofermentative hexose metabolism of *Fr. sanfranciscensis* based on the literature of enzymes and their by-products.

In situ monitoring of the expression of gene-coding enzymes is one of the most recent innovative approaches. A novel study focused, for the first time, on gene regulation in sourdough fermented with *Fr. sanfranciscensis* S1. Based on the results of total RNA isolation from sourdough, the transcriptional regulation library was structured using *Fr. sanfranciscensis* TMW 1.1304 as a model. In contrast to genes associated with glutamate metabolism, the gene profile of the carbohydrate degradation system (e.g. phosphoglycerate kinase, phosphoglycerate mutase, phosphopyruvate hydratase phosphoglucomutase, and pyruvate kinase) was promoted after 12 hours of fermentation (Liu et al. 2020).

Fr. sanfranciscensis uses fructose in the presence of oxygen and produces acetate with the help of acetate kinase (AK). It has been shown that acetyl phosphatase in *F. sanfranciscensis* contributes to acetate synthesis. The *ack* gene responsible for the expression of AK of *F. sanfranciscensis* DSM20451 has been expressed in *E. coli* and characterized as stable at 60°C for 30 min and when frozen at −20°C for at

TABLE 7.1
Characterization of Enzymes Associated with Sourdough Microorganisms and Additional Enzymes to Sourdough Matrix

No	Species	Enzyme	MW	Optimum temperature	Optimum pH	Km	Vm	References
1.	B. pseudocatenulatum ATCC 27919 and B. longum subsp. infantis ATCC 15697	Phytase[a]	—	50°C	5.5	—	—	(Tamayo-Ramos et al. 2012)
2.	B. pseudocatenulatum ATCC 27919	Phytase	—	60°C	6.5–7.0	—	—	(Haros, Bielecka, and Sanz 2005)
3.	P. kudriavzevii OG32	Phytase	—	37°C	4.6	—	—	(Ogunremi, Agrawal, and Sanni 2020)
	C. tropicalis BOM21	Phytase	—	20°C	3.6	—	—	(Ogunremi, Agrawal, and Sanni 2020)
4.	Lpb. pentosus CECT 4023	Acid phosphatase	69 kDa	50°C	5.0	2.44 $\mu mol\ l^{-1}$	820 $\mu mol\ h^{-1}$	(Palacios et al. 2005)
5.	Fr. sanfranciscensis CB1	Phytase	50 kDa	45°C	4.0	—	—	(De Angelis et al. 2003)
6.	Fr. sanfranciscensis CB1	ADI OTC CK	46 kDa 39 kDa 37 kDa	30°C 30°C 40°C	5.0 6.5 6.0	— — —	— — —	(De Angelis, Mariotti, et al. 2002)
7.	S. cerevisiae LC3	Phytase	48 kDa	40°C	5.0	—	—	(Caputo, Visconti, and De Angelis 2015)
8.	Lpb. plantarum CRL1964	Phytase, PhyLP	55 kDa	55°C	4.5	—	—	(Sandez Penidez et al. 2020)

(Continued)

TABLE 7.1 (CONTINUED)
Characterization of Enzymes Associated with Sourdough Microorganisms and Additional Enzymes to Sourdough Matrix

No	Species	Enzyme	MW	Optimum temperature	Optimum pH	Km	Vm	References
9.	*Lpb. pentosus* CFR3	Phytate-degrading enzyme	70 kDa	55–60°C	5	—	—	(Amritha, Halami, and Venkateswaran 2017)
10.	*Fr. sanfranciscensis* TMW 1.392	Levansucrase	90 kDa	35–45°C	5.4	13.1 ± 3.4 mM sucrose	206 ± 19 μmol mg^{-1} min^{-1}	(Tieking et al. 2005)
11.	*Lc. citreum* CW28	Inulosucrase[a]	165 kDa for IslA,	35°C	6.5	38 mM sucrose	—	(Del Moral et al. 2008)
12.	*W. confusa* VTT E-90392	Dextransucrase[b]	~160 kDa	35°C	5.4	14.7 mM sucrose	8.2 μmol mg^{-1} min^{-1}	(Kajala et al. 2015)
13.	*F. rossiae* DSM 15814	β-D-xylosidase[b]	—	40°C	6.0	—	—	(Pontonio et al. 2016)
14.	*Lim. fermentum* Mw2 *Lim. fermentum* Ogi E1	Amylase	—	35°C 45°C	4.5 5.0	—	—	(Agati et al. 1998)
15.	*Lim. reuteri* E81	Glucansucrase[a]	115 kDa	50°C	7	7.5 mM sucrose	1.49 IU mg^{-1}	(Ispirli et al. 2019)
16.	*Lat. curvatus* TMW 1.624 *Lim. reuteri* TMW 1.106 *Lb. animalis* TMW 1.971	Glucosyltransferases[a]	183.63 kDa — —	31°C 45°C 42°C and 53°C	4.4 4.4 5.8–6.0	— — —	— — —	(Rühmkorf et al. 2013)
17.	*W. confusa* LBAE C39-2	Dextransucrase[a]	154 kDa	35–40°C	5.4	8.6 mM	20 mg^{-1} min^{-1}	(Amari et al. 2013)

(*Continued*)

TABLE 7.1 (CONTINUED)
Characterization of Enzymes Associated with Sourdough Microorganisms and Additional Enzymes to Sourdough Matrix

No	Species	Enzyme	MW	Optimum temperature	Optimum pH	Km	Vm	References
18.	Lb. panis TMW1.648	Levansucrase[a]	87.802 kDa	45°C	4.0–4.6	29.9 mM for overall activity, 17.0 mM and 22.5 for hydrolysis and transfer reactions,	69.9 μmol mg^{-1} min^{-1} for overall activity, 44.0 μmol mg^{-1} min^{-1} and 21.3 μmol mg^{-1} min^{-1} for hydrolysis and transfer reactions	(Waldherr, Meissner, and Vogel 2008)
19.	Lb. crispatus DSM29598	AbnA	72.11 kDa	40°C	6.0	–	–	(Li et al. 2020)
		AbnB	58.56 kDa	40°C	7.5	–	–	
20.	Lim. reuteri CRL 1098	AP	–	37°C	6.0	–	–	(Rollán and de Valdez 2001)
		DP	–	37°C	8.0	–	–	
		TP	–	45°C	6.5	–	–	
21.	Lim. reuteri CRL 1098	ADI	–	65°C	5.5	–	–	(Rollan, Lorca, and de Valdez 2003)
		CK	–	37°C	5.5	–	–	
		OTC	–	40°C	~5.5–7.0	–	–	
22	Fr. sanfranciscensis CB1	PepX	~49.2 and 56 kDa	30°C	6.0	0.6 mM for Gly-Pro-pNA	99.6 nmol mg^{-1} min^{-1}	(Gallo et al. 2005)
23.	Fr. sanfranciscensis CB1	Cell envelope proteinase	57 kDa	40°C	7.0	–	–	(Gobbetti, Smacchi, and Corsetti 1996)
		Cytoplasmic dipeptidase	65 kDa	30–35°C	7.5	0.70 mM on Leu-Leu	–	
		Cytoplasmic aminopeptidase	75 kDa	30–35°C	7.5	0.44 mM on Leu-pNA	–	

(*Continued*)

TABLE 7.1 (CONTINUED)
Characterization of Enzymes Associated with Sourdough Microorganisms and Additional Enzymes to Sourdough Matrix

No	Species	Enzyme	MW	Optimum temperature	Optimum pH	Km	Vm	References
24.	Lim. reuteri DSM 20016	Cystathionine γ-lyase	~160 kDa	35°C	8.0	No access to this information	No access to this information	(De Angelis, Curtin, et al. 2002)
25.	Lpb. plantarum CRL 759 and CRL 778	Aminopeptidase	–	37°C	7.0 and 6.0	–	–	(Rollan et al. 2005)
		Dipeptidase	–	45°C and 37°C	8.0 and 6.0	–	–	
		Tripeptidase	–	37°C	8.0 for both	–	–	
		Endopeptidase	–	37°C	6.5 and 6.0	–	–	
26.	Fr. sanfranciscensis TMW1.392	MDH	44–53 kDa	35°C	5.8 and 8.0	24 mM fructose and 78 mM mannitol for reduction and oxidation	29 U mg^{-1} and 20 U mg^{-1} for reduction and oxidation	(Korakli and Vogel 2003)
27.	F. sanfranciscensis DSM20451	phosphotransacetylase[a]	35.5 kDa	49–58°C	8.1–9.1	1.3 and 0.1 mM^1 0.6, 6.7 mM^2	194 U mg^{-1} and 38 U mg^{-1}	(Knorr, Ehrmann, and Vogel 2001a)

(*Continued*)

TABLE 7.1 (CONTINUED)
Characterization of Enzymes Associated with Sourdough Microorganisms and Additional Enzymes to Sourdough Matrix

No	Species	Enzyme	MW	Optimum temperature	Optimum pH	Km	Vm	References
28.	Lim. reuteri subsp. rodentium LTH2584 and Lim. reuteri subsp. murium LTH5448	Glc1[a]	51.10 kDa	50°C	9.0	0.11 mM[c]; 1.54 mM[d]	Compared to Gcl1 and Gcl2, Vmax values were higher for Gcl3 with cysteine and glutamate	(Xie and Gänzle 2021)
	Lim. reuteri subsp. rodentium LTH2584 and Lim. reuteri subsp. murium LTH5448	Glc2[a]	59.29 kDa	50°C	9.0	0.10 mM[c]; 1.46 mM[d]		
	Lim. reuteri subsp. rodentium LTH2584	Glc3[a]	49.34 kDa	50°C	9.0	0.18 mM[c]; 1.56 mM[d]		
29.	Lim. fermentum TMW 1.890	Cinnamoyl esterases[a]	27–29 kDa	–	7.0–8.0	8.54 ± 1.27 mmol/L	76.98 ± 8.58 mmol mg^{-1} min^{-1}	(Fritsch et al. 2017)
	Lpb. plantarum TMW 1.460		27–29 kDa	20–30°C	7.0–8.0	5.25 ± 1.29 mmol/L	91.21 ± 14.71 mmol mg^{-1} min^{-1}	
30.	Streptomyces werraensis	FA esterase type D	48 kDa	40°C	7.0–8.0	–	–	(Schulz et al. 2018)

[a]Enzyme expressed in E. coli; [b]Enzyme expressed in L. lactis; [c]for cysteine; [d]for glutamate
MW: Molecular weight, Km: Michaelis–Menten type kinetic constants, Vm: The maximum velocity

least 2 months; also, it was inactivated at higher temperatures (Knorr, Ehrmann, and Vogel 2001b). The first report on the characterization of phosphotransacetylase of heterofermentative LAB was carried out after the expression of *pta* gene of *F. sanfranciscensis* DSM20451 in *E. coli*. The authors demonstrated that phosphotransacetylase was not affected by $MgCl_2$ (10 mM) or KCl (100 mM), stable at basic pH (7–10) and lost its activity after 15 min at 60°C (Knorr, Ehrmann, and Vogel 2001a). Furthermore, another study researched the *pta* gene using polymerase chain reaction (PCR) in *Fr. sanfranciscensis* TMW 1.392 (Tieking et al. 2005).

Metabolism of maltose in *Fr. sanfranciscensis* DSM20451 was molecularly characterized as the maltose phosphorylase and phosphoglucomutase producers, *mapA* and *pgmA* genes; the production of the *mapA* gene was found 87.5 kDa by sodium dodecyl sulfate polyacrylamide gel electrophoresis (SDS-PAGE) analysis for the first time for this species. Activity of maltose phosphorylase of *Fr. sanfranciscensis* DSM20451T expressed in *E.coli* was analyzed by measuring the glucose released from maltose by the spectrophotometer and high-performance liquid chromatography (HPLC)-refractive index system (Ehrmann and Vogel 1998).

Wheat and rye are great sources for non-starch polysaccharides with 1.5–3% and 7–8% arabinoxylans (AX) content, respectively (Gänzle 2014). Arabinoxylans, degraded by several enzyme activities, are necessary such as arabinosidases and endo-β-xylanases, glucuronidases, and α-L-arabinofuranosidases. Formation of pentosans from the action of additional enzymes, such as α-L-arabinofuranosidase or endo-xylanase in bread production, improved bread qualities (Gobbetti et al. 2000). Activities of endo-xylane, β-xylosidase, and arabinofuranosidase were found optimum in ungerminated rye at pH 4.5 and 40°C, 70°C and 60°C, respectively (Rasmussen et al. 2001). Recently, a novel study identified and characterized the multiple putative extracellular endo-arabinofuranosidases or arabinanases encoding operon in Lactobacillaceae for the first time. The operon of sourdough isolate *Lactobacillus crispatus* included genes that encode extracellular arabinan utilization enzymes AbnA (72.11 kDa) and AbnB (58.56 kDa) as well as enzymes for metabolism of arabinose. It was also shown that Mn^{2+}, Mg^{2+}, Fe^{2+}, and Fe^{3+} increased activity of AbnA, whereas EDTA lowered its activity. The additives did not alter the activity of AbnB (Li and Gänzle 2020).

The addition of enzymes to improve dough characteristics is a general strategy. Low MW Xylanase from *Aspergillus niger* IBT-90 were characterized with optimum activity at pH 5.5 and temperature 50°C, and it was inhibited by Ag^+, Fe^{3+}, and NBS and promoted by DTT and Na^+ (Romanowska, Polak, and Bielecki 2006). The addition of this enzyme to bread formulation enhanced the size, crumb structure, and storage time of wheat–rye and whole meal bread (Romanowska, Polak, and Bielecki 2006).

Another study researched the impact of substrate on the xly cluster (*xylE*, *xylA*, *xynT*, *xylT*, *xylI*, *xylK*, and *xylR* genes) and the *ara* cluster of *Furfurilactobacillus rossiae* cDSM 15814 with RNAseq analysis and determined the organization of these genes. Transcriptome analyses showed that XOS or D-xylose significantly increased expression of genes up to approximately 250-fold, and the *ara* gene cluster was significantly activated with strain growth of L-arabinose. Furthermore, the β-D-xylosidase enzyme was expressed in *Lc. lactis subsp. cremoris* NZ9000 by cloning *xylA* gene. The utilization of XOS as an enzyme substrate was explored by high-performance anion exchange chromatography with pulsed amperometric detection (HPAEC-PAD) system. The optimum pH and temperature of β-D-xylosidase was assessed in spectrophotometer by using p-nitrophenyl-β-D-xylopyranoside (pNPX) (Pontonio et al. 2016).

Heterofermentative lactobacilli metabolizes maltose or glucose to CO_2, lactate, and ethanol and produces acetate instead of ethanol from fructose, citrate, fumarate, or malate. Conversion of mannitol from fructose could occur with the aid of a phosphatase in homofermentative LAB or mannitol dehydrogenase (MDH) in heterofermentative LAB by oxidation of NADH to NAD^+ (nicotinamide adenine dinucleotide) and promotes the competitiveness of LAB (Korakli and Vogel 2003, Vrancken et al. 2008). The action of enzymes is indicated by quantification of mannitol production by (HPAEC-PAD)) (Vrancken et al. 2008). An MDH in *Fr. sanfranciscensis* TMW1.392 (isogenic with LTH2590) was characterized with optimal pH 5.8 for reduction of fructose and pH 8.0 for oxidation of mannitol by monitoring absorbance of NAD^+ production at 340 nm using spectrophotometer. MDH exhibited limited substrate specificity for oxidation against sorbitol, arabitol, and xylitol and it showed no reduction activity against glucose, arabinose, xylose, or mannose (Korakli and Vogel 2003). The starter culture potential of *Limosilactobacillus reuteri* CRL 1100 for fermentation of wheat dough has been proven previously (Gerez et al. 2008). High MDH activity of *Lim. reuteri* CRL 1101 was characterized in the presence of fructose in the log phase, and this high activity was associated with the expression of *mdh* gene by proteomic and genomic methods. Furthermore, two proteins of arginine deiminase (ADI) pathways pathway were also detected (Ortiz et al. 2017). Fermentation with *Lc. citreum* TR116 has been applied as a novel microbial approach for reducing sugar in sourdough burger buns, based on its role in the conversion of fructose to mannitol (Sahin et al. 2019). The next study of the authors explored genes that contribute to carbohydrate, organic acid, and amino acid metabolism in *Lc. citreum* TR116 isolated from yellow pea sourdough. Real time-polymerase chain reaction (RT-qPCR) analyses showed that fructose increased expression of *mdh* gene and transporter gene *manX* by 7.3 ± 1.2 and 3.4 ± 0.6 fold in this isolate, respectively (Sahin, Rice, and Coffey 2021).Moreover, the growth of yeast on the agar medium containing arbutin as the only carbon source was also preferred for screening β-glucosidase synthesis (Perricone et al. 2014). *Lactobacillus fermentum* Ogi E1 isolated from maize sourdough did not exhibit glucoamylase and maltose-phosphorylase activities and produced minimal α-Glucosidase activity except for the cell extract (Santoyo et al. 2003).

7.2.2 EPS Producer Enzymes

Glycosyltransferases (Gtfs) or fructosyltransferases (Ftfs) produce glucans and fructans as a result of their action on sucrose. In addition, fructosyltransferases also act on raffinose. The glucan-producing Gtf enzyme class consists of dextransucrase, alternansucrase, mutan, and reuteransucrase, whereas, Ftfs produce fructans based on activities of inulosucrase and levansucrase (van Hijum et al. 2006). Reuteran, dextran (Chen, Levy, and Gänzle 2016), inulin (Van Hijum, Van Der Maarel, and Dijkhuizen 2003), and levan (Ua-Arak, Jakob, and Vogel 2017) have been detected as metabolites of sourdough- and cereal-associated LAB due to the action of carbohydrate metabolism.

Similar to other enzyme screening techniques, agar plate assay has been widely used in detection of EPS producer microorganisms (Van der Meulen et al. 2007, Perri et al. 2021, Galli et al. 2020, Di Cagno et al. 2006, Coda et al. 2018, Milanović et al. 2020, Valerio et al. 2020). Additionally, EPS production analyses can be monitored by chromatographic and spectrophotometric techniques, such as HPLC analysis, phenol-sulfuric method, and enzymatic kit assays (Di Cagno et al. 2006, Galli et al. 2020).

A study revealed that sucrose metabolism of *Fr. sanfranciscensis* TMW 1.392 and formation of fructan and 1-kestose are dependent on the activity of a single enzyme, levansucrase. Molecular characterization of the levansucrase gene showed that it encoded levansucrase (879 aa protein) with an isoelectric point (pI) value of 4.69; it was highly similar to *Lim. reuteri* originated levansucrases in BLAST analysis. Overall, hydrolyzation and transferase activities of enzymes due to released glucose, fructose levels, and difference between them, hydrolase activity of enzymes was predominant at sucrose concentrations lower than 200 mM, and transferase activity increased when sucrose concentration increased (Tieking et al. 2005).

Similarly, EPS production by *Fr. sanfranciscensis* LTH2590 (TMW1.392) in fermented sourdough was determined by HPLC analysis. *In situ* EPS production using carbon isotope labeling was also used to verify formation of EPS and associated substrate preference (Korakli et al. 2001). Likewise, *Lim. reuteri* LTH5448 and TMW1.106 were investigated for the impact of sucrose-triggering glycosyltransferase and fructosyltransferase genes. Based on RT-PCR analysis, sucrose presence was ineffective on the expression of *inu* and *gtfA*, while it promoted expression of *ftfA*. Through HPLC analysis, the authors also confirmed activity of enzymes by detecting oligosaccharides (Schwab et al. 2007).

Although molecular screening is applied in numerous studies, it can be insufficient to detect all genes associated with related enzymes. For example, a study used gel permeation chromatography (GPC) to detect EPS production to confirm screening results of PCR analysis targeting glucansucrase, fructansucrase, *epsA*, and *epsB* genes. EPS production was detected in 10 out of 174 dairy and sourdough strains, and seven of them gave positive PCR results (Van der Meulen et al. 2007). Similarly, a study explored 111 strains from *Lactobacillus* and *Weissella* by two different GPC steps. Next, confirmation with PCR analysis using primers targeting conserved regions in genes demonstrated that the levansucrase gene was not detected in glucan producers or non-EPS producers, and only six of 15 fructan producers had a levansucrase gene (Tieking et al. 2003). More recently, spot inoculation in agar plate assay gave positive results in 28 strains, whereas six of them did not exhibit positive results in PCR analyses (Milanović et al. 2020). Likewise, a previous study detected that only 23 or 13 of 177 various sourdough LAB strains produced EPS in Chalmers agar with only sucrose or with a mixture of sugars, respectively; PRC analysis with primers for homopolysaccharide (*gtf* and *lev*) and heteropolysaccharide (*epsA*, *epsB*, *epsD*, *epsE*, and *epsEFG*) genes revealed that 13 of 14 strains harbored at least one EPS-producing gene, but no *epsEFG* primer targeting glycosyltransferase existed in any strain (Dubois et al. 1956, Palomba et al. 2012).

The first complete dextransucrase gene for *W. confusa* species were identified in *W. confusa* C39-2. It was also very similar to dextransucrase of soya-originated *W. confusa* based on BLAST analysis. Theoretical molecular weight (MW) and pI values of the C39-2 and K39 strains were calculated as 154 kDa, 159 kDa with a pI of 4.5 and 4.6, respectively (Amari et al. 2013). After detection of the EPS-forming ability of *Lb. panis* TMW 1.648 on sucrose containing agar plates, PCR and inverse PCR was used to detect and sequence the complete *ftf* gene synthesizing levansucrase. Predicted pI was determined as pI of 4.5. BLAST analysis showed that it shared 70% identity with levansucrase of *Fr. sanfranciscensis*. Its activity was reduced to half at temperatures above 50°C. In addition to its overall activity, K_M and V_{max} were detected as 17.0 mM, 22.5 mM, and 44.0 µmol mg^{-1} min^{-1}, 21.3 µmol mg^{-1} min^{-1}, respectively, for hydrolysis and transfer reactions. Presence of calcium was found necessary to repair the inhibitor effect of EDTA, whereas other bivalent cations and monovalent cations were insufficient (Waldherr, Meissner, and Vogel 2008). A detailed study characterized Gtf enzymes from 3 lactobacilli by their molecular and biochemical levels. Activities of Gtf were promoted in the presence of Ca^{2+}. Optimum pH range were found to be higher than in *Lim. reuteri* and *Lat. curvatus*. Also, Gtf from *Lb. animalis* was effective in a broad pH range (4.8–7.2) (Rühmkorf et al. 2013). Interestingly, the impact of temperature on the dextran-producing ability of *Weissella cibaria* 10M showed that optimum temperature for growth is not parallel with dextran production. The dextran-forming ability of this strain was optimum at 20°C or less, whereas its growth temperature was 30°C. RT-qPCR analysis revealed the dextransucrase producing *dsrM* gene, SDS-PAGE analysis resulted in a 162 kDa protein band with an increased band intensity after cold shift fermentation (incubation at 30°C and shift to 6°C to induce dextransucrase expression). Also, the oligosaccharides profiles from thin layer chromatography (TLC) confirmed the stimulating effect of a cold shift (Hu and Gänzle 2018).

Recently, a glucansucrase of *Lim. reuteri* E81 was expressed in *E. coli*. BLAST analysis of the glucansucrase gene and biochemical characterization of the enzyme showed high similarities with other glucansucrase genes and enzymes from *Lim. reuteri* strains. Also, TLC, HPLC, Liquid Chromatography-Mass Spectrometer (LC-MS) and nuclear magnetic resonance (NMR) spectroscopy techniques were used to confirm glucansucrase activity on sucrose and maltose (İspirli et al. 2019).

In another study, phylogenetic analysis exhibited that catalytic domain of levansucrase genes from *Lc. mesenteroides* LBAE-G15 and Lm 17 were highly conserved between *Leuconostoc* spp. Cloned levansucrase genes from strain Lm 17 and strain LBAE-G15 shared high sequence homology (97%), similar theoretical pI (4.6), and MW (113 kDa) (Iliev et al. 2018). In the literature, a limited number of *Lc. citreum* from sourdough (LBAE E16, LBAE C10, and LBAE C11) and pozol (CW28) has been explored for their draft genomes (Laguerre et al. 2012, Olvera et al. 2017). In addition, inulosucrase (IslA) activity of *Lc. citreum* CW28 has been reported (Olivares-Illana et al. 2002). Molecular analysis results showed a high similarity between C-terminal domain of IslA of *Lc. citreum* CW28 and alternansucrase (Asr) of *Lc. mesenteroides* NRRL B-1355. (Del Moral et al. 2008). Moreover, based on molecular and biochemical analyses, *Lc. citreum* LBAE-C11 has been identified as an alternan-producing sourdough strain (Amari et al. 2015).

Genomic and phenotypic characteristics of carbohydrate metabolism of sourdough-isolated L. reuteri strains LTH2584, LTH5448, TMW1.656, and TMW1.112 were researched by RT-qPCR using various primers for carbohydrate transporters (Zhao and Gänzle 2018). The results showed that *Lim. reuteri* contained transporter genes for maltose (*malT*), sucrose (*srcT*), and lactose or raffinose (*lacT*), respectively. *Lim. reuteri* had licheninase (EC 3.2.1.73) and endo-1, 4-β-galactosidase (EC 3.2.1.89) but not a phosphotransferase system (PTS) (Zhao and Gänzle 2018).

Another study used several methods to detect EPS-producing enzymes and their encoding genes in *Lc.*, *Lc. mesenteroides*, *W. cibaria*, *W. confusa*, *Fr. Sanfranciscensis*, and *Levilactobacillus brevis*, such as screening by agar plate assay, 3,5-dinitrosalicylic acid (DNS), and HPLC and PCR methods (Bounaix et al. 2009). In the next study, nine *Lc. citreum* and six *Lc. mesenteroides* were screened by the DNS method. Dextransucrase and glycansucrase activities in the cell-associated fractions and supernatants were also detected by SDS-PAGE analyses. Furthermore, the fermentation pattern of various sugars was used as another technique to classify the activity of enzymes. Dextran-producing strains gave almost 180 kDa bands, whereas strains revealed more than one single band in general. In addition, *Lc. citreum* strains produced 230 kDa bands similar to alternansucrase and the reference strain NRRL B-1355. Also, NMR analyses researched the structure of polymers formed by *Leuconostoc* strains. In conclusion, the study evaluated the enzyme from *Lc. citreum* G15 as levansucrase based on 130 kDa protein band from raffinose fermentation (Bounaix et al. 2010).

In another study, enzyme activity of *W. confusa* VTT E-90392 gene expressed in *L. lactis* was characterized by using the ^{14}C-sucrose radioisotope method. Also, the optimum pH of the enzyme was determined using Nelson-Somogyi method. Kinetic characteristics of the enzyme were revealed using both methods (Kajala et al. 2015).

Application of rye bran, originated from *Lc. citreum* FDR24, to sourdough was researched for EPS production and their associated genes. RT-qPCR analysis showed that at least five GH70 enzyme-encoding genes were harbored in *Lc. citreum* FDR241 similar to other *Lc. citreum* strains. Dextran production from *Lc. citreum* FDR24 was controlled by the *drsB* gene based on its significant increase during sourdough fermentation (Coda et al. 2018). Moreover, EPS-forming enzyme activities of 249 isolates from 12 Turkish sourdoughs were preliminarily screened by plate agar assay. Screening of EPS producers from each species by PCR analyses using *gtf*, *lev*, *epsA*, *epsB*, and *p-gtf* genes revealed that isolates harbored at least one EPS-associated gene in their genomes. Also, the yield of EPS was higher at 37°C rather than 30°C (Dertli et al. 2016). Apple-originated *Weissella cibaria* JAG8 has been proposed for a cereal application (Rao and Goyal 2013). Optimum pH and temperature of the purified enzyme was found to be 5.4 and 35°C. On the other hand, it was almost deactivated at 40°C. On the contrary of EDTA and urea, Mg^{2+}, Co^{2+}, and Ca^{2+} improved its activity (Mohan Rao and Goyal 2013).

A detailed research was undertaken on 255 sourdough strains focused on EPS production, using several techniques such as agar plate assay, phenol-sulfuric acid method, and enzymatic assay. PCR products of *Lactiplantibacillus plantarum* LP9 and *W. cibaria* CW4 shared high similarity with glucosyltransferase (GTF) of *Lpb. plantarum* WCFS1 and putative GTF (*epsD*) of *B. cereus* G9241, respectively. Also staining with Schiff's reagent of enzymes resulted in two bands in the range of 180–200 kDa (Di Cagno et al. 2006).

7.3 ENZYMES IN PROTEIN METABOLISM

7.3.1 PROTEINASE AND PEPTIDASES

The action of sourdough microbiota is promoted by the activity of endogenous wheat or rye flour enzymes. Proteolysis during the sourdough fermentation is associated with wheat enzymes, specifically aspartic proteinases. Decreasing the pH during the sourdough formation activates these endogenous proteinases and leads to formation of peptides to be degraded further by their strain-specific peptidases of sourdough microbiota (Gänzle, Loponen, and Gobbetti 2008).

A wide range of qualitative and quantitative methods has been developed to investigate microbial protease activities, which differ in their in-analysis time, simplicity, and limit of detection. During dough formation, free amino acids were profiled in many sourdough studies as a biochemical indicator of proteolytic activity (Chiş et al. 2019, Lhomme et al. 2015). In previous studies, degradation of fluorescent proteins such as fluorescein isothiocyanate (FITC)-casein, FITC-glutenin, or gliadins (Thiele, Gänzle, and Vogel 2003) have been monitored to determine proteolytic activities. Another study used azocasein to determine protease activity of isolates at 440 nm (Nakamura et al. 2007). On the other hand, agar plate assay is extensively used as a simple and inexpensive technique to qualitatively determine proteolytic activity. Agar medium can be formed from modified media with 2% skim milk for yeast and lactobacilli strains, respectively (Palla et al. 2017). Different methods, such as the trinitrobenzenesulfonic acid method (Adler-Nissen 1979, Zotta, Ricciardi, and Parente 2007, Reale et al. 2020, Reale et al. 2021) ando-phthaldialdehyde (OPA) (Church et al. 1983, Schettino et al. 2020, De Bellis et al. 2019, Perri et al. 2021, Dingeo et al. 2020, Maidana et al. 2020, Pepe et al. 2003, Coda et al. 2017) have been applied for spectrophotometric determination of proteolytic activities of sourdough microorganisms. Proteolytic activities during sourdough fermentation can be checked by SDS-PAGE analysis (Pepe et al. 2003). Free amino acid content due to enzymatic activities can be monitored by an amino acid analyzer (De Bellis et al. 2019, Coda et al. 2017, Montemurro et al. 2019, Nionelli et al. 2018). The effect of proteolytic activity of LAB on gluten and gliadin can be measured using the International Association for Cereal Science and Technology (ICC) method 137 and the enzyme-linked immunosorbent assay (ELISA) method, respectively (M'hir et al. 2008). Aminopeptidase (AP) and endopeptidase (EP) activities of sourdough LAB can be determined by measuring the absorbance of liberated p-nitroaniline content using p-nitroaniline (p-NA), N-succinyl l-phenyl-alanine-p-NA (Suphepa), and N-glutaryl l-phenyl-alanine-p-NA (Gluphepa) as substrates, respectively (Rollán and de Valdez 2001). Substrate specificity and activator/inhibitor of enzymes can be various; peptidase of *Lpb. plantarum* CRL 778 is 152% more efficient on Lys-p-NA than Leu-p-NA, and DTT act as an activator and inhibitor for the DP enzymes of *Lpb. plantarum* 759 and 778, respectively (Rollan et al. 2005). Wheat albumins and globulins can be used as a substrate in SDS-PAGE cell-envelope-associated proteinase activity detection by SDS-PAGE assay (De Angelis et al. 2007, Rizzello et al. 2012, Nionelli et al. 2018). The Cd–ninhydrin method at 510 nm has been selected to determine dipeptidase (EC 3.4.13.11; PepV) and tripeptidase (EC 3.4.11.4; PepT) activities using Leu-Leu, Gly-Gly, Leu-Pro, and Gly-Tyr and Leu-Leu-Leu, Gly-Gly-Gly, dl-Leu-Gly-dl-Phe, and Leu-Gly-Gly as substrates (Rollán and de Valdez 2001, De Angelis et al. 2007). This method has been applied in the measurement of prolidase (EC 3.4.13.9; PepQ) and prolinase (EC 3.4.13.8; PepR) using Val–Pro and Pro–Gly substrates (De Angelis et al. 2007). Rizzello et al. (2015) studied the peptidases (PepN, PepA, PepO, PepT, and PepX) of *Lpb. plantarum* C48 and *Lev. brevis* AM7 to be used for sourdough from legumes. Hu et al. (2011) researched PepN, PepX, PepO, and PepQ activities of *Lim. Reuteri* TMW 1.106, *Lim. Reuteri* LTH 5448, *F. rossiae* 34J, and *Lpb. plantarum* FUA 3002. De Angelis et al. (2007) screened 50 *Fr. sanfranciscensis* isolates for PepN, PepI, PepX, PepQ, PepR, PepV, PepT, and proteinase activities. Studies focused on the reduction of allergenicity of wheat explore the peptidase and proteolytic activities of starter cultures or sourdough microbial isolates. Fu et al. (2020) demonstrated that intracellular activities of PepX, PepI, and PepN were higher than extracellular activities, whereas extracellular proteinase activities were higher than intracellular ones in most of the LAB and yeast strains in the study.

Falasconi et al. (2020) investigated *W. cibaria* UC4051, *W. confusa* UC4052, and *Limosilactobacillus fermentum* UC3641 species using various bioinformatic tools, and their results showed that *W. cibaria* and *Lim. fermentum* harbored genes belonging to the EPS cluster (i.e. *epsB*, *epsC*, *epsD*, and *epsE*) and Pep system (*pepA*, *pepC*, *pepD*, *pepF*, *pepI*, *pepN*, *pepO*, *pepQ*, *pepS*, *pepT*, *pepV*, and *pepX*) were almost complete in three strains. A characterization study revealed that *Lim. reuteri* CRL 1098 produced metalloenzyme peptidases. Dipeptidase activity was enhanced in the presence of Zn^{2+}, Ca^{2+}, Mn^{2+} (0.1 mM), Mg^{2+}, and Co^{2+}; while tripeptidase activity was increased by Fe^{2+}, Mg^{2+}, Zn^{2+}, Co^{2+}, and Mn^{2+}. Mn^{2+} acts as a stabilizer for the AP (Rollán and de Valdez 2001). Screening of the *prt* gene in *Fr. sanfranciscensis* TMW1.53 (ATCC 27651^T, DSM 20451^T) using various primers showed that it did not share any similarities with other *prt* genes. Genes coding Opp, DtpT, PepT, PepC, PepR, and PepN were also monitored during sourdough fermentation by *Fr. sanfranciscensis* TMW1.53. Although microorganisms harbored the PepX-associated gene, it was not expressed in sourdough (Vermeulen et al. 2005). Purifed PepX from *Fr. sanfranciscensis* CB1 preserved 70% and 50–80% of its activity in the ranges of 25–30°C, pH 6.5–7.5 and pH 5.0–5.5. It cleaved Pro from N-terminus and did not possess any prolidase, aminopeptidase, or endopeptidase activities. On the other hand, PepX from dairy-originated microbial isolates had higher optimal pH and temperature values in the ranges of pH 6.5–7.5 and 40–55 (Gallo et al. 2005). Also, activity of PepX is dependent on the substrate; Zotta, Ricciardi, and Parente (2007) found that sourdough strains exhibited higher PepX activity on Gly-Pro-p-NA than on Arg-Pro-p-NA, except for several strains.

7.3.2 ACE Inhibitor Activity

Proteolytic systems of sourdough microorganisms and endogenous enzymes of flours contribute to the formation of angiotensin converting enzyme (ACE)-inhibitory peptides (Hu et al. 2011). Also, increased fermentation time contributes to lysis of bacterial cells, and liberated intracellular peptidases support proteolysis in rye-malt sourdoughs (Gänzle, Loponen, and Gobbetti 2008, Stromeck et al. 2011, Zhao et al. 2013). Gänzle, Loponen, and Gobbetti (2008) provided an excellent summary of the enzymes in the most used flours in sourdough production and highlighted the cleavage mechanism of endopeptidase and exopeptidases of sourdough LAB on protein sequences.

Nakamura et al. (2007) isolated the first ACE-inhibitory peptide (ACEIP) from sourdough. Hu et al. (2011) reported the ACEIP in rye malt sourdoughs via LC-MS/MS and related increased LQP and LLP content in sourdough with low PepO and high PepN activities of LAB. Furthermore, Zhao et al. (2013) studied the change in the ACEIP during sourdough fermentation, dough preparation, and baking steps. Reverse phase-HPLC and reverse-phase fast protein liquid chromatography techniques can be used to quantify the ACE-inhibitory activities by monitoring the formation of hippuric acid by ACE (Torino et al. 2013, Rizzello et al. 2008). In addition to identification of ACEIP by chromatographic techniques, monitoring the ACE inhibition by spectrophotometer using HHL as a substrate can be used as an indicator of proteolytic enzymes in sourdough (Diowksz et al. 2020).

In contrast, while ACEIP can be formed due to proteolytic action, other compounds such as phenolics and flavonoids can contribute to ACE inhibitor activities via their associated enzymatic actions (Zieliński et al. 2020). A study showed that metabolization of phenolic acids and flavonoids during digestion also enhanced ACE inhibitor activity in biscuits from fermented buckwheat flour (Zieliński et al. 2020).

7.3.3 Other Bioactive Peptide-Forming Activities

A limited number of studies focused on the formation of antioxidative, antimicrobial, and anticancer peptides formed during sourdough fermentation. The combination of *Companilactobacillus alimentarius*, *Lev. Brevis*, *Fr. Sanfranciscensis*, and *Lentilactobacillus hilgardii* for fermentation of various cereal sourdoughs increased the antioxidative peptide content up to 2–3-fold compared to chemically acidified doughs (Coda et al. 2012). Another research studied the antioxidative peptide by-products due to enzymatic action in quinoa sourdough (Rizzello et al. 2017). The anticancer potential of wholemeal wheat, soybean, barley, amaranth, and rye flour sourdoughs was analyzed after elimination of 40 LAB sourdough-based peptidase activities (PepN, PepA, PepT, PepX, and PepO). In addition, SDS-PAGE technique was used to detect proteinase activity in water-soluble extracts of sourdoughs. The authors confirmed the highest lunasin- and peptide-forming enzymatic activities were detected in *Lat. curvatus* SAL33 and *Lev. brevis* AM7 in all sourdoughs by spectrophotometry, Reverse phase-HPLC and nano-Liquid chromatography/electrospray ionization mass spectrometry (nano-LC/ESI-MS) analyses (Rizzello et al. 2012). Furthermore, the activities of peptidases from *Lpb. plantarum* C48 and *Lev. brevis* AM7 involved in the formation of anticancer lunasin-like polypeptides from native proteins in sourdough were studied and confirmed by western-blot analysis (Rizzello et al. 2015). Proteolysis products formed in sourdough fermented wheat and rye bread were profiled by using LC-MS technique, and peptides with antidiabetic activities were detected (Koistinen et al. 2018).

7.4 ENZYMES IN PHENOLIC METABOLISM

In cereals, the most abundant phenolic compounds (PCs) are phenolic acids, classified into hydroxybenzoic acids and hydroxycinnamic acids. Ferulic acid (FA) accounts for 50–90% of the phenolic acids in wheat and rye and contributes to the production of aroma compounds in bakery goods (Ripari, Bai, and Gänzle 2019, Boudaoud et al. 2021). During sourdough fermentation, the metabolism of PC by LAB occurs through the action of several enzymes such as decarboxylases (PAD), reductases (PAR), esterases, and/or glycosidases (Rollán, Gerez, and LeBlanc 2019). Bounded phenolic acid can be liberated by esterase enzyme, their structure can be altered by glucosidase activities, and they can be used by LAB and yeasts, depending on their enzyme activities. Meanwhile, other studies have pointed out that these enzymatic activities of LAB on phenolics depends on the phenolic content of the cereal and flour type, and the metabolism of FA in the medium is stain-dependent (Antognoni et al. 2019, Ravisankar, Dizlek, and Awika 2021, Konopka et al. 2014).

The characteristics of sourdough-isolated FA esterase are highly limited. In a recent study, isolated and characterized cinnamoyl esterases from sourdough strains *Lpb. plantarum* and *Lim. fermentum* were partially characterized for the first time (Fritsch et al. 2017). More recently, a study focused on the characterization of the FA esterase type D from *Streptomyces werraensis* and its impact on the volume of wheat dough pastries. FA esterase addition of 0.03 U or 0.3 U kg^{-1} to dough increased volume of patisseries up to 8.0 or 9.7%, respectively. These results revealed the desired impact of FA esterase on baking for the first time (Schulz et al. 2018).

On the other hand, several studies have been focused on the enzymes associated with phenolic metabolism. Yeast strains of *Kazachstania humilis*, *Kazachstania bulderi*, and *S. cerevisiae*, LAB strains of *Fru. Sanfranciscensis*, *Lactiplantibacillus plantarum*, *Lactiplantibacillus xiangfangensis*, *Levilactobacillus hammesii*, *Latilactobacillus curvatus*, and *Latilactobacillus sakei* were studied in stimulated sourdough medium for their enzyme activities. Free FA and its derivatives were measured in terms of FA decarboxylase and reductase activities. In addition, the bioinformatic research on the associated genes of FA reductase (*hcrF* and *par1*) and decarboxylase (*pdc1*) in *Fr. sanfranciscensis* has been conducted (Boudaoud et al. 2021). In another study, 110 *Bacillus* strains isolated from fermentative dough product Kambu koozh (from pearl millet dough) were investigated for their feruloyl esterase enzyme activities with agar plate assay and FA content *via* RP-HPLC with diode array detector at 280 nm. Feruloyl esterase producer strains were grown in modified media, and the crude enzyme from the cells were explored by measuring the PNP via spectrophotometer at 400 nm using 4-nitrophenyl-ferulate as a substrate (Palaniswamy and Govindaswamy 2016). LAB isolates from Daqu (fermented grain blend of barley, pea, and wheat) have been investigated for esterase activities by measuring p-nitrophenol in spectrophotometer at 540 nm using p-nitrophenyl acetate as a substrate and demonstrated that approximately 60% of isolates had esterase activity, and almost none of the *Weissella* isolates showed no activity. (Huang et al. 2021). Another research focused on the effect of spontaneous fermentation or fermentation with *Lev. brevis* L62 or *Lpb. plantarum* L73 or baker's yeast. The authors reported that yeast fermentation is the main reason for the high free FA content for rye or germinated rye fermentation. They attributed these results to microbial or grain enzymes (e.g. cinnamoyl esterase) produced or activated by increased pH (4.5–6.0) during yeast fermentation (Katina et al. 2007).

Most of the studies use the agar plate technique since it is easy to perform and relatively fast. Donaghy, Kelly, and McKay (1998) developed an improved agar plate assay by using 0.3 ml ethyl ferulate (10% v/v in dimethylformamide) as the only carbon source in each agar plate. This method has been applied in numerous studies as a selection criterion for further investigation of LAB species (Palaniswamy and Govindaswamy 2016, Xu et al. 2017, Wang et al. 2017, Hole et al. 2012). In another study, eight LAB strains were selected from a total of 56 LAB strains by determination of clear zone in MRS agar with ethyl ferulate and were used in the fermentation of whole wheat and oat flours. Based on the results obtained by HPLC-diode array detector with a C8 column, fermentation of flours by *Lb. johnsonii* LA1, *Lim. reuteri* SD2112, and *Lb. acidophilus* LA-5 boosted free FA content in the medium up to almost 20-fold (Hole et al. 2012). Ultra-performance liquid chromatography–MS for investigation of sourdough fermentation of whole grain wheat, sorghum, and teff flours (Ravisankar, Dizlek, and Awika 2021) and HPLC-UV for investigation of fermentation of wheat after extraction of free, soluble-conjugated, insoluble-bound phenolic extracts has been applied (Antognoni et al. 2019). Free FA content implying the FA esterase activity was monitored by HPLC-diode array detector in rye flour and five breads (Dynkowska, Cyran, and Ceglińska 2015). Another study measured absorbance of released FA content from ethyl ferulate at 325 nm to determine FA esterase activity of *K. marxianus* during pre-fermentation of wheat bran dough (Zhang et al. 2019). A more comprehensive evaluation on the detection techniques of for feruloyl esterase activity and the substrates of the enzymes used in these methods has been explained by Ramos-de-la-Peña and Contreras-Esquivel (2016).

7.5 PHYTASE

Phytate (myo-inositol hexaphosphate (InsP6)) is the major phosphorus source in legumes and cereals. It shows anticancer effects due to antioxidative properties (Kumar et al. 2010). However, multivalent cations such as Ca^{2+}, Mg^{2+}, Fe^{2+}, Zn^{2+}, and Mn^{2+} is converted to their insoluble forms. Thus, PA reduces the bioavailabilities of these minerals and trace elements. Phytases degrade phytic acid to lower myo-inositol phosphates by removing phosphate (Sharma et al. 2020). Sodium phytate, calcium phytate, or dipotassium phytate added modified media have been selected as primary techniques to identify phytase producer strains in numerous publications (Anastasio et al. 2010, Nuobariene, Hansen, and Arneborg 2012, Palla et al. 2017, Palla et al. 2020). In addition, these substrates have been utilized in spectrophotometric approaches to quantify phytase production by sourdough microorganisms (Anastasio et al. 2010, Nuobariene, Hansen, and Arneborg 2012, Palla et al. 2017, Palla et al. 2020). Detection of phytic acid levels by RP-HPLC has been used as another technique (Lopez et al. 2000, De Angelis et al. 2003). Furthermore, several reports determined the level of minerals as a result of phytase activity by inductively coupled plasma and atomic absorption spectroscopy (AAS) spectroscopy (Karaman, Sagdic, and Durak 2018, Sharma et al. 2018).

In general, LAB-originated phytases can be effective in a broad pH range of 4.0–7.5 and temperature range of 40–80°C except for thermostable phytases. For activities of yeast phytases, a limited acidic pH range (4.5–5.5) is more suitable (Sharma et al. 2020). Based on metagenomic analyses on gut microbiota, protein tyrosine phosphatase-like phytases or cysteine phytases are the main phytases. The first molecular data of protein tyrosine phosphatase from genus *Lactobacillus* were reported in *Lim. fermentum* NKN51 cheese isolate. The genetic locus-encoding phytase gene *(phyLf)* was demonstrated by several *in silico* analyses. Protein tyrosine phosphatase was unaffected with the addition of Ca^{2+}, Hg^{2+}, Mg^{2+}, and Co^{2+}, stabilized with EDTA and promoted with presence of Ag^{1+} (Sharma et al. 2018). In the next part of the study, phytase gene *(phyLs)* was cloned from *Lim. fermentum* NKN51, expressed in *E. coli* and molecular weight of enzyme was characterized by zymographic analysis. The change in the mineral levels of PhyLf enzyme added to durum wheat and finger millet doughs was monitored by atomic absorption spectrophotometer (Sharma et al. 2018).

Another study on enzyme characterization focused on phosphatase PhypA from human-originated *Bifidobacterium. infantis* ATCC 15697 and phosphatases PhylA from *Bifidobacterium pseudocatenulatum* ATCC 27919 after their overexpression in *E. coli*. Phytate was the major substrate for PhypA and phylA. The enzymes exhibited 73% and 44% activity at 80°C after 15 min. In addition, PhylA were more effective at higher pH (6.5) and temperatures (55°C) (Tamayo-Ramos et al. 2012).

Several studies on sourdough fermented by these strains focused on the phytate degradation. Addition of wheat sourdough fermented with *B. pseudocatenulatum* ATCC 27919 to doughs prepared with wheat, rye, and a mixture of wheat and rye flours resulted in significant reduction in the concentration phytate (García-Mantrana, Monedero, and Haros 2015). Further study investigated whole grain sourdough bread fermented with *Lacticaseibacillus casei* BL23-containing phytase gene belonging to *B. pseudocatenulatum* ATCC27919 and *B. longum spp. infantis* ATCC15697. MW of phytase in *Lcb. casei* was detected to confirm phytase activity by zymogram analysis. *Lcb. casei*-containing phytase gene of *B. pseudocatenulatum* ATCC27919 effectively hydrolyzed phytate to different lower myo-inositol phosphates (García-Mantrana et al. 2016). Another study screened LAB to explore phytase and phosphatase activities and characterized an acid phosphatase of *Lpb. pentosus* CECT 4023; Co^{2+} and L-ascorbic acid and EDTA stimulated the enzyme, whereas iodoacetic acid, phenylmethylsulphonyl fluoride, disodium pyrophosphate, and Ca^{2+} decreased its activity. Acid phosphatase was not efficient in the broad range of pH; since it kept 50% activity at pH 4.5 and 5.5, it protects its 40% of activity in the temperature range from 37°C to 60°C. It exhibited activity on phytase and other phosphorylated substrates, and p-nitrophenyl phosphate (p-NPP) was used for kinetic characterization of phosphatase (Palacios et al. 2005).

A recent study screened yeasts isolated from cereal-based traditional Nigerian fermented food. Supernatants of *Pichia kudriavzevii* OG32 and *Candida tropicalis* BOM21 were characterized without purification. Crude phytase of *C. tropicalis* BOM21 and *P. kudriavzevii* OG32 showed the highest activities at pH 3.6 and pH 4.6. Both phytases in supernatants preserved almost 80% of activities in the range of 20–50°C (Ogunremi, Agrawal, and Sanni 2020). Phytase characterization from *Fru. sanfranciscensis* suggested its nutritional quality enhancement in cereal products due to its phytase activities for the first time (De Angelis et al. 2003). The authors screened several microbial isolates from sourdough for their phytase activities and studied *Fru. sanfranciscensis* CB1 with the highest activity for phytase characterization. Even a 0.5-point increase or decrease from its optimum pH reduced enzyme activity sharply, and it only preserved 40% of phytase activity of *Fru. sanfranciscensis* CB1 at 30°C and 60°C. Furthermore, phenylmethylsulfonyl fluoride, Hg^{2+}, and Fe^{2+} inhibited activity of the enzyme. It was thermostable in the presence of Na-phytate at 60°C for 30 min. but lost almost all activity at 80°C for 10 min. 2D electrophoresis determined pI of enzyme as 5.0 (De Angelis et al. 2003). In another study, 16 yeast sourdough isolates were evaluated due to measurement of absorbance of liberated inorganic orthophosphate from the FA by phytase at 700 nm. EDTA, Cu^{2+}, and Fe^{2+} were ineffective on phytase, while phenylmethylsulfonyl fluoride, Zn^{2+}, Hg^{2+}, Mn^{2+}, and Co^{2+} reduced enzyme activity. Wheat sourdough fermentation with *S. cerevisiae* LC3 lowered phytate and increased free Ca^{2+}, Zn^{2+}, Fe^{2+}, and Mg^{2+} and inorganic orthophosphate concentration compared to doughs made with baker's yeast and *Lpb. plantarum* LP4 (Caputo, and De Angelis 2015).

Furthermore, phytase of *Lpb. plantarum* CRL196 showed the highest activity in a wide LAB isolate group isolated from cereals, pseudocereals, and sourdoughs (Sandez Penidez et al. 2020). Its enzyme (PhyLP) was partially purified at the end of molecular exclusion chromatography. Molecular mass of purified enzyme was determined by SDS-PAGE and zymogram analysis. PhyLP protected its activity in the range of pH 2.0–6.0 and high temperature conditions. PhyLP was identified as true phytase instead of a non-specific phosphatase, based on low activity ($15.99 \pm 1.54\%$) on p-NPP and the highest activity ($100.00 \pm 3.47\%$) on sodium phytate (Sandez Penidez et al. 2020).

Another study characterized a phytase from *Lpb. pentosus* CFR3 and used the isolate with finger millet flour to investigate its phytase degradation ability after screening of 42 LAB by agar plate assay. Non-specific phosphatase containing phytase activity was shown based on low substrate specificity on various substrates (Amritha, Halami, and Venkateswaran 2017). In another study, sourdough bread was produced by *Lev. brevis* HEB33, *Lpb. plantarum* ELB78, and *S. cerevisiae* TGM38 alone or together in different temperatures. The results of measurement of phytic acid by mineral content in these sourdoughs indicated that phytase activity was higher at 25°C rather than 30°C and 35°C (Yildirim and Arici 2019).

7.6 ENZYMES IN STRESS METABOLISM

During fermentation of sourdough, LAB are exposed to both cold stress during refrigerated storage and acid stress because of the increased acid content as an end product of sourdough fermentation. Environmental stress changes both the physiological activities of LAB and their production efficiency. The activities of microorganisms as a response to stress factors enhance the flavor and texture of sourdough and competition of LAB in sourdough. Thus, sourdough-starter strains with high stress tolerant mechanisms improve the potential of the strain in industrial applications (Serrazanetti et al. 2009, Zhang et al. 2012). LAB can be adapted to these harsh environmental conditions *via* several pathways, such as glutamate decarboxylase (GAD) and ADI pathways (Gobbetti et al. 2005, Serrazanetti et al. 2009).

7.6.1 GLUTAMATE DECARBOXYLASE

In addition to ACEIP, various studies focus on the potential antihypertensive content of sourdoughs by quantification of both GABA and ACE inhibitor activity (Rizzello et al. 2008, Peñas et al. 2015, Limón et al. 2015). GABA, a non-protein amino acid is another bioactive metabolite as a consequence of bioconversion by microorganisms during sourdough fermentation (Rizzello et al. 2008). It is a product of the stress metabolism of microorganisms and activated by environmental conditions such as acidic circumstances. It is produced by microorganisms with the action of the GAD enzyme which catalyzes the conversion of L-glutamate (or its salts) onto GABA through α-decarboxylation (Xu, Wei, and Liu 2017). GABA can contribute to acid tolerance and competitiveness in sourdough. Although during rapid fermentation of sourdough, non-GABA-producer strains were found to be dominant, GABA-producer strains can be more competitive after extended fermentation phases (Zheng et al. 2015). GAD activities are verified by different techniques, such as molecular characterization of *gad* gene or the GAD operon of strain (Fraberger, Ammer, and Domig 2020), TLC (Bhanwar et al. 2013), HPLC (Venturi et al. 2019), or spectrophotometry (Villegas et al. 2016).

To our knowledge, no GAD enzyme originating from a sourdough strain has been characterized until now. Molecular characterization of *acid resistance of Lim. reuteri* 100-23 showed that the GAD operon contained one glutamate decarboxylase gene (*gadB*) and two antiporter genes (*gadC1* and *gadC2*) in the presence of *gls3* (Su, Schlicht, and Gänzle 2011). The authors demonstrated the contribution of *gadB* gene during long sourdough fermentation time by RT-qPCR using associated primers and probes. A recent study screened 179 LAB isolates for their *gad* gene by using primers and PCR products, and the isolates were positive for 25% of sourdough strains including *Lev. Brevis*, *Loigolactobacillus coryniformis*, *Lpb. plantarum*, *Levilactobacillus senmaizukei*, *Levilactobacillus hammesii*, and *Lcb. paracasei* species (Fraberger, Ammer, and Domig 2020).

Companilactobacillus farciminis was used in a bakery product for GABA synthesis for the first time; the higher GABA production was associated with higher glutamic acid content of wheat flour compared to amaranth flour (Venturi et al. 2019). Similarly, white wheat sourdough contained higher GABA concentration (258.71 mg/kg) compared to rye flour when fermented with various LAB mixtures (Rizzello et al. 2008). A comprehensive study investigated the effect of fermentation by *Lpb. plantarum* C48 and *L. lactis subsp. lactis* PU1 on GABA production in various sourdough types (Coda, Rizzello, and Gobbetti 2010). GABA concentration due to the fermentation was promoted when a mixture of chickpea, amaranth, quinoa, and buckwheat flours were selected. The GABA-producing abilities of microbial isolates were explored by using amino acid analyzer, and the results were confirmed with molecular detection of *gad* genes with a PCR product of ca. 540 bp. In an optimization study of *Lev. brevis* CRL 1942, TLC and HPLC with a fluorescence detector and spectrophotometric GABase methods were used to screen 19 LAB strains. The optimum conditions for maximum GABA production were 270 mM of MSG and 30°C for 48 h for *Lev. brevis* CRL 1942 (Villegas et al. 2016). In one of the latest studies, impairment of the GAD system using pentoses as the only carbon source was reported for the first time by the GABase method. Highest GABA production (265 mM) of *Lev. brevis* CRL 2013 was obtained in MRS medium containing 267 mM MSG supplementation at pH 6.5 and at 30°C and 37°C for 72 h (Cataldo et al. 2020). Another research found that 8 of 15 sourdough microorganisms contained *gad* genes, whereas all isolates were identified as GABA producers based on TLC and HPLC analyses. Interestingly, isolates from rye sourdough had lower GABA production capacity (Demirbaş et al. 2017). In addition, the use of pear and orange to produce sourdough increased GABA production in breads up to 2-fold (Yu et al. 2018). Several LAB strains were investigated for their capacity to form GABA using the *gad* gene of *Lim. reuteri* 100-23 as a template primer in PCR analysis. Based on the presence of genes, *Lim. reuteri* TMW1.106 or LTH5448 were used to produce rye malt sourdough, and time-dependent increases in GABA and glutamate concentration were detected by HPLC after o-phthaldialdehyde derivatization (Stromeck et al. 2011). During the 96-h fermentation, the survival of GAD-positive strains compared to GAD-negative strains indicated the positive influence of the GAD system on acid resistance of *Lim. reuteri* (Stromeck et al. 2011).

Whole-genome microarray is a novel methodology research gene expression during sourdough fermentation. It was applied in back-slopping sourdough to monitor decarboxylation genes *gadB* of glutamate decarboxylase and *tyrDC* gene of tyrosine decarboxylase. Fermentation by *Lpb. plantarum* and *L. lactis* resulted in moderate *gadB* gene at the last and the first days of fermentation, respectively. The study found that *tryDC* gene increased during the first days of fermentation in sourdoughs, except the D13S spelt type (Weckx et al. 2011).

7.6.2 GLUTAMINASE AND ADI PATHWAY ENZYMES

Deamination of glutamine by glutamine-amidotransferases or glutaminases aids the acid tolerance of LAB and produces metabolites such as glutamate that can be transformed to GABA and γ-glutamyl peptides (Li et al. 2020). The acid tolerance mechanism of *Lim. reuteri* 100-23 has been characterized based on glutamine, glutamate, and arginine-based metabolisms. The quantification of expression of *gls1*, *gls2*, *gls3*, *adi*, and *gadB* genes associated with the synthesis of glutaminase, ADI pathway, and GAD enzymes in modified MRS media and sourdough were evaluated by RT-qPCR. The outcomes of RT-qPCR showed the contribution of acid stress on increased expression of *gls3* and *gadB* and the main role of the *gls3–gadB* operon in *Lim. reuteri* in glutamine metabolism (Teixeira et al. 2014). Moreover, the growth of these strains under low pH conditions showed that glutaminases improved acid tolerance of *Lim. reuteri* 100-23. Acid resistance of *Lim. reuteri* was enhanced by glutamine deamidation (Teixeira et al. 2014). Furthermore, the GAD and glutaminase operon in the genome of *Lim. reuteri* 100-23 were identified and the protective effect of the glutaminase mechanism has been explored under acid stress (Su, Schlicht, and Gänzle 2011, Teixeira et al. 2014). In addition, the results of a recent research indicates that acid resistance of *L. acidophilus* FUA3066 was enhanced, but its glutamine metabolism was not promoted substantially by glutaminases (Li et al. 2020).

Earliest study on glutaminase activity in lactobacilli during growth in sourdough was observed by the amination of glutamine to glutamate by spectrophotometer at 492 nm using a commercially available glutamate test kit. The results indicated glutaminase activities in both *F. sanfranciscensis* DSM20451 and *Lim. reuteri* TMW1.106. Moreover, the existence of glutamine in a modified MRS medium enabled them to survive under acidic conditions (Vermeulen, Gänzle, and Vogel 2007).

Arginine metabolism through the activities of ADI pathway enzymes (arginine deiminase (ADI), ornithine transcarbamoylase (OTC), and carbamate kinase (CK)) in the sourdough isolate *Lim. reuteri* CRL 1098 increases the resistance to acidic conditions based on the results of spectrophotometric analysis (Rollan, Lorca, and de Valdez 2003). Furthermore, *Lim. fermentum* IMDO 130101, an isolate from a laboratory rye sourdough, was monitored for conversion products of enzymes associated with ADI pathways (arginine, citrulline, and ornithine) through electron spray ionization mass spectrometry, and their concentration was interpreted under salt and temperature stress conditions (Vrancken, Rimaux, Wouters, et al. 2009). The study also identified the ADI operon as *arcBCAD* with an additional A/O antiporter *in Lim. fermentum* IMDO 130101 and found upregulation under salt and temperature stress (Vrancken, Rimaux, Weckx, et al. 2009). Additionally, gene expression of arginine/ornithine antiporter (*arcD*) and asparagine synthase (glutamine hydrolyzing) (*asnB*) were promoted in *Lim. reuteri* LTH5531 during type II sourdough fermentation (Dal Bello et al. 2005).

A comprehensive study focused on the arginine metabolism of sourdough LAB demonstrated that out of 70 LAB isolates from sourdough, only seven (*Fr. sanfranciscensis* CB1; *Lev. brevis* AM1, AM8, 10A; *Len. hilgardii* 51B; *Lb. fructivorans* DD3 and DA106) secreted all of the enzymes involved in the ADI pathway (De Angelis, Mariotti, et al. 2002). Activities of enzymes were recorded as absorbance of enzyme at 460 nm after addition of an acid mixture and boiling in the lightness conditions for ADI, in potassium phosphate as conversion of the ornithine to citrulline after 2 h at 37°C for OTC, and the conversion of carbamoyl phosphate to ammonia for CK. Characterization analysis revealed that ADI, OTC, and CK had similar optimum pH and temperature requirements similar to sourdough fermentation conditions (De Angelis, Mariotti, et al. 2002).

The earliest study on glutaminase activity in lactobacilli during growth in sourdough was observed by the amination of glutamine to glutamate by using a commercially available glutamate test kit. The results indicated glutaminase activities in both *F. sanfranciscensis* DSM20451 and *Lim. reuteri* TMW1.106. Moreover, the existence of glutamine in modified MRS medium enabled them to survive under acidic conditions (Vermeulen, Gänzle, and Vogel 2007).

Furthermore, an exclusive research employed metagenetics, metagenomics, metatranscriptomics, metaproteomics, meta-metabolomics, and genomic methodologies to explore the existence of various important genes associated with numerous enzymes, such as glutathione reductase, hexose-pentose, and phenolic metabolism enzymes (e.g. tannase/tannin acyl hydrolase, phenolic acid, and decarboxylase), ADI pathway enzymes, EPS-producing enzymes (e.g. glucansucrase, levansucrase, and inulosucrase), acid tolerance enzymes (e.g. gad and glutaminase) on the genome of 41 LAB from sourdough in the National Center for Biotechnology Information (NCBI) database (Weckx, Van Kerrebroeck, and De Vuyst 2019).

7.6.3 Glutathione Reductase

Glutamate, cysteine, and glycine-containing glutathione tripeptide (GSH) possess protective activity in the cell against environmental conditions. The expression of *gshR* gene-coding glutathione reductase (GshR) in sourdough was evaluated by PCR analysis, and the activity of GshR of *Fr. sanfranciscensis* TMW1.53 was confirmed by measuring the conversion of oxidized glutathione to reduced glutathione by HPLC-UV at 210 nm (Vermeulen et al. 2006). Another study focused on the effect of GSH on *Fr. sanfranciscensis* DSM 20451 during cold stress by monitoring activities of glyceraldehyde-3-phosphate dehydrogenase and pyruvate kinase of glycolysis metabolism using spectrophotometer at 340 nm. Also, 2D gel electrophoresis results showed that GSH producer cells promoted different protein expressions, such as β-phosphoglucomutase, GSH-peroxidase, and phosphate acetyltransferase in cold-treated cells (Zhang et al. 2012). Another study investigated the influence of GshR from *Fr. sanfranciscensis* DSM20451^T on oxygen tolerance and thiol formation by producing sourdough with both wild type and *Fr. sanfranciscensis* DSM20451^{T}ΔgshR strains together. The authors demonstrated that 446-amino-acid protein (GshR) with a MW 48.614 Da and pI 4.79 was expressed by the activity of *gshR* gene in *Fr. sanfranciscensis* DSM2045. Degenerate primers deg/gshRV and deg/gshRR were used by PCR to confirm the presence of *gshR* genes in several LAB strains. Deficiency of GshR activity was also tested by SDS-PAGE analysis. On the contrary, the results of gene screening with associated primers and enzymatic activity detections showed that cystathionine-γ-lyase activity was found in the absence of GshR activity (Jänsch et al. 2007). The following study focused on the impact of GshR activity on sourdough bread volume by using *Fr. sanfranciscensis* DSM20451^T and DSM20451^{T}ΔgshR. GSH, dimeric oxidized glutathione, and γ-glutamyl-cysteine content was determined by HPLC system together with liquid chromatography–tandem mass spectrometry. Fermentation with *Fr. sanfranciscensis* DSM20451^T reduced the volume, increased γ-glutamyl-cysteine, and did not alter the saltiness of bread (Tang, Zhao, and Gänzle 2017).

7.7 OTHER ENZYMATIC FORMATION OF BIOACTIVE COMPOUNDS IN SOURDOUGH

In their study Garofalo et al. (2012) reported antifungal activities of the *F. rossiae* LD108 and *Lb. paralimentarius* PB127 against *Aspergillus*, *Eurotium*, and *Penicillium* species. These activities of *F. rossiae* and *Lb. paralimentarius* sourdoughs in bread and panettone were attributed to bioactive peptides formed from gluten proteolysis and characterized by matrix-assisted laser desorption ionization time-of-flight mass spectrometry (Garofalo et al. 2012). Similarly, antifungal peptides were observed in sourdough fermented wheat germ (Rizzello et al. 2011). Bacteriocin, such as bavaricin A, plantaricin ST31, and bacteriocin-like inhibitory substance C57 from LAB have been explored (Corsetti, Settanni, and Van Sinderen 2004).

After the first characterization of antibiotic reutericyclin production by sourdough isolate *Lim. reuteri* LTH2584 (Höltzel et al. 2000), its production in wheat sourdough fermented with *Lim. reuteri* LTH2584 or *Fr. sanfranciscensis* LTH2581 was investigated (Gänzle and Vogel 2003). Furthermore, molecular detection and bioinformatic analysis of *Lim. reuteri* TMW1.112, LTH5448, LTH2584, and TMW1.656 have identified genomic islands associated in the reutericyclin synthesis by *Lim. reuteri* with the action of polyketide synthases and nonribosomal peptide synthetases (Lin et al. 2015). Another study revealed the ADI pathway-stimulating effect of low pH on *Lpb. plantarum* IMDO 130201 in wheat sourdough simulation medium by the microarray method. Increased gene expressions of *plnJ*, *plnL*, and *plnC* participating in a bacteriocin (plantaricin) peptide production was detected in wheat sourdough simulation medium under decreased pH conditions (Vrancken et al. 2011).

Activities of γ-glutamyl transferase/transpeptidase, glutaminase, glutathione synthetase, and γ-glutamyl cysteine ligase can form γ-Glutamyl peptides which exhibit anti-inflammatory bioactivities and enrich organoleptic characteristics of food matrix by giving kokumi activity (Xie and Gänzle 2021). In a recent study, quantification of γ-glutamyl peptides synthesis by LC-MS/MS in sourdough fermented with four different *Lim. reuteri* strains indicated the activities of γ-Glu-Cys synthetase during fermentation and its strain-specific activity (Zhao and Ganzle 2016). Sourdough fermentation with *Lim. reuteri* LTH5448 and its *glc*-deficient mutant revealed the contribution of *glc* genes on the formation of γ-glutamyl peptides, which were monitored by LC–MS/MS (Yan et al. 2018). Also, genomic approaches revealed the encoding genes of γ-Glu-Cys synthetase in the genome of *Lim. reuteri* LTH5448 and 100-23, while glutathione synthetase and glutamyl transferase were not determined (Zheng et al. 2015, Zhao and Ganzle 2016).

Furthermore, the extracts of *Fr. sanfranciscensis* A4, *H. alvei* ATCC-51815, and *D. hansenii* DSMZ-70590 were explored for their glutamate dehydrogenase and cystathionine γ-lyase activities based on quantifying the NADP (nicotinamide adenine dinucleotide phosphate)- or NAD-dependent reduction of glutamate at 340 nm and measuring the amount of keto acids, ammonia, and free thiols by the action of cystathionine γ-lyase on cystathionine (Cavallo et al. 2017).

7.8 CONCLUSION

According to all the reviewed studies, biochemical identification and/or gene level investigations are not enough for the discovery and characterization of sourdough enzymes. Screening should be conducted with specific primer design at the species level for all sourdough enzymes.

For sourdough enzymes, various quantitative methods and analytical methods, e.g. spectrophotometry and chromatography, were successfully used. Using only genotypic characterization have not been found very effective since the identified genes can be in the form of pseudogenes, and this type of genes cannot play a role in the expression of enzymes. On the other hand, it has been demonstrated that sourdough bacteria can produce enzymes in agar plate assays even though the gene responsible for the specific enzyme has not been detected by genetic approaches. Therefore, rather than only using genetic approaches, other screening methods should be considered, and EPS producer enzymes and phytases from sourdough are the most obvious target choices. The research on enzyme metabolism regarding bacteriocins is quite scarce. Few bioinformatic studies exist that compare sourdough enzyme gene structure to similar enzymes from other sources. Finally, enzyme (e.g. protease) activity has been widely studied, while investigations on enzyme characterization are limited. Therefore, this chapter emphasizes the importance and the need to discover new sourdough enzymes, develop and implement characterization strategies, and establish informational databases.

REFERENCES

Adler-Nissen, Jens. 1979. "Determination of the degree of hydrolysis of food protein hydrolysates by trinitrobenzenesulfonic acid." *Journal of Agricultural and Food Chemistry* 27 (6):1256–1262.

Agati, V, Jean-Pierre Guyot, Juliette Morlon-Guyot, Pascale Talamond, and DJ Hounhouigan. 1998. "Isolation and characterization of new amylolytic strains of Lactobacillus fermentum from fermented maize doughs (mawe and ogi) from Benin." *Journal of Applied Microbiology* 85 (3):512–520.

Amari, Myriam, Luisa Fernanda Gomez Arango, Valérie Gabriel, Hervé Robert, Sandrine Morel, Claire Moulis, Bruno Gabriel, Magali Remaud-Siméon, and Catherine Fontagné-Faucher. 2013. "Characterization of a novel dextransucrase from Weissella confusa isolated from sourdough." *Applied Microbiology and Biotechnology* 97 (12):5413–5422.

Amari, Myriam, Valerie Gabriel, Robert Hervé, Sandrine Morel, Claire Moulis, Bruno Gabriel, Magali Remaud-Simeon, and Catherine Fontagné-Faucher. 2015. "Overview of the glucansucrase equipment of Leuconostoc citreum LBAE-E16 and LBAE-C11, two strains isolated from sourdough." *FEMS Microbiology Letters* 362 (1):1–8.

Amritha, Girish K, Prakash M Halami, and G Venkateswaran. 2017. "Phytate dephosphorylation by Lactobacillus pentosus CFR 3." *International Journal of Food Science & Technology* 52 (7):1552–1558.

Anastasio, Marilena, Olimpia Pepe, Teresa Cirillo, Simona Palomba, Giuseppe Blaiotta, and Francesco Villani. 2010. "Selection and use of phytate-degrading LAB to improve cereal-based products by mineral solubilization during dough fermentation." *Journal of Food Science* 75 (1):M28–M35.

Antognoni, Fabiana, Roberto Mandrioli, Giulia Potente, Danielle Laure Taneyo Saa, and Andrea Gianotti. 2019. "Changes in carotenoids, phenolic acids and antioxidant capacity in bread wheat doughs fermented with different lactic acid bacteria strains." *Food Chemistry* 292:211–216.

Bhanwar, Seema, Meenakshi Bamnia, Moushumi Ghosh, and Abhijit Ganguli. 2013. "Use of Lactococcus lactis to enrich sourdough bread with γ-aminobutyric acid." *International Journal of Food Sciences and Nutrition* 64 (1):77–81.

Boudaoud, Sonia, Chahinez Aouf, Hugo Devillers, Delphine Sicard, and Diego Segond. 2021. "Sourdough yeast-bacteria interactions can change ferulic acid metabolism during fermentation." *Food Microbiology* 98:103790.

Bounaix, Marie-Sophie, Valerie Gabriel, Sandrine Morel, Herve Robert, Philippe Rabier, Magali Remaud-Simeon, Bruno Gabriel, and Catherine Fontagne-Faucher. 2009. "Biodiversity of exopolysaccharides produced from sucrose by sourdough lactic acid bacteria." *Journal of Agricultural and Food Chemistry* 57 (22):10889–10897.

Bounaix, Marie-Sophie, Valérie Gabriel, Hervé Robert, Sandrine Morel, Magali Remaud-Siméon, Bruno Gabriel, and Catherine Fontagné-Faucher. 2010. "Characterization of glucan-producing Leuconostoc strains isolated from sourdough." *International Journal of Food Microbiology* 144 (1):1–9.

Brandt, Markus J. 2007. "Sourdough products for convenient use in baking." *Food Microbiology* 24 (2):161–164.

Caputo, L, A Visconti, and M De Angelis. 2015. "Selection and use of a Saccharomyces cerevisae strain to reduce phytate content of wholemeal flour during bread-making or under simulated gastrointestinal conditions." *LWT-Food Science and Technology* 63 (1):400–407.

Cataldo, Pablo G, Josefina M Villegas, Graciela Savoy de Giori, Lucila Saavedra, and Elvira M Hebert. 2020. "Enhancement of γ-aminobutyric acid (GABA) production by Lactobacillus brevis CRL 2013 based on carbohydrate fermentation." *International Journal of Food Microbiology* 333:108792.

Cavallo, Noemi, Maria De Angelis, Maria Calasso, Maurizio Quinto, Annalisa Mentana, Fabio Minervini, Stefan Cappelle, and Marco Gobbetti. 2017. "Microbial cell-free extracts affect the biochemical characteristics and sensorial quality of sourdough bread." *Food Chemistry* 237:159–168.

Chen, Xiao Yan, Clemens Levy, and Michael G Gänzle. 2016. "Structure-function relationships of bacterial and enzymatically produced reuterans and dextran in sourdough bread baking application." *International Journal of Food Microbiology* 239:95–102.

Chiş, Maria Simona, Adriana Păucean, Laura Stan, Ramona Suharoschi, Sonia-Ancuţa Socaci, Simona Maria Man, Carmen Rodica Pop, and Sevastiţa Muste. 2019. "Impact of protein metabolic conversion and volatile derivatives on gluten-free muffins made with quinoa sourdough." *CyTA-Journal of Food* 17 (1):744–753.

Church, Frank C, Harold E Swaisgood, David H Porter, and George L Catignani. 1983. "Spectrophotometric assay using o-phthaldialdehyde for determination of proteolysis in milk and isolated milk proteins." *Journal of Dairy Science* 66 (6):1219–1227.

Coda, Rossana, Carlo Giuseppe Rizzello, and Marco Gobbetti. 2010. "Use of sourdough fermentation and pseudo-cereals and leguminous flours for the making of a functional bread enriched of γ-aminobutyric acid (GABA)." *International Journal of Food Microbiology* 137 (2–3):236–245.

Coda, Rossana, Carlo Giuseppe Rizzello, Daniela Pinto, and Marco Gobbetti. 2012. "Selected lactic acid bacteria synthesize antioxidant peptides during sourdough fermentation of cereal flours." *Applied and Environmental Microbiology* 78 (4):1087–1096.

Coda, Rossana, Jutta Varis, Michela Verni, Carlo G Rizzello, and Kati Katina. 2017. "Improvement of the protein quality of wheat bread through faba bean sourdough addition." *LWT: Food Science and Technology* 82:296–302.

Coda, Rossana, Yan Xu, David Sàez Moreno, Dominik Mojzita, Luana Nionelli, Carlo G Rizzello, and Kati Katina. 2018. "Performance of Leuconostoc citreum FDR241 during wheat flour sourdough type I propagation and transcriptional analysis of exopolysaccharides biosynthesis genes." *Food Microbiology* 76:164–172.

Corsetti, Aldo, L Settanni, and D Van Sinderen. 2004. "Characterization of bacteriocin-like inhibitory substances (BLIS) from sourdough lactic acid bacteria and evaluation of their in vitro and in situ activity." *Journal of Applied Microbiology* 96 (3):521–534.

Dal Bello, Fabio, Jens Walter, Stefan Roos, Hans Jonsson, and Christian Hertel. 2005. "Inducible gene expression in Lactobacillus reuteri LTH5531 during type II sourdough fermentation." *Applied and Environmental Microbiology* 71 (10):5873–5878.

De Angelis, M, R Di Cagno, G Gallo, M Curci, S Siragusa, C Crecchio, E Parente, and M Gobbetti. 2007. "Molecular and functional characterization of Lactobacillus sanfranciscensis strains isolated from sourdoughs." *International Journal of Food Microbiology* 114 (1):69–82.

De Angelis, Maria, Aine C Curtin, PAUL LH McSWEENEY, Michele Faccia, and Marco Gobbetti. 2002. "Lactobacillus reuteri DSM 20016: Purification and characterization of a cystathionine γ-lyase and use as adjunct starter in cheesemaking." *Journal of Dairy Research* 69 (2):255–267.

De Angelis, Maria, Giovanna Gallo, Maria Rosaria Corbo, Paul LH McSweeney, Michele Faccia, Marinella Giovine, and Marco Gobbetti. 2003. "Phytase activity in sourdough lactic acid bacteria: Purification and characterization of a phytase from Lactobacillus sanfranciscensis CB1." *International Journal of Food Microbiology* 87 (3):259–270.

De Angelis, Maria, Liberato Mariotti, Jone Rossi, Maurizio Servili, Patrick F Fox, Graciela Rollán, and Marco Gobbetti. 2002. "Arginine catabolism by sourdough lactic acid bacteria: Purification and characterization of the arginine deiminase pathway enzymes from Lactobacillus sanfranciscensis CB1." *Applied and Environmental Microbiology* 68 (12):6193–6201.

De Bellis, Palmira, Carlo Giuseppe Rizzello, Angelo Sisto, Francesca Valerio, Stella Lisa Lonigro, Amalia Conte, Valeria Lorusso, and Paola Lavermicocca. 2019. "Use of a selected Leuconostoc citreum strain as a starter for making a "yeast-free" bread." *Foods* 8 (2):70.

Del Moral, Sandra, Clarita Olvera, Maria Elena Rodriguez, and Agustin Lopez Munguia. 2008. "Functional role of the additional domains in inulosucrase (IslA) from Leuconostoc citreum CW28." *BMC Biochemistry* 9 (1):1–10.

Demirbaş, Fatmanur, Hümeyra İspirli, Asena Ayşe Kurnaz, Mustafa Tahsin Yilmaz, and Enes Dertli. 2017. "Antimicrobial and functional properties of lactic acid bacteria isolated from sourdoughs." *LWT: Food Science and Technology* 79:361–366.

Dertli, Enes, Emin Mercan, Muhammet Arıcı, Mustafa Tahsin Yılmaz, and Osman Sağdıç. 2016. "Characterisation of lactic acid bacteria from Turkish sourdough and determination of their exopolysaccharide (EPS) production characteristics." *LWT: Food Science and Technology* 71:116–124.

Di Cagno, Raffaella, Maria De Angelis, Antonio Limitone, Fabio Minervini, Paola Carnevali, Aldo Corsetti, Michael Gaenzle, Roberto Ciati, and Marco Gobbetti. 2006. "Glucan and fructan production by sourdough Weissella cibaria and Lactobacillus plantarum." *Journal of Agricultural and Food Chemistry* 54 (26):9873–9881.

Dingeo, Cinzia, Graziana Difonzo, Vito Michele Paradiso, Carlo Giuseppe Rizzello, and Erica Pontonio. 2020. "Teff type-I sourdough to produce gluten-free muffin." *Microorganisms* 8 (8):1149.

Diowksz, Anna, Alicja Malik, Agnieszka Jaśniewska, and Joanna Leszczyńska. 2020. "The inhibition of amylase and ACE enzyme and the reduction of immunoreactivity of sourdough bread." *Foods* 9 (5):656.

Donaghy, J, PF Kelly, and AM McKay. 1998. "Detection of ferulic acid esterase production by Bacillus spp. and lactobacilli." *Applied Microbiology and Biotechnology* 50 (2):257–260.

Dubois, Michel, Kyle A Gilles, Jean K Hamilton, PA t Rebers, and Fred Smith. 1956. "Colorimetric method for determination of sugars and related substances." *Analytical Chemistry* 28 (3):350–356.

Dynkowska, Wioletta M, Malgorzata R Cyran, and Alicja Moniuszko. 2015. "Soluble and cell wall-bound phenolic acids and ferulic acid dehydrodimers in rye flour and five bread model systems: Insight into mechanisms of improved availability." *Journal of the Science of Food and Agriculture* 95 (5):1103–1115.

Ehrmann, Matthias A, and Rudi F Vogel. 1998. "Maltose metabolism of Lactobacillus sanfranciscensis: Cloning and heterologous expression of the key enzymes, maltose phosphorylase and phosphoglucomutase." *FEMS Microbiology Letters* 169 (1):81–86.

Falasconi, Irene, Alessandra Fontana, Vania Patrone, Annalisa Rebecchi, Guillermo Duserm Garrido, Laura Principato, Maria Luisa Callegari, Giorgia Spigno, and Lorenzo Morelli. 2020. "Genome-assisted characterization of Lactobacillus fermentum, Weissella cibaria, and Weissella confusa strains isolated from sorghum as starters for sourdough fermentation." *Microorganisms* 8 (9):1388.

Fraberger, Vera, Claudia Ammer, and Konrad J Domig. 2020. "Functional properties and sustainability improvement of sourdough bread by lactic acid bacteria." *Microorganisms* 8 (12):1895.

Fritsch, Caroline, André Jänsch, Matthias A Ehrmann, Simone Toelstede, and Rudi F Vogel. 2017. "Characterization of cinnamoyl esterases from different Lactobacilli and Bifidobacteria." *Current Microbiology* 74 (2):247–256.

Fu, Wenhui, Wentong Xue, Chenglong Liu, Yang Tian, Ke Zhang, and Zibo Zhu. 2020. "Screening of lactic acid bacteria and yeasts from sourdough as starter cultures for reduced allergenicity wheat products." *Foods* 9 (6):751.

Galli, Viola, Manuel Venturi, Rossana Coda, Ndegwa Henry Maina, and Lisa Granchi. 2020. "Isolation and characterization of indigenous Weissella confusa for in situ bacterial exopolysaccharides (EPS) production in chickpea sourdough." *Food Research International* 138:109785.

Gallo, Giovanna, Maria De Angelis, Paul LH McSweeney, Maria Rosaria Corbo, and Marco Gobbetti. 2005. "Partial purification and characterization of an X-prolyl dipeptidyl aminopeptidase from Lactobacillus sanfranciscensis CB1." *Food Chemistry* 91 (3):535–544.

Gänzle, Michael G. 2014. "Enzymatic and bacterial conversions during sourdough fermentation." *Food Microbiology* 37:2–10.

Gänzle, Michael G, and Rudi F Vogel. 2003. "Contribution of reutericyclin production to the stable persistence of Lactobacillus reuteri in an industrial sourdough fermentation." *International Journal of Food Microbiology* 80 (1):31–45.

Gänzle, Michael G, Nicoline Vermeulen, and Rudi F Vogel. 2007. "Carbohydrate, peptide and lipid metabolism of lactic acid bacteria in sourdough." *Food Microbiology* 24 (2):128–138.

Gänzle, Michael G, Jussi Loponen, and Marco Gobbetti. 2008. "Proteolysis in sourdough fermentations: Mechanisms and potential for improved bread quality." *Trends in Food Science & Technology* 19 (10):513–521.

García-Mantrana, Izaskun, Vicente Monedero, and Monika Haros. 2015. "Myo-inositol hexakisphosphate degradation by Bifidobacterium pseudocatenulatum ATCC 27919 improves mineral availability of high fibre rye-wheat sour bread." *Food Chemistry* 178:267–275.

García-Mantrana, Izaskun, María J Yebra, Monika Haros, and Vicente Monedero. 2016. "Expression of bifidobacterial phytases in Lactobacillus casei and their application in a food model of whole-grain sourdough bread." *International Journal of Food Microbiology* 216:18–24.

Garofalo, Cristiana, Emanuele Zannini, Lucia Aquilanti, Gloria Silvestri, Olga Fierro, Gianluca Picariello, and Francesca Clementi. 2012. "Selection of sourdough lactobacilli with antifungal activity for use as biopreservatives in bakery products." *Journal of Agricultural and Food Chemistry* 60 (31):7719–7728.

Gerez, Carla Luciana, Silvia Cuezzo, G Rollán, and G Font de Valdez. 2008. "Lactobacillus reuteri CRL 1100 as starter culture for wheat dough fermentation." *Food Microbiology* 25 (2):253–259.

Gobbetti, Marco, E Smacchi, and Aldo Corsetti. 1996. "The proteolytic system of Lactobacillus sanfrancisco CB1: Purification and characterization of a proteinase, a dipeptidase, and an aminopeptidase." *Applied and Environmental Microbiology* 62 (9):3220–3226.

Gobbetti, Marco, P Lavermicocca, F Minervini, M De Angelis, and Aldo Corsetti. 2000. "Arabinose fermentation by Lactobacillus plantarum in sourdough with added pentosans and α α -L-arabinofuranosidase: A tool to increase the production of acetic acid." *Journal of Applied Microbiology* 88 (2):317–324.

Gobbetti, M, M De Angelis, Aldo Corsetti, and R Di Cagno. 2005. "Biochemistry and physiology of sourdough lactic acid bacteria." *Trends in Food Science & Technology* 16 (1–3):57–69.

Guzmán-Ortiz, Fabiola Araceli, Javier Castro-Rosas, Carlos Alberto Gómez-Aldapa, Rosalva Mora-Escobedo, Adriana Rojas-León, María Luisa Rodríguez-Marín, Reyna Nallely Falfán-Cortés, and Alma Delia Román-Gutiérrez. 2019. "Enzyme activity during germination of different cereals: A review." *Food Reviews International* 35 (3):177–200.

Hammes, Walter P, and MG Gänzle. 1998. "Sourdough breads and related products." In *Microbiology of Fermented Foods*, 199–216. Springer.

Haros, Monica, Maria Bielecka, and Yolanda Sanz. 2005. "Phytase activity as a novel metabolic feature in Bifidobacterium." *FEMS Microbiology Letters* 247 (2):231–239.

Hole, Anastasia S, Ida Rud, Stine Grimmer, Stefanie Sigl, Judith Narvhus, and Stefan Sahlstrøm. 2012. "Improved bioavailability of dietary phenolic acids in whole grain barley and oat groat following fermentation with probiotic Lactobacillus acidophilus, Lactobacillus johnsonii, and Lactobacillus reuteri." *Journal of Agricultural and Food Chemistry* 60 (25):6369–6375.

Höltzel, Alexandra, Michael G Gänzle, Graeme J Nicholson, Walter P Hammes, and Günther Jung. 2000. "The first low molecular weight antibiotic from lactic acid bacteria: Reutericyclin, a new tetramic acid." *Angewandte Chemie International Edition* 39 (15):2766–2768.

Hu, Ying, and Michael G Gänzle. 2018. "Effect of temperature on production of oligosaccharides and dextran by Weissella cibaria 10 M." *International Journal of Food Microbiology* 280:27–34.

Hu, Ying, Achim Stromeck, Jussi Loponen, Daise Lopes-Lutz, Andreas Schieber, and Michael G Gänzle. 2011. "LC-MS/MS quantification of bioactive angiotensin I-converting enzyme inhibitory peptides in rye malt sourdoughs." *Journal of Agricultural and Food Chemistry* 59 (22):11983–11989.

Huang, Xiaoning, Yi Fan, Jiao Meng, Shanfeng Sun, Xiaoyong Wang, Jingyu Chen, and Bei-Zhong Han. 2021. "Laboratory-scale fermentation and multidimensional screening of lactic acid bacteria from Daqu." *Food Bioscience* 40:100853.

Iliev, Ilia, Tonka Vasileva, Veselin Bivolarski, Ayshe Salim, Sandrine Morel, Philippe Rabier, and Valérie Gabriel. 2018. "Optimization of the expression of levansucrase L17 in recombinant E. coli." *Biotechnology & Biotechnological Equipment* 32 (2):477–486.

İspirli, Hümeyra, Mustafa Onur Yüzer, Christopher Skory, Ian J Colquhoun, Osman Sağdıç, and Enes Dertli. 2019. "Characterization of a glucansucrase from Lactobacillus reuteri E81 and production of malto-oligosaccharides." *Biocatalysis and Biotransformation* 37 (6):421–430.

İspirli, Hümeyra, Duygu Özmen, Mustafa Tahsin Yılmaz, Osman Sağdıç, and Enes Dertli. 2020. "Impact of glucan type exopolysaccharide (EPS) production on technological characteristics of sourdough bread." *Food Control* 107:106812.

Jänsch, André, Maher Korakli, Rudi F Vogel, and Michael G Gänzle. 2007. "Glutathione reductase from Lactobacillus sanfranciscensis DSM20451T: Contribution to oxygen tolerance and thiol exchange reactions in wheat sourdoughs." *Applied and Environmental Microbiology* 73 (14):4469–4476.

Kajala, Ilkka, Qiao Shi, Antti Nyyssölä, Ndegwa Henry Maina, Yaxi Hou, Kati Katina, Maija Tenkanen, and Riikka Juvonen. 2015. "Cloning and characterization of a Weissella confusa dextransucrase and its application in high fibre baking." *PloS One* 10 (1):e0116418.

Karaman, Kevser, Osman Sagdic, and M Zeki Durak. 2018. "Use of phytase active yeasts and lactic acid bacteria isolated from sourdough in the production of whole wheat bread." *LWT* 91:557–567.

Katina, Kati, K-H Liukkonen, Anu Kaukovirta-Norja, Herman Adlercreutz, S-M Heinonen, A-M Lampi, J-M Pihlava, and Kaisa Poutanen. 2007. "Fermentation-induced changes in the nutritional value of native or germinated rye." *Journal of Cereal Science* 46 (3):348–355.

Knorr, Ruth, Matthias A Ehrmann, and Rudi F Vogel. 2001a. "Cloning of the phosphotransacetylase gene from Lactobacillus sanfranciscensis and characterization of its gene product." *Journal of Basic Microbiology: An International Journal on Biochemistry, Physiology, Genetics, Morphology, and Ecology of Microorganisms* 41 (6):339–349.

Knorr, Ruth, Matthias A Ehrmann, and Rudi F Vogel. 2001b. "Cloning, expression, and characterization of acetate kinase from Lactobacillus sanfranciscensis." *Microbiological Research* 156 (3):267–277.

Koistinen, Ville M, Outi Mattila, Kati Katina, Kaisa Poutanen, Anna-Marja Aura, and Kati Hanhineva. 2018. "Metabolic profiling of sourdough fermented wheat and rye bread." *Scientific Reports* 8 (1):1–11.

Konopka, Iwona, Małgorzata Tańska, Alicja Faron, and Sylwester Czaplicki. 2014. "Release of free ferulic acid and changes in antioxidant properties during the wheat and rye bread making process." *Food Science and Biotechnology* 23 (3):831–840.

Korakli, Maher, and Rudi F Vogel. 2003. "Purification and characterisation of mannitol dehydrogenase from Lactobacillus sanfranciscensis." *FEMS Microbiology Letters* 220 (2):281–286.

Korakli, Maher, Andreas Rossmann, Michael G Gänzle, and Rudi F Vogel. 2001. "Sucrose metabolism and exopolysaccharide production in wheat and rye sourdoughs by Lactobacillus sanfranciscensis." *Journal of Agricultural and Food Chemistry* 49 (11):5194–5200.

Kumar, Vikas, Amit K Sinha, Harinder PS Makkar, and Klaus Becker. 2010. "Dietary roles of phytate and phytase in human nutrition: A review." *Food Chemistry* 120 (4):945–959.

Laguerre, Sandrine, Myriam Amari, Marlène Vuillemin, Hervé Robert, Valentin Loux, Christophe Klopp, Sandrine Morel, Bruno Gabriel, Magali Remaud-Siméon, and Valérie Gabriel. 2012. "Genome sequences of three Leuconostoc citreum strains, LBAE C10, LBAE C11, and LBAE E16, isolated from wheat sourdoughs." *American Society for Microbiology* 2012:1610–1611.

Lhomme, Emilie, Anna Lattanzi, Xavier Dousset, Fabio Minervini, Maria De Angelis, Guylaine Lacaze, Bernard Onno, and Marco Gobbetti. 2015. "Lactic acid bacterium and yeast microbiotas of sixteen French traditional sourdoughs." *International Journal of Food Microbiology* 215:161–170.

Li, Qing, and Michael G Gänzle. 2020. "Characterization of two extracellular arabinanases in Lactobacillus crispatus." *Applied Microbiology and Biotechnology* 104 (23):10091–10103.

Li, Qing, QianYing Tao, Jaunana S Teixeira, Marcia Shu-Wei Su, and Michael G Gänzle. 2020. "Contribution of glutaminases to glutamine metabolism and acid resistance in Lactobacillus reuteri and other vertebrate host adapted lactobacilli." *Food Microbiology* 86:103343.

Limón, Rocio I, Elena Peñas, M Inés Torino, Cristina Martínez-Villaluenga, Montserrat Dueñas, and Juana Frias. 2015. "Fermentation enhances the content of bioactive compounds in kidney bean extracts." *Food Chemistry* 172:343–352.

Lin, Xiaoxi B, Christopher T Lohans, Rebbeca Duar, Jinshui Zheng, John C Vederas, Jens Walter, and Michael Gänzle. 2015. "Genetic determinants of reutericyclin biosynthesis in Lactobacillus reuteri." *Applied and Environmental Microbiology* 81 (6):2032–2041.

Liu, Tongjie, Yang Li, Yuanyi Yang, Huaxi Yi, Lanwei Zhang, and Guoqing He. 2020. "The influence of different lactic acid bacteria on sourdough flavor and a deep insight into sourdough fermentation through RNA sequencing." *Food Chemistry* 307:125529.

Lopez, Hubert W, Ariane Ouvry, Elisabeth Bervas, Christine Guy, Arnaud Messager, Christian Demigne, and Christian Remesy. 2000. "Strains of lactic acid bacteria isolated from sour doughs degrade phytic acid and improve calcium and magnesium solubility from whole wheat flour." *Journal of Agricultural and Food Chemistry* 48 (6):2281–2285.

M'hir, Sana, Jean-Marc Aldric, Thami El-Mejdoub, Jaqueline Destain, Mondher Mejri, Moktar Hamdi, and Phillippe Thonart. 2008. "Proteolytic breakdown of gliadin by Enterococcus faecalis isolated from Tunisian fermented dough." *World Journal of Microbiology and Biotechnology* 24 (12):2775–2781.

Maidana, Stefania Dentice, Cecilia Aristimuño Ficoseco, Daniela Bassi, Pier Sandro Cocconcelli, Edoardo Puglisi, Graciela Savoy, Graciela Vignolo, and Cecilia Fontana. 2020. "Biodiversity and technological-functional potential of lactic acid bacteria isolated from spontaneously fermented chia sourdough." *International Journal of Food Microbiology* 316:108425.

Maske, Bruna L, Gilberto V de Melo Pereira, Alexander da S Vale, Dão Pedro de Carvalho Neto, Susan Grace Karp, Jéssica A Viesser, Juliano De Dea Lindner, Maria Giovana Pagnoncelli, Vanete Thomaz Soccol, and Carlos R Soccol. 2021. "A review on enzyme-producing lactobacilli associated with the human digestive process: from metabolism to application." *Enzyme and Microbial Technology* 149:109836.

Milanović, Vesna, Andrea Osimani, Cristiana Garofalo, Luca Belleggia, Antonietta Maoloni, Federica Cardinali, Massimo Mozzon, Roberta Foligni, Lucia Aquilanti, and Francesca Clementi. 2020. "Selection of cereal-sourced lactic acid bacteria as candidate starters for the baking industry." *PLoS One* 15 (7):e0236190.

Mohan Rao, T Jagan, and Arun Goyal. 2013. "Purification, optimization of assay, and stability studies of dextransucrase isolated from Weissella cibaria JAG8." *Preparative Biochemistry and Biotechnology* 43 (4):329–341.

Montemurro, Marco, Erica Pontonio, Marco Gobbetti, and Carlo Giuseppe Rizzello. 2019. "Investigation of the nutritional, functional and technological effects of the sourdough fermentation of sprouted flours." *International Journal of Food Microbiology* 302:47–58.

Nakamura, Toshihide, Ayako Yoshida, Noriko Komatsuzaki, Toshiyuki Kawasumi, and Jun Shima. 2007. "Isolation and characterization of a low molecular weight peptide contained in sourdough." *Journal of Agricultural and Food Chemistry* 55 (12):4871–4876.

Nionelli, Luana, Marco Montemurro, Erica Pontonio, Michela Verni, Marco Gobbetti, and Carlo Giuseppe Rizzello. 2018. "Pro-technological and functional characterization of lactic acid bacteria to be used as starters for hemp (Cannabis sativa L.) sourdough fermentation and wheat bread fortification." *International Journal of Food Microbiology* 279:14–25.

Nuobariene, Lina, Åse S Hansen, and Nils Arneborg. 2012. "Isolation and identification of phytase-active yeasts from sourdoughs." *LWT: Food Science and Technology* 48 (2):190–196.

Ogunremi, Omotade Richard, Renu Agrawal, and Abiodun Sanni. 2020. "Production and characterization of volatile compounds and phytase from potentially probiotic yeasts isolated from traditional fermented cereal foods in Nigeria." *Journal of Genetic Engineering and Biotechnology* 18:1–8.

Olivares-Illana, V, C Wacher-Rodarte, S Le Borgne, and A López-Munguía. 2002. "Characterization of a cell-associated inulosucrase from a novel source: A Leuconostoc citreum strain isolated from Pozol, a fermented corn beverage of Mayan origin." *Journal of Industrial Microbiology and Biotechnology* 28 (2):112–117.

Olvera, Clarita, Rosa I Santamaría, Patricia Bustos, Cristina Vallejo, Juan J Montor, Carmen Wacher, and Agustín Lopez Munguia. 2017. "Draft genome sequence of Leuconostoc citreum CW28 isolated from pozol, a pre-hispanic fermented corn beverage." *Genome Announcements* 5 (48):e01283-17.

Ortiz, Maria Eugenia, Juliana Bleckwedel, Silvina Fadda, Gianluca Picariello, Elvira M Hebert, Raúl R Raya, and Fernanda Mozzi. 2017. "Global analysis of mannitol 2-dehydrogenase in Lactobacillus reuteri CRL 1101 during mannitol production through enzymatic, genetic and proteomic approaches." *PLoS One* 12 (1):e0169441.

Palacios, MC, M Haros, CM Rosell, and Y Sanz. 2005. "Characterization of an acid phosphatase from Lactobacillus pentosus: Regulation and biochemical properties." *Journal of Applied Microbiology* 98 (1):229–237.

Palaniswamy, Sakthi Kumaran, and Vijayalakshmi Govindaswamy. 2016. "In-vitro probiotic characteristics assessment of feruloyl esterase and glutamate decarboxylase producing Lactobacillus spp. isolated from traditional fermented millet porridge (kambu koozh)." *LWT: FOOD SCIENCE and Technology* 68:208–216.

Palla, Michela, Massimo Blandino, Arianna Grassi, Debora Giordano, Cristina Sgherri, Mike Frank Quartacci, Amedeo Reyneri, Monica Agnolucci, and Manuela Giovannetti. 2020. "Characterization and selection of functional yeast strains during sourdough fermentation of different cereal wholegrain flours." *Scientific Reports* 10 (1):1–15.

Palla, Michela, Caterina Cristani, Manuela Giovannetti, and Monica Agnolucci. 2017. "Identification and characterization of lactic acid bacteria and yeasts of PDO Tuscan bread sourdough by culture dependent and independent methods." *International Journal of Food Microbiology* 250:19–26.

Palomba, Simona, Silvana Cavella, Elena Torrieri, Alessandro Piccolo, Pierluigi Mazzei, Giuseppe Blaiotta, Valeria Ventorino, and Olimpia Pepe. 2012. "Wheat sourdough from Leuconostoc lactis and Lactobacillus curvatus exopolysaccharide-producing starter culture: Polyphasic screening, homopolysaccharide composition and viscoelastic behavior." *Applied and Environmental Microbiology*.

Peñas, Elena, Marina Diana, Juana Frías, Joan Quílez, and Cristina Martínez-Villaluenga. 2015. "A multistrategic approach in the development of sourdough bread targeted towards blood pressure reduction." *Plant Foods for Human Nutrition* 70 (1):97–103.

Pepe, O, F Villani, D Oliviero, T Greco, and S Coppola. 2003. "Effect of proteolytic starter cultures as leavening agents of pizza dough." *International Journal of Food Microbiology* 84 (3):319–326.

Perri, Giuseppe, Rossana Coda, Carlo Giuseppe Rizzello, Giuseppe Celano, Marco Ampollini, Marco Gobbetti, Maria De Angelis, and Maria Calasso. 2021. "Sourdough fermentation of whole and sprouted lentil flours: In situ formation of dextran and effects on the nutritional, texture and sensory characteristics of white bread." *Food Chemistry* 355:129638.

Perricone, Marianne, Antonio Bevilacqua, Maria Rosaria Corbo, and Milena Sinigaglia. 2014. "Technological characterization and probiotic traits of yeasts isolated from Altamura sourdough to select promising microorganisms as functional starter cultures for cereal-based products." *Food Microbiology* 38:26–35.

Pontonio, Erica, Jennifer Mahony, Raffaella Di Cagno, Mary O'Connell Motherway, Gabriele Andrea Lugli, Amy O'Callaghan, Maria De Angelis, Marco Ventura, Marco Gobbetti, and Douwe van Sinderen. 2016. "Cloning, expression and characterization of a β-D-xylosidase from Lactobacillus rossiae DSM 15814 T." *Microbial Cell Factories* 15 (1):1–12.

Ramos-de-la-Peña, Ana Mayela, and Juan Carlos Contreras-Esquivel. 2016. "Methods and substrates for feruloyl esterase activity detection, a review." *Journal of Molecular Catalysis B: Enzymatic* 130:74–87.

Rao, T, Jagan Mohan, and Arun Goyal. 2013. "A novel high dextran yielding Weissella cibaria JAG8 for cereal food application." *International Journal of Food Sciences and Nutrition* 64 (3):346–354.

Rasmussen, Claus V, Hanne Boskov Hansen, Åse Hansen, and Lone Melchior Larsen. 2001. "pH-, temperature-and time-dependent activities of endogenous endo-β-D-xylanase, β-D-xylosidase and α-L-arabinofuranosidase in extracts from ungerminated rye (Secale cereale L.) grain." *Journal of Cereal Science* 34 (1):49–60.

Ravisankar, Shreeya, Halef Dizlek, and Joseph M Awika. 2021. "Changes in extractable phenolic profile during natural fermentation of wheat, sorghum and teff." *Food Research International* 145:110426.

Reale, Anna, Teresa Zotta, Rocco G Ianniello, Gianfranco Mamone, and Tiziana Di Renzo. 2020. "Selection criteria of lactic acid bacteria to be used as starter for sweet and salty leavened baked products." *LWT* 133:110092.

Reale, Anna, Luigia Di Stasio, Tiziana Di Renzo, Salvatore De Caro, Pasquale Ferranti, Gianluca Picariello, Francesco Addeo, and Gianfranco Mamone. 2021. "Bacteria do it better! Proteomics suggests the molecular basis for improved digestibility of sourdough products." *Food Chemistry* 359:129955.

Ripari, Valery, Yunpeng Bai, and Michael G Gänzle. 2019. "Metabolism of phenolic acids in whole wheat and rye malt sourdoughs." *Food Microbiology* 77:43–51.

Rizzello, Carlo Giuseppe, Angela Cassone, Rossana Coda, and Marco Gobbetti. 2011. "Antifungal activity of sourdough fermented wheat germ used as an ingredient for bread making." *Food Chemistry* 127 (3):952–959.

Rizzello, Carlo Giuseppe, Luana Nionelli, Rossana Coda, and Marco Gobbetti. 2012. "Synthesis of the cancer preventive peptide lunasin by lactic acid bacteria during sourdough fermentation." *Nutrition and Cancer* 64 (1):111–120.

Rizzello, Carlo Giuseppe, Blanca Hernández-Ledesma, Samuel Fernández-Tomé, José Antonio Curiel, Daniela Pinto, Barbara Marzani, Rossana Coda, and Marco Gobbetti. 2015. "Italian legumes: Effect of sourdough fermentation on lunasin-like polypeptides." *Microbial Cell Factories* 14 (1):1–20.

Rizzello, Carlo Giuseppe, Anna Lorusso, Vito Russo, Daniela Pinto, Barbara Marzani, and Marco Gobbetti. 2017. "Improving the antioxidant properties of quinoa flour through fermentation with selected autochthonous lactic acid bacteria." *International Journal of Food Microbiology* 241:252–261.

Rizzello, CG, A Cassone, R Di Cagno, and M Gobbetti. 2008. "Synthesis of angiotensin I-converting enzyme (ACE)-inhibitory peptides and γ-aminobutyric acid (GABA) during sourdough fermentation by selected lactic acid bacteria." *Journal of Agricultural and Food Chemistry* 56 (16):6936–6943.

Rollán, G, and G Font de Valdez. 2001. "The peptide hydrolase system of Lactobacillus reuteri." *International Journal of Food Microbiology* 70 (3):303–307.

Rollan, G, GL Lorca, and G Font de Valdez. 2003. "Arginine catabolism and acid tolerance response in Lactobacillus reuteri isolated from sourdough." *Food Microbiology* 20 (3):313–319.

Rollan, G, M De Angelis, M Gobbetti, and G Font De Valdez. 2005. "Proteolytic activity and reduction of gliadin-like fractions by sourdough lactobacilli." *Journal of Applied Microbiology* 99 (6):1495–1502.

Rollán, Graciela C, Carla L Gerez, and Jean G LeBlanc. 2019. "Lactic fermentation as a strategy to improve the nutritional and functional values of pseudocereals." *Frontiers in Nutrition* 6:98.

Romanowska, Irena, Jacek Polak, and Stanisław Bielecki. 2006. "Isolation and properties of Aspergillus niger IBT-90 xylanase for bakery." *Applied Microbiology and Biotechnology* 69 (6):665–671.

Rühmkorf, Christine, Christian Bork, Petra Mischnick, Heinrich Rübsam, Thomas Becker, and Rudi F Vogel. 2013. "Identification of Lactobacillus curvatus TMW 1.624 dextransucrase and comparative characterization with Lactobacillus reuteri TMW 1.106 and Lactobacillus animalis TMW 1.971 dextransucrases." *Food Microbiology* 34 (1):52–61.

Sahin, Aylin W, Tom Rice, Emanuele Zannini, Kieran M Lynch, Aidan Coffey, and Elke K Arendt. 2019. "The incorporation of sourdough in sugar-reduced biscuits: A promising strategy to improve techno-functional and sensory properties." *European Food Research and Technology* 245 (9):1841–1854.

Sahin, Aylin W, Tom Rice, and Aidan Coffey. 2021. "Genomic analysis of Leuconostoc citreum TR116 with metabolic reconstruction and the effects of fructose on gene expression for mannitol production." *International Journal of Food Microbiology* 354:109327.

Sandez Penidez, Sergio H, Marina A Velasco Manini, Carla L Gerez, and Graciela C Rollán. 2020. "Partial characterization and purification of phytase from Lactobacillus plantarum CRL1964 isolated from pseudocereals." *Journal of Basic Microbiology* 60 (9):787–798.

Santoyo, M Calderon, G Loiseau, R Rodriguez Sanoja, and JP Guyot. 2003. "Study of starch fermentation at low pH by Lactobacillus fermentum Ogi E1 reveals uncoupling between growth and α-amylase production at pH 4.0." *International Journal of Food Microbiology* 80 (1):77–87.

Schettino, Rosa, Erica Pontonio, Marco Gobbetti, and Carlo Giuseppe Rizzello. 2020. "Extension of the shelf-life of fresh pasta using chickpea flour fermented with selected lactic acid bacteria." *Microorganisms* 8 (9):1322.

Schulz, Kathrin, Annabel Nieter, Ann-Karolin Scheu, José L Copa-Patiño, David Thiesing, Lutz Popper, and Ralf G Berger. 2018. "A type D ferulic acid esterase from Streptomyces werraensis affects the volume of wheat dough pastries." *Applied Microbiology and Biotechnology* 102 (3):1269–1279.

Schwab, Clarissa, Jens Walter, Gerald W Tannock, Rudi F Vogel, and Michael G Gänzle. 2007. "Sucrose utilization and impact of sucrose on glycosyltransferase expression in Lactobacillus reuteri." *Systematic and Applied Microbiology* 30 (6):433–443.

Serrazanetti, Diana I, Maria Elisabetta Guerzoni, Aldo Corsetti, and Rudi Vogel. 2009. "Metabolic impact and potential exploitation of the stress reactions in lactobacilli." *Food Microbiology* 26 (7):700–711.

Sharma, Neha, Steffy Angural, Monika Rana, Neena Puri, Kanthi Kiran Kondepudi, and Naveen Gupta. 2020. "Phytase producing lactic acid bacteria: Cell factories for enhancing micronutrient bioavailability of phytate rich foods." *Trends in Food Science & Technology* 96:1–12.

Sharma, Rekha, Piyush Kumar, Vandana Kaushal, Rahul Das, and Naveen Kumar Navani. 2018. "A novel protein tyrosine phosphatase like phytase from Lactobacillus fermentum NKN51: Cloning, characterization and application in mineral release for food technology applications." *Bioresource Technology* 249:1000–1008.

Stolz, Peter, Rudi F Vogel, and Walter P Hammes. 1995. "Utilization of electron acceptors by lactobacilli isolated from sourdough." *Zeitschrift für Lebensmittel-Untersuchung und Forschung* 201 (4):402–410.

Stromeck, Achim, Ying Hu, Lingyun Chen, and Michael G Gänzle. 2011. "Proteolysis and bioconversion of cereal proteins to glutamate and γ-aminobutyrate (GABA) in rye malt sourdoughs." *Journal of Agricultural and Food Chemistry* 59 (4):1392–1399.

Su, Marcia S, Sabine Schlicht, and Michael G Gänzle. 2011. "Contribution of glutamate decarboxylase in Lactobacillus reuteri to acid resistance and persistence in sourdough fermentation." *Microbial Cell Factories* 10:1–12, BioMed Central.

Tamayo-Ramos, Juan Antonio, Juan Mario Sanz-Penella, María J Yebra, Vicente Monedero, and Monika Haros. 2012. "Novel phytases from Bifidobacterium pseudocatenulatum ATCC 27919 and Bifidobacterium longum subsp. infantis ATCC 15697." *Applied and Environmental Microbiology* 78 (14):5013–5015.

Tang, Kai Xing, Cindy J Zhao, and Michael G Gänzle. 2017. "Effect of glutathione on the taste and texture of type I sourdough bread." *Journal of Agricultural and Food Chemistry* 65 (21):4321–4328.

Teixeira, Januana S, Arisha Seeras, Alma Fernanda Sanchez-Maldonado, Chonggang Zhang, Marcia Shu-Wei Su, and Michael G Gänzle. 2014. "Glutamine, glutamate, and arginine-based acid resistance in Lactobacillus reuteri." *Food Microbiology* 42:172–180.

Thiele, Claudia, Michael G Gänzle, and Rudi F Vogel. 2003. "Fluorescence labeling of wheat proteins for determination of gluten hydrolysis and depolymerization during dough processing and sourdough fermentation." *Journal of Agricultural and Food Chemistry* 51 (9):2745–2752.

Tieking, Markus, Maher Korakli, Matthias A Ehrmann, Michael G Gänzle, and Rudi F Vogel. 2003. "In situ production of exopolysaccharides during sourdough fermentation by cereal and intestinal isolates of lactic acid bacteria." *Applied and Environmental Microbiology* 69 (2):945–952.

Tieking, Markus, Matthias A Ehrmann, Rudi F Vogel, and Michael G Gänzle. 2005. "Molecular and functional characterization of a levansucrase from the sourdough isolate Lactobacillus sanfranciscensis TMW 1.392." *Applied Microbiology and Biotechnology* 66 (6):655–663.

Torino, Maria Inés, Rocío I Limón, Cristina Martínez-Villaluenga, Sari Mäkinen, Anne Pihlanto, Concepción Vidal-Valverde, and Juana Frias. 2013. "Antioxidant and antihypertensive properties of liquid and solid state fermented lentils." *Food Chemistry* 136 (2):1030–1037.

Ua-Arak, Tharalinee, Frank Jakob, and Rudi F Vogel. 2017. "Influence of levan-producing acetic acid bacteria on buckwheat-sourdough breads." *Food Microbiology* 65:95–104.

Valerio, Francesca, Anna Rita Bavaro, Mariaelena Di Biase, Stella Lisa Lonigro, Antonio Francesco Logrieco, and Paola Lavermicocca. 2020. "Effect of amaranth and quinoa flours on exopolysaccharide production and protein profile of liquid sourdough fermented by Weissella cibaria and Lactobacillus plantarum." *Frontiers in Microbiology* 11:967.

Van der Meulen, Roel, Silvia Grosu-Tudor, Fernanda Mozzi, Frederik Vaningelgem, Medana Zamfir, Graciela Font de Valdez, and Luc De Vuyst. 2007. "Screening of lactic acid bacteria isolates from dairy and cereal products for exopolysaccharide production and genes involved." *International Journal of Food Microbiology* 118 (3):250–258.

van Hijum, Sacha AFT, Slavko Kralj, Lukasz K Ozimek, Lubbert Dijkhuizen, and Ineke GH van Geel-Schutten. 2006. "Structure-function relationships of glucansucrase and fructansucrase enzymes from lactic acid bacteria." *Microbiology and Molecular Biology Reviews* 70 (1):157–176.

Van Hijum, SAFT, MJEC Van Der Maarel, and L Dijkhuizen. 2003. "Kinetic properties of an inulosucrase from Lactobacillus reuteri 121." *FEBS Letters* 534 (1–3):207–210.

Venturi, Manuel, Viola Galli, Niccolò Pini, Simona Guerrini, and Lisa Granchi. 2019. "Use of selected lactobacilli to increase γ-Aminobutyric acid (GABA) content in sourdough bread enriched with amaranth flour." *Foods* 8 (6):218.

Vermeulen, Nicoline, Melanie Pavlovic, Matthias A Ehrmann, Michael G Gänzle, and Rudi F Vogel. 2005. "Functional characterization of the proteolytic system of Lactobacillus sanfranciscensis DSM 20451T during growth in sourdough." *Applied and Environmental Microbiology* 71 (10):6260–6266.

Vermeulen, Nicoline, Jan Kretzer, Hetty Machalitza, Rudi F Vogel, and Michael G Gänzle. 2006. "Influence of redox-reactions catalysed by homo-and hetero-fermentative lactobacilli on gluten in wheat sourdoughs." *Journal of Cereal Science* 43 (2):137–143.

Vermeulen, Nicoline, Michael G Gänzle, and Rudi F Vogel. 2007. "Glutamine deamidation by cereal-associated lactic acid bacteria." *Journal of Applied Microbiology* 103 (4):1197–1205.

Villegas, Josefina M, Lucia Brown, Graciela Savoy de Giori, and Elvira M Hebert. 2016. "Optimization of batch culture conditions for GABA production by Lactobacillus brevis CRL 1942, isolated from quinoa sourdough." *LWT: Food Science and Technology* 67:22–26.

Vrancken, Gino, Tom Rimaux, Luc De Vuyst, and Frederic Leroy. 2008. "Kinetic analysis of growth and sugar consumption by Lactobacillus fermentum IMDO 130101 reveals adaptation to the acidic sourdough ecosystem." *International Journal of Food Microbiology* 128 (1):58–66.

Vrancken, Gino, Tom Rimaux, Stefan Weckx, Luc De Vuyst, and Frederic Leroy. 2009a. "Environmental pH determines citrulline and ornithine release through the arginine deiminase pathway in Lactobacillus fermentum IMDO 130101." *International Journal of Food Microbiology* 135 (3):216–222.

Vrancken, Gino, Tom Rimaux, Dorrit Wouters, Frederic Leroy, and Luc De Vuyst. 2009b. "The arginine deiminase pathway of Lactobacillus fermentum IMDO 130101 responds to growth under stress conditions of both temperature and salt." *Food Microbiology* 26 (7):720–727.

Vrancken, Gino, Luc De Vuyst, Tom Rimaux, Joke Allemeersch, and Stefan Weckx. 2011. "Adaptation of Lactobacillus plantarum IMDO 130201, a wheat sourdough isolate, to growth in wheat sourdough simulation medium at different pH values through differential gene expression." *Applied and Environmental Microbiology* 77 (10):3406–3412.

Waldherr, Florian W, Daniel Meissner, and Rudi F Vogel. 2008. "Genetic and functional characterization of Lactobacillus panis levansucrase." *Archives of Microbiology* 190 (4):497–505.

Wang, Xiaomei, Yajun Bai, Yujie Cai, and Xiaohui Zheng. 2017. "Biochemical characteristics of three feruloyl esterases with a broad substrate spectrum from Bacillus amyloliquefaciens H47." *Process Biochemistry* 53:109–115.

Weckx, Stefan, Joke Allemeersch, Roel Van der Meulen, Gino Vrancken, Geert Huys, Peter Vandamme, Paul Van Hummelen, and Luc De Vuyst. 2011. "Metatranscriptome analysis for insight into whole-ecosystem gene expression during spontaneous wheat and spelt sourdough fermentations." *Applied and Environmental Microbiology* 77 (2):618–626.

Weckx, Stefan, Simon Van Kerrebroeck, and Luc De Vuyst. 2019. "Omics approaches to understand sourdough fermentation processes." *International Journal of Food Microbiology* 302:90–102.

Xie, Jin, and Michael G Gänzle. 2021. "Characterization of γ-glutamyl cysteine ligases from Limosilactobacillus reuteri producing kokumi-active γ-glutamyl dipeptides." *Applied Microbiology and Biotechnology* 105:1–13.

Xu, Zhenshang, Huiying He, Susu Zhang, Tingting Guo, and Jian Kong. 2017. "Characterization of feruloyl esterases produced by the four lactobacillus species: L. amylovorus, L. acidophilus, L. farciminis and L. fermentum, isolated from ensiled corn stover." *Frontiers in Microbiology* 8:941.

Yan, Bowen, Yuan Yao Chen, Weilan Wang, Jianxin Zhao, Wei Chen, and Michael Gänzle. 2018. "γ-Glutamyl cysteine ligase of Lactobacillus reuteri synthesizes γ-glutamyl dipeptides in sourdough." *Journal of Agricultural and Food Chemistry* 66 (46):12368–12375.

Yildirim, Rusen Metin, and Muhammet Arici. 2019. "Effect of the fermentation temperature on the degradation of phytic acid in whole-wheat sourdough bread." *LWT* 112:108224.

Yu, Y., Wang, L., Qian, H., Zhang, H., and Qi, X. (2018). Contribution of spontaneously-fermented sourdoughs with pear and navel orange for the bread-making. *LWT* 89: 336–343.

Zhang, Binle, Yang Wendan, Feng Wang, Jacob Ojobi Omedi, Ruoshi Liu, Jinxin Huang, Luan Zhang, Qibo Zou, Weining Huang, and Shaolei Li. 2019. "Use of Kluyveromyces marxianus prefermented wheat bran as a source of enzyme mixture to improve dough performance and bread biochemical properties." *Cereal Chemistry* 96 (1):142–153.

Zhang, Juan, Yin Li, Wei Chen, Guo-Cheng Du, and Jian Chen. 2012. "Glutathione improves the cold resistance of Lactobacillus sanfranciscensis by physiological regulation." *Food Microbiology* 31 (2):285–292.

Zhao, Cindy J, and Michael G Ganzle. 2016. "Synthesis of taste-active γ-glutamyl dipeptides during sourdough fermentation by Lactobacillus reuteri." *Journal of Agricultural and Food Chemistry* 64 (40):7561–7568.

Zhao, Cindy J, Ying Hu, Andreas Schieber, and Michael Gänzle. 2013. "Fate of ACE-inhibitory peptides during the bread-making process: Quantification of peptides in sourdough, bread crumb, steamed bread and soda crackers." *Journal of Cereal Science* 57 (3):514–519.

Zhao, Xin, and Michael G Gänzle. 2018. "Genetic and phenotypic analysis of carbohydrate metabolism and transport in Lactobacillus reuteri." *International Journal of Food Microbiology* 272:12–21.

Zheng, Jinshui, Xin Zhao, Xiaoxi B Lin, and Michael Gänzle. 2015. "Comparative genomics Lactobacillus reuteri from sourdough reveals adaptation of an intestinal symbiont to food fermentations." *Scientific Reports* 5 (1):1–11.

Zieliński, Henryk, Joanna Honke, Joanna Topolska, Natalia Bączek, Mariusz Konrad Piskuła, Wiesław Wiczkowski, and Małgorzata Wronkowska. 2020. "ACE inhibitory properties and phenolics profile of fermented flours and of baked and digested biscuits from buckwheat." *Foods* 9 (7):847.

Zotta, Teresa, Annamaria Ricciardi, and Eugenio Parente. 2007. "Enzymatic activities of lactic acid bacteria isolated from Cornetto di Matera sourdoughs." *International Journal of Food Microbiology* 115 (2):165–172.

8 Enzyme Production from Sourdough

Mensure Elvan and Sebnem Harsa

CONTENTS

8.1 Introduction 199
8.2 Production of Enzymes Discovered from Sourdough 202
    8.2.1 Proteases 203
    8.2.2 Amylase 206
    8.2.3 Phytase 207
    8.2.4 Xylanase 210
    8.2.5 Tannase and Gallate Decarboxylase 215
    8.2.6 Glycosidase 215
    8.2.7 Esterase 216
    8.2.8 Glucansucrase and Fructansucrase 216
    8.2.9 Lipoxygenase 217
    8.2.10 FODMAP Hydrolysis 217
    8.2.11 Arginine Deiminase 218
8.3 Conclusion 218
References 219

8.1 INTRODUCTION

Enzymes are biological catalyzers, applied in numerous areas such as pharmaceuticals, cosmetics, textiles, feed, and food. Enzymes are taking an important place in these industrial applications, and their use is due to the recent scientific developments in the biochemical engineering and microbiological fields (Whitehurst and Van Oort 2010). Industrial enzymes derive from various sources, e.g., animal, plant, or microbial origin. Microbial production offers advantages, therefore most of the production is carried out by using various microorganisms. The recovery of intracellular enzymes, such as glucose oxidase, is rather difficult since these molecules usually remain connected to the cells. Extracellular enzymes, such as proteinases and lipases, account for most of the total sales. These biocatalysts are readily available from renewable resources, are biodegradable, and are active under mild conditions (e.g., temperature and pH) and selectivity in substrate and product stereochemistry; thus, they are superior alternatives to chemical catalysts (Jemli et al. 2016). The most used enzymes are derived from food and represent the largest share in the enzyme market (Miguel et al. 2013).

DOI: 10.1201/9781003141143-10

The global enzyme market is currently growing, being valued in 10.6 billion dollars in 2020; with an annual growth rate of 7.1%, this will mean an increased value market to 14.9 billion dollars by 2027. Enzymes are expected to be highly in demand for the global food and beverage industry due to the increasing consumption of functional foods, as a result of healthy lifestyle and rising consumer awareness of and demand for healthy foods (GrandViewResearch 2021). Furthermore, according to Novozymes, food, beverages, and human health enzymes represented 20% of total enzyme sales in 2020, this demand being mainly raised by market penetration and the success of suitable solutions in baking (Novozymes 2021). While carbohydrase and protease revenues make up about 70% of the present industrial enzyme market, lipase and phytase revenues make up about 8% and 7%, respectively. In addition, transferases, lactases, isomerases, and redox enzymes also contribute to this market. The increase in microbial enzyme production volume together with the increasing demand and use of enzymes are indicators of an increasing global enzyme market. The general processes for the industrial production of enzymes are shown in Figure 8.1.

Major enzyme manufacturers, such as Novozymes, DSM, DuPont, and BASF represent almost 75% of the industrial enzyme market (Arbige, Shetty, and Chotani 2019). There are two ways to use enzymes. In the first, raw materials are converted into main products by using enzymes; and in the second, functional attributes of products can be altered by the action of enzymes as food additives (Illanes 2008). Increasing consumer demand for qualified food products and large-scale production of them have resulted in using food additives such as enzymes (Joye, Lagrain, and Delcour 2009). Industrial enzymes and their applications in the food sector are shown in Figure 8.2 and Figure 8.3.

Enzymes are important additives applied in many bakery products mainly due to the limitations of chemicals or chemically derived additives in bread and other fermented foods during their manufacturing processes; thus, enzymes have gained great relevance in this sector (Cauvain and Young 2007). In the bakery industry, three sources of enzymes are routinely used: flour-based enzymes, added exogenous enzymes, and secreted enzymes by dominant microorganisms in the dough (Di Cagno et al. 2003). The exogenous enzymes are generally applied to modify

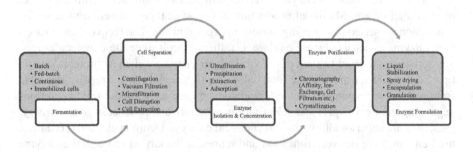

FIGURE 8.1 Main steps for enzyme production at industrial level.

Enzyme Production from Sourdough 201

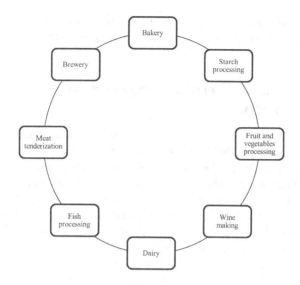

FIGURE 8.2 Enzymes and their applications in the food industries.

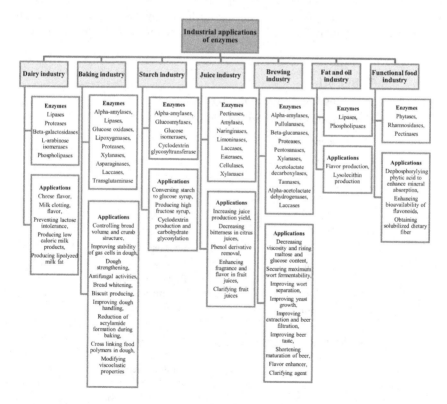

FIGURE 8.3 Industrial applications of enzymes.

the rheology of dough, crumb softness, and gas retention during bread production as well as to lower formation of acrylamide in bakery products (Cauvain and Young 2007).

8.2 PRODUCTION OF ENZYMES DISCOVERED FROM SOURDOUGH

Sourdough, an important resource of genetic diversity is also the source of a wide variety of enzymatic activities for biotechnological and industrial applications (Pepe et al. 2004) (see Table 8.1). The production of various industrial enzymes, such as amylase, protease, and glucosidase has been reported from by many microorganisms and cereals from sourdough. These enzymes have been applied extensively in food industries for food additive production, food processing and preservation (such as bread making), functional food production; they are also applied in the pharmaceutical industries. In addition, enzymes, i.e., xylanase and phytase, play an effective role in both the technological and nutritional potential of foods (Poutanen, Flander, and Katina 2009).

Activity of cereal enzymes and metabolism of sourdough microorganisms are interdependent. Acidifying modifies the cereal's enzymatic activity and the solubility of the substrates (De Vuyst and Neysens 2005). Enzymes from heterofermentative lactic acid bacteria (LAB) metabolism were affected by reducing the sourdough's oxidation-reduction potential, and by the accumulation of low-molecular-weight

TABLE 8.1
Enzymes Used to Metabolize Different Compounds during Sourdough Fermentation

Enzymatic metabolisms	Enzymes
Carbohydrate metabolism	Alpha-amylase
	Beta-amylase
	Glucoamylase
	Glycosidases
	Xylanase
	Alpha-galactosidase
	Fructanase
Protein metabolism	Proteases
	Peptidases
	Aspartic proteases
Lipid metabolism	Lipoxygenase
	Lyase
Phenolic compound metabolism	Glycosidases
	Esterase
	Feruloyl esterase
	Tannase
	Gallate decarboxylases

thiol substances (Capuani, Behr, and Vogel 2012, Jänsch et al. 2007). In turn, cereal enzymes degrade complex carbohydrates into available substrates for microbial growth (Hammes and Gänzle 1998).

8.2.1 Proteases

Proteases constitute one of the largest industrial enzyme groups, and they have an important role in industrial biotechnology, especially in food, medicine, detergent, and pharmaceutical production (Zhang, Wang, and Jiang 2020). These enzymes are used by different food sectors, such as cheese production, meat tenderization, and prevention of haze in beer (Damodaran, Parkin, and Fennema 2007) as well as to produce breads, crackers, and sweet bakery products (Bombara, Anon, and Pilosof 1997) (see Figure 8.3). The wide range of applications of proteases has led to an increasing demand for these enzymes that ideally have high enzymatic activity and can be produced from cheap substrates (Zhang, Wang, and Jiang 2020). Studies have shown that sourdough is a low-cost and natural carbon source for protease production.

Proteases hydrolyze peptide bonds of protein molecules. These enzymes are divided into two main groups based on the action side: endopeptidases and exopeptidases. Primary proteolysis is based on endogenous proteinases (Gänzle, Loponen, and Gobbetti 2008, Thiele, Grassl, and Gänzle 2004). These enzymes are applicable for reducing the kneading of dough, ensuring dough homogeneity, modifying gluten strength, controlling texture, and improving flavor (Di Cagno et al. 2003, Goesaert et al. 2005). Gluten proteins improve gas retention and dough hydration (Wieser 2007).

The breakdown of proteins is based on the metabolic activity of microorganisms and cereal enzymes during sourdough fermentation (Gänzle, Loponen, and Gobbetti 2008). Acidification and low-molecular-weight thiol accumulation enhance the solubilization of gluten and, thus, they become susceptible to enzymatic degradation (Jänsch et al. 2007, Thiele, Grassl, and Gänzle 2004). In addition, aspartic proteases play a major role in resting rye and wheat grains during sourdough fermentation where the optimum pH has been provided (Bleukx et al. 1998, Brijs, Bleukx, and Delcour 1999).

Proteolysis is restricted to breakdown of the cereal proteins in rye and wheat sourdoughs; high degradation of protein involves the addition of fungal or malt enzymes (Gänzle, Loponen, and Gobbetti 2008). Previously, it has been reported that proteolytic activity by LAB during sourdough fermentation was found to be higher than doughs prepared without yeast and starter culture addition (Gobbetti et al. 1994). Besides, amino acid concentrations have been predominantly raised because of strain-specific intracellular peptidase activities by *Lactobacillus subsp.* during simulated sourdough fermentation (Di Cagno et al. 2002, Gobbetti, Smacchi, and Corsetti 1996). Sourdough LAB possess a wide range of peptidases. *Fructilactobacillus sanfranciscensis* CB1, *Fructilactobacillus lindneri* A79, *Levilactobacillus brevis* H12, *Limosilactobacillus fermentum* CC5, *Fructilactobacillus fructivorans* DD1, *Lactiplantibacillus plantarum subsp. plantarum* DC400, *Companilactobacillus farciminis* CC1, *Lactobacillus acidophilus* BF4, and *Lactobacillus delbrueckii subsp.*

delbrueckii B5V, based on wheat sourdough, were used for their proteolytic activity. Based on the findings, all species were able to secrete proteinase, aminopeptidase, iminopeptidase, dipeptidase, tripeptidase, endopeptidase activities. Moreover, *F. fructivorans* DD1 and *L. plantarum* DC400 also possessed carboxypeptidase activity. The activity of proteinase was evaluated on gluten and casein. Degradation of gluten was greater compared to casein. On the other hand, *L. acidophilus* BF4, *L. brevis* H12, and *L. plantarum* DC400 demonstrated proteinase activity at the highest level (3.9–6.0 units/mg) on casein, among other species (Gobbetti, Smacchi, and Corsetti 1996). According to the findings of a new study, *Pediococcus acidilactici* XZ31 showed high proteolytic activity on both wheat protein and casein and has been reported to have a strong capability to lower wheat protein allergenicity (Fu et al. 2020). In a study, protein metabolisms of sourdough LAB were determined directly, by evaluating peptidase and proteinase activities of isolates and indirectly, by quantifying free amino acids released in the medium. The isolated strains demonstrated a widespread proteolytic activity, free amino acids were released (1.87 mmol Leu/L) by *Weissella paramesenteroides* CRL-2191, exhibiting the greatest capacity among strains. Twelve of the 18 isolates showed proteinase enzyme activity (Sáez et al. 2018). In addition, according to the results obtained from the analysis of sourdoughs using different starter cultures, higher activity of protease was detected in *Latilactobacillus curvatus* and *Pediococcus pentosaceus, Enterococcus pseudoavium*, and *P. pentosaceus* sourdoughs (Bartkiene et al. 2017). Aminopeptidases from several LAB have been isolated and characterized (Tan, Poolman, and Konings 1993). Active peptidase systems and proteinase affinity for both gluten and casein informed the selection of useful enzymatic characteristics (Gobbetti, Smacchi, and Corsetti 1996).

Bread quality can be improved to a great extent when the flour protein fraction has been changed during sourdough fermentation. Quality attributes, such as dough rheology, gas retention, and bread texture and volume, were regulated by the protein composition in wheat dough (Vermeulen et al. 2005). Furthermore, proteolytic activities offer substrates, e.g., amino acids for microbial development and the transformation of these into flavoring compounds and antifungal substances during sourdough fermentation (Lavermicocca et al. 2000, Thiele, Gänzle, and Vogel 2002). Proteolysis activity during sourdough fermentation comprises the protein hydrolysis to polypeptides, and consequent release of free amino acids were enabled by LAB peptidase (Gobbetti et al. 2005, Thiele, Gänzle, and Vogel 2002). Free amino acids, once released, contribute to flavor and odor directly or are prone to chemical changes during baking process or enzymatic catabolism (Kieronczyk et al. 2001), resulting in the release or synthesis of volatile flavoring compounds (Corsetti 2013). Free amino acid catabolism in sourdough LAB involves the expression of the arginine deiminase pathway (De Angelis et al. 2002), which increases the biomass growth and resistance to acidic environments; in particular, it enhances the synthesis of ornithine, the 2-acetyl-pyrroline precursor accountable for the roasted smell of wheat bread crust (Gobbetti et al. 2005). Thus, proteolysis is thought to improve the organoleptic and nutritional properties of bakery products (Rizzello et al. 2014). Moreover, the release of free amino acids has been reported to be effective in reducing the addition of salt without affecting the consistency of foods (Gobbetti et al. 2019). In a study, *F. sanfranciscensis* DSM 20451 was characterized, and it did not have proteolytic activity on proteinaceous substrates. However, *F. sanfranciscensis* utilizes peptides formed as a result of protein hydrolysis by endogenous proteinases based on flour during sourdough fermentation. The peptide metabolism of *F. sanfranciscensis* may lead to the accumulation of small peptides and amino acids that provide direct or indirect contribution to the quality attributes of sourdough bread (Vermeulen et al. 2005).

Protein hydrolyzation by exogenous peptidases and proteases from sourdough *Lactobacillus* strains or endogenous cereal proteases beneficially affects the intolerance and allergy responses of cereal-sensitive people. This potential can be seen in the production of bakery products with wheat flour. During sourdough fermentation, wheat gluten is fully degraded, therefore it would be safe to consume these products by celiac patients (Gobbetti et al. 2014). The level of wheat proteins that cause cereal intolerance was lowered by LAB isolated from sourdough (Di Cagno et al. 2002). In sourdough breads produced using *F. sanfranciscensis* (LS40, LS41) and *L. plantarum* CF1 cultures, the gluten concentration in bread was reduced by 95% (Di Cagno et al. 2008). Likewise, a previous research was conducted to determine whether the catalysts of gluten degradation in wheat sourdough are endogenous wheat proteases or microbial proteases. Based on the results, it was observed that activity of cereal enzymes and dough pH are associated with the proteolytic breakdown and depolymerization of glutens during sourdough fermentation. In addition, sourdough-isolated *Limosilactobacillus pontis* and fungal protease demonstrated a synergistic effect on proteolytic activity (Thiele, Gänzle, and Vogel 2003). Furthermore, while microbial enzymes do not greatly affect the overall hydrolysis of proteins and amino acid accumulation, the degradation of individual wheat proteins was influenced by fermentation with *Lactobacillus* strains having high proteolytic enzyme activity (Di Cagno et al. 2002). In a novel study, it was revealed that *Bacillus* spp. isolated from wheat sourdough showed high gluten-degrading activity. According to the results, the gluten concentration ranged between 254.17 and 267.59 mg/kg after 24 hours of sourdough fermentation, while the content decreased to 98–112 mg/kg after 48 hours of fermentation (Rashmi et al. 2020).

Proteolytic activity is also effective increasing the shelf life of bread. With the proteolytic activity of LAB on wheat proteins, different antifungal peptides have been characterized in water-soluble extracts of fermented wheat flour (Coda et al. 2011, Coda et al. 2008). Peptides biosynthesized by LAB in wheat-based media allow bread to be stored for long periods with a broad inhibitory effect against certain species that frequently contaminate bakery products and bakeries (Coda et al. 2011, Coda et al. 2008, Rizzello et al. 2011).

Bioactive peptides are chains of amino acids that perform various functions in humans after being released by microbial enzyme activities. Cereals and pseudocereals are important bioactive peptides sources with antioxidant, anti-inflammatory, anticancer, and cardiovascular protective properties (Malaguti et al. 2014). As a result of sourdough fermentation with proteolytic LAB and flours of whole wheat, barley, soybean, rye, and amaranth, the concentration of lunasin, which is an anticancer bioactive peptide, increased (Rizzello et al. 2012). In another study, selected

LAB demonstrated the ability to release antioxidant bioactive peptides from kamut and spelt during sourdough fermentation (Gobbetti et al. 2014).

Protease enzyme is used in many food fields such as bakery, beverage, brewery, and dairy industries. This enzyme, which is mainly responsible for the hydrolysis of proteins, plays a major role in the production of bioactive peptides, antifungal agents, and flavoring volatile compounds. Moreover, protease prevents cereal intolerance by providing the degradation of gluten protein. A high level of free amino acids is provided during sourdough fermentation compared to baker's yeast (Gobbetti 1998). Generally, higher protease activity implies that in practical applications a lower amount of enzyme can be used, or the enzymatic reaction time may be shorter (Zhang, Wang, and Jiang 2020). Thus, sourdoughs can be used to economically prepare high-level proteases.

8.2.2 Amylase

Amylase has a key role for the industrial hydrolyzation of starch. Amylase is used to control crumb structure and volume of breads in the bakery industry, to clarify cloudy juice in the beverage industry, and to lower viscosity and increase glucose and maltose content in the brewery industry (see Figure 8.3).

There are multiple amylases that can be used at the industrial level. Alpha-amylase is an intracellular enzyme that hydrolyzes α-1,4 glycosidic linkages of starch, glycogen, and cyclodextrins in a random manner, cleaving amylose and amylopectin-releasing dextrins and oligosaccharides. Beta-amylase is an extracellular enzyme that acts on α-1,4 glycosidic linkages, releasing maltose from non-reducing ends in polysaccharides. Glucoamylase or amyloglucosidase is an extracellular enzyme that cleaves α-1,6 glycosidic linkages and releases glucose units from non-reducing ends of polysaccharide chains (Damodaran, Parkin, and Fennema 2007). All these enzymes are produced by cereal, bacterial, and fungal sources (Jòzef 2007). Resting grains of rye and wheat include α-amylase, β-amylase, and amyloglucosidase activities. Amylase enzyme activities in rye and wheat sourdoughs release glucose, maltose, and maltodextrin during fermentation (Röcken and Voysey 1995). According to a previous study, amylolytic *L. fermentum* OgiE1 and *L. fermentum* Mw2 strains were first isolated from starchy fermented foods, such as Benin maize sourdough (Agati et al. 1998).

Microbial and malt α-amylases have been commonly applied in the bakery industry (Gupta et al. 2003). α-amylases from fungal or malt sources are often added to optimize the flour amylase activity for increasing the level of reducing and fermentable sugars. Due to their lower thermostability, fungal α-amylases are more suitable for flour standardization than malt amylases. α- and β-amylases have complementary functions in the bread manufacturing process (Martin and Hoseney 1991). Added α-amylase hydrolyzes starch into low-molecular-weight dextrins in the dough stage; whereas β-amylase degrades these oligosaccharides and releases maltose, which is utilized by sourdough bacteria or yeasts (Goesaert et al. 2005, Jòzef 2007). The increased maltose level, as reducing sugars, result in the production of Maillard reaction products, improving crust color and bread flavor. Furthermore, amylase enzymes alter texture and volume of the bakery products, and they show anti-staling activity (Cauvain and Young 2007, Goesaert et al. 2009, Poutanen 1997). Moreover, the reducing sugars' content is also related to acrylamide formation (Nguyen, Peters, and Van Boekel 2016). According to a study by Bartkiene et al. (2017), LAB strains show different carbohydrate metabolisms and reduce the amount of acrylamide in bread samples in all cases. Acidification and proteolysis of dough can affect both amino acid and sugar composition directly and indirectly, and the low pH value in the dough prevents acrylamide formation (Bartkiene et al. 2013). In one study, different combinations of sourdough-isolated microorganisms were used to determine their effects on the quality of wheat bread and formation of acrylamide. Based on the findings, higher amylolytic enzyme activity was found in *L. curvatus* and *L. plantarum subsp. plantarum* sourdoughs, while lower amylolytic enzyme activity was established in *L. brevis* and *P. pentosaceus* sourdoughs (Bartkiene et al. 2017).

Amylase hydrolyzes starches into dextrins, increasing the levels of fermentable sugars in the dough, and hence, the high CO_2 generation by yeasts (Kim, Maeda, and Morita 2006), which increases the bread volume. Nevertheless, increased gas levels cannot be maintained if the gluten matrix of dough is not strong enough. The addition of glucose oxidase creates additional linkages in the gluten network, resulting in a strengthening effect on the dough (Bonet et al. 2006). These two enzyme combinations showed beneficial effects on dough and sourdough breads produced using teff flour (Alaunyte et al. 2012). Likewise, some researchers have used enzymes alone or in combination to improve the bread quality (Laurikainen et al. 1998, Katina et al. 2006). Previous studies show that the use of combined enzymes can have more noteworthy influences on the properties of bread. The main reason for this is that enzymes that act on several flour components may be used at the same time, and some enzymes can reduce the negative effects of each other (Caballero, Gómez, and Rosell 2007a, b). The combined enzymes such as amylase-glucose oxidase and xylanase-glucose oxidase combinations have demonstrated positive effects on teff sourdough breads such as improved loaf volume, taste characteristics, crumb structure, and firmness (Alaunyte et al. 2012).

LAB such as *L. fermentum, Lentilactobacillus buchneri subsp. buchneri, L. plantarum, Limosilactobacillus reuteri*, and *L. brevis* also have α-galactosidase activity and, thus, the ability to hydrolyze raffinose (LeBlanc et al. 2004, Connes et al. 2004). The raffinose concentration in sourdough fermented wheat germ reduced to 45% compared to raw wheat germ (Rizzello et al. 2010). *Lactococcus garvieae* CRL-2199 and *Enterococcus durans* CRL-2194 isolated from chickpea sourdoughs demonstrated α-galactosidase activity (Sáez et al. 2018).

The amylase enzymes are used in many food fields, such as the starch industry and beverage, bakery, and alcoholic beverage industries. Therefore, industrial production of amylases is of great importance, and the cereals and microorganisms in sourdough can play a part in the production of these enzymes.

8.2.3 PHYTASE

Phytase is the enzyme involved in the dephosphorylation of phytic acid, being of great relevance to improve the absorption of minerals, i.e., iron, calcium, zinc, potassium, magnesium, and manganese and the digestibility of proteins. With the increasing demand for functional foods and the development of this food industry in recent years, industrial production of phytase enzyme has gained importance.

Phytic acid is naturally found in legumes, cereals, and pseudocereals. It produces insoluble mineral complexes, decreasing their bioavailability (Kumar et al. 2010, Martinez et al. 1996). Moreover, insoluble phytate–protein complexes are also produced at low pH, also decreasing the digestibility of protein (Kies et al. 2006). However, the beneficial effects of phytate as an anticancer, antidiabetic, or antioxidant agent have been reported (Kumar et al. 2010). Phytases can hydrolyze phytic acid and release soluble phosphate. Hydrolysis of phytate in dough is mainly associated with cereal phytases (Fretzdorff and Brummer 1992), in contrast, the sourdough LAB has also been reported to show phytase enzyme activity (De Angelis et al. 2003). While some studies reveal that LAB strains demonstrate intracellular phytase activity (De Angelis et al. 2003, Lopez et al. 2000, Reale et al. 2004), other studies have shown that LAB strains in sourdough exhibit extracellular phytase activities (Cizeikiene et al. 2015, Karaman, Sagdic, and Durak 2018, Yildirim and Arici 2019). On the other hand, cereals have endogenous phytase activities that can be increased by lowering the pH during sourdough fermentation during the bread making process; however, cereal phytase alone is not sufficient for breaking down phytate (Greiner and Konietzny 2006, Sanz-Penella, Tamayo-Ramos, and Haros 2012). Different studies focused on determining phytic acid or mineral residue content in bread and dough exposed to sourdough fermentation, to assess their bioavailability. An important increase in mineral bioavailability occurred due to sourdough acidification; therefore, endogenous phytases and microbial enzyme activities of the flour were indirectly activated (Larsson and Sandberg 1991). Phytase supplementation can increase protein digestibility by preventing complex formation or allowing the complex to dissolve faster (Kies et al. 2006).

Phytase enzymes are found in cereals, yeast (Türk et al. 2000), and LAB isolated from sourdough (Lopez et al. 2003). During the bread making process, phytase enzymes have the role of reducing phytic acid (myo-inositol hexaphosphate) to myo-inositol phosphate esters and lowering esters that are less probable to bind minerals and form mineral complexes (Persson et al. 1998, Sandberg et al. 1999). Phytate is found in the cereal structure as a salt of cations such as K^+, Mg^{2+}, and Ca^{2+} (Deshpande and Cheryan 1984). Previous studies have shown that sourdough fermentation was more effective than baker's yeast fermentation at reducing phytic acid content in bread (Leenhardt et al. 2005, Lopez et al. 2001). Furthermore, strains of *P. pentosaceus* (KTU05-8 and KTU05-9) demonstrated high activity of extracellular phytase (Cizeikiene et al. 2015); similarly, the combination of *P. pentosaceus* and *Saccharomyces cerevisiae* showed high phytate breakdown (Karaman, Sagdic, and Durak 2018). In addition, phytase activity was found in water/salt soluble sourdough extracts when the strains of *P. pentosaceus* F16A and *L. plantarum* 13I were used as starter cultures (Coda et al. 2011). A recent study confirmed the phytase activity of *P. pentosaceus*; breads made with the sourdough isolate *P. pentosaceus* had significantly less phytic acid contents compared to those of the control group (Ebrahimi, Sadeghi, and Mortazavi 2020).

Enzyme Production from Sourdough

Microorganisms, animals, and plants can produce phytases, however, microorganism-based phytases are promising for industrial production and cereal-based food applications. Microbial phytases are extracellularly synthesized in a medium. Genetically modified or non-modified natural species of bacteria, yeasts, and molds are used to produce microbial phytases such as *Bacillus amyloliquefaciens, Bacillus subtilis, Escherichia coli, Klebsiella sp., Hansenula polymorpha, Schwanniomyces occidentalis, Schwanniomyces castellii,* and *Rhodotorula gracilis, Aspergillus ficuum,* and *Aspergillus niger* (Pandey et al. 2001).

Lactobacillus species are involved in various fermentation processes. In many studies, phytase producing *Lactobacillus* species have been identified from sourdough (Anastasio et al. 2010, Chaoui, Faid, and Belhcen 2003, De Angelis et al. 2003). Yeasts generally demonstrate lower phytase activity compared to LAB, and the growing conditions must support the expression of phytase genes for high yeast enzyme activity to occur (Andlid, Veide, and Sandberg 2004). According to the systematic review of Arora et al. (2020), there are approximately 150 LAB and yeast strains that are probable phytase activity sources, 18 of them belonging to the *Lactobacillus* genus. The most common Italian sourdough microorganisms, such as *F. sanfranciscensis* CB1, *Companilactobacillus alimentarius* 5Q, *L. fermentum* 6E, *L. brevis* 14G, *Leuconostoc citreum* 23B, *Weissella confusa* 14A, *Lactococcus lactis* subsp. *lactis* 11M, *L. plantarum* DC400, *F. fructivorans* DA106, *L. acidophilus* 16A, *Lentilactobacillus hilgardii* 52B, *and C. farciminis* I2 (Corsetti et al. 2001, Gobbetti 1998) were evaluated for their phytase activity. All sourdough LAB had phytase activity; however, *F. sanfranciscensis* had the highest capacity to break down sodium phytate. In another study, sourdough LAB strains *Leuconostoc mesenteroides* subsp. *mesenteroides, L. acidophilus,* and *L. plantarum* demonstrated phytase activity (Lopez et al. 2000). Sourdough-like species isolated from spontaneously fermented wheat bran, such as *L. plantarum* (CE42, CE84), *L. brevis* (CE94, CE85), *L. curvatus* CE83, *L. mesenteroides* (CE52, CE48), *Latilactobacillus sakei* subsp. *sakei* CE47, *L. citreum* (CE88, CE54), *Lacticaseibacillus casei* Q11, and *P. pentosaceus* (CE65, CE23) demonstrated the capability to breakdown sodium phytate (Manini et al. 2016). Phytase active strains isolated from wheat and rye sourdough, including *L. fermentum, Limosilactobacillus panis, L. reuteri, P. pentosaceus,* and *L. panis* GH1, demonstrated a higher volumetric extracellular enzyme activity compared to others (Nuobariene et al. 2015). Resistance to high temperature is considered a useful and important principle for industrial use of phytase enzymes. Furthermore, the phytase of *F. sanfranciscensis* CB1 (phytase activity 420.8 U/mL) retained 70% of its activity after 30 minutes of exposure to 70°C; in other words, it can remain active during the initial stage of dough baking (De Angelis et al. 2003).

In one study, traditional wheat sourdough isolates *Furfurilactobacillus rossiae* ED1, *Companilactobacillus paralimentarius* E106, *Lactiplantibacillus paraplantarum* N15, *L. plantarum* ED-10, and *L. brevis* E25 have shown higher phytase activities when compared to the other isolates *L. curvatus* N19, *Weissella cibaria* N9, *W. paramesenteroides* N7, *Leuconostoc pseudomesenteroides* N13, and *L. mesenteroides* N6. Their phytase activities ranged from 0.12 U/mg to 0.61 U/mg (Demirbaş et al. 2017). In a novel study, *L. plantarum* and *Lactobacillus amylovorus* showed extracellular phytase enzyme activities with 125–146 U/mL values (Cizeikiene et al. 2020). Phytase activities of LAB and yeasts isolated from sourdough samples taken from different places were found to be 703.1–1153.8 U/mL and 352.2–943.4 U/mL, respectively (Karaman, Sagdic, and Durak 2018). Previous studies have indicated that the phytase activities of LAB are strain-specific (Demirbaş et al. 2017, Palacios et al. 2008) (see Table 8.2). In another study, the phytase activity in sourdough fermented wheat germ was 2.78 ± 0.08 U, while the activity in raw wheat germ was 0.77 ± 0.02 U. Therefore, sourdough fermented wheat germ extract included relatively high levels of free Fe^{++}, Mn^{++}, Ca^{++}, Zn^{++}, and K^+ (Rizzello et al. 2010). Spontaneously fermented amaranth and quinoa sourdough isolates of LAB strains possessed extracellular phytase activities ranging between 0.02 and 0.36 U/10Log CFU (Sandez Penidez et al. 2020). Moreover, the application of Boza, a Turkish cereal-based fermented beverage, as a starter culture for sourdough production resulted in high phytase enzyme yields. According to the results, sourdough isolates *P. pentosaceus* EK1 and *L. lactis* B12 showed high levels of extracellular phytase activity, 1285.5 U/mL and 943.1 U/mL, respectively (Doğan and Tekiner 2020). Additionally, in a novel study, an Iranian sourdough yeast isolate *Kluyveromyces marxianus* showed relatively higher phytase activity (1.64 U/mL) than the control *S. cerevisiae* (Fekri et al. 2020).

The phytase enzyme, which has gained importance with the development of the functional food industry, plays an effective role in both biological access to minerals and digestibility of proteins. Phytate naturally found in cereals is dephosphorylated by phytase, also found in cereals, or secreted by local microorganisms. LAB strains exhibiting phytate degradation ability can be used in fermented foods as starter cultures to improve the nutritional quality and mineral bioavailability of food products rich in phytic acid (Manini et al. 2016, Rollán, Gerez, and LeBlanc 2019, Sumengen, Dincer, and Kaya 2013), or they can be directly incorporated into foods to provide phytate degradation.

8.2.4 Xylanase

Xylanase catalyzes the hydrolyzation of 1,4-β-D-xylosidic bonds in xylan and arabinoxylan. Xylanases, also called endoxylanases, were originally called pentosanases (Collins, Gerday, and Feller 2005). The production of a wide diversity of xylanases from many microorganisms consisting of bacteria, fungi, and archaea has been reported (Verma and Satyanarayana 2012). Xylanases are frequently used in the bakery industry in combination with lipases, amylases, and numerous oxidoreductases to improve the rheological properties of dough and to enhance the organoleptic properties of bread (Collins et al. 2006). In addition, these enzymes have also been applied for enhancing the quality attributes of biscuits, cakes, and other bakery products (Poutanen 1997). Using the xylanase enzyme, a more flexible, stable, and easy-to-process dough is obtained, causing a larger loaf volume, a softer crumb (Heldt-Hansen 2006), and an extended shelf life (Poutanen, Flander, and Katina 2009).

TABLE 8.2
Selected LAB Strains Isolated from Sourdough and Their Enzymatic Activities

LAB strain	Amylase activity	Protease activity	Phytase activity	Xylanase activity	References
L. acidophilus DSM 20079	1.33±0.004 U/g	(5.5±0.1) x10⁻³ U/g	10±0.2 U/g	370±10 U/g	Cizeikiene et al. (2020)
L. bulgaricus MI	1.35±0.002 U/g	(5.5±0.1) x10⁻³ U/g	12.9±0.2 U/g	170±10 U/g	
F. rossiae GL14	1.30±0.003 U/g	(5.4±0.1) x10⁻³ U/g	10.4±0.3 U/g	120±10 U/g	
L. plantarum CRL 1964			312.97±2.09 U/mg		Sandez Penidez et al. (2020)
L. casei B21			594.6 U/mL		Doğan and Tekiner (2020)
L. casei B31A			682.7 U/mL		
L. casei K11			487.3 U/mL		
L. casei K22			506.4 U/mL		
L. casei K32			635.3 U/mL		
L. fermentum B1A			743.7 U/mL		
L. pentosus B1			678.5 U/mL		
L. pentosus B31			763.0 U/mL		
L. pentosus B33			634.4 U/mL		
L. pentosus B33A			714.7 U/mL		
L. lactis B11			463.6 U/mL		
L. lactis B12			943.1 U/mL		
L. lactis B32			810.5 U/mL		
P. pentosaceus EK1			1285.5 U/mL		

(Continued)

TABLE 8.2 (CONTINUED)
Selected LAB Strains Isolated from Sourdough and Their Enzymatic Activities

LAB strain	Amylase activity	Protease activity	Phytase activity	Xylanase activity	References
P. pentosaceus EK2			559.4 U/mL		
P. pentosaceus EK3			576.3 U/mL		
P. pentosaceus K33			603.6 U/mL		
P. pentosaceus NB1			497.6 U/mL		
P. pentosaceus NB32			521.7 U/mL		
P. pentosaceus NB34			532.8 U/mL		
E. faecium NB32A			548.2 U/mL		Karaman, Sagdic, and Durak (2018)
LAB strains (undefined)			703.1–1153.8 U/mL		Demirbaş et al. (2017)
L. mesenteroides N6			0.12 ± 0.008 U/mg		
W. paramesenteroides N7			0.19 ± 0.004 U/mg		
W. cibaria N9			0.15 ± 0.001 U/mg		
L. pseudomesenteroides N13			0.19 ± 0.001 U/mg		
L. paraplantarum N15			0.61 ± 0.001 U/mg		
L. curvatus N19			0.12 ± 0.001 U/mg		
F. rossiae ED1			0.55 ± 0.001 U/mg		
L. plantarum ED10			0.59 ± 0.006 U/mg		
L. brevis E25			0.54 ± 0.001 U/mg		
C. paralimentarius E106			0.46 ± 0.004 U/mg		
L. sakei KTU05-6			25 ± 0 U/mL		Cizeikiene et al. (2015)
P. acidilactici KTU05-7			11 ± 1 U/mL		
P. pentosaceus KTU05-10			21 ± 4 U/mL		
P. pentosaceus KTU05-8			32 ± 1 U/mL		
P. pentosaceus KTU05-9			54 ± 6 U/mL		

(Continued)

TABLE 8.2 (CONTINUED)
Selected LAB Strains Isolated from Sourdough and Their Enzymatic Activities

LAB strain	Amylase activity	Protease activity	Phytase activity	Xylanase activity	References
P. pentosaceus LUHS183	145.8±5.7 AU	204.9±4.9 AU			Bartkiene et al. (2018)
P. acidilactici LUHS29	160.1±6.4 AU	211.7±2.8 AU			
L. paracasei LUHS244	224.4±9.1 AU	190.8±6.0 AU			
L. brevis LUHS173	158.3±5.2 AU	196.7±4.8 AU			
L. mesenteroides LUHS242	143.4±6.4 AU	185.0±5.2 AU			
L. plantarum LUHS135	204.2±3.9 AU	195.3±6.2 AU			
P. pentosaceus LUHS183 × L. mesenteroides LUHS225	260.6±9.3 AU	250.5±6.9 AU			Bartkiene et al. (2017)
P. pentosaceus LUHS183 × L. brevis LUHS173	234.7±7.2 AU	255.1±7.3 AU			
P. pentosaceus LUHS183 × E. pseudoavium LUHS234	251.2±3.9 AU	299.9±8.5 AU			
P. pentosaceus LUHS183 × L. curvatus LUHS51	269.9±6.5 AU	290.5±9.3 AU			
L. plantarum LUHS135 × L. curvatus LUHS51	311.1±8.1 AU	254.8±4.8 AU			
L. plantarum LUHS135 × P. pentosaceus LUHS183	263.3±7.4 AU	213.8±6.2 AU			
P. pentosaceus 1.2			43±2 U/mL		Nuobariene et al. (2015)
L. panis GH1			141±10 U/mL		
L. reuteri 2.3			6±1 U/mL		
L. reuteri 8.1			19±2 U/mL		
L. fermentum 5.1			46±2 U/mL		
L. fermentum 5.2			3±0 U/mL		

(Continued)

TABLE 8.2 (CONTINUED)
Selected LAB Strains Isolated from Sourdough and Their Enzymatic Activities

LAB strain	Amylase activity	Protease activity	Phytase activity	Xylanase activity	References
L. fermentum 6.1			11±3 U/mL		De Angelis et al. (2003)
L. fermentum 7.2			21±2 U/mL		
C. alimentarius 5Q			230.7 U/mL		
C. farciminis 2I			116.3 U/mL		
L. citreum 23B			172.6 U/mL		
L. acidophilus 16A			89.4 U/mL		
L. plantarum DC400			110.6 U/mL		
L. fermentum 6E			130.6 U/mL		
L. hilgardii 52B			159.1 U/mL		
F. fructivorans DA106			295.3 U/mL		
L. lactis subsp. lactis 11M			249.2 U/mL		
W. confusa 14A			259.6 U/mL		
F. sanfranciscensis CB1			420.8 U/mL		
F. sanfranciscensis 22E			7.6 U/mL		
L. brevis 14G			105.4 U/mL		

Abbreviations of table units: U (enzyme unit, μmol/min) and AU (enzyme activity unit).

Rye and wheat flours include 7–8% and 1.5–7% arabinoxylan, respectively, but only a small part of arabinoxylan is soluble in water (Gebruers et al. 2010, Shewry et al. 2010). The solubility of arabinoxylan in water improves the foam stability and dough hydration of rye and wheat dough. On the other hand, water-insoluble arabinoxylans slow down the formation of gluten during dough mixing and interrupt gas cells (Goesaert et al. 2005). Water-soluble arabinoxylans contribute to rye bread quality during sourdough fermentation, and the content of these compounds increases during simulated sourdough fermentation of rye and wheat (Hansen et al. 2002, Korakli et al. 2001). The breakdown of rye flour arabinoxylans in sourdough causes the solubilization of polysaccharides with high molecular weight (Loponen et al. 2009). The degradation of arabinoxylans into xylose and arabinose was detected in rye malt sourdoughs, but the degradation was observed after the addition of pentosanases in wheat sourdough (Gobbetti et al. 2000, Loponen et al. 2009).

Furthermore, xylanase addition to dough may positively affect human health by decreasing these compounds (Bhat 2000). Water-extractable arabinoxylan increases are associated with delayed carbohydrate ingestion and absorption rates and decreased insulinemic and glycemic responses (Lu et al. 2004, Lu et al. 2000, Möhlig et al. 2005). Furthermore, the soluble oligosaccharides obtained from the hydrolysis of non-water-extractable arabinoxylans exhibit beneficial effects on health with prebiotic, antioxidant, anticarcinogenic, and hypocholesterolemic properties (Broekaert et al. 2011).

8.2.5 TANNASE AND GALLATE DECARBOXYLASE

Tannins found in many plants are undesirable in food and feed, from a nutritional point of view, because these compounds inhibit digestive enzymes, affecting the use of minerals and vitamins. The enzyme tannase yields gallic acids from hydrolyzable tannins, while gallate decarboxylase enzyme decarboxylates gallic acids to pyrogallol. Both tannase and gallate decarboxylase enzymes can be involved in the elimination of tannins and in the liberation of bioactive phenolics (Muñoz et al. 2017). Tannin hydrolyzation activities in fermented foods have been stated for *L. plantarum subsp. plantarum* (Muñoz et al. 2017), *Lactiplantibacillus pentoses*, and *L. paraplantarum* (Osawa et al. 2000). Activities of gallate decarboxylase enzyme were also discovered in *W. cibaria* CRL2195 and *P. pentosaceus* CRL2145 isolated from chickpea sourdough (Sáez et al. 2018), and in *Weissella* species isolated from beans (Sáez et al. 2017).

8.2.6 GLYCOSIDASE

Glycoside hydrolases, also known as glycosidases, are enzymes that catalyze the glycosidic bond hydrolysis in complex carbohydrates. These enzymes are widely used in the breakdown of polysaccharides such as cellulose, hemicellulose, and starch. Together with glycosyltransferases, glycosidases are involved in the main catalytic mechanism in the synthesis and breaking of glycosidic bonds (Bourne and Henrissat 2001, Henrissat and Davies 1997).

Phenolic acids are covalently bonded to plant cell walls or present in their soluble forms in the cytoplasm. Microbial enzymes and acidification provide biological conversion of polyphenols to more available bioactive compounds (Lee, Lo, and Pan 2013). Glycosidases of *Lacticaseibacillus rhamnosus* and *L. plantarum* subsp. *plantarum* have been used to break the plant–cell wall matrix, resulting in the release of phenolic compounds and depolymerization of high molecular weight phenolics (Filannino, Di Cagno, and Gobbetti 2018). This enzymatic activity is important, as the release of extractable phenolic compounds also results in an increased antioxidant capacity of the products (Đorđević, Šiler-Marinković, and Dimitrijević-Branković 2010). Another study stated that *L. casei* and *L. plantarum*-secreted enzymes are liberated flavonoids and isoflavone aglycones from related glycosides, and then *p*-coumaric and ferulic acid were released by aglycone metabolism by LAB (Filannino, Di Cagno, and Gobbetti 2018). During sorghum sourdough fermentation, even though endogenic cereal enzymes have the potential to hydrolyze flavonoid glycosides and phenolic acid esters, glucosidase, esterase, phenolic acid reductase, and phenolic acid decarboxylase enzymes synthesized by lactobacilli exhibited activity to transform phenolic compounds (Gänzle 2014).

8.2.7 Esterase

Phenolic compounds play a leading role in the formation of aroma during bread making (Czerny and Schieberle 2002). The most important phenolics in rye and wheat are alkylresorcinols and phenolic acids. Ferulic acid is the major phenolic acid in rye (Shewry et al. 2010). Feruloyl esterase hydrolyzes feruloylated sugar esters. Rye flour has feruloyl esterase activity (Hansen et al. 2002); this enzyme becomes active during the germination stage, and ferulic acid is liberated from the cell walls of barley or wheat during malt mashing (Coghe et al. 2004, Sancho et al. 1999).

Some studies revealed that cereal enzymes and starter cultures used in sourdough fermentation are effective in the metabolism of phenolic compounds. It has been reported that *Lactobacillus* species from sorghum fermentations and *Levilactobacillus hammesii* DSM 16381 isolated from wheat sourdough influence phenolic acid metabolism (Svensson et al. 2010, Valcheva et al. 2005). An increased free ferulic acid content was observed in simulated sourdough fermentation of wholemeal rye (Hansen et al. 2002), and increases in phenolic acid were detected in simulated sourdough fermentation of whole grain barley and oats (Hole et al. 2012). The study showed that using feruloyl esterase-active starter cultures significantly increased the free phenolic acid contents specific to the strains (Hole et al., 2012). In another study, flavonoid glucosides from red sorghum sourdough were converted into flavonoids during lactic acid fermentation (Svensson et al. 2010).

8.2.8 Glucansucrase and Fructansucrase

Glucansucrases and fructansucrases are involved in the production of exopolysaccharides (EPS) (Korakli and Vogel 2006, van Hijum et al. 2006). Formation of EPS has beneficial effects on bread quality, increasing bread volume, improving texture,

Enzyme Production from Sourdough

and increasing the content of dietary fiber (Galle and Arendt 2014). Substrates with high buffering capacity preserve the medium within the optimum pH range (4.5–5.5) for efficient glucansucrase activity. According to a study, this favorable medium allowed *L. reuteri* TMW 1.106 to produce high amounts of EPS during sourdough fermentation (Kaditzky and Vogel 2008). High concentration of maltose, a glycosyl acceptor LAB glucansucrase, alters glucansucrase activity from polysaccharides to oligosaccharides synthesis (Galle et al. 2010, van Hijum et al. 2006). Fermentation of maize, sorghum, and pearl millet lacking beta-amylase activity allows a high production of exopolysaccharides (Galle et al. 2010, RŘhmkorf et al. 2012).

8.2.9 Lipoxygenase

Lipoxygenase is an endogenous cereal enzyme that oxidizes linoleic acid into hydroperoxy acids that are converted into the hydroxy fatty acids (HFA) with potential biological activities (Kim and Oh 2013).

LAB species can hydrolyze oleic and linoleic acids into HFA. Linoleic acid is converted into 13-hydroxy-9-octadecenoic acid or 10-hydroxy-12-octadecenoic acid, the latter showing antifungal activity (Ogawa et al. 2001). HFA formation during *L. hammesii* growth in sourdough has resulted in mold-free products and an extended bread shelf life (Black et al. 2013). Aldehydes, which are significant flavor compounds, are formed during the lipid oxidation in flour and by the activity of enzymes hydroperoxide lyase and lipoxygenase that occur naturally in the wheat endosperm (Birch, Petersen, and Hansen 2014).

8.2.10 FODMAP Hydrolysis

Irritable bowel syndrome (IBS) is an intestinal disorder resulting in bloating, abdominal pain, constipation, gas, and diarrhea and altered gut microbiota (Cozma-Petruţ et al. 2017). Fermentable oligosaccharides, disaccharides, monosaccharides, and polyols (FODMAPs) are short-chain carbohydrates (Gibson and Shepherd 2005). A low-FODMAP diet might be able to alleviate digestive symptoms in IBS patients (Rao, Yu, and Fedewa 2015). Therefore, FODMAP hydrolysis has gained attention, and sourdough biotechnology is applicable for that. During bread making, extended sourdough fermentation can reduce the level of FODMAPs by up to 90% (Ziegler et al. 2016). Essentially, the process technique is more critical than simply selecting low-FODMAP cereals. In contrast to baker's yeast bread, sourdough reduces amylase/trypsin inhibitors – especially FODMAPs – without baking enhancers, by prolonged fermentation of more than 12 hours (Laatikainen et al. 2017). Sourdough yeast species can break down fructans. Sucrose breaks down into glucose and fructose by the activity of invertase enzymes produced by *S. cerevisiae* and those present in wheat flour (De Vuyst and Neysens 2005). The enzyme invertase secreted by *S. cerevisiae* can hydrolyze fructans, and *K. marxianus* strains that secrete the inulinase enzyme, lowered the fructan level (Struyf et al. 2017). Additionally, *Lactobacillus crispatus* possess extracellular fructanase, which shows fructan hydrolyzing activity during rye and wheat sourdough fermentation (Li, Loponen, and Ganzle 2020). Even though most fructans have been hydrolyzed, fructose accumulation still presents risks for IBS patients. *Apilactobacillus kunkeei* (B23I, PF16, and PLA21) completely metabolized fructose during sourdough fermentation (Acín Albiac et al. 2020). Moreover, the application of the traditional sourdough isolate *Lacticaseibacillus paracasei* subsp. *paracasei* FJSSZ3L1 as a starter culture reduced the FODMAP content in Chinese steamed bread, indicating that *L. paracasei* has great potential to produce low-FODMAP baked goods (Fang et al. 2021).

8.2.11 ARGININE DEIMINASE

The arginine deiminase pathway provides adenosine triphosphate (ATP) for the growth of microorganisms under various environmental conditions, particularly when carbohydrate is absent or in low concentration, providing carbamoyl phosphate for the biosynthesis of pyrimidines or citrulline, and protecting bacteria against damage triggered, for example, by acid and/or starvation stresses (Casiano-Colón and Marquis 1988). Hence, expression of the arginine deiminase pathway in LAB and other microorganisms can have great relevance for industrial applications (De Angelis et al. 2002). This pathway consists of three enzymes: arginine deiminase, which breaks down arginine into ammonia and citrulline; ornithine transcarbamylase, which degrades citrulline into ornithine and carbamoyl phosphate; and carbamate kinase, which dephosphorylates carbamoyl phosphate into ammonia, carbon dioxide, and ATP (De Angelis et al. 2002).

Arginine degradation by sourdough microorganisms was investigated. Obligate heterofermentative sourdough bacteria, such as *F. sanfranciscensis* CB1, *L. brevis* (AM1, AM8, 10A), *L. hilgardii* 51B, and *F. fructivorans* (DD3, DA106) demonstrated all the arginine deiminase, ornithine transcarbamylase, and carbamate kinase enzymatic activities (De Angelis et al. 2002). Arginine deiminase pathway expression by *F. sanfranciscensis* CB1 during sourdough fermentation improved microbial growth and survival and increased tolerance to environmental stressors, such as acid, and greater ornithine production, which improves the organoleptic properties of sourdough (De Angelis et al. 2002).

8.3 CONCLUSION

Enzymes isolated from sourdough are of extreme relevance in the food industry both in terms of technological improvement of products as well as to increase their shelf life and healthy attributes. Briefly, phytic acid is degraded by phytase, thus increasing mineral and protein bioavailability (Milanović et al. 2020); hydrolysis of gluten by protease makes it possible to produce gluten-free or low-gluten foods (Rashmi et al. 2020); the breakdown of FODMAP by fructanase may address non-celiac wheat sensitivity and may be a solution for IBS patients. In addition, it has been emphasized by recent studies that fructanase and amylase enzymes play a significant role in reducing the sugar content of products (Li, Loponen, and Ganzle 2020, Müller et al. 2021). Reducing the amount of sugar in foods will play an important role in the prevention of diseases, such as diabetes and obesity, especially in developed countries.

In addition, since the protease enzyme increases the amount of free amino acids, this can be an effective strategy in reducing salt consumption (Gobbetti et al. 2019). Considering the technological effects of enzymes during sourdough fermentation, amylase, protease, and xylanase enzymes improve dough texture and organoleptic properties (Alaunyte et al. 2012, Collins et al. 2006, Rizzello et al. 2014), lipoxygenase extends shelf life (Black et al. 2013), esterase plays a leading role in aroma/ taste formation (Hole et al. 2012), and glucansucrase and fructansucrase are effective in EPS production, thus increasing dietary fiber content and bread volume (Galle and Arendt 2014).

Chemicals and/ or enzymes are used in food formulations to improve product quality and shelf life. Nowadays, due to an increased awareness of the serious influences of chemicals in food products and human health, there is an increasing trend that products need to be bio- or organic, and free of or with minimal amounts of chemicals. In this respect, the use of enzymes is of great interest. Enzyme production from a product with a rich microflora, such as sourdough is of great industrial importance within the current scenario in the food sector. Microorganisms are recognized as the main source of enzymes that play a major role in the food industry improving the physicochemical and microbiological quality of food products as well as their shelf life and digestibility. Studies have shown that high levels of proteinase, phytase, amylase, and xylanase enzymes, which are widely used in the food industry, are produced at high yields from sourdoughs and thus, sourdoughs' microflora can be considered as warehouses or enzyme factories with promising applications in the food industry at the industrial level.

REFERENCES

Acín Albiac, Marta, Raffaella Di Cagno, Pasquale Filannino, Vincenzo Cantatore, and Marco Gobbetti. 2020. "How fructophilic lactic acid bacteria may reduce the FODMAPs content in wheat-derived baked goods: A proof of concept." *Microbial Cell Factories* 19 (1). https://doi.org/10.1186/s12934-020-01438-6.

Agati, V, Jean-Pierre Guyot, Juliette Morlon-Guyot, Pascale Talamond, and DJ Hounhouigan. 1998. "Isolation and characterization of new amylolytic strains of Lactobacillus fermentum from fermented maize doughs (mawe and ogi) from Benin." *Journal of Applied Microbiology* 85 (3):512–520.

Alaunyte, Ieva, Valentina Stojceska, Andrew Plunkett, Paul Ainsworth, and Emma Derbyshire. 2012. "Improving the quality of nutrient-rich Teff (Eragrostis tef) breads by combination of enzymes in straight dough and sourdough breadmaking." *Journal of Cereal Science* 55 (1):22–30.

Anastasio, Marilena, Olimpia Pepe, Teresa Cirillo, Simona Palomba, Giuseppe Blaiotta, and Francesco Villani. 2010. "Selection and use of phytate-degrading LAB to improve cereal-based products by mineral solubilization during dough fermentation." *Journal of Food Science* 75 (1):M28–M35.

Andlid, Thomas A, Jenny Veide, and Ann-Sofie Sandberg. 2004. "Metabolism of extracellular inositol hexaphosphate (phytate) by Saccharomyces cerevisiae." *International Journal of Food Microbiology* 97 (2):157–169.

Arbige, Michael V, Jay K Shetty, and Gopal K Chotani. 2019. "Industrial enzymology: The next chapter." *Trends in Biotechnology* 37 (12):1355–1366.

Arora, Kashika, Hana Ameur, Andrea Polo, Raffaella Di Cagno, Carlo Giuseppe Rizzello, and Marco Gobbetti. 2020. "Thirty years of knowledge on sourdough fermentation: A systematic review." *Trends in Food Science & Technology* 108:71–83.

Bartkiene, Elena, Ida Jakobsone, Grazina Juodeikiene, Daiva Vidmantiene, Iveta Pugajeva, and Vadims Bartkevics. 2013. "Study on the reduction of acrylamide in mixed rye bread by fermentation with bacteriocin-like inhibitory substances producing lactic acid bacteria in combination with Aspergillus niger glucoamylase." *Food Control* 30 (1):35–40.

Bartkiene, Elena, Vadims Bartkevics, Vita Krungleviciute, Iveta Pugajeva, Daiva Zadeike, and Grazina Juodeikiene. 2017. "Lactic acid bacteria combinations for wheat sourdough preparation and their influence on wheat bread quality and acrylamide formation." *Journal of Food Science* 82 (10):2371–2378.

Bartkiene, Elena, Vadims Bartkevics, Vita Lele, Iveta Pugajeva, Paulina Zavistanaviciute, Ruta Mickiene, Daiva Zadeike, and Grazina Juodeikiene. 2018. "A concept of mould spoilage prevention and acrylamide reduction in wheat bread: Application of lactobacilli in combination with a cranberry coating." *Food Control* 91:284–293.

Bhat, MK14538100. 2000. "Cellulases and related enzymes in biotechnology." *Biotechnology Advances* 18 (5):355–383.

Birch, Anja Niehues, Mikael Agerlin Petersen, and Åse Solvej Hansen. 2014. "Aroma of wheat bread crumb." *Cereal Chemistry* 91 (2):105–114.

Black, Brenna A, Emanuele Zannini, Jonathan M Curtis, and Michael G Gï. 2013. "Antifungal hydroxy fatty acids produced during sourdough fermentation: Microbial and enzymatic pathways, and antifungal activity in bread." *Applied and Environmental Microbiology* 79 (6):1866–1873.

Bleukx, Wouter, Kristof Brijs, Sophie Torrekens, Fred Van Leuven, and Jan Arsène Delcour. 1998. "Specificity of a wheat gluten aspartic proteinase." *Biochimica et Biophysica Acta (BBA)-Protein Structure and Molecular Enzymology* 1387 (1–2):317–324.

Bombara, N, MC Anon, and AMR Pilosof. 1997. "Functional properties of protease modified wheat flours." *LWT: Food Science and Technology* 30 (5):441–447.

Bonet, Arturo, Cristina M Rosell, Pedro A Caballero, Manuel Gómez, Isabel Pérez-Munuera, and María Angeles Lluch. 2006. "Glucose oxidase effect on dough rheology and bread quality: A study from macroscopic to molecular level." *Food Chemistry* 99 (2):408–415.

Bourne, Yves, and Bernard Henrissat. 2001. "Glycoside hydrolases and glycosyltransferases: Families and functional modules." *Current Opinion in Structural Biology* 11 (5):593–600.

Brijs, Kristof, Wouter Bleukx, and Jan A Delcour. 1999. "Proteolytic activities in dormant rye (Secale cereale L.) grain." *Journal of Agricultural and Food Chemistry* 47 (9):3572–3578.

Broekaert, Willem F, Christophe M Courtin, Kristin Verbeke, Tom Van de Wiele, Willy Verstraete, and Jan A Delcour. 2011. "Prebiotic and other health-related effects of cereal-derived arabinoxylans, arabinoxylan-oligosaccharides, and xylooligosaccharides." *Critical Reviews in Food Science and Nutrition* 51 (2):178–194.

Caballero, Pedro A, Manuel Gómez, and Cristina M Rosell. 2007a. "Bread quality and dough rheology of enzyme-supplemented wheat flour." *European Food Research and Technology* 224 (5):525–534.

Caballero, Pedro A, Manuel Gómez, and Cristina M Rosell. 2007b. "Improvement of dough rheology, bread quality and bread shelf-life by enzymes combination." *Journal of Food Engineering* 81 (1):42–53.

Capuani, Alessandro, Jürgen Behr, and Rudi F Vogel. 2012. "Influence of lactic acid bacteria on the oxidation–reduction potential of buckwheat (Fagopyrum esculentum Moench) sourdoughs." *European Food Research and Technology* 235 (6):1063–1069.

Casiano-Colón, Aida, and Robert E Marquis. 1988. "Role of the arginine deiminase system in protecting oral bacteria and an enzymatic basis for acid tolerance." *Applied and Environmental Microbiology* 54 (6):1318–1324.

Cauvain, Stanley P, and Linda S Young. 2007. *"Technology of breadmaking."*(2nd ed.). Springer

Chaoui, A, M Faid, and R Belhcen. 2003. "Effect of natural starters used for sourdough bread in Morocco on phytate biodegradation." *EMHJ-Eastern Mediterranean Health Journal*, 9 (1–2): 141–147.

Cizeikiene, Dalia, Grazina Juodeikiene, Elena Bartkiene, Jonas Damasius, and Algimantas Paskevicius. 2015. "Phytase activity of lactic acid bacteria and their impact on the solubility of minerals from wholemeal wheat bread." *International Journal of Food Sciences and Nutrition* 66 (7):736–742.

Cizeikiene, Dalia, Jolita Jagelaviciute, Mantas Stankevicius, and Audrius Maruska. 2020. "Thermophilic lactic acid bacteria affect the characteristics of sourdough and wholegrain wheat bread." *Food Bioscience* 38:100791.

Coda, Rossana, Carlo G Rizzello, Franco Nigro, Maria De Angelis, Philip Arnault, and Marco Gobbetti. 2008. "Long-term fungal inhibitory activity of water-soluble extracts of Phaseolus vulgaris cv. Pinto and sourdough lactic acid bacteria during bread storage." *Applied and Environmental Microbiology* 74 (23):7391–7398.

Coda, Rossana, Angela Cassone, Carlo G Rizzello, Luana Nionelli, Gianluigi Cardinali, and Marco Gobbetti. 2011. "Antifungal activity of Wickerhamomyces anomalus and Lactobacillus plantarum during sourdough fermentation: Identification of novel compounds and long-term effect during storage of wheat bread." *Applied and Environmental Microbiology* 77 (10):3484–3492.

Coghe, Stefan, Koen Benoot, Filip Delvaux, Bart Vanderhaegen, and Freddy R Delvaux. 2004. "Ferulic acid release and 4-vinylguaiacol formation during brewing and fermentation: Indications for feruloyl esterase activity in Saccharomyces cerevisiae." *Journal of Agricultural and Food Chemistry* 52 (3):602–608.

Collins, Tony, Charles Gerday, and Georges Feller. 2005. "Xylanases, xylanase families and extremophilic xylanases." *FEMS Microbiology Reviews* 29 (1):3–23.

Collins, Tony, Anne Hoyoux, Agnes Dutron, Jacques Georis, Bernard Genot, Thierry Dauvrin, Filip Arnaut, Charles Gerday, and Georges Feller. 2006. "Use of glycoside hydrolase family 8 xylanases in baking." *Journal of Cereal Science* 43 (1):79–84.

Connes, Cristelle, Aurelio Silvestroni, Jean Guy Leblanc, Vincent Juillard, Graciela Savoy de Giori, Fernando Sesma, and Jean-Christophe Piard. 2004. "Towards probiotic lactic acid bacteria strains to remove raffinose-type sugars present in soy-derived products." *Le Lait* 84 (1–2):207–214.

Corsetti, Aldo. 2013. "Technology of sourdough fermentation and sourdough applications." In *Handbook on Sourdough Biotechnology*, 85–103. Springer.

Corsetti, Aldo, P Lavermicocca, M Morea, F Baruzzi, N Tosti, and Marco Gobbetti. 2001. "Phenotypic and molecular identification and clustering of lactic acid bacteria and yeasts from wheat (species Triticum durum and Triticum aestivum) sourdoughs of Southern Italy." *International Journal of Food Microbiology* 64 (1–2):95–104.

Cozma-Petruţ, Anamaria, Felicia Loghin, Doina Miere, and Dan Lucian Dumitraşcu. 2017. "Diet in irritable bowel syndrome: What to recommend, not what to forbid to patients!" *World Journal of Gastroenterology* 23 (21):3771. https://doi.org/10.3748/wjg.v23.i21.3771.

Czerny, Michael, and Peter Schieberle. 2002. "Important aroma compounds in freshly ground wholemeal and white wheat flour identification and quantitative changes during sourdough fermentation." *Journal of Agricultural and Food Chemistry* 50 (23):6835–6840.

Damodaran, Srinivasan, Kirk L Parkin, and Owen R Fennema. 2007. *Fennema's Food Chemistry*: CRC Press.

De Angelis, Maria, Liberato Mariotti, Jone Rossi, Maurizio Servili, Patrick F Fox, Graciela Rollán, and Marco Gobbetti. 2002. "Arginine catabolism by sourdough lactic acid bacteria: Purification and characterization of the arginine deiminase pathway enzymes from Lactobacillus sanfranciscensis CB1." *Applied and Environmental Microbiology* 68 (12):6193–6201.

De Angelis, Maria, Giovanna Gallo, Maria Rosaria Corbo, Paul LH McSweeney, Michele Faccia, Marinella Giovine, and Marco Gobbetti. 2003. "Phytase activity in sourdough lactic acid bacteria: Purification and characterization of a phytase from Lactobacillus sanfranciscensis CB1." *International Journal of Food Microbiology* 87 (3):259–270.

De Vuyst, Luc, and Patricia Neysens. 2005. "The sourdough microflora: Biodiversity and metabolic interactions." *Trends in Food Science & Technology* 16 (1–3):43–56.

Demirbaş, Fatmanur, Hümeyra İspirli, Asena Ayşe Kurnaz, Mustafa Tahsin Yilmaz, and Enes Dertli. 2017. "Antimicrobial and functional properties of lactic acid bacteria isolated from sourdoughs." *LWT: Food Science and Technology* 79:361–366.

Deshpande, SS, and Munir Cheryan. 1984. "Effects of phytic acid, divalent cations, and their interactions on α -amylase activity." *Journal of Food Science* 49 (2):516–519.

Di Cagno, Raffaella, Maria De Angelis, Paola Lavermicocca, Massimo De Vincenzi, Claudio Giovannini, Michele Faccia, and Marco Gobbetti. 2002. "Proteolysis by sourdough lactic acid bacteria: Effects on wheat flour protein fractions and gliadin peptides involved in human cereal intolerance." *Applied and Environmental Microbiology* 68 (2):623–633.

Di Cagno, R, M De Angelis, Aldo Corsetti, P Lavermicocca, P Arnault, P Tossut, G Gallo, and M Gobbetti. 2003. "Interactions between sourdough lactic acid bacteria and exogenous enzymes: Effects on the microbial kinetics of acidification and dough textural properties." *Food Microbiology* 20 (1):67–75.

Di Cagno, Raffaella, Carlo G Rizzello, Maria De Angelis, Angela Cassone, Giammaria Giuliani, Anna Benedusi, Antonio Limitone, Rosalinda F Surico, and Marco Gobbetti. 2008. "Use of selected sourdough strains of Lactobacillus for removing gluten and enhancing the nutritional properties of gluten-free bread." *Journal of Food Protection* 71 (7):1491–1495.

Doğan, Murat, and İsmail Hakkı Tekiner. 2020. "Extracellular phytase activites of lactic acid bacteria in sourdough mix prepared from traditionally produced boza as starter culture." *Food Health* 6(2):117–127.

Đorđević, Tijana M, Slavica S Šiler-Marinković, and Suzana I Dimitrijević-Branković. 2010. "Effect of fermentation on antioxidant properties of some cereals and pseudo cereals." *Food Chemistry* 119 (3):957–963.

Ebrahimi, Maryam, Alireza Sadeghi, and Seyed Ali Mortazavi. 2020. "The use of cyclic dipeptide producing LAB with potent anti-aflatoxigenic capability to improve technofunctional properties of clean-label bread." *Annals of Microbiology* 70:1–12.

Fang, Siyi, Bowen Yan, Fengwei Tian, Huizhang Lian, Jianxin Zhao, Hao Zhang, Wei Chen, and Daming Fan. 2021. "β-fructosidase FosE activity in Lactobacillus paracasei regulates fructan degradation during sourdough fermentation and total FODMAP levels in steamed bread." *LWT* 145:111294.

Fekri, Arezoo, Mohammadali Torbati, Ahmad Yari Khosrowshahi, Hasan Bagherpour Shamloo, and Sodeif Azadmard-Damirchi. 2020. "Functional effects of phytate-degrading, probiotic lactic acid bacteria and yeast strains isolated from Iranian traditional sourdough on the technological and nutritional properties of whole wheat bread." *Food Chemistry* 306:125620.

Filannino, Pasquale, Raffaella Di Cagno, and Marco Gobbetti. 2018. "Metabolic and functional paths of lactic acid bacteria in plant foods: Get out of the labyrinth." *Current Opinion in Biotechnology* 49:64–72.

Fretzdorff, B, and J-M Brummer. 1992. "Reduction of phytic acid during breadmaking of whole-meal breads." *Cereal Chemistry* 69 (3):266–270.
Fu, Wenhui, Wentong Xue, Chenglong Liu, Yang Tian, Ke Zhang, and Zibo Zhu. 2020. "Screening of lactic acid bacteria and yeasts from sourdough as starter cultures for reduced allergenicity wheat products." *Foods* 9 (6):751.
Galle, Sandra, and Elke K Arendt. 2014. "Exopolysaccharides from sourdough lactic acid bacteria." *Critical Reviews in Food Science and Nutrition* 54 (7):891–901.
Galle, Sandra, Clarissa Schwab, Elke Arendt, and Michael Gänzle. 2010. "Exopolysaccharide-forming Weissella strains as starter cultures for sorghum and wheat sourdoughs." *Journal of Agricultural and Food Chemistry* 58 (9):5834–5841.
Gänzle, Michael G. 2014. "Enzymatic and bacterial conversions during sourdough fermentation." *Food Microbiology* 37:2–10.
Gänzle, Michael G, Jussi Loponen, and Marco Gobbetti. 2008. "Proteolysis in sourdough fermentations: Mechanisms and potential for improved bread quality." *Trends in Food Science & Technology* 19 (10):513–521.
Gebruers, Kurt, Emmie Dornez, Zoltan Bedo, Mariann Rakszegi, Christophe M Courtin, and Jan A Delcour. 2010. "Variability in xylanase and xylanase inhibition activities in different cereals in the HEALTHGRAIN diversity screen and contribution of environment and genotype to this variability in common wheat." *Journal of Agricultural and Food Chemistry* 58 (17):9362–9371.
Gibson, Peter Raymond, and Susan J Shepherd. 2005. "Personal view: Food for thought–western lifestyle and susceptibility to Crohn's disease. The FODMAP hypothesis." *Alimentary Pharmacology & Therapeutics* 21 (12):1399–1409.
Gobbetti, M. 1998. "The sourdough microflora: Interactions of lactic acid bacteria and yeasts." *Trends in Food Science & Technology* 9 (7):267–274.
Gobbetti, Marco, MS Simonetti, J Rossi, L Cossignani, Aldo Corsetti, and P Damiani. 1994. "Free D-and L-amino acid evolution during sourdough fermentation and baking." *Journal of Food Science* 59 (4):881–884.
Gobbetti, Marco, E Smacchi, and Aldo Corsetti. 1996. "The proteolytic system of Lactobacillus sanfrancisco CB1: Purification and characterization of a proteinase, a dipeptidase, and an aminopeptidase." *Applied and Environmental Microbiology* 62 (9):3220–3226.
Gobbetti, Marco, P Lavermicocca, F Minervini, M De Angelis, and Aldo Corsetti. 2000. "Arabinose fermentation by Lactobacillus plantarum in sourdough with added pentosans and α α -L-arabinofuranosidase: A tool to increase the production of acetic acid." *Journal of Applied Microbiology* 88 (2):317–324.
Gobbetti, M, M De Angelis, Aldo Corsetti, and R Di Cagno. 2005. "Biochemistry and physiology of sourdough lactic acid bacteria." *Trends in Food Science & Technology* 16 (1–3):57–69.
Gobbetti, Marco, Carlo G Rizzello, Raffaella Di Cagno, and Maria De Angelis. 2014. "How the sourdough may affect the functional features of leavened baked goods." *Food Microbiology* 37:30–40.
Gobbetti, Marco, Maria De Angelis, Raffaella Di Cagno, Maria Calasso, Gabriele Archetti, and Carlo Giuseppe Rizzello. 2019. "Novel insights on the functional/nutritional features of the sourdough fermentation." *International Journal of Food Microbiology* 302:103–113.
Goesaert, Hans, Kristof Brijs, WS Veraverbeke, CM Courtin, Kurt Gebruers, and JA Delcour. 2005. "Wheat flour constituents: How they impact bread quality, and how to impact their functionality." *Trends in Food Science & Technology* 16 (1–3):12–30.
Goesaert, Hans, Louise Slade, Harry Levine, and Jan A Delcour. 2009. "Amylases and bread firming–an integrated view." *Journal of Cereal Science* 50 (3):345–352.

GrandViewResearch. 2021. "Enzymes Market Size, Share & Trends Analysis Report By Application (Industrial Enzymes, Specialty Enzymes), By Product (Carbohydrase, Proteases, Lipases), By Source, By Region, And Segment Forecasts, 2020–2027." https://www.grandviewresearch.com/industry-analysis/enzymes-industry

Greiner, Ralf, and Ursula Konietzny. 2006. "Phytase for food application." *Food Technology & Biotechnology* 44 (2):125–140.

Gupta, Rani, Paresh Gigras, Harapriya Mohapatra, Vineet Kumar Goswami, and Bhavna Chauhan. 2003. "Microbial α-amylases: A biotechnological perspective." *Process Biochemistry* 38 (11):1599–1616.

Hammes, Walter Peter, and MG Gänzle. 1998. "Sourdough breads and related products." In *Microbiology of Fermented Foods*, 199–216. Springer.

Hansen, H Boskov, M Andreasen, M Nielsen, L Larsen, Bach K Knudsen, A Meyer, L Christensen, and Å Hansen. 2002. "Changes in dietary fibre, phenolic acids and activity of endogenous enzymes during rye bread-making." *European Food Research and Technology* 214 (1):33–42.

Heldt-Hansen, Hans Peter. 2006. "Macromolecular interactions in enzyme applications for food products." *Food Science and Technology-New York-Marcel Dekker* 154:363.

Henrissat, Bernard, and Gideon Davies. 1997. "Structural and sequence-based classification of glycoside hydrolases." *Current Opinion in Structural Biology* 7 (5):637–644.

Hole, Anastasia S, Ida Rud, Stine Grimmer, Stefanie Sigl, Judith Narvhus, and Stefan Sahlstrøm. 2012. "Improved bioavailability of dietary phenolic acids in whole grain barley and oat groat following fermentation with probiotic Lactobacillus acidophilus, Lactobacillus johnsonii, and Lactobacillus reuteri." *Journal of Agricultural and Food Chemistry* 60 (25):6369–6375.

Illanes, Andrés. 2008. "Enzyme biocatalysis." *Principles and Applications*, 1–56. Editorial Springer-Verlag New York Inc.

Jänsch, André, Maher Korakli, Rudi F Vogel, and Michael G Gänzle. 2007. "Glutathione reductase from Lactobacillus sanfranciscensis DSM20451T: Contribution to oxygen tolerance and thiol exchange reactions in wheat sourdoughs." *Applied and Environmental Microbiology* 73 (14):4469–4476.

Jemli, Sonia, Dorra Ayadi-Zouari, Hajer Ben Hlima, and Samir Bejar. 2016. "Biocatalysts: Application and engineering for industrial purposes." *Critical Reviews in Biotechnology* 36 (2):246–258.

Joye, Iris J, Bert Lagrain, and Jan A Delcour. 2009. "Use of chemical redox agents and exogenous enzymes to modify the protein network during breadmaking: A review." *Journal of Cereal Science* 50 (1):11–21.

Jòzef, Synowiecki. 2007. "The use of starch processing enzymes in the food industry." In *Industrial Enzymes*, 19–34. Springer.

Kaditzky, Susanne, and Rudi F Vogel. 2008. "Optimization of exopolysaccharide yields in sourdoughs fermented by lactobacilli." *European Food Research and Technology* 228 (2):291–299.

Karaman, Kevser, Osman Sagdic, and M Zeki Durak. 2018. "Use of phytase active yeasts and lactic acid bacteria isolated from sourdough in the production of whole wheat bread." *LWT* 91:557–567.

Katina, Kati, Marjatta Salmenkallio-Marttila, Riitta Partanen, Pirkko Forssell, and Karin Autio. 2006. "Effects of sourdough and enzymes on staling of high-fibre wheat bread." *LWT-Food Science and Technology* 39 (5):479–491.

Kieronczyk, A, S Skeie, K Olsen, and T Langsrud. 2001. "Metabolism of amino acids by resting cells of non-starter lactobacilli in relation to flavour development in cheese." *International Dairy Journal* 11 (4–7):217–224.

Kies, Arie K, Leon H De Jonge, Paul A Kemme, and Age W Jongbloed. 2006. "Interaction between protein, phytate, and microbial phytase. In vitro studies." *Journal of Agricultural and Food Chemistry* 54 (5):1753–1758.
Kim, Ji Hyun, Tomoko Maeda, and Naofumi Morita. 2006. "Effect of fungal α-amylase on the dough properties and bread quality of wheat flour substituted with polished flours." *Food Research International* 39 (1):117–126.
Kim, Kyoung-Rok, and Deok-Kun Oh. 2013. "Production of hydroxy fatty acids by microbial fatty acid-hydroxylation enzymes." *Biotechnology Advances* 31 (8):1473–1485. https://doi.org/10.1016/j.biotechadv.2013.07.004.
Korakli, Maher, and Rudi F Vogel. 2006. "Structure/function relationship of homopolysaccharide producing glycansucrases and therapeutic potential of their synthesised glycans." *Applied Microbiology and Biotechnology* 71 (6):790–803.
Korakli, Maher, Andreas Rossmann, Michael G Gänzle, and Rudi F Vogel. 2001. "Sucrose metabolism and exopolysaccharide production in wheat and rye sourdoughs by Lactobacillus sanfranciscensis." *Journal of Agricultural and Food Chemistry* 49 (11):5194–5200.
Kumar, Vikas, Amit K Sinha, Harinder PS Makkar, and Klaus Becker. 2010. "Dietary roles of phytate and phytase in human nutrition: A review." *Food Chemistry* 120 (4):945–959.
Laatikainen, Reijo, Jari Koskenpato, Sanna-Maria Hongisto, Jussi Loponen, Tuija Poussa, Xin Huang, Tuula Sontag-Strohm, Hanne Salmenkari, and Riitta Korpela. 2017. "Pilot study: Comparison of sourdough wheat bread and yeast-fermented wheat bread in individuals with wheat sensitivity and irritable bowel syndrome." *Nutrients* 9 (11):1215.
Larsson, Marie, and A-S Sandberg. 1991. "Phytate reduction in bread containing oat flour, oat bran or rye bran." *Journal of Cereal Science* 14 (2):141–149.
Laurikainen, Taru, Helena Härkönen, Karin Autio, and Kaisa Poutanen. 1998. "Effects of enzymes in fibre-enriched baking." *Journal of the Science of Food and Agriculture* 76 (2):239–249.
Lavermicocca, Paola, Francesca Valerio, Antonio Evidente, Silvia Lazzaroni, Aldo Corsetti, and Marco Gobbetti. 2000. "Purification and characterization of novel antifungal compounds from the sourdough Lactobacillus plantarum strain 21B." *Applied and Environmental Microbiology* 66 (9):4084–4090.
LeBlanc, JG, Marisa Selva Garro, Aurelio Silvestroni, Cristelle Connes, J-C Piard, F Sesma, and G Savoy de Giori. 2004. "Reduction of α-galactooligosaccharides in soyamilk by Lactobacillus fermentum CRL 722: In vitro and in vivo evaluation of fermented soyamilk." *Journal of Applied Microbiology* 97 (4):876–881.
Lee, Bao-Hong, Yi-Hsuan Lo, and Tzu-Ming Pan. 2013. "Anti-obesity activity of Lactobacillus fermented soy milk products." *Journal of Functional Foods* 5 (2):905–913.
Leenhardt, Fanny, Marie-Anne Levrat-Verny, Elisabeth Chanliaud, and Christian Rémésy. 2005. "Moderate decrease of pH by sourdough fermentation is sufficient to reduce phytate content of whole wheat flour through endogenous phytase activity." *Journal of Agricultural and Food Chemistry* 53 (1):98–102.
Li, Qing, Jussi Loponen, and Michael G Ganzle. 2020. "Characterization of the extracellular fructanase FruA in Lactobacillus crispatus and its contribution to fructan hydrolysis in breadmaking." *Journal of Agricultural and Food Chemistry* 68 (32):8637–8647.
Lopez, Hubert W, Ariane Ouvry, Elisabeth Bervas, Christine Guy, Arnaud Messager, Christian Demigne, and Christian Remesy. 2000. "Strains of lactic acid bacteria isolated from sour doughs degrade phytic acid and improve calcium and magnesium solubility from whole wheat flour." *Journal of Agricultural and Food Chemistry* 48 (6):2281–2285.

Lopez, HW, V Krespine, C Guy, A Messager, C Demigne, and C Remesy. 2001. "Prolonged fermentation of whole wheat sourdough reduces phytate level and increases soluble magnesium." *Journal of Agricultural and Food Chemistry* 49 (5):2657–2662.

Lopez, Hubert W, Virgile Duclos, Charles Coudray, Virginie Krespine, Christine Feillet-Coudray, Arnaud Messager, Christian Demigné, and Christian Rémésy. 2003. "Making bread with sourdough improves mineral bioavailability from reconstituted whole wheat flour in rats." *Nutrition* 19 (6):524–530.

Loponen, Jussi, Paivi Kanerva, Chonggang Zhang, Tuula Sontag-Strohm, Hannu Salovaara, and Michael G Gänzle. 2009. "Prolamin hydrolysis and pentosan solubilization in germinated-rye sourdoughs determined by chromatographic and immunological methods." *Journal of Agricultural and Food Chemistry* 57 (2):746–753.

Lu, Zhong X, Karen Z Walker, Jane G Muir, Tom Mascara, and Kerin O'Dea. 2000. "Arabinoxylan fiber, a byproduct of wheat flour processing, reduces the postprandial glucose response in normoglycemic subjects." *The American Journal of Clinical Nutrition* 71 (5):1123–1128.

Lu, Zhong X, Karen Zell Walker, Jane Grey Muir, and Kerin O'Dea. 2004. "Arabinoxylan fibre improves metabolic control in people with Type II diabetes." *European Journal of Clinical Nutrition* 58 (4):621–628.

Malaguti, Marco, Giovanni Dinelli, Emanuela Leoncini, Valeria Bregola, Sara Bosi, Arrigo FG Cicero, and Silvana Hrelia. 2014. "Bioactive peptides in cereals and legumes: Agronomical, biochemical and clinical aspects." *International Journal of Molecular Sciences* 15 (11):21120–21135.

Manini, F, MC Casiraghi, K Poutanen, M Brasca, D Erba, and C Plumed-Ferrer. 2016. "Characterization of lactic acid bacteria isolated from wheat bran sourdough." *LWT-Food Science and Technology* 66:275–283.

Martin, ML, and RC Hoseney. 1991. "A mechanism of bread firming. II, Role of starch hydrolyzing enzymes." *Cereal Chemistry* 68 (5):503–507.

Martinez, CRGP, G Ros, MJ Periago, G Lopez, J Ortuno, and F Rincon. 1996. "Phytic acid in human nutrition." *Food Science and Technology International* 2 (4):201–210.

Miguel, Ângelo Samir Melim, Tathiana Souza Martins-Meyer, EVDC Figueiredo, Bianca Waruar Paulo Lobo, and Gisela Maria Dellamora-Ortiz. 2013. "Enzymes in bakery: Current and future trends." *Food Industry*:278–321.

Milanović, Vesna, Andrea Osimani, Cristiana Garofalo, Luca Belleggia, Antonietta Maoloni, Federica Cardinali, Massimo Mozzon, Roberta Foligni, Lucia Aquilanti, and Francesca Clementi. 2020. "Selection of cereal-sourced lactic acid bacteria as candidate starters for the baking industry." *Plos One* 15 (7):e0236190.

Möhlig, M, C Koebnick, MO Weickert, W Lueder, B Otto, J Steiniger, M Twilfert, F Meuser, AFH Pfeiffer, and HJ Zunft. 2005. "Arabinoxylan-enriched meal increases serum ghrelin levels in healthy humans." *Hormone and Metabolic Research* 37 (05):303–308.

Müller, Denise Christina, Ha Nguyen, Qing Li, Regine Schönlechner, Susanne Miescher Schwenninger, Wendy Wismer, and Michael Gänzle. 2021. "Enzymatic and microbial conversions to achieve sugar reduction in bread." *Food Research International* 143:110296.

Muñoz, R, B de las Rivas, F López de Felipe, I Reverón, L Santamaría, M Esteban-Torres, JA Curiel, H Rodríguez, and JM Landete. 2017. "Biotransformation of phenolics by Lactobacillus plantarum in fermented foods." In *Fermented Foods in Health and Disease Prevention*, 63–83. Elsevier.

Nguyen, Ha T, Ruud JB Peters, and Martinus AJS Van Boekel. 2016. "Acrylamide and 5-hydroxymethylfurfural formation during baking of biscuits: Part I: Effects of sugar type." *Food Chemistry* 192:575–585.

Novozymes. 2021. "The Novozymes Report 2020." https://report2020.novozymes.com/#Facts-and-figures

Nuobariene, Lina, Dalia Cizeikiene, Egle Gradzeviciute, Åse S Hansen, Søren K Rasmussen, Grazina Juodeikiene, and Finn K Vogensen. 2015. "Phytase-active lactic acid bacteria from sourdoughs: Isolation and identification." *LWT: Food Science and Technology* 63 (1):766–772.

Ogawa, Jun, Kenji Matsumura, Shigenobu Kishino, Yoriko Omura, and Sakayu Shimizu. 2001. "Conjugated linoleic acid accumulation via 10-hydroxy-12-octadecaenoic acid during microaerobic transformation of linoleic acid by Lactobacillus acidophilus." *Applied and Environmental Microbiology* 67 (3):1246–1252.

Osawa, RO, Keiko Kuroiso, Satoshi Goto, and Akira Shimizu. 2000. "Isolation of tannin-degrading lactobacilli from humans and fermented foods." *Applied and Environmental Microbiology* 66 (7):3093–3097.

Palacios, María Consuelo, Monica Haros, Yolanda Sanz, and Cristina M Rosell. 2008. "Selection of lactic acid bacteria with high phytate degrading activity for application in whole wheat breadmaking." *LWT-Food Science and Technology* 41 (1):82–92.

Pandey, Ashok, George Szakacs, Carlos R Soccol, Jose A Rodriguez-Leon, and Vanete T Soccol. 2001. "Production, purification and properties of microbial phytases." *Bioresource technology* 77 (3):203–214.

Pepe, Olimpia, Guiseppe Blajotta, Marilena Anastasio, Giancarlo Moschetti, Danilo Ercolini, and Francesco Villani. 2004. "Technological and molecular diversity of Lactobacillus plantarum strains isolated from naturally fermented sourdoughs." *Systematic and Applied Microbiology* 27 (4):443–453.

Persson, Hans, Maria Türk, Margareta Nyman, and Ann-Sofie Sandberg. 1998. "Binding of Cu2+, Zn2+, and Cd2+ to inositol tri-, tetra-, penta-, and hexaphosphates." *Journal of agricultural and food chemistry* 46 (8):3194–3200.

Poutanen, Kaisa. 1997. "Enzymes: An important tool in the improvement of the quality of cereal foods." *Trends in Food Science & Technology* 8 (9):300–306.

Poutanen, Kaisa, Laura Flander, and Kati Katina. 2009. "Sourdough and cereal fermentation in a nutritional perspective." *Food Microbiology* 26 (7):693–699.

Rao, Satish Sanku Chander, S Yu, and A Fedewa. 2015. "Systematic review: Dietary fibre and FODMAP-restricted diet in the management of constipation and irritable bowel syndrome." *Alimentary Pharmacology & Therapeutics* 41 (12):1256–1270.

Rashmi, Bennur Somashekharaiah, Devaraja Gayathri, Mahanthesh Vasudha, Chakra Siddappa Prashantkumar, Chidanandamurthy Thippeswamy Swamy, Kumar S Sunil, Palegar Krishnappa Somaraja, and Patil Prakash. 2020. "Gluten hydrolyzing activity of Bacillus spp isolated from sourdough." *Microbial Cell Factories* 19 (1):1–11.

Reale, Anna, Luisa Mannina, Patrizio Tremonte, Anatoli P Sobolev, Mariantonietta Succi, Elena Sorrentino, and Raffaele Coppola. 2004. "Phytate degradation by lactic acid bacteria and yeasts during the wholemeal dough fermentation: A 31P NMR study." *Journal of Agricultural and Food Chemistry* 52 (20):6300–6305.

Rizzello, Carlo Giuseppe, Luana Nionelli, Rossana Coda, Maria De Angelis, and Marco Gobbetti. 2010. "Effect of sourdough fermentation on stabilisation, and chemical and nutritional characteristics of wheat germ." *Food Chemistry* 119 (3):1079–1089.

Rizzello, Carlo Giuseppe, Angela Cassone, Rossana Coda, and Marco Gobbetti. 2011. "Antifungal activity of sourdough fermented wheat germ used as an ingredient for bread making." *Food Chemistry* 127 (3):952–959.

Rizzello, Carlo Giuseppe, Luana Nionelli, Rossana Coda, and Marco Gobbetti. 2012. "Synthesis of the cancer preventive peptide lunasin by lactic acid bacteria during sourdough fermentation." *Nutrition and Cancer* 64 (1):111–120.

Rizzello, Carlo Giuseppe, José Antonio Curiel, Luana Nionelli, Olimpia Vincentini, Raffaella Di Cagno, Marco Silano, Marco Gobbetti, and Rossana Coda. 2014. "Use of fungal proteases and selected sourdough lactic acid bacteria for making wheat bread with an intermediate content of gluten." *Food Microbiology* 37:59–68.

Rollán, Graciela C, Carla L Gerez, and Jean G LeBlanc. 2019. "Lactic fermentation as a strategy to improve the nutritional and functional values of pseudocereals." *Frontiers in Nutrition* 6:98.

Röcken, W, and PA Voysey. 1995. "Sour-dough fermentation in bread making." *Journal of Applied Bacteriology*. Oxford[J. APPL. BACTERIOL.]. 79.

RŘhmkorf, Christine, Sarah Jungkunz, Maria Wagner, and Rudi F Vogel. 2012. "Optimization of homoexopolysaccharide formation by lactobacilli in gluten-free sourdoughs." *Food Microbiology* 32 (2):286–294.

Sáez, Gabriel D, Elvira M Hébert, Lucila Saavedra, and Gabriela Zárate. 2017. "Molecular identification and technological characterization of lactic acid bacteria isolated from fermented kidney beans flours (Phaseolus vulgaris L. and P. coccineus) in northwestern Argentina." *Food Research International* 102:605–615.

Sáez, Gabriel D, Lucila Saavedra, Elvira M Hebert, and Gabriela Zárate. 2018. "Identification and biotechnological characterization of lactic acid bacteria isolated from chickpea sourdough in northwestern Argentina." *LWT* 93:249–256.

Sancho, Ana I, Craig B Faulds, Begoña Bartolomé, and Gary Williamson. 1999. "Characterisation of feruloyl esterase activity in barley." *Journal of the Science of Food and Agriculture* 79 (3):447–449.

Sandberg, Ann-Sofie, Mats Brune, Nils-Gunnar Carlsson, Leif Hallberg, Erika Skoglund, and Lena Rossander-Hulthén. 1999. "Inositol phosphates with different numbers of phosphate groups influence iron absorption in humans." *The American Journal of Clinical Nutrition* 70 (2):240–246.

Sandez Penidez, Sergio H, Marina A Velasco Manini, Carla L Gerez, and Graciela C Rollán. 2020. "Partial characterization and purification of phytase from Lactobacillus plantarum CRL1964 isolated from pseudocereals." *Journal of Basic Microbiology* 60 (9):787–798.

Sanz-Penella, Juan Mario, Juan Antonio Tamayo-Ramos, and Monika Haros. 2012. "Application of bifidobacteria as starter culture in whole wheat sourdough breadmaking." *Food and Bioprocess Technology* 5 (6):2370–2380.

Shewry, Peter R, Vieno Piironen, Anna-Maija Lampi, Minnamari Edelmann, Susanna Kariluoto, Tanja Nurmi, Rebeca Fernandez-Orozco, Annica AM Andersson, Per Åman, and Anna Fras. 2010. "Effects of genotype and environment on the content and composition of phytochemicals and dietary fiber components in rye in the HEALTHGRAIN diversity screen." *Journal of Agricultural and Food Chemistry* 58 (17):9372–9383.

Struyf, Nore, Jitka Laurent, Joran Verspreet, Kevin J Verstrepen, and Christophe M Courtin. 2017. "Saccharomyces cerevisiae and Kluyveromyces marxianus cocultures allow reduction of fermentable oligo-, di-, and monosaccharides and polyols levels in whole wheat bread." *Journal of Agricultural and Food Chemistry* 65 (39):8704–8713.

Sumengen, Melis, Sadik Dincer, and Aysenur Kaya. 2013. "Production and characterization of phytase from Lactobacillus plantarum." *Food Biotechnology* 27 (2):105–118.

Svensson, Louise, Bonno Sekwati-Monang, Daise Lopes Lutz, Andreas Schieber, and Michael G Ganzle. 2010. "Phenolic acids and flavonoids in nonfermented and fermented red sorghum (Sorghum bicolor (L.) Moench)." *Journal of Agricultural and Food Chemistry* 58 (16):9214–9220.

Tan, PS Tjwan, Bert Poolman, and Wil N Konings. 1993. "Proteolytic enzymes of Lactococcus lactis." *Journal of Dairy Research* 60 (2):269–286.

Thiele, Claudia, Michael G Gänzle, and Rudi F Vogel. 2002. "Contribution of sourdough lactobacilli, yeast, and cereal enzymes to the generation of amino acids in dough relevant for bread flavor." *Cereal Chemistry* 79 (1):45–51.
Thiele, Claudia, Michael G Gänzle, and Rudi F Vogel. 2003. "Fluorescence labeling of wheat proteins for determination of gluten hydrolysis and depolymerization during dough processing and sourdough fermentation." *Journal of Agricultural and Food Chemistry* 51 (9):2745–2752.
Thiele, Claudia, Simone Grassl, and Michael Gänzle. 2004. "Gluten hydrolysis and depolymerization during sourdough fermentation." *Journal of Agricultural and Food Chemistry* 52 (5):1307–1314.
Türk, Maria, Ann-Sofie Sandberg, Nils-Gunnar Carlsson, and Thomas Andlid. 2000. "Inositol hexaphosphate hydrolysis by Baker's yeast. Capacity, kinetics, and degradation products." *Journal of Agricultural and Food Chemistry* 48 (1):100–104.
Valcheva, Rosica, Maher Korakli, Bernard Onno, Herve Prevost, Iskra Ivanova, Matthias A Ehrmann, Xavier Dousset, Michael G Gänzle, and Rudi F Vogel. 2005. "Lactobacillus hammesii sp. nov., isolated from French sourdough." *International Journal of Systematic and Evolutionary Microbiology* 55 (2):763–767.
van Hijum, Sacha AFT, Slavko Kralj, Lukasz K Ozimek, Lubbert Dijkhuizen, and Ineke GH van Geel-Schutten. 2006. "Structure-function relationships of glucansucrase and fructansucrase enzymes from lactic acid bacteria." *Microbiology and Molecular Biology Reviews* 70 (1):157–176.
Verma, Digvijay, and T Satyanarayana. 2012. "Molecular approaches for ameliorating microbial xylanases." *Bioresource Technology* 117:360–367.
Vermeulen, Nicoline, Melanie Pavlovic, Matthias A Ehrmann, Michael G Gänzle, and Rudi F Vogel. 2005. "Functional characterization of the proteolytic system of Lactobacillus sanfranciscensis DSM 20451T during growth in sourdough." *Applied and Environmental Microbiology* 71 (10):6260–6266.
Whitehurst, Robert J, and Maarten Van Oort. 2010. *Enzymes in Food Technology*. Vol. 388: Wiley Online Library.
Wieser, Herbert. 2007. "Chemistry of gluten proteins." *Food Microbiology* 24 (2):115–119.
Yildirim, Rusen Metin, and Muhammet Arici. 2019. "Effect of the fermentation temperature on the degradation of phytic acid in whole-wheat sourdough bread." *LWT* 112:108224.
Zhang, Shuhang, Jingjing Wang, and Hong Jiang. 2020. "Microbial production of value-added bioproducts and enzymes from molasses, a by-product of sugar industry." *Food Chemistry*:128860.
Ziegler, Jochen U, Deborah Steiner, C Friedrich H Longin, Tobias Würschum, Ralf M Schweiggert, and Reinhold Carle. 2016. "Wheat and the irritable bowel syndrome–FODMAP levels of modern and ancient species and their retention during bread making." *Journal of Functional Foods* 25:257–266.

9 Biotechnological Applications of Sourdough Lactic Acid Bacteria
A Source for Vitamins Fortification and Exopolysaccharides Improvement

*Pasquale Russo, Kenza Zarour,
María Goretti Llamas-Arriba,
Norhane Besrour-Aouam, Vittorio Capozzi,
Nicola De Simone, Paloma López, and
Giuseppe Spano*

CONTENTS

9.1 Introduction	232
9.2 Sourdough Fermentation by Lactic Acid Bacteria Producing B-Group Vitamins	233
9.2.1 Folate Production by Lactic Acid Bacteria	234
9.2.2 Riboflavin Production by Lactic Acid Bacteria	237
9.2.3 Cobalamin Production by Lactic Acid Bacteria	239
9.3 Exopolysaccharides Produced by Lactic Acid Bacteria	240
9.4 Dextrans Produced by Lactic Acid Bacteria	243
9.5 Prebiotics in Bakery Products	248
Conclusion	251
Acknowledgments	251
References	252

DOI: 10.1201/9781003141143-11

9.1 INTRODUCTION

In recent years, consumers have moved away from a simple need to satisfy hunger toward consuming food to maintain well-being and reduce the risk of disease. Thus, there is a growing interest in the development of foods with functional properties, and among these foods, there is bread, which is a daily staple consumed all over the world (Siró et al. 2008). Eating bread can have a significant nutritional impact (Crucean et al. 2019) because, apart from providing macronutrients, wheat bread represents an important source of micronutrients such as minerals and vitamins. A variety of B vitamins is detected in wheat and bread, such as thiamine (B_1), riboflavin (B_2), niacin (B_3), pantothenic acid (B_5), pyridoxine (B_6), biotin (B_8), and folates (B_9). These water-soluble molecules are key substances in carbohydrate, protein, and fat metabolism (Batifoulier et al. 2006; Kurek et al. 2017). The concentrations of B vitamins in grain flour and bread are affected by several factors, starting with (i) the variety of wheat, (ii) the growing locality, (iii) the soil type, (iv) the use of fertilizers, and ending with (v) the storage conditions of wheat, (vi) milling process, and (vii) bread making (Mihhalevski et al. 2013). Therefore, it is important to control the loss of vitamins during the manufacturing stages. The decrease of daily B vitamins intake could be related to a decrease of bread consumption (Batifoulier et al. 2005), as well as to the consumption of refined products and the attenuation of the use of cereals in the daily diet (Rohi et al. 2013). Among the proposed keys for ensuring vitamins, the administration of fortified products can be a medium-term solution (Czeizel and Merhala 1998; Povoroznyuk et al. 2015). The incorporation of such vital molecules in bakery products, like bread and biscuits, ensures, in addition to better nutritional quality, novel textural and sensorial properties. The traditional and original way to enrich bread composition is food to food fortification, using herbs, spices, millets, oilseeds, and pulses, which contain significant amounts of proteins, fat, calcium, potassium, iron, and vitamins (Agrahar-Murugkar 2020; Boukid et al. 2019). Fortification with exopolysaccharides (EPS), in particular dextran, provides doughs and breads with important rheological and textural properties while improving their quality and acceptability. This biopolymer also has anti-infectious (Zarour, Llamas, et al. 2017) and antiviral activities (Nácher-Vázquez et al. 2015) or antioxidant, anti-inflammatory, and cholesterol-lowering properties (Mohd Nadzir et al. 2021). These characteristics confer on EPS a high potential as bioingredients with nutritional and health interests.

The incorporation of sourdough-containing microorganisms, such as lactic acid bacteria (LAB), which are able to produce many metabolic products, constitutes a novel technological approach to (i) ensure the sourdough fermentation and (ii) to fortify it by producing several end products like B vitamins, dextran, polyols, oligosaccharides, organic acids, CO_2, and/or antimicrobial substances (Moreno de LeBlanc et al. 2018; Novotni, Gänzle, and Rocha 2021).

Sourdoughs are a mixture of cereal flour (mainly wheat and rye) and water, which are spontaneously fermented by indigenous yeasts and LAB and are normally used as leavening agents (type I sourdoughs) or as starters (type II sourdoughs) for bread elaboration (Gänzle and Zheng 2019; Brandt 2019). Dough is usually dried (type III sourdoughs) for easy storage and utilization (Lau et al. 2021), and this type is used as acidifier supplements and aroma carriers (Meroth et al. 2003). In general, type I sourdoughs are employed to produce artisanal breads and they are maintained by sequential refreshments (backslopping), while type II sourdoughs are used in the industrial process (Brandt 2019). However, both sourdoughs usually contain one or two different yeast species and two or three different species of LAB. It is very common to find *Fructilactobacillus sanfranciscensis* (formerly *Lactobacillus sanfranciscensis*) populating type I sourdoughs, while type II sourdoughs are rich in *Limosilactobacillus reuteri* (previously named *Lactobacillus reuteri*) or *Lactobacillus delbrueckii* (Gänzle and Zheng 2019). Traditionally, the fermentation during bread making is performed by a simple mixture, and efforts are made to diversify the type of breads and to improve their organoleptic properties. Furthermore, in recent years, research has increased to improve microbiological, technological, and nutritional profiles, thereby creating healthier bread.

In this context, this book chapter reports the main role of LAB in the sourdough used in bread making, based on the beneficial effect exerted by their metabolites with significant technological interest: vitamins, dextran, and prebiotics, used as fortification sources.

9.2 SOURDOUGH FERMENTATION BY LACTIC ACID BACTERIA PRODUCING B-GROUP VITAMINS

Several strains of LAB can produce different bioactive compounds, including vitamins that, when accumulated in the food matrix, contribute to enhancing the nutritional value of fermented foods. In recent years, the concept of *in situ* fortification exploiting bacterial fermentation has been proposed as a low-cost and eco-friendly approach to increase the level of B-group vitamins in some foods (Levit et al. 2021). Due to the high incidence of these deficiencies, current fortification programs and recent investigation projects aim to improve the levels of riboflavin (vitamin B_2), folate (vitamin B_9), and cobalamin (vitamin B_{12}) (Burgess, Smid, and van Sinderen 2009). Cereal-based fermented foods are major contributors to energy intake in both developed and low-income countries (Guyot 2012). Therefore, sourdough fermentation by LAB producing B-group vitamins has a worldwide potential in the field of functional/fortified cereal products (Capozzi et al. 2012). However, LAB's contribution to the vitamin enrichment of fermented foods is often negligible since, under physiological conditions, they do not need to synthesize high concentrations of these micronutrients. In addition, the occurrence of exogenous vitamins in complex vegetable matrices, such as sourdough, could inhibit the expression of the pathways responsible for their production. These considerations are partially in contrast with the concept of *in situ* biofortification, limiting the application of tailored bio-based strategies.

Therefore, there is a growing interest to develop innovative approaches aiming to select vitamin-overproducing strains for application in the food industry. It is important to point out that though metabolic engineering is a suitable strategy to increase the production of secondary metabolites, the introduction of genetically modified microorganisms into the food chain may be limited by ethical and regulatory concerns (Bekaert et al. 2008). In contrast, several food-grade strategies can be adopted to increase the level of vitamin production during fermentation, mainly including (i) the isolation and selection of new strains and/or species from different ecological niches able to produce high levels of B vitamins; (ii) the employment of co-culture in mixed fermentations; (iii) the selection of strains producing more stable forms of the vitamin (called vitamers); (iv) the exposure to the selective pressure of toxic analogues and/or compounds in the absence of the vitamin of interest; (v) the induction of random mutations by exposure to mutagenic agents; and (vi) the optimization of fermentation conditions.

Recent applications of LAB strains isolated from cereal- and/or pseudocereal-based matrices able to synthesize B-group vitamins in order to produce bio-enriched foods by sourdough fermentation are summarized in Table 9.1.

9.2.1 Folate Production by Lactic Acid Bacteria

Folate is involved in essential biological functions such as DNA replication, repair and methylation, and synthesis of nucleotides and some vitamins and amino acids (e.g. methionine). Folate deficiency in humans causes megaloblastic anemia, Alzheimer's, heart disease, osteoporosis, increased risk of breast and colorectal cancers, hearing loss, congenital malformations, neural tube defects. The vitamin B_9 recommended dietary allowance (RDA) is 400 mg of dietary folate equivalent (DFE) per day for adults. In general, cereals are good folate sources that are the most abundant vitamin of B-group in their grains.

Fermentation has been reported to increase the nutritional value of sourdough and, in particular, its folate content (Saubade et al. 2017). However, several factors can impact the level of folate in sourdough, including the cereal matrix, the microbial strains used as starter cultures, and the fermentation conditions. For example, in multigrain products, the folate concentration increased with the addition of rye or hulled oats (Aprodu, Bolea, and Banu 2019). The same authors reported that the amount of folate was lower in sourdoughs fermented by starter cultures than in spontaneous fermentation, also suggesting that folate production could be a pro-technological criterion to select new strains as starter cultures for sourdough fermentation (Aprodu, Bolea, and Banu 2019). Accordingly, it has been reported that selected yeast strains increased folates in white wheat bread compared to doughs leavened by commercial baker's cultures, and this feature was attributed to the cultivation procedure prior to food fermentation (Hjortmo et al. 2008). In general, baker's yeast contributes to the sourdough's folate enrichment by synthesizing folates to a greater extent than LAB (Kariluoto et al. 2004). Indeed, by using *S. cerevisiae* as starter cultures, the folate level was increased up to seven-fold compared to a sourdough fermentation driven by various LAB strains or spontaneous fermentation (Katina et al. 2007). Moreover, the sourdough's complex microbial biodiversity also includes several non-*Saccharomyces* and endogenous bacteria that have been investigated for

TABLE 9.1
LAB Strains Isolated from Cereal and/or Pseudocereal-Based Matrices Able to Synthesize B-Group Vitamins and Their Employment to Produce Bio-enriched Foods by Sourdough Fermentation

LAB strain	Source of isolation	Vitamin production (culture media)	Vitamin bio-enrichment	Bio-enriched food	Reference
L. plantarum CRL 2106	Andean amaranth sourdough and grains	138 ng/mL (B_9)	-	-	Carrizo et al. 2017
L. plantarum CRL 2107		158 ng/mL (B_2)			
L. fermentum 8.2	Ben-saalga	97 ng/mL (B_9)	-	-	Greppi et al. 2017
L. plantarum 6.2		93 ng/mL (B_9)			
L. plantarum CRL 1973	Quinoa sourdough	143 ng/mL (B_9)	-	-	Carrizo et al. 2016
L. rhamnosus CRL 1963		360 ng/mL (B_2)			
L. pentosus ES124	Raw cereal grains	61 ng/mL (B_9)	-	-	Salvucci et al. 2016
L. plantarum ES137		57 ng/mL (B_9)			
L. reuteri CRL1098	Sourdough	0.5 mg/L (B_{12})	—	–	Taranto et al. 2003
L. plantarum CRL 2107 + L. plantarum CRL 1964	Quinoa sourdough and grains	—	1.6 µg/g (B_9) 5.1 µg/g (B_2)	Quinoa pasta	Carrizo et al. 2020
L. fermentum 8.2 + L. plantarum A6	Ben-saalga	—	7.3 µg/100 g fresh matter (B_9)	Cereal-based fermented porridge	Bationo et al. 2019
L. fermentum PBCC 11.5	Sourdough	1.2 µg/L (B_2)	6.66 µg/g (B_2)	Bread	Russo et al. 2014
L. plantarum UNIFG104 + L. plantarum UNIFG209	Sourdough	642 µg/L (B_2) 586 µg/L (B_2)	6.81 µg/g (B_2) 4.41 µg/g (B_2)	Bread Pasta (dried)	Capozzi et al. 2011

their folate production capability in cereal-based products (Herranen et al. 2010; Korhola et al. 2014).

LAB strains are known to produce folate during dough fermentation, and folate genomic determinants have been detected by using metagenomic approaches to investigate the nutritional potential of the microbiota of cereal-based fermented African products (Turpin, Humblot, and Guyot 2011; Saubade et al. 2017).

However, this apparent ability of LAB strains is quite controversial. Some authors reported that LAB could be detrimental in sourdough fermentation since they could deplete the folate synthesized by yeasts and/or that occurring in the cereal matrix (Kariluoto et al. 2006). Moreover, lactic fermentation might indirectly impact negatively on the folate content of sourdough, lowering the pH level and inhibiting the growth of folate-producer yeasts, and inducing the degradation and/or interconversion of some folate vitamers, since most of them are less stable at low pH (Gujska, Michalak, and Klepacka 2009). Yeasts and LAB mainly produce 5-methyl-tetrahydrofolate, and tetrahydrofolate as the major vitamers, the latter being more sensitive to acidic conditions. Thus, a strategy to enrich sourdough in folate could be the selection of starter cultures mainly producing vitamers stable in acidic environments.

However, folate production by autochthonous and/or starter LAB cultures is a topic deserving further investigation since only a few studies report on their contribution to the bio-enrichment in sourdough. A recent screening of lactobacilli isolated from cereal-based fermented foods showed that folate production was a shared trait in several strains of *Lactiplantibacillus plantarum* (formerly *Lactobacillus plantarum*) and *Limosillactobacillus fermentum* (formerly *Lactobacillus fermentum*) (Greppi et al. 2017). In particular, two of the best producers were able to increase the folate content of *ben-saalga*, an African traditional cereal-based fermented porridge (Bationo et al. 2019), underlining that biofortification of traditional staple foods is a way to make new folate-rich food available in low-income countries, where a high rate of folate deficiency has been reported (Bationo et al. 2020).

There is also a growing interest in sourdoughs obtained by non-wheat cereal-based fermentations as well as in fermented pseudocereals, such as amaranth, quinoa, and chia sourdoughs, as sources of new strains of biotechnological interest (Coda et al. 2014; Petrova and Petrov 2020).

Thus, LAB strains have been isolated from quinoa and amaranth sourdoughs and characterized based on some technological (i.e. ability to grow in the dough and to acidify it, production of EPS) and nutritional (i.e. production of vitamins of the B-group and degradation of antinutritional compounds) properties in order to select suitable starters for new functional foods (Carrizo et al. 2016; Carrizo et al. 2017). In particular, *Leuconostoc mesenteroides* subsp. *mesenteroides* CRL 2131, *Enterococcus durans* CRL 2122, and *L. plantarum* CRL 2106 from the amaranth sourdough (Carrizo et al. 2017) and *L. plantarum* CRL 1973 and CRL 1970, *L. rhamnosus* CRL 1972 and *L. sakei* CRL 1978 from quinoa sourdough produced elevated concentrations of folate with the strain CRL 1973 producing the highest concentration (Carrizo et al. 2016). Furthermore, quinoa sourdough fermented by these *L. plantarum* strains prevented vitamin deficiencies in an *in vivo* murine model, indicating that bio-enrichment of quinoa pasta using LAB could be a novel strategy to increase vitamin and mineral bioavailability in cereal/pseudocereal-derived foods (Carrizo et al. 2020).

Moreover, some LAB strains isolated from different raw cereal grains (namely *Lactobacillus pentosus* ES124, *L. plantarum* ES137, and *Enterococcus mundtii* ES63) were able to increase the folate level in oat bran and rye sourdoughs as well as showing other excellent technological properties (Salvucci, LeBlanc, and Pérez 2016). This suggests that selection of starter cultures within the autochthonous microbiota is strongly recommended since indigenous cultures ensure better performance compared to allochthonous/commercial strains.

It was reported that folate overproduction confers resistance against the folate antagonist methotrexate in *L. plantarum* WCFS1, and screening for methotrexate-resistant isolates has been proposed as an effective strategy to select natural folate-overproducing strains (Wegkamp, De Vos, and Smid 2009). However, robust folate production reduced the growth rate of the overproducing *L. plantarum* WCFS1, thus limiting the interest of selected strains for use in the food industry (Wegkamp et al. 2010).

9.2.2 Riboflavin Production by Lactic Acid Bacteria

Riboflavin is the precursor of the coenzymes flavin mononucleotide and flavin adeninedinucleotide, which are mainly involved as electron carriers in redox reactions. Riboflavin is also required for essential biochemical reactions, including the metabolism of some aminoacids, energy production, and activation of other coenzymes. Riboflavin deficiency is associated with eye-related problems, stomatitis, cheilosis, rash of the mouth and tongue, cardiac risk, preeclampsia, anemia, liver damage, and changes in cerebral glucose metabolism (LeBlanc et al. 2011). According to the European Food Information Council, the riboflavin RDA is 1.6 mg per day (Turck et al. 2017). In general, cereal grains are poor sources of vitamin B_2.

Riboflavin biosynthesis is a strain-dependent feature distributed among LAB species and mediated through the genes of the *rib* operon (Averianova et al. 2020). Due to their importance in the food industry, biotechnological applications of LAB as cell factories for riboflavin production have been recently reviewed (Thakur, Tomar, and De 2016). In general, riboflavin-producing LAB strains are selected based on their ability to grow in a corresponding vitamin-free medium (Russo et al. 2021). This is a fast and low-cost way to identify spontaneous riboflavin producers from ecological niches of biological interest. However, the ability to synthesize riboflavin is a widely shared trait in LAB since 79% of the strains isolated from amaranth sourdough and grains were able to grow in a riboflavin-free medium (Carrizo et al. 2017). In particular, *L. plantarum* from sourdough and *E. durans* and *Leuc. mesenteroides* from grains were the best riboflavin producers. However, in this study, *L. plantarum* CRL 2106 and CRL 2107 and *L. mesenteroides* subsp. *mesenteroides* CRL 2131 were selected for further analysis based on different criteria, including the ability to produce both vitamins B_2 and B_9 (Carrizo et al. 2017). Similarly, in a screening of functional LAB isolated from quinoa sourdough, *L. rhamnosus* was found as the species producing high levels of riboflavin (> 270 ng/mL), with strain

CRL 1963 producing the highest amounts (360 ng/mL) (Carrizo et al. 2016). Strains from this screen were used as starters to produce a vitamin B_2 and B_9 bio-enriched quinoa dough, able to prevent nutritional deficiencies in mice, demonstrating that this novel biotechnological process could be an alternative to exogenous fortification to counteract vitamin deficiencies (Carrizo et al. 2020). Interestingly, these LAB species are also typical inhabitants of cereal sourdoughs, suggesting that they could also be employed as functional starters for these matrices.

However, even though some of the above strains were able to produce somewhat elevated levels of vitamin B_2, there are biotechnological techniques to obtain food-grade vitamin B_2 overproducing strains. In Gram-positive bacteria, including LAB, riboflavin biosynthesis is regulated by the flavin mononucleotide (FMN) riboswitch-mediated transcriptional attenuation (Abbas and Sibirny 2011). Roseoflavin, a toxic analogue of riboflavin, inhibits the transcription of the *rib* operon by binding to the RFN regulator (Ott et al. 2009). It was reported that spontaneous roseoflavin-resistant strains carried out deletions or mutations in the RFN sequence that can impair the proper regulatory activity of the FMN riboswitch resulting in riboflavin overproduction (Ripa et al. 2022). Thus, the exposure of riboflavin-producing strains to the selective pressure of roseoflavin has been reported as a promising approach to select vitamin B_2-overproducing LAB strains belonging to the species *Lactococcus lactis*, *L. plantarum*, *L. fermentum*, and *L. mesenteroides* (Burgess et al. 2004; Burgess et al. 2006; Capozzi et al. 2011; Russo et al. 2014; Yépez et al. 2019). This strategy has been successfully employed to select spontaneous roseoflavin-resistant riboflavin-overproducing derivatives of *L. plantarum* and *L. fermentum* isolated from traditional Italian sourdoughs (Capozzi et al. 2011; Russo et al. 2014) and *Weissella cibaria* strains isolated from Spanish mother doughs made from rye (Hernández-Alcántara et al. 2022; Diez-Ozaeta et al. 2023). The use of the best riboflavin producers as starters for sourdough fermentation resulted in bread with a vitamin B_2 content enhanced by two- or three-fold when fermented by *L. fermentum* or *L. plantarum* in co-inoculum with *Saccharomyces cerevisiae*, respectively (Capozzi et al. 2011; Russo et al. 2014) and ten-fold when fermented by *W. cibaria* in the absence of yeast (Hernández-Alcántara et al 2022). In addition, Russo et al. (2014) reported a positive effect of extending the fermentation time with lactobacilli, proposing that 16 hours is adequate to obtain a bread containing a vitamin B_2 concentration enough to meet approximately 50% of the RDA for an adult human.

In recent years, sourdough fermentation ingredients have been proposed to enhance the nutritional and functional properties of various cereal-based products, and the employment of sourdough for pasta fortification has also been investigated, including fortification in B-group vitamins (Montemurro, Coda, and Rizzello 2019). Thus, the introduction of a novel sourdough fermentation step has been proposed as an innovation that might lead to an advancement in pasta production (Capozzi et al. 2012). Interestingly, sourdough fermentation for 16 hours with two overproducing *L. plantarum* strains produced a vitamin B_2 enriched pasta (Capozzi et al. 2011). This dough was enriched about two-fold compared to the control. This difference was maintained throughout extrusion and drying, though cooking negatively impacted the B_2 content (Capozzi et al. 2011).

9.2.3 COBALAMIN PRODUCTION BY LACTIC ACID BACTERIA

Cobalamin belongs to the cobalamin group playing a pivotal role in cellular metabolism, especially in DNA synthesis, methylation, and mitochondrial metabolism. Vitamin B_{12} deficiency is responsible for hematological and neurological illnesses and for the serious disease, pernicious anemia (Green et al. 2017). It has been estimated that vitamin B_{12} intakes 3.8–20.7 µg/day are needed to prevent deficiency in healthy adults and elderly people (Sobczyńska-Malefora et al. 2021). Cereal grains are not nutritional sources of this vitamin.

Unlike riboflavin and folate, the ability of LAB to encode complete *de novo* biosynthetic pathways of cobalamin has been scarcely reported since most species are auxotrophic for this vitamin. Nevertheless, this biotechnological feature was reported for the first time in *L. reuteri* CRL1098, a strain isolated from sourdough (Taranto et al. 2003). It has been demonstrated that *L. reuteri* requires this vitamin for glycerol co-fermentation in the synthesis of reuterin (Sriramulu et al. 2008). Transcriptional analysis showed the expression of different genes involved in cobalamin biosynthesis during sourdough fermentation, suggesting that this pathway could be involved in the adaptation to the ecological niche of sourdough (Hüfner et al. 2008). Over the years, other LAB species belonging to *L. fermentum* (Basavanna and Prapulla 2013), *Furfurilactobacillus rossiae* (De Angelis et al. 2014), *Loigolactobacillus coryniformis* (Torres et al. 2016), and *L. plantarum* (Bhushan, Tomar, and Mandal 2016) have been identified as vitamin B_{12} producers. Of these, *L. rossiae* DSM 15814 was the only strain of sourdough origin (De Angelis et al. 2014). However, the other cobalamin-producing species are versatile microbes considered as typical inhabitants of sourdough and contribute to the overall quality of sourdough fermentation (Alfonzo et al. 2017).

Some of these strains have been proposed to *in situ* enrich the cobalamin content of fermented soy-milk products (Gu et al. 2015; Bhushan, Tomar, and Chauhan 2017). However, to the best of our knowledge, no reported studies have investigated the *in situ* fortifications of sourdough by LAB. This could be partially explained by the fact that *L. reuteri* strains have been reported to produce only intracellular vitamin B_{12} (Taranto et al. 2003), while *L. coryniformis* and *L. plantarum* synthesize only low amounts of extracellular cobalamin (Masuda et al. 2012). However, the selection of strains able to produce high levels of extracellular cobalamin has been recently reported for *L. plantarum* LZ95 and CY2 strains, and this could be a promising approach for the *in situ* fortification of fermented matrices (Li et al. 2017).

Moreover, recent advances in spectroscopy can discriminate between cobalamin and other corrinoid compounds. This differentiation between vitamin and pseudovitamin is crucial to the interpretation of human metabolism, and it has recently been questioned since the corresponding transporter in the human gastrointestinal tract has a very low affinity to the pseudovitamin, making it virtually unavailable to humans (Varmanen et al. 2016). Therefore, it is noteworthy that different biosynthetic pathways of cobalamin-type corrinoid compounds in *Lactobacillus* strains have been recently reported (Torres et al. 2018), as well the first demonstration of a partial metabolic shift to produce vitamin B_{12}, instead of pseudo-B_{12}, by a *Lactobacillus* strain in a purine-free medium (Torres et al. 2020).

Cobalt exposition generates oxidative stress to bacterial cells, and it has been proposed as an effective strategy to select cobalt-resistant vitamin B_{12}-overproducing strains of *Propionibacterium* spp. and *Lactobacillus* spp. (Seidametova et al. 2004; Bhushan, Tomar, and Mandal 2016). Unlike LAB, *Propionibacterium*, which is known to produce cobalamin to different extents, has been proposed to increase the *in situ* active vitamin B_{12} production in wheat flour and bran (Xie et al. 2018). Interestingly, mixed fermentation of *P. freudenreichii* and *Levilactobacillus brevis* ATCC 14869 (formerly *Lactobacillus brevis*) were successfully employed for the bio-enrichment in cobalamin and the control of indigenous microorganisms in non-sterile cereal and pseudocereal sourdoughs, indicating their mutualistic role as an effective strategy in fortifying cereal-based food with vitamin B_{12} (Xie et al. 2019; Xie et al. 2021).

9.3 EXOPOLYSACCHARIDES PRODUCED BY LACTIC ACID BACTERIA

LAB can produce a range of beneficial metabolites apart from vitamins, which can promote human health, e.g. the production of short-chain fatty acids (SCFA), polyols such as mannitol, antimicrobial molecules, organic acids (lactic acid, acetic acid, etc.), different amino acids, and EPS (Oerlemans et al. 2020; Hwang and Lee 2019; Stoyanova, Ustyugova, and Netrusov 2012).

Bacterial polysaccharides can form capsular polysaccharides (CPS) or can be released into the medium as EPS (Figure 9.1). Production of EPS by LAB has been extensively studied because of their beneficial properties as immunomodulators (Zarour et al. 2017), antivirals (Montserrat Nácher-Vázquez et al. 2015), positive influencers in adhesion, and colonization of the producing strains (Besrour-Aouam et al. 2019); or due to their negative effects when appearing in alcoholic beverages (Puertas et al. 2018) or in meat products (Iulietto et al. 2015), causing wide economic losses.

Some EPS confer a very characteristic phenotype to the producing colony when growing on surface media, known as the "ropy" phenotype (Ruas-Madiedo and De Los Reyes-Gavilán 2005). These mucous colonies form a filament when touched with a loop. EPS can also be observed surrounding the producing bacteria using electromicroscopic techniques (Zarour et al. 2017).

According to their chemical composition and the pathway used for their synthesis, EPS can be divided into homopolysaccharides (HoPS) and heteropolysaccharides (HePS). The former are composed of a monosaccharide, typically glucose or fructose, giving rise to α-glucans, β-glucans, or β-fructans, depending on the position of the carbon involved in the linkage (Torino, Font de Valdez, and Mozzi 2015; Zeidan et al. 2017; Castro-Bravo et al. 2018). In addition, polygalactans (constituted by galactose) have also been reported (Mozzi et al. 2006). The latter, HePS, comprise repeating units of different monosaccharides, the most abundant being composed of glucose, galactose, and rhamnose. However, sometimes acetylated or phosphorylated

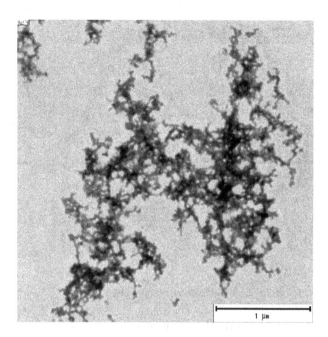

FIGURE 9.1 EPS from *Leuconostoc carnosum* CUPV411 released to the medium and observed by transmission electron microscopy.

monosaccharides, glycerol, fucose, or glucuronic acids can also be present in HePS (Castro-Bravo et al. 2018; Torino, Font de Valdez, and Mozzi 2015; Sanalibaba and Çakmak 2016). HoPS and HePS are biosynthesized by different mechanisms. HePS biosynthesis is more complex than that of HoPS, due to the various genes participating in the process, which are organized in *eps*-clusters with a conserved functional operon-like structure with the genes oriented in one direction. Genes coding for glycosyltransferases (GTF), proteins related to polymerization and export, and genes with unknown functions as well as mobile elements are some of the genes found in the *eps*-clusters (Figure 9.2A). Currently, four pathways are described, two for HePS biosynthesis: the Wzx/Wzy-dependent pathway and the ABC transporter-dependent pathway; and the other two for HoPS biosynthesis: the synthase-dependent pathway and the extracellular biosynthesis by the sucrase protein (Nadzir et al. 2021; Castro-Bravo et al. 2018).

- Wzx/Wzy-dependent pathway: this mechanism is very common for the biosynthesis of HePS containing highly diverse sugar units. The process is represented in Figure 9.2B and starts with the priming-GTF enzyme, which links an activated nucleotide sugar to a lipid carrier (undecaprenyl phosphate C55-P) located in the cytoplasmic membrane. Then, different GTFs add sugar moieties sequentially by catalyzing the glycosidic linkage until

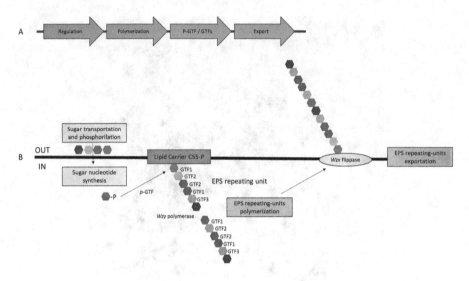

FIGURE 9.2 Operons involved in HePS biosynthesis by LAB.

the repeating unit is formed. Each repeating unit is secreted and polymerized extracellularly by the "flippase"–polymerase complex. The Wzx "flips" the repeating units across the membrane, and the Wzy polymerase adds those units to the forming chain (Castro-Bravo et al. 2018; Mohd Nadzir et al. 2021; Lebeer et al. 2009).

- ABC transporter-dependent pathway: ABC transporter homologue genes have been reported for the *Bifidobacterium* genus. It is believed that these genes participate in the export of HePS in those taxa where the priming-GTF has not been detected. Once the repeating unit is attached to the lipid carrier, an efflux pump complex in the inner membrane translocates it to the exocellular membrane (Mohd Nadzir et al. 2021; Castro-Bravo et al. 2018; Ferrario et al. 2016).
- Synthase-dependent pathway: some bacterial EPS such as cellulose are synthesized by this pathway. Here, an activated nucleotide sugar is assembled by a membrane synthase, and then a membrane transporter exports the EPS (Nadzir et al. 2021).
- Extracellular biosynthesis by sucrase proteins: dextrans, alternans, levans, reuterans, etc. are synthesized by this shunt. The process starts with the extracellular hydrolysis of sucrose into glucose and fructose by a sucrase, and this enzyme is also responsible for the polymerization of the forming chain with the monosaccharidic moieties of glucose or fructose with different branches (Figure 9.3). These enzymes take their name from the EPS produced: dextransucrase, alternansucrase, mutansucrase, etc. (Sanalibaba and Çakmak 2016; Mohd Nadzir et al. 2021; Guérin et al. 2020).

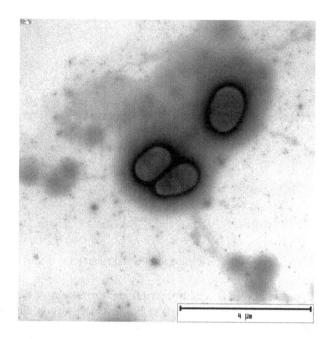

FIGURE 9.3 *Weissella cibaria* AV2ou colonies visualized by transmission electron microscopy surrounded by membrane-attached dextran.

9.4 DEXTRANS PRODUCED BY LACTIC ACID BACTERIA

Among the HoPS, dextrans have worldwide economic importance. They are widely used in industries where their applications range from cryoprotectants, moisturizers, or thickeners to plasma substitutes or as coating for chromatographic columns (Naessens et al. 2005; Xu et al. 2019). They are used in the bakery industry and also in ice-creams, milk shakes, and other beverages to enhance palatability (Kothari et al. 2014). Dextrans are α-glucans synthesized by the dextransucrase enzyme using sucrose as a substrate. Depending on the linkage specificity of the dextransucrase, dextrans will have different types of glucosidic linkages and different extents of branching. They are composed of glucose units linked mainly by α-(1→6) glycosidic bonds with branches at C3. In addition, partially branched dextrans in O-4 (Llamas-Arriba et al. 2019) and O-2 (Bozonnet et al. 2002) positions have also been reported. Many LAB belonging to different genera produce dextrans (Figure 9.4): *Leuconostoc* (Zarour et al. 2017; Llamas-Arriba et al. 2019), *Weissella* (Besrour-Aouam et al. 2021), *Lactobacillus* (Nácher-Vázquez et al. 2017), and *Streptococcus* or *Oenococcus* (Dimopoulou et al. 2018; Vuillemin et al. 2018). Moreover, some LAB genomes produce more than one gene encoding for a glucan synthase, thus being able to produce more than one HoPS (Lynch, Coffey, and Arendt 2018).

Dextrans have been reported to be synthesized under stress conditions for the producing bacteria, presumably protecting them from hazardous environments (Yan

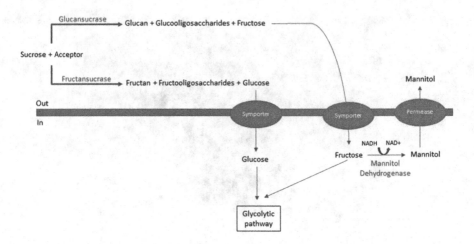

FIGURE 9.4 Biosynthetic pathways leading to EPS and mannitol synthesis in LAB.

et al. 2016; Schmid et al. 2021). The transcriptional expression of dextransucrases in different conditions has also been studied. For instance, several authors confirm the activation of the enzyme in the presence of sucrose though not with other carbon sources. One example is the DsrD from *L. mesenteroides* Lcc4, whose biosynthesis increased 10-fold in the presence of 20 g/L of sucrose (Neubauer, Bauche, and Mollet 2003). However, the DsrLS from *L. sakei* MN1 not only expressed in the presence of sucrose but also when glucose was the carbon source (Nácher-Vázquez et al. 2017). A recent study of the expression of the Dsr of *Weissella* strains in the presence of glucose and sucrose has shown that, for *Weissella confusa* V30, the enzyme was synthesized in the presence of both carbon sources. In contrast, for *W. cibaria* AV2ou and *W. confusa* FS54, Dsr was only expressed in the presence of sucrose (Besrour-Aouam et al. 2021). Moreover, the dextransucrase DsrLL from *L. lactis* AV1n was also expressed at low temperatures, producing 10-fold higher levels of dextran at 20°C than at 37°C. In addition, Hu and Gänzle (2018) found that the maximal expression and activity of the dextransucrase from *W. cibaria* 10 M was at 15°C.

Dextran production by LAB, is important to the food industry. They have the potential to improve dough rheology and bread texture (Wang et al. 2021). The analysis of the microbiota of 35 sourdoughs revealed that most of them contained at least one strain able to produce HoPS using sucrose as a substrate (Ripari 2019; Ripari, Bai, and Gänzle 2019). Thus, EPS production seems to be an important feature in the sourdough microbial composition.

The European Commission has approved the use of dextrans in baked goods, up to the level of 5% (Kothari et al. 2014). Incorporating dextran in bakery can replace or reduce hydrocolloids currently used as bread improvers (Tieking and Gänzle 2005), and adding 5g/kg dextran to the wheat flour improved viscoelastic properties of wheat doughs and the volume of the corresponding breads. These results were more important than those observed after adding other EPS (Tieking et al. 2003). Finally, research results showed that dextran produced *in situ* was more functional than when dextran was added externally (Perri et al. 2021; Tieking et al. 2003; Wang et al. 2018).

In the bakery industry, *in situ* production of dextran in sourdough has become a strategy to improve shelf life, volume, and the nutritional value of bread, without using additives (Xu et al. 2019). Nonetheless, the metabolic activity of dextran-producing bacteria can also hinder the positive impact of EPS, due to the acidification of the sourdough. Therefore, research in this field is necessary to reach an equilibrium for the manufacture of tasteful bread with good technological properties.

Apart from improving textural properties of bread, sensorial amelioration has also been reported for dextrans. For instance, *in situ* dextran production by *W. confusa* A16 in sorghum sourdough fermentation masked sour flavor, bitter taste, and aftertaste, lowering the flavor intensity perception (Wang et al. 2020). In addition, it produced bread that was more elastic, foldable, moist, cohesive, soft, flexible, and smooth compared to control sorghum sourdough or native sorghum breads (Wang et al. 2020). In another study using the same strain, the results showed that *in situ* dextran production increased waste bread slurry viscosity and yielded residual fructose or glucose that could effectively replace the sugar added for yeast leavening, while no improvement was detected in bread containing β-glucan produced *in situ* by *Pediococcus claussenii* E-032355T (Immonen et al. 2020).

Wheat bread supplemented with 30% (w/w) sourdough-containing dextran-producing *W. confusa* SLA4 showed increased specific volume and decreased crumb hardness and staling rate than the control wheat bread. This incorporation increased total and soluble fiber content, and the aroma profile (Perri et al. 2021).

Analysis of bread generated with high molecular weight (Mw) dextran supplementation and fermented with *L. mesenteroides* ATCC 8239^T, prior to baking, showed that both dextran and weak acidification improved bread quality, especially crumb softness (Zhang et al. 2019). Dextran decreased moisture migration due to high water-binding capacity, and the weak acidification contributed to the amylopectin recrystallization retardation. A covalent synergy has been detected between dextran and weak acidification. Dough rheology demonstrated that weak acidification was fundamental for dextran-induced positive dough viscoelasticity. This synergy could be a promising strategy to prolong the shelf life of wheat bread. For instance, using 43% of *Weissella confusa* VTT E-90392 dextran-containing wheat sourdough results in mild acidification of bread (Katina et al. 2009), reaching a pH 0.5 lower than the EPS-negative control bread. Moreover, VTT E-90392 dextran meant a 5-fold increase in the viscosity of the sourdough, which when used for bread baking improved volume (up to 10%) and crumb softness (25–40%) during six days of storage. In another study, the dextran-producing *W. confusa* VTT E-143403 was used to produce *in situ* dextran-rich sourdoughs. Viscoelastic properties of the dough were improved, and in final breads, the specific volume was increased (21%), while the crumb hardness reduced (12%) (Wang et al. 2018). The use of dextran-producing *W. confusa* TT E–143403 (E3403) in faba bean sourdough improved the viscoelastic properties of the dough. Also, it increased the specific volume of the bread produced. At the same time, no similar positive effect was detected after the incorporation of dextran-producing *L. pseudomesenteroides*

DSM 20193, which may be explained by the high acidity produced during the fermentation process (Wang et al. 2018). These results have been confirmed by several research works. In situ dextran production by W. confusa altered the nutritional value of millet, leading to an increase of antioxidant activity and free phenolic content. This fermentation also improved protein digestibility and lowered the bread's predicted glycemic index (Wang et al. 2019).

However, not all dextrans exert the same good technological impact on bread quality. Comparative study of the effect of dextrans of different Mw (T10, T70, T250, T750, T2000) on the wheat bread quality showed that there is a covalent relationship between the M_w of dextran and the improvement of bread qualities (Zhang et al. 2018). Dextran with high M_w values has the potential for improving the quality of wheat sourdough breads, especially extending shelf life. Dextran T2000 (i) presented the most inhibitory effects on the aging process, (ii) suppressed gelatinization of starch and bound a large amount of water, (iii) improved the elastic properties of sourdough-containing dough (Zhang et al. 2018), (iv) decreased the formation of amylose crystallites, and (v) inhibited amylopectin recrystallization (Zhang et al. 2018).

It has been reported that reuteran, from *L. reuteri* TMW1.656, obtained by directed mutagenesis, with a lower molecular mass than the wild type, exerted better textural properties in bread (Chen, Levy, and Gänzle 2016); therefore, the molecular weight of α-glucans becomes an important parameter to consider for the improvement of bread quality. Moreover, the same authors also found that the proportion and type of branching of reuterans can affect the quality of bread (Chen, Levy, and Gänzle 2016).

The study of lactobacilli, *Leuconostoc* and *Weissella* strains producing dextran, levan, and/or β-glucan in the carrot raw material had been carried out in order to replace hydrocolloid additives in the preparation of carrot puree as a model for a vegetable-based product (Juvonen et al. 2015). *L. lactis* E-032298 and *W. confusa* E-90392 showed the best result compared with the β-glucan-producing strains due to the production of low branching dextran, which gave a thick texture (not too elastic), accompanied by a pleasant smell and flavor. Also, a synergy has been detected between sourdoughs during bread making, since during *in situ* dextran production by LAB, the increase of glucose and fructose released from sucrose enhanced CO_2 production by yeasts in bread (Galle et al. 2012).

In the bread industry, the natural bioingredients produced in a co-culture of *W. confusa* 11GU-1 (dextran, lactate, and acetate) and *Propionibacterium freudenreichii Pf* JS15 (HePS, propionate, and others) have the potential to be applied as antifungal, texture-building, and anti-staling agents, consistent with consumer demands for "clean label" products (Tinzl-Malang et al. 2020; Tinzl-Malang et al. 2015).

Celiac disease is a pathology of immunological origin. In some people, ingestion of a protein found in flour (wheat, barley, and rye) – gluten – triggers an exaggerated reaction of the immune system, resulting in inflammation which destroys the villi of the intestinal mucosa. As a result, the absorption of nutrients is more or less reduced, depending on the severity and extent of the disease. The gluten-free diet is the only effective treatment available for celiac disease (Itzlinger et al. 2018). According to the World Gastroenterology Organization Global Guidelines, gluten can be defined as the rubbery protein mass that remains when the wheat dough is washed to remove starch. Gluten is a group of wheat grain storage proteins found in wheat, rye, and barley. It is consumed in the Western diet (10–20 g per person/day) (Cohen, Day, and Shaoul 2019). A significant technological challenge is presented by replacing gluten in bread and other foods (Kothari et al. 2014). Most gluten-free breads on the market are of poor sensory and textural quality. Dextran is used as a bioingredient produced during sourdough fermentation to improve gluten-free breads' technological and rheological properties and potentially replace hydrocolloids (Galle et al. 2012).

Up to now, all research done to improve gluten-free bread has been to imitate the gluten network. Less attention is granted to technological approaches that change the consistency of the dough. Alternative methods, such as using crosslinking enzymes and/or incorporating an alternative polymeric network, such as EPS produced by LAB, are proposed. Other new approaches, like high hydrostatic pressure, sourdough technology, and unconventional heating methods, are applied. Particular attention is paid to arabinoxylan, which forms a stable carbohydrate network in gluten-free pasta that can replace gluten (Bender and Schönlechner 2020). *In situ* production of dextran by *W. cibaria* MG1 in sorghum, a gluten-free cereal, showed an improvement in shelf life, better than that observed with reuteran produced by *L. reuteri* VIP or the fructan produced by *L. reuteri* Y2, although all three strains contributed to the nutritional enrichment of the gluten-free bread formed (Galle et al. 2012; Rühmkorf et al. 2012). Moreover, another study demonstrated that the *in situ* production of α-glucan by *L. reuteri* E81 improved the elasticity of the sourdough, but it did not confer any effect on the textural properties of bread (İspirli et al. 2020).

The use of *L. plantarum* (ATCC 14917) and *L. fermentum* (ATCC 9338) as starter cultures and olive leaf extract can improve antifungal properties and the quality of the bread regarding the taste, staleness, odor, chewability, softness, moisture content, pH, acidity, and texture. Therefore, incorporating these ingredients as alternatives to gluten may be employed to improve the final product and acceptability of the gluten-free bread (Moghaddam et al. 2020).

The incorporation of dextran-producing LAB into the fermentation of dough and the production of breads can make them functional foods as they tend to be consumed daily. In addition to the effect of dextran, discussed above, organic acids affect the protein and starch fractions of flour by increasing the protease and amylase activities, thus leading to a reduction in staling (Arendt, Ryan, and Dal Bello 2007). Dextran, produced *in situ*, should have a high Mw and few branched linkages for the application in sourdough to improve the bread quality (structural, thermal, and rheological properties). However, *in situ* production of this polymer during sourdough fermentation is challenged by simultaneous acidification due to metabolic activities of the bacteria, which may significantly diminish the positive technological impact of dextran (Katina et al. 2009). For this reason, and others, dextran-producing *Weissella* species, such as *W. confusa* or *W. cibaria*, are proposed as good candidates to make bread (Zhang et al. 2020).

9.5 PREBIOTICS IN BAKERY PRODUCTS

Prebiotics are involved in preserving human life quality against cancer, mental disorders, vascular diseases, and obesity (Davani-Davari et al. 2019). They have been given generally recognized as safe (GRAS) status (Chourasia et al. 2020), and the prebiotics concept was updated in 2016 by the International Scientific Association for Probiotics and Prebiotics (ISAPP), referring to "a substrate that is selectively utilised by host microorganisms conferring a health benefit" (Gibson et al. 2017). This definition expands the list of prebiotics by including new non-carbohydrate substances, applications other than the gastrointestinal tract, and by covering other categories in addition to food (Mohanty et al. 2018). Prebiotics were recently classified into three main groups: (i) polyols (lactulose, xylitol, mannitol, lactiol); (ii) oligosaccharides (fructooligosaccharides (FOS), galactooligosaccharides (GOS), xylooligosacharides (XOS), isomaltooligosaccharides (IMO), raffinose oligosaccharides (RFOs), isomaltulose, arabinoxylanoligosaccharides (AXOS), and inulin, among others); and (iii) fibers (cellulose, dextrins, pectins, and β-glucans) (de Paulo Farias et al. 2019; Mohanty et al. 2018). The use of microbial fermentation processes for the biotechnological synthesis of prebiotics is considered a good alternative (i) to increase the production scale; (ii) to improve the nutritional quality of the final food product (sensory properties, texture, physicochemical characteristics, fibers, and partial replacement of sugars and fats); and (iii) to reduce production costs, making the process economically feasible (de Paulo Farias et al. 2019).

Prebiotics can directly improve bakery goods and, indirectly, consumer health (Glibowski and Skrzypczak 2017). They can be formulated either as a powder or syrup and marketed as supplements or incorporated into food products such as yogurts and breads (Mohanty et al. 2018). Prebiotics act mainly as fat and sugar replacers in bakery products, improving their sensory acceptance (Morais et al. 2014). To be served as functional food bioingredients, prebiotics must (i) not deteriorate the organoleptic properties of the end product and (ii) be chemically stable in food processing treatments, such as heat, low pH and Maillard reaction conditions. If there is an alteration or degradation of the prebiotic compound, it will not be available for bacterial metabolism, nor will it provide the stimulation of LAB probiotics; therefore, it cannot exert its mechanism to improve consumer health (Neri-Numa and Pastore 2020).

LAB and yeasts present in the dough can produce key enzymes responsible for synthesizing several prebiotic groups. The formation of these compounds and their level in the sourdough depend on (i) the material's quality, (ii) the activity of the flour's natural microbiota and/or (iii) the selected microbial strains used to start the fermentation (Păcularu-Burada, Georgescu, and Bahrim 2020). Corn flour and wheat bran can be used as substrate sources to synthesize oligosaccharides prebiotic by LAB involved in bread fermentation (Rolim, Hu, and Gänzle 2019). Oligosaccharide prebiotic fermentation allows the production of short-chain fatty acids (Davani-Davari et al. 2019).

The incorporation of fibers in cereal bars is encouraged by nutritional guidelines recommending an increase in fiber intake since consumption of foods low in fiber constitutes a risk factor for human health (Voss, Campos, and Pintado 2021). Adding natural fibers like wheat bran and oat hulls for bread making has positive effects on the dough rheological proprieties. Also, when used as fiber sources, wheat, rice, oat, and barley have an influence on the farinograph characteristics, extensibility, breaking strength, and spread ratio of the biscuits (Padma Ishwarya and Prabhasankar 2014).

According to Vera and colleagues (Vera, Illanes, and Guerrero 2021), FOS and GOS are used in bakery products due to their functional properties. Given their characteristics and their low natural concentration in foods, they are manufactured on a large industrial scale. The addition of oligosaccharides FOS, GOS, and XOS can (i) inhibit the formation of starch crystal structures to a certain extent, (ii) reduce the damage from ice crystals to starch, and (iii) delay starch retrogradation.

FOS can be found naturally in wheat and barley (Al-Sheraji et al. 2013). In addition, they can be produced from sucrose by transferring one to three molecules of fructose (Prapulla, Subhaprada, and Karanth 2000). Fructosyl-transferase is a key enzyme-producing FOS, whose structure is a linear chain of fructose with β-$(2\rightarrow 1)$ linkage. Different microorganisms present the FOS enzyme, such as *Candida* spp. and *Saccharomyces cerevisiae* (Yun 1996). The latter can use glucose, fructose, and sucrose to produce carbon dioxide and ethanol. This fermentation profile probably allows it to compete with dextran-producing LAB for sucrose in bread making.

GOS is generated from lactose, by the action of β-galactosidases. The end product of this reaction is mainly a mixture of tri- to penta-saccharides containing galactose molecules with β-$(1\rightarrow 6)$, β-$(1\rightarrow 3)$, and β-$(1\rightarrow 4)$ linkages (Davani-Davari et al. 2019). Several microorganisms can produce β-galactosidases such as bifidobacteria and lactobacilli (Davani-Davari et al. 2019). Several lactobacilli species, such as *Lactobacillus sakei* and *Fructilactobacillus sanfranciscensis*, were proposed in different combinations to ferment wheat sourdough (Xu et al. 2019). Some yeasts, such as *S. cerevisiae*, have been used for producing recombinant forms of β-galactosidase, which can be used for GOS production (Demain and Vaishnav 2009).

The incorporation of dextran-producing LAB (*Leuconostoc*, *Weissella*, and *Lactobacillus*), allows conversion of sucrose to prebiotic oligosaccharides such as IMO, which occurs in many fermented foods, including bread (Marco et al. 2017). When an acceptor is present in the culture medium, maltose, isomaltose, and glucose, in addition to sucrose and dextransucrase, uses them to generate IMO at the expense of dextran synthesis (Hu et al. 2017). Part of the released D-glucopyranosyl residues is consumed to form these acceptor products, which decrease the dextran yield (Dols et al. 1997; Werning et al. 2012). Dextransucrase catalysis having oligosaccharides as a virtually exclusive product was shifted by the addition of equimolar sucrose and efficient acceptor molecules (Hu et al. 2017). An alternative way of IMO production involves two reactions: (i) the hydrolysis of starch-generating dextrins by an amylase, which can be also produced by some dextran-producing species (Khusniati et al. 2021), and (ii) conversion of dextrins to α-$(1\rightarrow 6)$-linked oligosaccharides by an α-D-glucosidase (Rolim, Hu, and Gänzle 2019). IMO are one of the most common oligosaccharides used in several countries as a functional food component due to their excellent nutritional properties and prebiotic activity

(Yan, Hu, and Gänzle 2018). In the global oligosaccharide market, IMO demand has developed rapidly and is used now as a low-calorie sweetener in food products, such as bakery goods (Huang et al. 2020; Rolim, Hu, and Gänzle 2019; Vera, Illanes, and Guerrero 2021). The beneficial coexistence of yeasts and LAB during bread making confers better organoleptic and nutritional properties to the final product (Gobbetti et al. 2019). However, this coexistence does not prevent competition for nutrients when both kinds of microorganisms can metabolize the same substrate. Maltose is the most abundant fermentable sugar available to yeasts in unsweetened bread doughs. Many of the traditional yeast fermentation industries require yeasts that can ferment maltose efficiently after hexose depletion to maintain a steady rate of fermentation (Hazell and Attfield 1999). This point can generate negative competition with ISO-producing LAB.

Mannitol, a hexitol polyol, is used as a low-calorie ingredient in the food industry. Some yeasts can produce it (Candida mannitofaciens, Candida zeylanoides, Candida magnoliae) and heterofermentative LAB belonging to lactobacilli and the *Leuconostoc/Oenococcus* genera (Zarour et al. 2017). It is produced directly from fructose in a reaction catalyzed by the mannitol 2-dehydrogenase (Gaspar et al. 2013). It has been shown that co-extrusion of probiotic *Lactobacillus acidophilus* NCFM with mannitol as prebiotic can enhance its growth and survival in an acidic environment (Yee et al. 2019). Mannitol, naturally produced by sourdough dextran-producing LAB, is a functional ingredient for multiple uses, such as (i) a prebiotic product and (ii) a substitute sugar to mimic the rheological properties exerted by sucrose in wheat dough systems. This combination revealed an improvement in the quality, flavor, and aroma of the bread (Arendt, Ryan, and Dal Bello 2007; Axel et al. 2015; Canesin and Cazarin 2021; Sahin et al. 2019), which are among the principal characteristics perceived by the consumer (Longin et al. 2020).

Today, more organoleptic, technological, microbiological, and nutritional properties are being studied in-depth, with the aim of understanding symbiotic relations in sourdough (prebiotic/probiotic and yeast/LAB), to industrialize and to select the LAB and yeasts involved in bread fermentation in order to offer the consumer better bakery products.

Although sourdoughs have been studied for many years, little is known about the relationship and interactions between yeasts and LAB in this particular environment. Carbonetto et al. suggested negative and positive interactions, depending on the effect one species has on the other, therefore, these interactions are categorized as: (i) competition (negative/negative interaction), (ii) amensalism (negative/neutral interaction), (iii) predation/parasitism (negative/positive interaction), (iv) commensalism (positive/neutral interaction), and (v) mutualism (positive/positive interaction) (Carbonetto et al. 2020). As reviewed in the previous section, positive interactions between yeasts and LAB appear when their metabolic requirements (e.g. maltose) are different. This was the case of *F. sanfranciscensis* and *Kazachstania humilis* (Carbonetto et al. 2020). Nevertheless, other relationships occur, such as the prevalence of low-pH adapted yeasts for survival in the presence of lactic acid produced by LAB, or the growth-promoting amino acid production by both yeasts and LAB (Carbonetto et al. 2020). Negative effects have also been described, and the reason why those LAB/yeasts are present in that hostile environment is that they are maintained with the sequential refreshments of the sourdoughs. In addition, sourdough fermentations are not very long in time, thus, the competition for nutrients does not exist, and this minoritarian species can survive (Fujimoto et al. 2019).

Considering positive relationships between yeasts and LAB in a sourdough environment, the presence of acetic acid bacteria in sourdoughs, too, has been described as beneficial because of their potential for aroma and dextran production (Comasio et al. 2020). Moreover, some species have been reported to produce mannitol from the fructose released by other species when hydrolyzing the sucrose present in flour (Paramithiotis et al. 2007). This would imply a strategy to manufacture low-sugar sourdoughs and breads, due to the sweetener capacity of mannitol. Moreover, new products with multiple healthy functionalities and that are attractive, tasteful, and biosafe for different population groups in developed and developing countries would be the next challenge. Considering that bread is a very important food in the Mediterranean diet, it could be considered perfect for nutrient delivery. Bread elaborated with these sourdoughs could be one solution for gluten-intolerances, vegans, and obesity.

CONCLUSION

LAB are a heterogeneous group of prokaryotes with relevant applications in modulating the safety and quality of fermented foods and beverages. The chapter proposes an overview of sourdough LAB as bio-producers for *in situ* improvements, directly in the edible matrices, of B-group vitamins, exopolysaccharides, dextrans, and prebiotics, biomolecules crucial in shaping nutritional, functional, and sensory properties. The overview underlines the importance of transdisciplinary studies that integrate bacteria genomics and physiology, biotechnological approaches, and food sciences to propose innovative paths in food biosciences. The proposed microbial-based solutions are consistent with sustainable economic, social, and environmental development objectives.

ACKNOWLEDGMENTS

We thank Dr. Stephen Elson for his critical reading of the manuscript. Pasquale Russo is the beneficiary of a grant by MIUR in the framework of "AIM: Attraction and International Mobility" (PON R&I2014-2020) (practice code D74I18000190001). This work was supported by the Spanish Ministry of Science, Innovation and Universities (grant RTI2018-097114-B-I00) and CSIC (grant 2022AEP028). In addition, this work is also based upon the work from European COST Action 18101, SOURDOMICS–Sourdough biotechnology network toward novel, healthier, and sustainable food and bioprocesses, supported by COST, where G.S., P.L., and P.R. are members. COST is a funding agency for research and innovation networks.

REFERENCES

Abbas, C. A., and A. A. Sibirny. 2011. Genetic Control of Biosynthesis and Transport of Riboflavin and Flavin Nucleotides and Construction of Robust Biotechnological Producers. *Microbiology and Molecular Biology Reviews* 75:321–60.

Agrahar-Murugkar, D. 2020. Food to Food Fortification of Breads and Biscuits with Herbs, Spices, Millets and Oilseeds on Bio-Accessibility of Calcium, Iron and Zinc and Impact of Proteins, Fat and Phenolics. *LWT* 130;109703.

Alfonzo, A., C. Miceli, A. Nasca, et al. 2017. Monitoring of Wheat Lactic Acid Bacteria from the Field until the First Step of Dough Fermentation. *Food Microbiology* 62:256–69.

Al-Sheraji, S. H., A. Ismail, M. Y. Manap, S. Mustafa, R. M. Yusof, and F. A. Hassan. 2013. Prebiotics as Functional Foods: A Review. *Journal of Functional Foods* 5:1542–53.

Aprodu, I., C. Bolea, and I. Banu. 2019. Effect of Lactic Fermentation on Nutritional Potential of Multigrain Flours Based on Wheat, Rye and Oat. *The Annals of the University Dunarea de Jos of Galati. Fascicle VI-Food Technology* 43:69–80.

Arendt, E. K., L. A. Ryan, and F. Dal Bello. 2007. Impact of Sourdough on the Texture of Bread. *Food Microbiology* 24:165–74.

Averianova, L. A., L. A. Balabanova, O. M. Son, A. B. Podvolotskaya, and L. A. Tekutyeva. 2020. Production of Vitamin B_2 (Riboflavin) by Microorganisms: An Overview. *Frontiers in Bioengineering and Biotechnology* 8:570828.

Axel, C., B. Röcker, B. Brosnan, et al. 2015. Application of *Lactobacillus amylovorus* DSM19280 in Gluten-Free Sourdough Bread to Improve the Microbial Shelf Life. *Food Microbiology* 47:36–44.

Basavanna, G., and S. G. Prapulla. 2013. Evaluation of Functional Aspects of *Lactobacillus fermentum* CFR 2195 Isolated from Breast Fed Healthy Infants' Fecal Matter. *Journal of Food Science and Technology* 50:360–6.

Batifoulier, F., M. -A. Verny, E. Chanliaud, C. Rémésy, and C. Demigné. 2005. Effect of Different Bread-making Methods on Thiamine, Riboflavin and Pyridoxine Contents of Wheat Bread. *Journal of Cereal Science* 42:101–8.

Batifoulier, F., M. -A. Verny, E. Chanliaud, C. Remesy, and C. Demigné. 2006. Variability of B Vitamin Concentrations in Wheat Grain, Milling Fractions and Bread Products. *European Journal of Agronomy* 25:163–19.

Bationo, F., L. T. Songré-Ouattara, F. Hama-Ba, et al. 2020. Folate Status of Women and Children in Africa: Current Situation and Improvement Strategies. *Food Reviews International* 36:1–14.

Bationo, F., L. T. Songré-Ouattara, Y. M. Hemery, et al. 2019. Improved Processing for the Production of Cereal-Based Fermented Porridge Enriched in Folate Using Selected Lactic Acid Bacteria and a Back Slopping Process. *LWT* 106 (June 1): 172–8.

Bekaert, S., S. Storozhenko, P. Mehrshahi, et al. 2008. Folate Biofortification in Food Plants. *Trends in Plant Science* 13:28–35.

Bender, D., and R. Schönlechner. 2020. Innovative Approaches towards Improved Gluten-Free Bread Properties. *Journal of Cereal Science* 91:102904.

Besrour-Aouam, N., I. Fhoula, A. M. Hernández-Alcántara, et al. 2021. The Role of Dextran Production in the Metabolic Context of *Leuconostoc* and *Weissella* Tunisian Strains. *Carbohydrate Polymers* 253:117254.

Besrour-Aouam, N., M. L. Mohedano, I. Fhoula, et al. 2019. Different Modes of Regulation of the Expression of Dextransucrase in *Leuconostoc lactis* AV1n and *Lactobacillus sakei* MN1. *Frontiers in Microbiology* 10:959.

Bhushan, B., S. K. Tomar, and A. Chauhan. 2017. Techno-Functional Differentiation of Two Vitamin B 12 Producing *Lactobacillus plantarum* Strains: An Elucidation for Diverse Future Use. *Applied Microbiology and Biotechnology* 101:697–709.

Bhushan, B., S. K. Tomar, and S. Mandal. 2016. Phenotypic and Genotypic Screening of Human-Originated Lactobacilli for Vitamin B 12 Production Potential: Process Validation by Micro-Assay and UFLC. *Applied Microbiology and Biotechnology* 100:6791–803.

Boukid, F., E. Zannini, E. Carini, and E. Vittadini. 2019. Pulses for Bread Fortification: A Necessity or a Choice? *Trends in Food Science & Technology* 88:416–28.

Bozonnet, S., M. Dols-Laffargue, E. Fabre, et al. 2002. Molecular Characterisation of DSR-E, an α-1,2 Linkage-Synthesizing Dextransucrase with Two Catalytic Domains. *Journal of Bacteriology* 184:5753–61.

Brandt, M. J. 2019. Industrial Production of Sourdoughs for the Baking Branch: An Overview. *International Journal of Food Microbiology* 302:3–7.

Burgess, C., M. O'connell-Motherway, W. Sybesma, J. Hugenholtz, and D. van Sinderen. 2004. Riboflavin Production in *Lactococcus lactis*: Potential for in Situ Production of Vitamin-Enriched Foods. *Applied and Environmental Microbiology* 70:5769–77.

Burgess, C. M., E. J. Smid, G. Rutten, and D. Van Sinderen. 2006. A General Method for Selection of Riboflavin-Overproducing Food Grade Micro-Organisms. *Microbial Cell Factories* 5:1–12.

Burgess, C. M., E. J. Smid, and D. van Sinderen. 2009. Bacterial Vitamin B_2, B_{11} and B_{12} Overproduction: An Overview. *International Journal of Food Microbiology* 133:1–7.

Canesin, M. R., and C. B. B. Cazarin. 2021. Nutritional Quality and Nutrient Bioaccessibility in Sourdough Bread. *Current Opinion in Food Science* 40:81–6.

Capozzi, V., V. Menga, A. M. Digesu, et al. 2011. Biotechnological Production of Vitamin B_2-Enriched Bread and Pasta. *Journal of Agricultural and Food Chemistry* 59:8013–20.

Capozzi, V., P. Russo, M. T. Dueñas, P. López, and G. Spano. 2012a. Lactic Acid Bacteria Producing B-Group Vitamins: A Great Potential for Functional Cereals Products. *Applied Microbiology and Biotechnology* 96:1383–94.

Capozzi, V., P. Russo, M. Fragasso, P. de Vita, D. Fiocco, and G. Spano. 2012b. Biotechnology and Pasta-Making: Lactic Acid Bacteria as a New Driver of Innovation. *Frontiers in Microbiology* 3:94.

Carbonetto, B., T. Nidelet, S. Guezenec, M. Perez, D. Segond, and D. Sicard. 2020. Interactions between Kazachstania Humilis Yeast Species and Lactic Acid Bacteria in Sourdough. *Microorganisms* 8:240.

Carrizo, S. L., C. E. Montes de Oca, J. E. Laiño, et al. 2016. Ancestral Andean Grain Quinoa as Source of Lactic Acid Bacteria Capable to Degrade Phytate and Produce B-Group Vitamins. *Food Research International* 89:488–94.

Carrizo, S. L., A. de Moreno de LeBlanc, J. G. LeBlanc, and G. C. Rollán. 2020. Quinoa Pasta Fermented with Lactic Acid Bacteria Prevents Nutritional Deficiencies in Mice. *Food Research International* 127:108735.

Carrizo, S. L., C. E. M. de Oca, M. E. Hébert, et al. 2017. Lactic Acid Bacteria from Andean Grain Amaranth: A Source of Vitamins and Functional Value Enzymes. *Microbial Physiology* 27:289–98.

Castro-Bravo, N., J. M. Wells, A. Margolles, and P. Ruas-Madiedo. 2018. Interactions of Surface Exopolysaccharides from Bifidobacterium and Lactobacillus within the Intestinal Environment. *Frontiers in Microbiology* 9:2426.

Chen, X. Y., C. Levy, and M. G. Gänzle. 2016. Structure-Function Relationships of Bacterial and Enzymatically Produced Reuterans and Dextran in Sourdough Bread Baking Application. *International Journal of Food Microbiology* 239:95–102.

Chourasia, R., L. C. Phukon, S. P. Singh, A. K. Rai, and D. Sahoo. 2020. Role of Enzymatic Bioprocesses for the Production of Functional Food and Nutraceuticals. In *Biomass, Biofuels, Biochemicals*, eds S. P. Singh, R. R. Singhania, Z. Li, A. Pandey, and C. Larroche, 309–34. Amsterdam: Elsevier.

Coda, R., R. Di Cagno, M. Gobbetti, and C. G. Rizzello. 2014. Sourdough Lactic Acid Bacteria: Exploration of Non-Wheat Cereal-Based Fermentation. *Food Microbiology* 37:51–8.

Cohen, I. S., A. S. Day, and R. Shaoul. 2019. Gluten in Celiac Disease: More or Less? *Rambam Maimonides Medical Journal* 10: e0007.

Comasio, A., M. Verce, S. Van Kerrebroeck, and L. De Vuyst. 2020. Diverse Microbial Composition of Sourdoughs From Different Origins. *Frontiers in Microbiology* 11:1212.

Crucean, D., G. Debucquet, C. Rannou, A. le-Bail, and P. le-Bail. 2019. Vitamin B4 as a Salt Substitute in Bread: A Challenging and Successful New Strategy. Sensory Perception and Acceptability by French Consumers. *Appetite* 134:17–25.

Czeizel, A. E., and Z. Merhala. 1998. Bread Fortification with Folic Acid, Vitamin B_{12}, and Vitamin B_6 in Hungary. *Lancet* 352:1225.

Davani-Davari, D., M. Negahdaripour, I. Karimzadeh, et al. 2019. Prebiotics: Definition, Types, Sources, Mechanisms, and Clinical Applications. *Foods* 8:92.

De Angelis, M., F. Bottacini, B. Fosso, et al. 2014. *Lactobacillus rossiae*, a Vitamin B12 Producer, Represents a Metabolically Versatile Species within the Genus *Lactobacillus*. *PloS One* 9:e107232.

Demain, A. L., and P. Vaishnav. 2009. Production of Recombinant Proteins by Microbes and Higher Organisms. *Biotechnology Advances* 27:297–306.

Diez-Ozaeta, I., L. Martín Loarte, M. L. Mohedano, et al. 2023. A Methodology for the Selection and Characterization of Riboflavin-overproducing *Weissella cibaria* Strains after Treatment with Roseoflavin. *Frontiers in Microbiology* 14:1154130.

Dimopoulou, M., J. Raffenne, O. Claisse, et al. 2018. Oenococcus Oeni Exopolysaccharide Biosynthesis, a Tool to Improve Malolactic Starter Performance. *Frontiers in Microbiology* 9:1276.

Dols, M., W. Chraibi, M. Remaud-Simeon, N. D. Lindley, and P. F. Monsan. 1997. Growth and Energetics of *Leuconostoc mesenteroides* NRRL B-1299 during Metabolism of Various Sugars and Their Consequences for Dextransucrase Production. *Applied and Environmental Microbiology* 63:2159–65.

Ferrario, C., C. Milani, L. Mancabelli, et al. 2016. Modulation of the Eps-Ome Transcription of Bifidobacteria through Simulation of Human Intestinal Environment. *FEMS Microbiology Ecology* 92:fiw056.

Fujimoto, A., K. Ito, N. Narushima, and T. Miyamoto. 2019. Identification of Lactic Acid Bacteria and Yeasts, and Characterisation of Food Components of Sourdoughs Used in Japanese Bakeries. *Journal of Bioscience and Bioengineering* 127:575–81.

Galle, S., C. Schwab, F. Dal Bello, A. Coffey, M. G. Gänzle, and E. K. Arendt. 2012. Influence of In-Situ Synthesized Exopolysaccharides on the Quality of Gluten-Free Sorghum Sourdough Bread. *International Journal of Food Microbiology* 155:105–12.

Gänzle, M. G., and J. Zheng. 2019. Lifestyles of Sourdough Lactobacilli–Do They Matter for Microbial Ecology and Bread Quality? *International Journal of Food Microbiology* 302:15–23.

Gaspar, P., A. L. Carvalho, S. Vinga, H. Santos, and A. R. Neves. 2013. From Physiology to Systems Metabolic Engineering for the Production of Biochemicals by Lactic Acid Bacteria. *Biotechnology Advances* 31:764–88.

Gibson, G. R., R. Hutkins, M. E. Sanders, et al. 2017. Expert Consensus Document: The International Scientific Association for Probiotics and Prebiotics (ISAPP) Consensus Statement on the Definition and Scope of Prebiotics. *Nature Reviews Gastroenterology & Hepatology* 14:491.

Glibowski, P., and K. Skrzypczak. 2017. Prebiotic and Synbiotic Foods. In *Microbial Production of Food Ingredients and Additives*, eds A. Grumezescu, and A. M. Holban, 155–88. Amsterdam: Elsevier.

Gobbetti, M., M. De Angelis, R. Di Cagno, M. Calasso, G. Archetti, and C. G. Rizzello. 2019. Novel Insights on the Functional/Nutritional Features of the Sourdough Fermentation. *International Journal of Food Microbiology* 302:103–13.

Green, R., L. H. Allen, A. -L. Bjørke-Monsen, et al. 2017. Vitamin B 12 Deficiency. *Nature Reviews Disease Primers* 3:1–20.

Greppi, A., Y. Hemery, I. Berrazaga, Z. Almaksour, and C. Humblot. 2017. Ability of Lactobacilli Isolated from Traditional Cereal-Based Fermented Food to Produce Folate in Culture Media under Different Growth Conditions. *LWT* 86:277–84.

Gu, Q., C. Zhang, D. Song, P. Li, and X. Zhu. 2015. Enhancing Vitamin B_{12} Content in Soy-Yogurt by *Lactobacillus reuteri*. *International Journal of Food Microbiology* 206:56–9.

Guérin, M., C. R. -D. Silva, C. Garcia, and F. Remize. 2020. Lactic Acid Bacterial Production of Exopolysaccharides from Fruit and Vegetables and Associated Benefits. *Fermentation* 6:115.

Gujska, E., J. Michalak, and J. Klepacka. 2009. Folates Stability in Two Types of Rye Breads during Processing and Frozen Storage. *Plant Foods for Human Nutrition* 64:129–34.

Guyot, J. -P. 2012. Cereal-Based Fermented Foods in Developing Countries: Ancient Foods for Modern Research. *International Journal of Food Science & Technology* 47:1109–14.

Hazell, B. W., and P. V. Attfield. 1999. Enhancement of Maltose Utilisation by Saccharomyces Cerevisiae in Medium Containing Fermentable Hexoses. *Journal of Industrial Microbiology and Biotechnology* 22:627–32.

Hernández-Alcántara, A.M., R. Chiva, M.L. Mohedano, et al. 2022. *Weissella cibaria* Riboflavin-overproducing and Dextran-Producing Strains Useful for the Development of Functional Bread. *Frontiers in Nutrition* 9:978831.

Herranen, M., S. Kariluoto, M. Edelmann, et al. 2010. Isolation and Characterisation of Folate-Producing Bacteria from Oat Bran and Rye Flakes. *International Journal of Food Microbiology* 142:277–85.

Hjortmo, S., J. Patring, J. Jastrebova, and T. Andlid. 2008. Biofortification of Folates in White Wheat Bread by Selection of Yeast Strain and Process. *International Journal of Food Microbiology* 127:32–6.

Hu, Y., and M. G. Gänzle. 2018. Effect of Temperature on Production of Oligosaccharides and Dextran by *Weissella cibaria* 10 M. *International Journal of Food Microbiology* 280:27–34.

Hu, Y., V. Winter, X. Y. Chen, and M. G. Gänzle. 2017. Effect of Acceptor Carbohydrates on Oligosaccharide and Polysaccharide Synthesis by Dextransucrase DsrM from *Weissella cibaria*. *Food Research International* 99:603–11.

Huang, S. -X., D. -Z. Hou, P. -X. Qi, et al. 2020. Enzymatic Synthesis of Non-Digestible Oligosaccharide Catalysed by Dextransucrase and Dextranase from Maltose Acceptor Reaction. *Biochemical and Biophysical Research Communications* 523:651–7.

Hüfner, E., R. A. Britton, S. Roos, H. Jonsson, and C. Hertel. 2008. Global Transcriptional Response of *Lactobacillus reuteri* to the Sourdough Environment. *Systematic and Applied Microbiology* 31:323–38.

Hwang, H., and J. -H. Lee. 2019. Evaluation of Metabolites Derived from Lactic Acid Bacteria Isolated from Kimchi. In *Chemistry of Korean Foods and Beverages*, Eds C. H. Do, A. M. Rimando, Y. Kim, 3–10. Washington, DC: ACS Publications.

Immonen, M., N. H. Maina, Y. Wang, R. Coda, and K. Katina. 2020. Waste Bread Recycling as a Baking Ingredient by Tailored Lactic Acid Fermentation. *International Journal of Food Microbiology* 327:108652.

İspirli, H., D. Özmen, M. T. Yılmaz, O. Sağdıç, and E. Dertli. 2020. Impact of Glucan Type Exopolysaccharide (EPS) Production on Technological Characteristics of Sourdough Bread. *Food Control* 107:106812.

Itzlinger, A., F. Branchi, L. Elli, and M. Schumann. 2018. Gluten-Free Diet in Celiac Disease: Forever and for All? *Nutrients* 10:1796.

Iulietto, M. F., P. Sechi, E. Borgogni, and B. T. Cenci-Goga. 2015. Meat Spoilage: A Critical Review of a Neglected Alteration Due to Ropy Slime Producing Bacteria. *Italian Journal of Animal Science* 14:4011.

Juvonen, R., K. Honkapää, N. H. Maina, et al. 2015. The Impact of Fermentation with Exopolysaccharide Producing Lactic Acid Bacteria on Rheological, Chemical and Sensory Properties of Pureed Carrots (Daucus Carota L.). *International Journal of Food Microbiology* 207:109–18.

Kariluoto, S., M. Aittamaa, M. Korhola, H. Salovaara, L. Vahteristo, and V. Piironen. 2006. Effects of Yeasts and Bacteria on the Levels of Folates in Rye Sourdoughs. *International Journal of Food Microbiology* 106:137–43.

Kariluoto, S., L. Vahteristo, H. Salovaara, K. Katina, K. -H. Liukkonen, and V. Piironen. 2004. Effect of Baking Method and Fermentation on Folate Content of Rye and Wheat Breads. *Cereal Chemistry* 81:134–9.

Katina, K., K. -H. Liukkonen, A. Kaukovirta-Norja, et al. 2007. Fermentation-Induced Changes in the Nutritional Value of Native or Germinated Rye. *Journal of Cereal Science* 46:348–55.

Katina, K., N. H. Maina, R. Juvonen, et al. 2009. In Situ Production and Analysis of *Weissella confusa* Dextran in Wheat Sourdough. *Food Microbiology* 26:734–43.

Khusniati, T., D. A. Trilestari, S. Yuningtyas, and Sulistiani. 2021. The Stability of Alpha-Amylase from *Leuconostoc mesenteroides* EN 17–11 and *Lactobacillus plantarum* B110 at Various Storage Times and Temperatures. In *AIP Conference Proceedings*, 2331:050015. AIP Publishing LLC.

Korhola, M., R. Hakonen, K. Juuti, et al. 2014. Production of Folate in Oat Bran Fermentation by Yeasts Isolated from Barley and Diverse Foods. *Journal of Applied Microbiology* 117:679–89.

Kothari, D., D. Das, S. Patel, and A. Goyal. 2014. Dextran and Food Application. In *Polysaccharides: Bioactivity and Biotechnology*, eds K. G. Ramawat, J. M. Merillon, 735–52. Cham: Springer International Publishing AG.

Kurek, M. A., J. Wyrwisz, S. Karp, and A. Wierzbicka. 2017. Particle Size of Dietary Fiber Preparation Affects the Bioaccessibility of Selected Vitamin B in Fortified Wheat Bread. *Journal of Cereal Science* 77:166–71.

Lau, S. W., A. Q. Chong, N. L. Chin, R. A. Talib, and R. K. Basha. 2021 Sourdough Microbiome Comparison and Benefits. *Microorganisms* 9:1355.

Lebeer, S., T. L. Verhoeven, G. Francius, et al. 2009. Identification of a Gene Cluster for the Biosynthesis of a Long, Galactose-Rich Exopolysaccharide in *Lactobacillus rhamnosus* GG and Functional Analysis of the Priming Glycosyltransferase. *Applied and Environmental Microbiology* 75:3554–63.

de LeBlanc, A. de M., T. D. Luerce, A. Miyoshi, V. Azevedo, and J. G. LeBlanc. 2018. Functional Food Biotechnology: The Use of Native and Genetically Engineered Lactic Acid Bacteria. In *Omics Technologies and Bio-Engineering*, eds D. Barh, V. Azevedo, 105–28. Amsterdam: Elsevier.

LeBlanc, J. G., J. E. Laiño, M. J. del Valle, et al. 2011. B-Group Vitamin Production by Lactic Acid Bacteria – Current Knowledge and Potential Applications. *Journal of Applied Microbiology* 111:1297–309.

Levit, R., G. S. de Giori, A. de M. de LeBlanc, and J. G. LeBlanc. 2021. Recent Update on Lactic Acid Bacteria Producing Riboflavin and Folates: Application for Food Fortification and Treatment of Intestinal Inflammation. *Journal of Applied Microbiology* 130:1412–24.

Li, P., Q. Gu, L. Yang, Y. Yu, and Y. Wang. 2017. Characterisation of Extracellular Vitamin B_{12} Producing *Lactobacillus plantarum* Strains and Assessment of the Probiotic Potentials. *Food Chemistry* 234:494–501.

Llamas-Arriba, M. G., A. I. Puertas, A. Prieto, et al. 2019. Characterization of Dextrans Produced by *Lactobacillus mali* CUPV271 and *Leuconostoc carnosum* CUPV411. *Food Hydrocolloids* 89:613–22.

Longin, F., H. Beck, H. Gütler, et al. 2020. Aroma and Quality of Breads Baked from Old and Modern Wheat Varieties and Their Prediction from Genomic and Flour-Based Metabolite Profiles. *Food Research International* 129:108748.

Lynch, K. M., A. Coffey, and E. K. Arendt. 2018. Exopolysaccharide Producing Lactic Acid Bacteria: Their Techno-Functional Role and Potential Application in Gluten-Free Bread Products. *Food Research International* 110:52–61.

Marco, M. L., D. Heeney, S. Binda, et al. 2017. Health Benefits of Fermented Foods: Microbiota and Beyond. *Current Opinion in Biotechnology* 44:94–102.

Masuda, M., M. Ide, H. Utsumi, T. Niiro, Y. Shimamura, and M. Murata. 2012. Production Potency of Folate, Vitamin B_{12}, and Thiamine by Lactic Acid Bacteria Isolated from Japanese Pickles. *Bioscience, Biotechnology, and Biochemistry* 76:2061–7.

Meroth, C. B., J. Walter, C. Hertel, M. J. Brandt, and W. P. Hammes. 2003. Monitoring The Bacterial Population Dynamics In Sourdough Fermentation Processes By Using Pcr-Denaturing Gradient Gel Electrophoresis. *Applied and Environmental Microbiology* 69:475–82.

Mihhalevski, A., I. Nisamedtinov, K. Hälvin, A. Ošeka, and T. Paalme. 2013. Stability of B-Complex Vitamins and Dietary Fiber during Rye Sourdough Bread Production. *Journal of Cereal Science* 57:30–8.

Moghaddam, M. F. T., H. Jalali, A. M. Nafchi, and L. Nouri. 2020. Evaluating the Effects of Lactic Acid Bacteria and Olive Leaf Extract on the Quality of Gluten-Free Bread. *Gene Reports* 21:100771.

Mohanty, D., S. Misra, S. Mohapatra, and P. S. Sahu. 2018. Prebiotics and Synbiotics: Recent Concepts in Nutrition. *Food Bioscience* 26:152–60.

Mohd Nadzir, M., R. W. Nurhayati, F. N. Idris, and M. H. Nguyen. 2021. Biomedical Applications of Bacterial Exopolysaccharides: A Review. *Polymers* 13:530.

Montemurro, M., R. Coda, and C. G. Rizzello. 2019. Recent Advances in the Use of Sourdough Biotechnology in Pasta Making. *Foods* 8:129.

Morais, E. C., A. G. Cruz, J. A. F. Faria, and H. M. A. Bolini. 2014. Prebiotic Gluten-Free Bread: Sensory Profiling and Drivers of Liking. *LWT-Food Science and Technology* 55:248–54.

Mozzi, F., F. Vaningelgem, E. M. Hébert, et al. 2006. Diversity of Heteropolysaccharide-Producing Lactic Acid Bacterium Strains and Their Biopolymers. *Applied and Environmental Microbiology* 72:4431–5.

Nácher-Vázquez, M., N. Ballesteros, Á. Canales, et al. 2015. Dextrans Produced by Lactic Acid Bacteria Exhibit Antiviral and Immunomodulatory Activity against Salmonid Viruses. *Carbohydrate Polymers* 124:292–301.

Nácher-Vázquez, M., I. Iturria, K. Zarour, et al. 2017. Dextran Production by *Lactobacillus sakei* MN1 Coincides with Reduced Autoagglutination, Biofilm Formation and Epithelial Cell Adhesion. *Carbohydrate Polymers* 168:22–31.

Nácher-Vázquez, M., J. A. Ruiz-Masó, M. L. Mohedano, G. Del Solar, R. Aznar, and P. López. 2017. Dextransucrase Expression Is Concomitant with That of Replication and Maintenance Functions of the PMN1 Plasmid in *Lactobacillus sakei* MN1. *Frontiers in Microbiology* 8:2281.

Naessens, M., A. Cerdobbel, W. Soetaert, and E. J. Vandamme. 2005. Leuconostoc Dextransucrase and Dextran: Production, Properties and Applications. *Journal of Chemical Technology and Biotechnology* 80:845–60.

Neri-Numa, I. A., and G. M. Pastore. 2020. Novel Insights into Prebiotic Properties on Human Health: A Review. *Food Research International* 131:108973.

Neubauer, H., A. Bauche, and B. Mollet. 2003. Molecular Characterisation and Expression Analysis of the Dextransucrase DsrD of *Leuconostoc mesenteroides* Lcc4 in Homologous and Heterologous *Lactococcus lactis* Cultures. *Microbiology* 149:973–82.

Novotni, D., M. Gänzle, and J. M. Rocha. 2021. Composition and Activity of Microbiota in Sourdough and Their Effect on Bread Quality and Safety. In *Trends in Wheat and Bread Making*, ed. C. M. Galanakis, 129–72. Cambridge: Academic Press.

Oerlemans, M. M., R. Akkerman, M. Ferrari, M. T. Walvoort, and P. de Vos. 2020. Benefits of Bacteria-Derived Exopolysaccharides on Gastrointestinal Microbiota, Immunity and Health. *Journal of Functional Foods* 76:104289.

Ott, E., J. Stolz, M. Lehmann, and M. Mack. 2009. The RFN Riboswitch of Bacillus Subtilis Is a Target for the Antibiotic Roseoflavin Produced by Streptomyces Davawensis. *RNA Biology* 6:276–80.

Păcularu-Burada, B., L. A. Georgescu, and G. -E. Bahrim. 2020. Current Approaches in Sourdough Production with Valuable Characteristics for Technological and Functional Applications. *The Annals of the University Dunarea de Jos of Galati. Fascicle VI-Food Technology* 44:132–48.

Padma Ishwarya, S., and P. Prabhasankar. 2014. Prebiotics: Application in Bakery and Pasta Products. *Critical Reviews in Food Science and Nutrition* 54:511–22.

Paramithiotis, S., A. Sofou, E. Tsakalidou, and G. Kalantzopoulos. 2007. Flour Carbohydrate Catabolism and Metabolite Production by Sourdough Lactic Acid Bacteria. *World Journal of Microbiology and Biotechnology* 23:1417–23.

de Paulo Farias, D., F. F. de Araújo, I. A. Neri-Numa, and G. M. Pastore. 2019. Prebiotics: Trends in Food, Health and Technological Applications. *Trends in Food Science & Technology* 93:23–35.

Perri, G., R. Coda, C. G. Rizzello, et al. 2021. Sourdough Fermentation of Whole and Sprouted Lentil Flours: In Situ Formation of Dextran and Effects on the Nutritional, Texture and Sensory Characteristics of White Bread. *Food Chemistry* 355:129638.

Petrova, P., and K. Petrov. 2020. Lactic Acid Fermentation of Cereals and Pseudocereals: Ancient Nutritional Biotechnologies with Modern Applications. *Nutrients* 12:1118.

Povoroznyuk, V., N. Balatska, V. Dotsenko, L. Synyeok, V. Havrysh, and O. Bortnichuk. 2015. Fortified Bread in Correction of Vitamin D Status. *Maturitas* 81:219.

Prapulla, S. G., V. Subhaprada, and N. G. Karanth. 2000. Microbial Production of Oligosaccharides: A Review. 47:299–343.

Puertas, A. I., I. Ibarburu, P. Elizaquivel, et al. 2018. Disclosing Diversity of Exopolysaccharide-Producing Lactobacilli from Spanish Natural Ciders. *LWT* 90:469–74.

Ripa, I., J. Á. Ruiz-Masó, N. De Simone, P. Russo, G. Spano, and G. del Solar. 2022. A single change in the aptamer of the *Lactiplantibacillus plantarum rib* operon riboswitch severely impairs its regulatory activity and leads to a vitamin B 2 - overproducing phenotype. *Microbial Biotechnology*. https://doi.org/10.1111/1751-7915.13919

Ripari, V. 2019. Techno-Functional Role of Exopolysaccharides in Cereal-Based, Yogurt-like Beverages. *Beverages* 5:16.

Ripari, V., Y. Bai, and M. G. Gänzle. 2019. Metabolism of Phenolic Acids in Whole Wheat and Rye Malt Sourdoughs. *Food Microbiology* 77:43–51.

Rohi, M., I. Pasha, M. S. Butt, and H. Nawaz. 2013. Variation in the Levels of B-Vitamins and Protein Content in Wheat Flours. *Pakistan Journal of Nutrition* 12:441–7.

Rolim, P. M., Y. Hu, and M. G. Gänzle. 2019. Sensory Analysis of Juice Blend Containing Isomalto-Oligosaccharides Produced by Fermentation with *Weissella cibaria*. *Food Research International* 124:86–92.

Ruas-Madiedo, P., and C. G. De Los Reyes-Gavilán. 2005. Methods for the Screening, Isolation, and Characterisation of Exopolysaccharides Produced by Lactic Acid Bacteria. *Journal of Dairy Science* 88:843–56.

Rühmkorf, C., H. Rübsam, T. Becker, et al. 2012. Effect of Structurally Different Microbial Homoexopolysaccharides on the Quality of Gluten-Free Bread. *European Food Research and Technology* 235:139–46.

Russo, P., V. Capozzi, M. P. Arena, et al. 2014. Riboflavin-Overproducing Strains of *Lactobacillus fermentum* for Riboflavin-Enriched Bread. *Applied Microbiology and Biotechnology* 98:3691–700.

Russo, P., N. De Simone, V. Capozzi, et al. 2021. Selection of Riboflavin Overproducing Strains of Lactic Acid Bacteria and Riboflavin Direct Quantification by Fluorescence. *Methods in Molecular Biology* 2280:3–14.

Sahin, A. W., T. Rice, E. Zannini, et al. 2019. *Leuconostoc Citreum* TR116: In-Situ Production of Mannitol in Sourdough and Its Application to Reduce Sugar in Burger Buns. *International Journal of Food Microbiology* 302:80–9.

Salvucci, E., J. G. LeBlanc, and G. Pérez. 2016. Technological Properties of Lactic Acid Bacteria Isolated from Raw Cereal Material. *LWT* 70:185–91.

Sanalibaba, P., and G. A. Çakmak. 2016. Exopolysaccharides Production by Lactic Acid Bacteria. *Applied Microbiology: Open Access* 2:1000115.

Saubade, F., Y. M. Hemery, J. -P. Guyot, and C. Humblot. 2017. Lactic Acid Fermentation as a Tool for Increasing the Folate Content of Foods. *Critical Reviews in Food Science and Nutrition* 57:3894–910.

Saubade, F., C. Humblot, Y. M. Hemery, and J. -P. Guyot. 2017. PCR Screening of an African Fermented Pearl-Millet Porridge Metagenome to Investigate the Nutritional Potential of Its Microbiota. *International Journal of Food Microbiology* 244:103–10.

Schmid, J., D. Wefers, R. F. Vogel, and F. Jakob. 2021. Analysis of Structural and Functional Differences of Glucans Produced by the Natively Released Dextransucrase of *Liquorilactobacillus hordei* TMW 1.1822. *Applied Biochemistry and Biotechnology* 193:96–110.

Seidametova, E. A., M. R. Shakirzyanova, D. M. Ruzieva, and T. G. Gulyamova. 2004. Isolation of Cobalt-Resistant Strains of Propionic Acid Bacteria, Potent Producers of Vitamin B 12. *Applied Biochemistry and Microbiology* 40:560–2.

Siró, I., E. Kápolna, B. KápolnA, and A.Lugasi.2008. Functional Food. Product Development, Marketing and Consumer Acceptance: A Review. *Appetite* 51:456–67.

Sobczyńska-Malefora, A., E. Delvin, A. McCaddon, K. R. Ahmadi, and D. J. Harrington. 2021. Vitamin B_{12} Status in Health and Disease: A Critical Review. Diagnosis of Deficiency and Insufficiency: Clinical and Laboratory Pitfalls. *Critical Reviews in Clinical Laboratory Sciences*. https://doi.org/10.1080/10408363.2021.1885339.

Sriramulu, D. D., M. Liang, D. Hernandez-Romero, et al. 2008. *Lactobacillus reuteri* DSM 20016 Produces Cobalamin-Dependent Diol Dehydratase in Metabolosomes and Metabolises 1,2-Propanediol by Disproportionation. *Journal of Bacteriology* 190:4559–67.

Stoyanova, L. G., E. A. Ustyugova, and A. I. Netrusov. 2012. Antibacterial Metabolites of Lactic Acid Bacteria: Their Diversity and Properties. *Applied Biochemistry and Microbiology* 48:229–43.

Taranto, M. P., J. L. Vera, J. Hugenholtz, G. F. De Valdez, and F. Sesma. 2003. *Lactobacillus reuteri* CRL1098 Produces Cobalamin. *Journal of Bacteriology* 185:5643–7.

Thakur, K., S. K. Tomar, and S. De. 2016. Lactic Acid Bacteria as a Cell Factory for Riboflavin Production. *Microbial Biotechnology* 9:441–51.

Tieking, M., and M. G. Gänzle. 2005. Exopolysaccharides from Cereal-Associated Lactobacilli. *Trends in Food Science & Technology* 16:79–84.

Tieking, M., M. Korakli, M. A. Ehrmann, M. G. Gänzle, and R. F. Vogel. 2003. In Situ Production of Exopolysaccharides during Sourdough Fermentation by Cereal and Intestinal Isolates of Lactic Acid Bacteria. *Applied and Environmental Microbiology* 69:945–52.

Tinzl-Malang, S. K., F. Grattepanche, P. Rast, P. Fischer, J. Sych, and C. Lacroix. 2020. Purified Exopolysaccharides from *Weissella confusa* 11GU-1 and *Propionibacterium freudenreichii* JS15 Act Synergistically on Bread Structure to Prevent Staling. *LWT* 127:109375.

Tinzl-Malang, S. K., P. Rast, F. Grattepanche, J. Sych, and C. Lacroix. 2015. Exopolysaccharides from Co-Cultures of *Weissella confusa* 11GU-1 and *Propionibacterium freudenreichii* JS15 Act Synergistically on Wheat Dough and Bread Texture. *International Journal of Food Microbiology* 214:91–101.

Torino, M. I., G. Font de Valdez, and F. Mozzi. 2015. Biopolymers from Lactic Acid Bacteria. Novel Applications in Foods and Beverages. *Frontiers in Microbiology* 6:834.

Torres, A. C., M. Elean, E. M. Hebert, L. Saavedra, and M. P. Taranto. 2020. Metabolic Shift in the Production of Corrinoid Compounds by *Lactobacillus coryniformis* in the Absence of Purines. *Biochimie* 168:185–9.

Torres, A. C., V. Vannini, J. Bonacina, G. Font, L. Saavedra, and M. P. Taranto. 2016. Cobalamin Production by *Lactobacillus coryniformis*: Biochemical Identification of the Synthetized Corrinoid and Genomic Analysis of the Biosynthetic Cluster. *BMC Microbiology* 16:240.

Torres, A. C., V. Vannini, G. Font, L. Saavedra, and M. P. Taranto. 2018. Novel Pathway for Corrinoid Compounds Production in *Lactobacillus*. *Frontiers in Microbiology* 9:2256.

Turck, D., J. -L. Bresson, B. Burlingame, et al. 2017. Dietary Reference Values for Riboflavin. *EFSA Journal* 15:e04919.

Turpin, W., C. Humblot, and J. -P. Guyot. 2011. Genetic Screening of Functional Properties of Lactic Acid Bacteria in a Fermented Pearl Millet Slurry and in the Metagenome of Fermented Starchy Foods. *Applied and Environmental Microbiology* 77:8722–34.

Varmanen, P. K., P. Deptula, B. S. Chamlagain, and V. I. Piironen. 2016. Letter to the Editor on "Enhancing Vitamin B12 Content in Soy-Yogurt by *Lactobacillus reuteri*. *International Journal of Food Microbiology* 206:56–9.

Vera, C., A. Illanes, and C. Guerrero. 2021. Enzymatic Production of Prebiotic Oligosaccharides. *Current Opinion in Food Science* 37:160–70.

Voss, G. B., D. A. Campos, and M. M. Pintado. 2021. Cereal Bars Added With Probiotics and Prebiotics. In *Probiotics and Prebiotics in Foods*, ed. A. Gomes da Cruz, C. S. Ranadheera, F. Nazzaro, and A. Mortazavian, 201–17. Cambridge: Academic Press.

Vuillemin, M., F. Grimaud, M. Claverie, et al. 2018. A Dextran with Unique Rheological Properties Produced by the Dextransucrase from Oenococcus Kitaharae DSM 17330. *Carbohydrate Polymers* 179:10–8.

Wang, Y., D. Compaoré-Sérémé, H. Sawadogo-Lingani, R. Coda, K. Katina, and N. H. Maina. 2019. Influence of Dextran Synthesised in Situ on the Rheological, Technological and Nutritional Properties of Whole Grain Pearl Millet Bread. *Food Chemistry* 285:221–30.

Wang, Y., N. H. Maina, R. Coda, and K. Katina. 2021. Challenges and Opportunities for Wheat Alternative Grains in Bread-making: Ex-Situ-versus in-Situ-Produced Dextran. *Trends in Food Science & Technology* 113:232–44.

Wang, Y., P. Sorvali, A. Laitila, N. H. Maina, R. Coda, and K. Katina. 2018. Dextran Produced in Situ as a Tool to Improve the Quality of Wheat-Faba Bean Composite Bread. *Food Hydrocolloids* 84:396–405.

Wang, Y., A. Trani, A. Knaapila, et al. 2020. The Effect of in Situ Produced Dextran on Flavour and Texture Perception of Wholegrain Sorghum Bread. *Food Hydrocolloids* 106:105913.

Wegkamp, A., W. M. De Vos, and E. J. Smid. 2009. Folate Overproduction in *Lactobacillus plantarum* WCFS1 Causes Methotrexate Resistance. *FEMS Microbiology Letters* 297:261–5.

Wegkamp, A., A. E. Mars, M. Faijes, et al. 2010. Physiological Responses to Folate Overproduction in *Lactobacillus plantarum* WCFS1. *Microbial Cell Factories* 9:100.

Werning, M. L., S. Notararigo, M. Nácher, P. Fernández de Palencia, R. Aznar, and P. López. 2012. Biosynthesis, Purification and Biotechnological Use of Exopolysaccharides Produced by Lactic Acid Bacteria. *Food Additives*, ed. Y. El-Samragy, 83–114. Rijeka: InTech Open.

Xie, C., R. Coda, B. Chamlagain, et al. 2018. In Situ Fortification of Vitamin B_{12} in Wheat Flour and Wheat Bran by Fermentation with Propionibacterium Freudenreichii. *Journal of Cereal Science* 81:133–9.

Xie, C., R. Coda, B. Chamlagain, et al. 2021. Fermentation of Cereal, Pseudo-Cereal and Legume Materials with *Propionibacterium freudenreichii* and *Levilactobacillus brevis* for Vitamin B_{12} Fortification. *LWT* 137:110431.

Xie, C., R. Coda, B. Chamlagain, P. Varmanen, V. Piironen, and K. Katina. 2019. Co-Fermentation of *Propionibacterium freudenreichii* and *Lactobacillus brevis* in Wheat Bran for in Situ Production of Vitamin B_{12}. *Frontiers in Microbiology* 10:1541.

Xu, D., Y. Zhang, K. Tang, Y. Hu, X. Xu, and M. G. Gänzle. 2019. Effect of Mixed Cultures of Yeast and Lactobacilli on the Quality of Wheat Sourdough Bread. *Frontiers in Microbiology* 10:2113.

Xu, Y., Y. Cui, F. Yue, et al. 2019. Exopolysaccharides Produced by Lactic Acid Bacteria and Bifidobacteria: Structures, Physiochemical Functions and Applications in the Food Industry. *Food Hydrocolloids* 94:475–99.

Yan, M., J. Han, X. Xu, et al. 2016. Gsy, a Novel Glucansucrase from *Leuconostoc mesenteroides*, Mediates the Formation of Cell Aggregates in Response to Oxidative Stress. *Scientific Reports* 6:1–13.

Yan, Y. L., Y. Hu, and M. G. Gänzle. 2018. Prebiotics, FODMAPs and Dietary Fiber: Conflicting Concepts in Development of Functional Food Products? *Current Opinion in Food Science* 20:30–7.

Yee, W. L., C. L. Yee, N. K. Lin, and P. L. Phing. 2019. Microencapsulation of *Lactobacillus acidophilus* NCFM Incorporated with Mannitol and Its Storage Stability in Mulberry Tea. *Ciência e Agrotecnologia* 43:005819.

Yépez, A., P. Russo, G. Spano, et al. 2019. In Situ Riboflavin Fortification of Different Kefir-like Cereal-Based Beverages Using Selected Andean LAB Strains. *Food Microbiology* 77:61–8.

Yun, J. W. 1996. Fructooligosaccharides: Occurrence, Preparation, and Application. *Enzyme and Microbial Technology* 19:107–17.

Zarour, K., M. G. Llamas, A. Prieto, et al. 2017. Rheology and Bioactivity of High Molecular Weight Dextrans Synthesised by Lactic Acid Bacteria. *Carbohydrate Polymers* 174:646–57.

Zarour, K., N. Vieco, A. Pérez-Ramos, M. Nácher-Vázquez, M. L. Mohedano, and P. López. 2017. Food Ingredients Synthesized by Lactic Acid Bacteria. In *Microbial Production of Food Ingredients and Additives*, eds A. M. Holban, and A. M. Grumezescu, 89–124. Cambridge: Academic Press.

Zeidan, A. A., V. K. Poulsen, T. Janzen, et al. 2017. Polysaccharide Production by Lactic Acid Bacteria: From Genes to Industrial Applications. *FEMS Microbiology Reviews* 41(Supplement 1):S168–S200.

Zhang, Y., L. Guo, D. Li, Z. Jin, and X. Xu. 2019. Roles of Dextran, Weak Acidification and Their Combination in the Quality of Wheat Bread. *Food Chemistry* 286:197–203.

Zhang, Y., L. Guo, D. Xu, et al. 2018. Effects of Dextran with Different Molecular Weights on the Quality of Wheat Sourdough Breads. *Food Chemistry* 256:373–9.

Zhang, Y., T. Hong, W. Yu, N. Yang, Z. Jin, and X. Xu. 2020. Structural, Thermal and Rheological Properties of Gluten Dough: Comparative Changes by Dextran, Weak Acidification and Their Combination. *Food Chemistry* 330:127154.

Zhang, Y., D. Li, N. Yang, Z. Jin, and X. Xu. 2018. Comparison of Dextran Molecular Weight on Wheat Bread Quality and Their Performance in Dough Rheology and Starch Retrogradation. *LWT* 98:39–45.

Section III

Innovative Applications of Sourdough Microbiota

10 Sourdough as a Source of Technological, Antimicrobial, and Probiotic Microorganisms

Vera Fraberger, Görkem Özülkü, Penka Petrova, Knežević Nada, Kaloyan Petrov, Domig Konrad Johann, and João Miguel F. Rocha

CONTENTS

Abbreviations 266
10.1 Microorganisms in Bread Making: An Introduction 267
10.2 Microbiota in Different Cereal Flours and Bread Making 271
10.3 Microbiota throughout Sourdough Fermentation 275
10.4 Production of Microbial Metabolites: Their Functions and Properties 278
     10.4.1 Low-Molecular-Weight Metabolites with Antimicrobial Activity: Functional Properties besides Flavor Improvement 278
     10.4.2 Bioactive Peptides with Antimicrobial Activity: Bacteriocins and BLIS 281
     10.4.3 LAB and EPS Production for Sourdough Bread with Improved Quality 283
10.5 Potential Applications of Sourdough Microorganisms and Their Metabolites 285
     10.5.1 Development of Sourdough Starters with Antifungal Activity 286
     10.5.2 Development of Functional Cereal Products 287
     10.5.3 Reducing the Amount of Anti-nutrients in Sourdough 289
     10.5.4 Fermentation of Phenolic Compounds for Increased Antioxidant Activity of Cereal Products 290
     10.5.5 Synthesis of Phytoestrogens 291
10.6 An Integrated Industrial Outlook 292
10.7 Conclusion 294
Acknowledgments 294
References 295

DOI: 10.1201/9781003141143-13

ABBREVIATIONS

AA	Acetic acid;
ABC	ATP-binding cassette;
ACE	angiotensin-converting enzyme;
ATI	amylase/trypsin inhibitors;
ATIs	α-amylase-trypsin-inhibitors;
ATP	adenosine triphosphate;
AXOS	arabinoxylans;
bfrA	β-fructosidase gene;
BLIS	bacteriocin-like inhibitory substances;
CFU	colony-forming units;
CO_2	carbon dioxide;
CSB	Chinese steamed bread;
Δp	membrane potential;
EC	European Commission;
EFSA	European Food Safety Authority;
EPS	exopolysaccharides;
EPS	Exopolysaccharides;
EU	European Union;
FA	ferulic acid;
FAA	free amino acids;
FAO	Food and Agriculture Organization of the United Nations;
FODMAPs	fermentable oligo-, di-, monosaccharides and polyols;
FOS	fructooligosaccharides;
FOS	fructooligosaccharides;
FQ	fermentation quotient;
GABA	γ-Aminobutyric acid;
GABA	γ-aminobutyric-acid;
GIT	gastrointestinal tract;
GOS	galactooligosaccharides;
GRAS	generally regarded as safe;
H_2O_2	hydrogen peroxide;
ILA	indolelactic;
IMO	isomaltooligosaccharides;
IPA	indole-3-propionic acid;
LA	lactic acid;
LAB	lactic acid bacteria;
MIC	minimum inhibitory concentration;
msmR	msm operon containing LacI regulator;
Mw	molecular weight;
NAD^+	nicotinamide adenine dinucleotide;
NCWS	non-celiac-wheat sensitivity;
•OH	hydroxyl radicals;
OH-PAA	hydroxyphenylacetic;

OH-PLA	4-hydroxyphenyllactic;
PCR	polymerase chain reaction;
PE	Phytoestrogens;
PLA	phenyllactic;
PTS	phosphoenolpyruvate phosphotransferase system;
QPS	qualified presumption of safety;
RS	resistant starch;
SCFA	short-chain fatty acids;
TTA	total titratable acidity;
UN	United Nations;
XOS	xylooligosaccharides;
YGNGV	tyrosine-glycine-asparagine-glycine-valine.

10.1 MICROORGANISMS IN BREAD MAKING: AN INTRODUCTION

For centuries bread has constituted a staple food in the human diet and is one of the most consumed traditional foods worldwide. Nevertheless, only *ca.* 55 years ago sourdough fermentation attracted the interest of the scientific community when Spicher and Stephan (1966) discussed the influence of sourdough-related lactic acid bacteria (LAB) on baking properties. This delayed onset of scientific interest in sourdough can be explained by the almost exclusive use of baker's yeast until the late 1900's; but due to consumer and industry awareness about the beneficial influence of sourdough on the quality of bakery goods, sourdough fermentation again gained broad acceptance and increased technological interest. Initially, research efforts were focused on the complex microbial ecosystem existing in baking sourdough and its impact on rheological and sensory properties, especially, but also on the shelf life of sourdough products (Hansen et al., 1989; Lotong et al., 2000; Venturi et al., 2016). In recent years, however, research activities have shifted toward the impact of sourdough biotechnology on the enhancement of nutritional and health properties of the sourdough-based baking goods.

Sourdough fermentation and production of bakery sourdoughs are often undertaken by artisanal (or homemade) and industrial bakers. Specific sourdoughs are often purchased from sourdough manufacturers to obtain a variety of sourdough bakery products (Brandt, 2019). A general definition given for sourdough is "a mixture of flour(s) and water (sometimes with salt) that are mixed and left to be spontaneously fermented by the natural microbiota present therein." The prevailing microorganisms after such natural fermentations mainly consist of a synergistic consortium of LAB and acid-tolerant yeast. Indeed, the continuous propagation of sourdough (or "mother dough") leads to the natural selection of such microorganisms that are beneficial for humans (Novotni et al., 2020; Rocha, 2011; Rocha and Malcata, 2012, 2016a, b).

The metabolic activity of the microorganisms during sourdough fermentation is responsible for the improvement of the quality of the end baking goods, in terms of technological, organoleptic, and nutritional properties. LAB and yeast can either be naturally present on sourdoughs or intentionally added to the dough as microbial (single or mixed/co-fermented) starter cultures. Thus, different types of sourdoughs (types I to III) can be distinguished based on the inoculation and processing technology undertaken (Novotni et al., 2020; Rocha, 2011; De Vuyst et al., 2017). Type I (traditional) sourdoughs differ from the others by the absence of microbial starter cultures and the frequent backslopping of sourdoughs, i.e. by using a portion of the fermented dough as a natural inoculum or leavening agent for the next fermentation batch. Furthermore, these doughs are typically fermented for less than 24 h at temperatures below 30°C, according to Van Kerrebroeck et al. (2017). Hence, the microbial diversity of type I sourdoughs is influenced by the ingredients – such as the type of flour or flours, tap water, and use of other ingredients, such as honey grape must – and processing factors – such as dough yield, fermentation time and temperature, and number of sourdough backsloppings (De Vuyst et al., 2002; Minervini et al., 2016; Minervini et al., 2018; Minervini et al., 2019; Novotni et al., 2020; Rocha and Malcata, 2012, 2016a, b). As a reference, the median values of LAB counts of type I sourdoughs are generally higher (log 8.5 colony-forming units per g, CFU/g) compared to those of yeast (log 6.5 CFU/g), resulting in ratios of 100:1 to 10:1, respectively (Arora et al., 2020).

Although the microbiota from the environment and additional ingredients can contribute significantly to the assembly of the microbial ecology, cereal flours and grains comprise the main source of microorganisms present in sourdoughs. Above all, heterofermentative or homofermentative LAB species of *Enterococcus*, *Lactococcus*, *Leuconostoc*, *Pediococcus*, *Streptococcus*, and *Weissella* (all from the formerly recognized genus *Lactobacillus*) have been frequently isolated (Figure 10.1). Homofermentative LAB convert hexoses into lactate as the major metabolic product, while heterofermentative LAB metabolize hexoses to yield lactate, acetate, ethanol, and carbon dioxide (CO_2) (Gänzle, 2015; Novotni et al., 2020; Rocha, 2011). Due to wide differences in the phenotypic, ecological, and genotypic traits of the former 261 species of *Lactobacillus*, a reclassification into 25 genera including *Lactobacillus* was suggested in 2020 (Zheng et al., 2020). This scheme included the addition of 12 genera frequently isolated from sourdoughs, namely *Companilactobacillus*, *Fructilactobacillus*, *Furfurilactobacillus*, *Latilactobacillus*, *Lacticaseibacillus*, *Lentilactobacillus*, *Levilactobacillus*, *Limosilactobacillus*, *Loigolactobacillus*, *Lactiplantibacillus*, *Paucilactobacillus*, and *Schleiferilactobacillus*. Within these genera, the most frequently isolated species in sourdoughs were reported to include the heterofermentative species *Lactiplantibacillus plantarum*, *Levilactobacillus brevis*, *Fructilactobacillus sanfranciscensis*, and *Limosilactobacillus fermentum* (Arora et al., 2020).

Regarding yeast, up to 80 different species have been isolated worldwide from sourdoughs, with species of the *Saccharomyces*, *Kazachstania*, *Candida*, *Torulaspora*, and *Pichia* genera being the most abundant (Arora et al., 2020; Fraberger et al., 2020c; Fujimoto et al., 2019; Palla et al., 2017; Palla et al., 2019; Urien et al., 2019; Yagmur et al., 2016; Rocha and Malcata, 1999). Yeast is primarily responsible for leavening the cereal dough and metabolizing fermentable carbohydrates. Notably, they even contribute to flavor and aroma formation in bakery products. The species

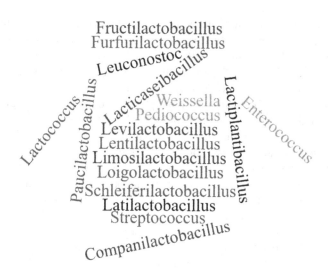

FIGURE 10.1 Most abundant LAB in sourdoughs.

most frequently identified in sourdough is *Saccharomyces cerevisiae*. Interestingly, the debate on the potential environmental cross contamination of baker's yeast or presence of wild strains continues.

The complex interaction between LAB and yeast affects the extent of the beneficial characteristics that microorganisms exert on the final product. LAB were particularly studied for a broad range of beneficial effects. They produce lactic and acetic acids but also substances like fatty acids, ethanol, and diacetyl, which delay or inhibit the growth of undesirable bacteria and fungi, thereby prolonging shelf life (Chen et al., 2019; Fraberger et al., 2020a; Küley et al., 2020; Păcularu-Burada et al., 2021, 2020). Furthermore, concentrations of hazardous compounds, like mycotoxins (Zadeike et al., 2021) and acrylamide (Bartkiene et al., 2017; Nachi et al., 2018), are decreased by the activity of LAB. Hence, sourdough fermentation exerts promising potentialities to increase shelf life and safety of bakery products. Besides the organic acids, a wide range of other complex aroma compounds, including alcohols, ketones, and aldehydes, contribute substantially to the improved flavor attained in sourdough products. Studies have also determined the presence of a more complex aroma profile in doughs and final baking products when co-fermenting processes with LAB and yeast are employed (Xu et al., 2019).

Sourdough-based fermentation products have also shown improved rheology characteristics when compared to those leavened with only baker's yeast (*Saccharomyces cerevisiae*). Texture, moisture retention, loaf volume, crumb structure, and color, as well as crust color, were the attributes particularly improved (Ma et al., 2021; Ryan et al., 2011). For example, sourdough-based fermented products showed a desirably higher number and smaller sizes of crumb holes. Specific microbial strains exhibit the ability to produce metabolites like exopolysaccharides (EPS), positively affecting bread quality criteria – like technological but also sensory properties – due to the increased loaf volume and reduced crumb hardness (Chen et al., 2016; Gezginc and Kara, 2019). Furthermore, functional attributes are enhanced, for instance, with the abovementioned production of EPS, which are classified as prebiotics (Dertli et al., 2016; Păcularu-Burada, 2020, 2021; Skendi et al., 2020). Other functional benefits include the increased bioavailability of minerals, vitamins, and phytochemicals due to the breakdown of the anti-nutritional phytic acid during sourdough fermentation (Yildirim and Arici, 2019) and also due to the synthesis of nutritionally active peptides and derivates of amino acids, like γ-aminobutyric-acid (GABA) (Diana et al., 2014; Villegas et al., 2016; Novotni et al., 2020; Rocha, 2011). Boudaoud et al. (2021) analyzed the metabolism of ferulic acid (FA) – which is mainly known for its antioxidant, antimicrobial, and anti-inflammatory properties – into components influencing nutritional and organoleptic properties, and with LAB and yeast associations promoting such reactions.

In recent years, research interest toward more digestible bakery products has increased significantly. New studies have shown that besides gluten, α-amylase-trypsin-inhibitors (ATIs) can induce an innate immune response in patients with wheat sensitivity. Furthermore, ATIs are assumed to be the main triggers of non-celiac wheat sensitivity (NCWS). During sourdough fermentation, however, these immunogenic proteins can be degraded (Fraberger et al., 2020b; Huang et al., 2020). Fermentable oligosaccharides, disaccharides, monosaccharides, and polyols (FODMAPs) are known to contribute to inflammatory reactions in patients with wheat sensitivity (Scherf, 2019). In this regard, yeasts play an important role, as they can hydrolyze FODMAPs due to invertase and inulinase activities (Fraberger et al., 2018; Menezes et al., 2019). Nevertheless, analyses of the impact of sourdough yeast communities are less frequent when compared to the research carried out with sourdough LAB, although the former play an essential role in textural properties due to their CO_2-building capacity. Indeed, yeast occurring in traditionally fermented sourdoughs help to strengthen the gluten network and produce health-promoting metabolites (De Vuyst et al., 2016; Rocha and Malcata, 1999, 2012). Aside from positively affecting the quality of the end product, yeast contribute to food safety by their antifungal and anti-aflatoxigenic activities (Böswald et al., 1995; De Vuyst et al., 2016; Pfliegler et al., 2015; Rocha, 2011). Table 10.1 summarizes the main impacts on final baking products when using sourdough fermentation.

In summary, the use of traditionally fermented sourdough exhibits numerous beneficial effects on the end baking products when compared to other leavening agents. Research studies comparing baker's yeast with sourdough-leavened goods have demonstrated that the latter exhibited improved rheological, organoleptic, and nutritional characteristics (Crowley et al., 2002). Furthermore, microbial starter cultures with a long-use tradition in food like backslopped sourdoughs need no qualified presumption of safety (QPS) status, as most starter cultures used in sourdough processing have already met the criteria. A restricted safety assessment for newly described species is therefore sufficient, and investigating the specific use of sourdough-based yeast and LAB strains to produce bakery products with concrete properties can be rewarding.

TABLE 10.1
Summary of the Main Beneficial Effects of LAB and Yeasts on Bakery Products

Characteristics	Impact on	References
Spoilage	Shelf life ↑	(Corsetti et al., 2000)
Nutrients	Gluten ↓	(Fraberger, Ladurner, et al., 2020b)
	ATIs ↓	(Fraberger, Ladurner, et al., 2020b; Huang et al., 2020)
	GABA ↑	(M. Venturi, Galli, Pini, Guerrini, & Granchi, 2019; Wu, Tun, Law, Khafipour, & Shah, 2017)
	FODMAPs ↓	(Fraberger et al., 2018)
	Minerals ↑	(Katina et al., 2005)
	Exopolysaccharide ↑	(Dertli et al., 2016; Liu et al., 2018)
Anti-nutritional factors	Phytic acid ↓	(Karaman, Sagdic, & Durak, 2018; Yildirim & Arici, 2019)
	Mycotoxins ↓	(Bartkiene et al., 2018)
	Acrylamide ↓	(Nachi et al., 2018)

Note: Arrow ↑: increasing effect; Arrow ↓: decreasing effect.

10.2 MICROBIOTA IN DIFFERENT CEREAL FLOURS AND BREAD MAKING

Cereals and pseudocereals are the major ingredients for bakery goods and, with water, constitute the sourdough matrix. Sourdough also consists of a microbial ecosystem harboring mainly LAB and yeast, which play a role in leavening, acidification, and flavor and aroma formation (De Vuyst et al., 2017; Novotni et al., 2020; Rocha and Malcata, 2012; Rocha, 2011). The cereal flour or flours, other dough ingredients, and environment in the bread making processes influence the sourdough microbiota. Geographical origin is another parameter which may influence the sourdough ecosystem (Suo et al., 2021). Cereal flours are a source of nutrients for the development of the microbiota and contain autochthonous LAB and yeast, among many other microorganisms (De Vuyst et al., 2017; Ercolini et al., 2013; Rocha and Malcata, 1999, 2012; Rocha, 2011).

Cereals, which are annual plants of family Poaceae, were the first plants to be processed by mankind. Wheat (*Triticum* spp.), maize (*Zea mays* L.), and rice (*Oryza sativa* L.) are cultivated worldwide and account for more than 50% of worldwide production. Wheat is widely used in sourdough fermentations. *Lactiplantibacillus plantarum* and *Pediococcus pentosaceus* are the dominant species found in Belgium wheat sourdough samples, which are followed by *Limosilactobacillus fermentum*. The LAB species *Leuconostoc mesenteroides*, *Levilactobacillus brevis*,

Latilactobacillus curvatus, Lc. Plantarum, and *Fructilactobacillus sanfranciscensis* have been found in French wheat sourdoughs but *Leuc. mesenteroides* is the most abundant LAB species of all. On the other hand, in German wheat sourdoughs, the dominant LAB species are *Lacticaseibacillus casei* and *Lp. Plantarum. Lev. Brevis, Li. Fermentum*, and *Fr. sanfranciscensis* are also found (De Vuyst et al., 2017; Huys et al., 2013).

Triticum durum and Triticum aestivum are the most common wheat species used in sourdough bread. Generally, the nutrient composition of *T. durum* – which contains the highest amount of maltose, glucose, fructose, and free amino acids (FAA) – provides a more suitable environment for obligate heterofermentative LAB than facultative heterofermentative LAB and yeasts. Moreover, in *T. aestivum* flour – which is characterized by lower concentrations of such nutrients – obligate heterofermentative LAB are less competitive than facultative and/or obligate homofermentative LAB. *Fr. sanfranciscensis* can be isolated and identified most frequently from both *T. aestivum* and *T. durum* flours, according to the study undertaken by Minervini et al., (2012), who investigated 19 Italian sourdoughs. *Lp. plantarum* and *Lactobacillus paralimentarius* are also abundant and the most commonly isolated yeast species in Italian sourdoughs is *Saccharomyces cerevisiae. Candida humilis* prevails in the protected designation of origin (PDO) sourdough bread *Pagnotta del Dittaino*, which is made of *T. durum* (Minervini et al., 2012). In Austrian wheat sourdough, the main genus found is *Lactobacillus* spp., followed by *Pediococcus* spp. and *Leuconostoc* spp. However, *Pediococcus* spp. and *Leuconostoc* spp. are only identified within Austrian wheat sourdough and not in Austrian rye sourdough (Fraberger et al., 2020c).

Jiaozi is a kind of Chinese traditional sourdough starter culture used to ferment wheat or maize flours. Chinese steamed bread (CSB) can be made from this typical traditional sourdough starter. *Lactobacillus, Weissella, Acetobacter, Sphingomonas*, and *Serratia* are reported as predominant LAB in CSB manufactured with Jiaozi. Species of *Saccharomyces, Candida*, and *Alternaria* are predominant yeast in this CSB. *Acetobacter tropicalis, Enterococcus durans, Lacp. Plantarum*, and *P. pentosaceus* are also isolated as dominant bacteria strains in CSB with Jiaozi in samples from Henan province in China. The predominant yeast species are *S. cerevisiae, W. anomalus, T. delbrueckii*, and *S. fibuligera* in samples from the same region. *Fr. sanfranciscensis* and *Limosilactobacillus pontis* are identified in CSB with Type I sourdoughs. *S. cerevisiae* and *Kazachstania humilis* are the most representative yeast in CSB with type I sourdough (Suo et al., 2021).

Rye (*Secale cereale*) is a cold-resistant cereal, very adaptable in Northern climates, and a good source for dietary fibers in Northern and Eastern Europe (Koistinen et al., 2018; Weckx et al., 2010). Sourdough fermentation is needed to inhibit α-amylase activity, to ensure good bread making performance of rye. *Lp. plantarum, Li. pontis, Lev. brevis, L. paralimentarius*, and *Li. fermentum*, in decreasing order of abundance, are encountered. Additionally, *S. cerevisiae, C. glabrata, C. humilis, K. unispora*, and *W. anomalus* are also dominant yeast species identified in rye sourdough (Weckx et al., 2010; Novotni et al., 2020; Rocha and Malcata, 2012; Rocha, 2011).

Maize flour is used to produce some traditional sourdoughs, such as *Pozol* from Mexico, *Masa Agria* from Colombia (Ben Omar and Ampe, 2000; Chaves Lopez et al., 2016) and *Broa*, an ancestral Portuguese sourdough bread made of regional maize and rye flours (Rocha and Malcata, 1999, 2012, 2016a, b; Rocha, 2011). *Enterococcus saccharolyticus, Lc. casei, Lactobacillus delbrueckii, Li. fermentum, Lp. plantarum,* and *Streptococcus bovis* were identified in a study of Nuraida et al. (1995). *L. alimentarius, Lc. casei, L. delbrueckii, Lp. plantarum, Lactococcus lactis,* and *S. suis* were also detected by Ben Omar and Ampe (2000). *Masa Agria* is a kind of naturally fermented maize dough and popular in the traditional gastronomy of Colombia. *Lp. Plantarum* and *Acetobacter fabarum*, followed by *Li. Fermentum, L. vaccinostercus,* and *Pediococcus argentinicus* are the most common species in *Masa Agria* (Chaves Lopez et al., 2016). Figure 10.2 shows the most common LAB species found in sourdoughs from wheat, rye, and maize flours.

Rice is one of the main ingredients of gluten-free products and requires sourdough fermentation to improve palatability and nutritional value (Moroni et al., 2009). The dominant microbiota found in rice sourdoughs using spontaneous fermentation is *Lc. paracasei, L. paralimentarius, L. perolens, L. Spicheri,* and *S. cerevisiae* according to the study of Meroth et al. (2003).

Barley (*Hordeum vulgare* L.), oat (*Avena sativa* L.), millet (*Pennisetum glaucum*), and sorghum (*Sorghum bicolor* L. Moench) have gained enormous interest in recent years due to the healthy food that can be manufactured with these cereals (De Vuyst et al., 2017; Coda et al., 2014). Barley flour may also need sourdough fermentation due to its negative effects on bread technological characteristics. *Lactobacillus* spp., *Leuconostoc* spp., and *Weissella* spp. were detected in barley sourdough by Harth et al. (2016) under different experimental conditions. *Li. fermentum, Lp.*

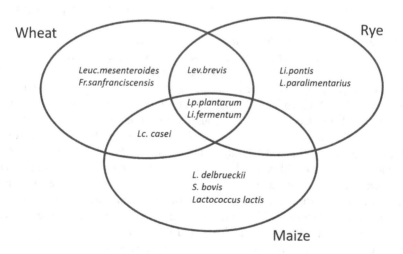

FIGURE 10.2 Common lactic acid bacteria (LAB) species found in sourdoughs from wheat, rye, and maize flours.

Plantarum, and *Lev. brevis* were dominant species isolated after ten backsloppings of barley sourdough. When using eight backsloppings in bread making conditions, *Leuconostoc citreum*, *Leu. mesenteroides*, *Weissella confusa*, and *W. cibaria* were detected. Among yeast, *S. cerevisiae* dominates both laboratory and bakery sourdough bread making conditions (Harth et al., 2016). Zannini et al. (2009) employed a co-cultured starter consisting of selected LAB and yeast from wheat sourdoughs and reported the dominance of *Lev. brevis*, *Lp. Plantarum*, and *S. cerevisiae* in barley sourdough.

The potential health benefits of oat, millet, and sorghum make these flours compatible with the production of bread that has been adjusted to the modern consumer – who increasingly demands food products with improved nutritional quality and free of additives. In a study from Hüttner et al., (2010), *Leu. argentinum*, *P. pentosaceus*, *W. cibaria*, and *L. coryniformis* prevail in oat sourdough. In millet sourdough, *Li. fermentum*, *L. helveticus*, and *Li. pontis* are encountered in descending order of abundance (Vogelmann et al., 2009). *Lc. casei*, *L. coryniformis*, *L. Parabuchneri*, and *L. harbinensis* are found predominantly in *ting*, which is a traditional fermented sorghum product used in Botswana. Additionally, these LAB are not detected frequently in wheat or rye sourdoughs (Sekwati-Monang et al., 2012). *Li. fermentum* and *P. pentosaceus* were also reported in fermented sorghum dough bread and spontaneous fermentation of the sorghum flour, respectively (Hamad et al., 1992; Mohammed et al., 1991).

Sourdoughs produced in bakeries may sometimes contain different ingredients such as fruits, flowers, or rumen cuts to stabilize sourdough microbiota. Sourdough microbiota start with plant materials (mustard flowers, speedwell flowers, myrtle berries, etc) that rapidly develop fermentation and are dominated by *Fr. sanfranciscensis*, *Lp. plantarum*, *L. graminis* or *L. rossiae*, and *S. cerevisiae*. *Acetobacter* spp. was identified only in sourdoughs using apple flowers or apple pulp as ingredients (Ripari et al., 2016).

Across the globe, pseudocereals, such as amaranth, buckwheat, and quinoa, have attracted increased attention, since they are gluten-free and have a high nutritional profile. *Lp. plantarum* RTa12, *L. sakei* RTa14, and *P. pentosaceus* RTa11 were detected as dominant strains in a spontaneous fermentation of amaranth flour from India, Peru, Mexico, and Germany (Sterr et al., 2009). The microbiota of quinoa sourdough produced by spontaneous fermentation consists of *Lactobacillus*, *Enterococcus*, *Leuconostoc*, *Lactococcus*, *Pediococcus*, and *Weissella* according to a study from Franco et al. (2020). In a study from Rizzello et al. (2016), *Lp. plantarum* and *L. rossiae* were identified in quinoa flour, while *Lp. plantarum* and *P. pentosaceus* were detected in quinoa dough (fermented at 30°C for 16 h). The spontaneous microbiota of buckwheat sourdough is composed of *L. crispatus*, *Li. Fermentum*, *L. gallinarum*, *L. graminis*, *Lp. Plantarum*, *L. sakei*, *L. vaginalis*, *Leuc. Holzapfelii*, *P. Pentosaceus*, and *W. cibaria* for LAB, and *K. barnettii* for yeast. *Li. pontis* was identified as a dominant species in teff sourdough besides *Li. fermentum*, *L. gallinarum*, *L. vaginalis*, *Leuc. holzapfelii*, and *P. Pentosaceus*, whereas *C. glabrata* and *S. cerevisiae* yeast species prevailed in teff sourdough (Moroni et al., 2011).

The enzymatic, microbiological, nutritional, and physicochemical quality as well as the type of cereal flours used in bread making is of the greatest importance.

Cereal flours as a key raw material in bread making is the main source of nutrients and endogenous enzymes – which have an influence on microbial stability of the developed sourdoughs (Novotni et al., 2020; Rocha and Malcata, 2012; Rocha, 2011). The quality of flour is not a controllable parameter, but its monitoring is an important part of the bread making process since it determines the microbial stability of mature sourdoughs (De Vuyst et al., 2017) as well as the technological traits of the final bakery products. Common wheat and durum wheat beside spelt, einkorn, kamut, and/or rye flours are used to prepare most sourdoughs. Other cereals (maize, barley, oat, etc) and pseudocereals (quinoa, amaranth, and buckwheat) are also commonly used in sourdough bread making. On the other hand, the composition of type I sourdoughs may be changed to type II ones when using other cereals and pseudocereals. This may result in the inhibition of *Fructilactobacillus sanfranciscensis* prevalence (Rocha and Malcata, 2016a; Novotni et al., 2020).

The inhibition of some microorganisms is another typical phenomenon observed in sourdough fermentations. Prevention of microbial spoilage caused by molds and undesirable bacteria is also performed by sourdough fermentation. Banishment of *Bacillus subtilis, B. licheniformis, B. amyloliquefaciens,* and *B. velezensis* as ropey microorganisms is carried out via acidification in sourdough fermentation by LAB. *Bacillus* endospores as seed endophytes can be found in baking ingredients, especially flours (Rocha and Malcata, 1999, 2012, 2016a, b). Additionally, another effect of acidification by LAB in sourdough fermentation is the disappearance of (among other microorganisms) the undesirable Gram-negative rods, such as *Acinetobacter, Comamonas, Enterobacter, Erwinia, Pantoea, Pseudomonas,* and *Sphingomonas* (Dinardo et al., 2019; Rocha and Malcata, 1999).

Discoloration, decomposition, off-flavor, and activation of some pathogenic and allergenic effects are some types of food spoilage caused by mold growth that threatens food safety. LAB belonging to the genera of *Lactobacillus, Lactococcus, Pediococcus,* and *Leuconostoc* show antifungal activity and, concomitantly, increase the shelf life of sourdough bread. Acetic acid produced by heterofermentative LAB has primarily more effective fungistatic activity than lactic acid. Production of organic acids, such as acetic acid, should be controlled in bread dough before baking since shelf life extension is fundamentally attributed to fermentation conditions and ecological parameters of sourdough (Novotni et al., 2020; Rocha and Malcata, 2012). Furthermore, cereal and pseudocereal flours as the main ingredient (or ingredients) of sourdough have a significant effect on sourdough microbiota. Alteration of these ingredients and other ecological factors further change sourdough microbiota. As a result of metabolic activities of this microbiota, sourdough has a nutritive, flavoring, and shelf life-extending impact. It also contains metabolites that ensure food safety.

10.3 MICROBIOTA THROUGHOUT SOURDOUGH FERMENTATION

Sourdough fermentation is a distinct natural process when compared to other food fermentations. Yeast and LAB are the main microbiota found in mature sourdoughs. However, cereal flours contain a diversity of undesirable microorganisms including those from the genera *Acinetobacter, Pantoea, Pseudomonas, Comamonas,*

Enterobacter, Erwinia, and *Sphingomonas*. The dominant microbiota in sourdoughs depends on the flour or flour mixtures and many other factors, such as fermentation time and temperature, the use or not of a mother dough (a piece of sourdough kept between batches to be used as natural starter culture), fermentation quotient, number of backslopping steps, etc (Novotni et al., 2020; Rocha, 2011; Rocha and Malcata, 2012). In general, undesirable genera of microorganisms can be mostly inactivated after one day of refreshment (*ca.* fermentation of 8 h at 25°C) except for the Enterobacteriaceae family. The phylum *Firmicutes*, including LAB, become dominant after one day of refreshment, although they appear in low to intermediate viable count levels in the cereal flours. *Firmicutes* still dominate after five and ten days of propagation according to both DNA and RNA based analyses (Ercolini et al., 2013). Weckx et al. (2010) also investigated the microbial ecosystem of rye sourdoughs during backslopping and stated that the major changes occurred from day 0 to day 4. During the first two days, *Lactococcus lactis, Leuconostoc citreum*, and *Weissella confusa* were detected. After the second day of fermentation, *L. curvatus* and *Pediococcus pentosaceus* were detected and dominated the sourdough ecosystem on the third day. From day 3 onward, *Li. fermentum* and *Lp. plantarum* were found.

The main genera of microorganisms in rye dough before fermentation are *Pantoea* (25.5%), *Pseudomonas* (19.7%), *Weissella* (19.7%), and *Acinetobacter* (8.3%). However, it was reported that *Comamonas, Sphingomonas, Staphylococcus, Erwinia, Chryseobacterium*, and *Luteibacter* were also found to have low incidence. *Weissella* (94.3%) started to become dominant in rye dough after the first fermentation and caused remarkable changes in the bacterial profile. The prevalence of *Weissella* is almost constant during further sourdough propagation steps. After ten days of propagation, *Weissella* is still encountered with high relative abundance (55.6%), beside the presence of *Lactobacillus* (32.5%) and *Pediococcus* (6.3%) (Ercolini et al., 2013). A fermentation with ten backsloppings (once every 24 h) at 23°C were performed by Vrancken et al. (2011) for the type I wheat sourdough propagated in the laboratory, and *Leuconostoc citreum* was detected as the dominant species. *Li. fermentum* prevailed in sourdoughs fermented at 30°C and 37°C with backslopping every 24 h. In the longer backslopping times (every 48 h at 30°C), the combination of *Li. Fermentum* and *Lp. Plantarum* were observed.

Durum wheat dough before fermentation has also shown almost the same microbial diversity as rye dough. *Acinetobacter, Pantoea, Pseudomonas, Chryseobacterium, Comamonas, Staphylococcus, Erwinia*, and *Sphingomonas* were found in durum wheat dough before fermentation. Lactic acid bacteria such as *Weissella, Lactococcus, Leuconostoc*, and *Lactobacillus* were detected after the first fermentation. Also, *Enterobacter* and Enterobacteriaceae family are encountered, but they disappear after five days of propagation. *Lactobacillus* (56.4%) prevailed at ten days with a high proportion besides *Leuconostoc* (18.7%), *Lactococcus* (11.1%), and *Weissella* (8.8%) (Ercolini et al., 2013).

Similar microbial dynamics were observed between *T. aestivum* wheat, rye, and durum wheat sourdoughs. Moreover, LAB can be identified in *T. aestivum* wheat dough before fermentation unlike the other doughs (rye and durum wheat). From the early stage of propagation, *Weissella* spp. is the dominant genus in all three sourdoughs (*T. aestivum*, *T. durum*, and rye). Exceptionally, the *L. sakei* group and *Leuconostoc* spp. prevails in wheat sourdough in the late stage of the propagation step (Ercolini et al., 2013). Generally, plant-associated microbiota (Enterobacteriaceae including *Enterobacter* and *Cronobacter*) initiate sourdough fermentation, and these are replaced by *Leuconostoc* and *Weissella* spp. until *Lp. Plantarum*, *Li. Fermentum*, and *Lev. brevis* dominate (Novoni et al., 2020).

In addition to the above type of sourdoughs mentioned, Rocha and Malcata (2016a) studied *broa* sourdough, made of maize and rye flours, in terms of microbial ecology dynamics throughout a long-term sourdough fermentation (Rocha and Malcata, 2016a) and storage of maize and rye flours and mother dough (Rocha and Malcata, 2016b). *Broa* sourdough contains molds, Gram-negative rods, endospore-forming and endospore-nonforming Gram-positive rods, and catalase-positive and negative Gram-positive cocci, beside *Lactobacillus* and yeast in the different stages of the sourdough bread making process. Throughout spontaneous *broa* sourdough fermentation, the presumptive *Bacillus* genus was unveiled as ubiquitous. After kneading, large quantities of presumptive *Staphylococcus* spp. Were detected, whereas presumptive yeast, *Lactobacillus* and *Bacillus*, were found at low levels. Yeast and *Lactobacillus* became dominant by the end of fermentation and inhibited undesirable rods and *Staphylococcus*.

In the particular studies of *broa* sourdough, high viable counts of presumptive *Staphylococcus* and Enterobacteriaceae were observed, while low viable counts of presumptive yeast, Bacillaceae and *Lactobacillus* were detected after manual kneading and under fermentation conditions. Moreover, high total viable counts were reached by two days of fermentation, although presumptive *Staphylococcus* vanished by this time period. Three days of fermentation caused the disappearance of presumptive Enterobacteriaceae and *Clostridium*, whereas presumptive Pseudomonadaceae disappeared after seven days, and presumptive Gram-positive catalase-negative cocci after 14 days of long-term fermentation. Moreover, significant growth was observed for presumptive Pseudomonadaceae, *Lactobacillus*, endospore-forming Gram-positive rods, and Gram-positive catalase-negative cocci within the first 24 h. *Bacillus* and *Lactobacillus* typically appeared under increased aeration (Rocha and Malcata, 2016a). As already mentioned, the typical ratio between LAB and yeast in mature wheat and rye sourdoughs is 10:1, and this case was also observable in *broa* sourdough. Presumptive *Bacillus* was observed to multiply and largely stay at substantial levels throughout the long-term fermentation. This can be due the fact that *Bacillus* spp. may grow in pH value from 2 to 11, which explains their ubiquitous presence in nature (Rocha and Malcata, 2012, 2016a, b; Rocha, 2011).

Although interest in the role of *Bacillus* in spontaneous sourdough fermentation has waned, these organisms may be present at important levels in sourdoughs. In addition to mold spoilage, some species of *Bacillus* (e.g. *Bacillus subtilis*, *Bacillus coagulans*, and *Bacillus cereus*) are the reason for food spoilage, such as ropiness in bread and related foods. Furthermore, some strains of *B. subtilis* and *B. licheniformis* may cause foodborne illness if present to levels over 10^5 CFU/g. The consumption of ropey bread is not very likely due to the slimy appearance of the crumb but, still, loaves containing high levels of those species and no apparent ropiness may cause health symptoms, such as diarrhea and vomiting. Thus, special caution is required for their potential risks (Rocha and Malcata, 2016a).

The typical microbiota evolves during sourdough fermentation. Spontaneous sourdough fermentation starts with a complex microbial diversity mainly from flour or flours and becomes less diverse as microbial synergistic and competing phenomena happens. At the end of fermentation, sourdoughs become more like each other in terms of LAB and yeast diversity and number. The presence of spore-forming bacteria in spontaneous sourdough fermentation should also be kept in mind in terms of food safety.

10.4 PRODUCTION OF MICROBIAL METABOLITES: THEIR FUNCTIONS AND PROPERTIES

Sourdough contains several microbial metabolites that contribute to the unique taste, aroma, texture, and nutritional value found in the final bakery products (Novotni et al., 2020; Rocha, 2011; Rocha and Malcata, 2012). The most significant volatile organic compounds, such as aldehydes, ketones, and organic acids, are produced by the microbial starter cultures or the autochthonous microbiota engaged in spontaneous sourdough fermentations. Sourdough LAB contribute to a better digestibility of the cereal flours by reducing the levels of long-chain carbohydrates and stimulating the degradation of indigestible poly- and oligosaccharides (Turpin et al., 2011). They also enhance the uptake of trace elements, such as iron, zinc, and calcium (Chaves-López et al., 2020) and improve bread preservation and prolong its shelf life. Bio-preservatives naturally produced by LAB include antibacterial, antifungal, and antioxidant substances and can be classified into compounds with low and high molecular weight (Mw) (Šušković et al., 2010). The first group includes organic acids, alcohols and derivatives, and gasses; while the second encompasses bacteriocins, bacteriocin-like inhibitory substances (BLIS), and cyclic peptides. In this way, LAB combine the sensory improvement of bread with functional effects, such as consumer utility, extended shelf life, and microbiological safety. Figure 10.3 schematizes generically the metabolic pathways of LAB during fermentation of cereal flour constituents.

10.4.1 LOW-MOLECULAR-WEIGHT METABOLITES WITH ANTIMICROBIAL ACTIVITY: FUNCTIONAL PROPERTIES BESIDES FLAVOR IMPROVEMENT

A comprehensive study of Bartkiene et al. (2020) showed that sourdough is a source of LAB with exceptional antimicrobial and antifungal properties. Investigation of the LAB microbiota of spontaneous rye sourdough revealed 13 LAB species of particular industrial interest. Most notably, certain strains of *Lp. plantarum*, *Lc. casei*, *Latilactobacillus curvatus*, and *Lc. paracasei* showed marked inhibitory effect on the growth of 15 different strains of pathogenic bacteria, as well as antifungal activity against *Aspergillus nidulans*, *Penicilium funiculosum*, and *Fusarium poae*. Since

Sourdough as a Source of Functional Microorganisms 279

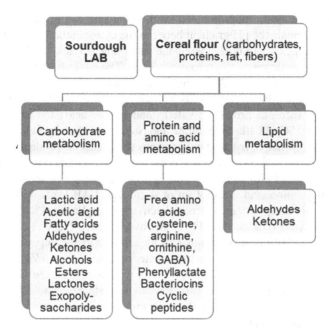

FIGURE 10.3 Carbon flow in sourdough lactic acid bacteria (LAB) and the resulting metabolic products.

heterofermentative LAB usually display a wider spectrum of antifungal activity, the authors postulated that the antimicrobial compounds in sourdough are metabolites produced by LAB during their carbohydrate metabolism. Several other studies show the potential benefits of LAB resulting from their antimicrobial and antioxidant capabilities (Trakselyte-Rupsiene et al., 2021; Păcularu-Burada et al., 2021, 2020; Zokaityte et al., 2020; Küley et al., 2020).

Sourdough LAB assimilate cereal carbohydrates (starch, maltose, sucrose) assisted by the action of flour amylases and a wide range of amylolytic enzymes (Petrova et al., 2013; Velikova et al., 2016). Directing the substrates toward homo- or heterofermentative pathways depends on the LAB species and predetermines the final products from sourdough fermentation. Examples of facultative heterofermentative species are *Lp. plantarum*, *Companilactobacillus alimentarius*, *Lc. casei*, and *Lc. paracasei*. Obligate heterofermentative lactobacilli are *Fr. sanfranciscensis*, *Li. pontis*, *Lev. Brevis*, and *Li. fermentum*. Most sourdoughs are dominated by a combination of obligate and facultative heterofermentative taxa, as common biochemical routes for hexose conversion in sourdoughs are the glycolytic and phosphogluconate pathways. The external electron acceptors (fructose or citrate) enhance energy generation and increase sourdough LAB competitiveness by alternative substrates uptake (Gobbetti et al., 2005; Zaunmüller et al., 2006) or by switching on the arginine deiminase pathway (De Angelis et al., 2002). In all cases, however, the main product of the fermentation is lactic acid (LA), a metabolite well-established by its antimicrobial activity. In its protonated form, lactic acid can permeate the plasma membrane and impair the pH gradient between the cytosol (alkaline) and the external environment (acidic), thus dissipating the membrane potential and destroying the cells of the pathogenic species (Arena et al., 2018). The total titratable acidity (TTA) in sourdough could reach 3.8–3.9, or up to 122 H^+ equivalents per kg (Guerzoni et al., 2011), thanks to other organic acids with antimicrobial effects, *viz.* acetic (AA), butyric, formic, succinic, propionic, valeric, and caproic acids and branched short-chain fatty acids (SCFA), like isobutyrate and isovalerate (Fazeli et al., 2004; Birt et al., 2013). The presence of fructose or pentosans stimulates the production of acetate in a ratio of 1:2.5 acetate: lactate, which is the optimal fermentation quotient (FQ) in sourdough (Röcken, 1996), while succinate is a product of the citrate pathway.

The specific amino acid metabolism of sourdough LAB leads to the formation of unique organic acids with antimicrobial activity: phenyllactic (PLA), indolelactic (ILA), 4-hydroxyphenyllactic (OH-PLA), and hydroxyphenylacetic (OH-PAA) (Van der Meulen et al., 2007). PLA is known as an antimicrobial compound with broad microbial activity against Gram-positive bacteria such as *Staphylococcus aureus*, *Enterococcus faecalis*, and *Bacillus cereus*, as well as against the Gram-negative *Salmonella enterica*, *Escherichia coli*, *Providencia stuartii*, and *Klebsiella oxytoca* (Mu et al., 2012). Importantly, PLA has high antifungal activity against yeast – including *Candida pulcherrima*, *C. parapsilosis*, and *Rhodotorula mucilaginosa* – and a wide range of mold species (i.e. filamentous fungi, including producers of mycotoxins) in cereal flours – such as *Asp. Ochraceus*, *P. roqueforti*, and *P. citrinum*, with minimum inhibitory concentration (MIC) value of 45 mM at pH 4.0 (Ström et al., 2002; Schwenninger et al., 2008). Lavermicocca et al. (2000) reported that PLA and OH-PLA produced by sourdough-isolated *Lp. plantarum* strains have anti-mold activity against *Aspergillus*, *Penicillium*, *Eurotium*, *Endomyces*, and *Monilia*, delaying the fungal contamination of bread with seven days of storage at room temperature. After examination of 23 different fungal strains, Lavermicocca et al. (2003) estimated the minimum fungicidal concentration of PLA. The concentration of PLA, which inhibited the mold's growth with 90% was 7.5 mg/ml, while fungicidal effect was exerted by 10 mg/ml PLA. These results are similar to the key effect of the caproic acid produced by *Fr. sanfranciscensis* CB1 against bread spoilage by *Fusarium*, *Penicillium*, *Aspergillus*, and *Monilia* (Corsetti et al., 1998). PLA, ILA, and OH-PLA produced by *Lp. plantarum* and *Lentilactobacillus buchneri* have been shown to inhibit the growth of *Penicillium nordicum* and the synthesis of mycotoxins (Guimarães et al., 2018). Other low-molecular-weight substances with inhibitory activity against fungi are benzoic acid, methylhydantoin, and mevalonolactone (Niku-Paavola et al., 1999).

Fermentation of cereal fibers results in the formation of indole-3-propionic acid (IPA), active against *Mycobacterium tuberculosis* (Kaufmann, 2018). IPA acts also as a "chemical chaperone" suppressing stress-induced cell death of neurons (Mimori et al., 2019), decreases the risk of type-2 diabetes (De Mello et al., 2017), acts as a neuro-protector in therapy for Alzheimer's disease because of its activity against a variety of toxins (Bendheim et al., 2002; Silva et al., 2019), and is a potent scavenger of hydroxyl radicals (•OH) (Zhang et al., 2016).

Other compounds with antioxidant activity in sourdough are phenolic acids. They can be derivatives of hydroxybenzoic acid (gallic, p-hydroxybenzoic, protocatechoic, syringic, and vanillic acids), and of hydroxycinnamic acid (caffeic, hydrocaffeic, ferulic, hydroferulic, p-coumaric, phloretic, and sinapic acids). They all function as free radical scavengers, thus increasing the overall sourdough antioxidant activity (Dordevic et al., 2010; Adebo and Medina-Meza, 2020).

γ-Aminobutyric acid (GABA) is an amino acid with valuable physiological functions, which is produced by sourdough LAB, such as *Lactococcus lactis* (Diana et al., 2014), *Lev. Brevis* (Huang et al., 2007), *Lc. Paracasei* (Komatsuzaki et al., 2005), *Furfurilactobacillus rossiae*, and *Comp. farciminis* (Venturi et al., 2019). GABA is an inhibitory neurotransmitter with a tranquilizing effect, a regulator of insomnia and depression, which acts efficiently just in the amounts provided by sourdough bread (Okada et al., 2000).

Among volatiles with antimicrobial activity are alcohols, diacetyl, acetaldehyde, acetoin, 2,3-butanediol, carbon dioxide and hydrogen peroxide (H_2O_2) (Tejero-Sariñena et al., 2012; Pétel et al., 2017; Pizarro et al., 2017). Carbonyls have activity against specific pathogens, such as *Escherichia coli*, *Listeria monocytogenes*, and *Staphylococcus aureus* (Lanciotti et al., 2003), while gasses play a role in extending the general bread durability. Carbon dioxide inhibits the enzymatic decarboxylation and increases the membrane permeability; whereas hydrogen peroxide damages the cellular structures by its oxidizing effect and disarranges the membrane redox potential (Reid, 2008).

10.4.2 Bioactive Peptides with Antimicrobial Activity: Bacteriocins and BLIS

Until recent years, protein metabolism in sourdough microbiota was primarily considered in the light of relieving celiac disease by breaking down the protein fractions of gliadins and glutenins in wheat, particularly due to the high proteolytic activity of LAB (Pepe et al., 2003; Novotni et al., 2020). Indeed, the anti-inflammatory effect of LAB probiotics in the gut has been linked with their ability to detoxify gluten. Other bioactive peptides in sourdough have angiotensin-converting enzyme (ACE)-inhibitory activity. According to Gänzle et al. (2008), their concentration in bread prepared with yeast dough is 100-fold lower than in sourdough bread. Sourdough LAB cause more than 50% inhibition of the ACE-activity and from 10% to 90% decrease of the amylase activity in wheat flour (Diowksz et al., 2020) – which indicates that sourdough bread consumption may alleviate cardiovascular and diabetic disorders.

However, the most important functional compounds produced by LAB in sourdough are the bacteriocins, the heat- and acid-resistant oligopeptides with antimicrobial activity against foodborne pathogenic bacteria and fungi. Besides, they can prevent the formation of mycotoxins. Bacteriocins are ribosomally synthesized peptides harmless to the human body, hence acting as effective bio-preservative agents. The most recent classification divides bacteriocins into four classes according to their nucleotide sequence and amino acid composition, molecular weight, thermal stability, and mode of action: (i) molecules less than 5 kDa in size, heat-stable, and lanthionine-containing (lantibiotics, labyrinthopeptins, and sanctibiotics); (ii) bacteriocins below 10 kDa in size, heat-stable, and non-lanthionine-containing (pediocin-like, dimers, circular, and linear, non-pediocin-like bacteriocins); (iii) proteins with molecular weights higher than 30 kDa, heat-sensitive (helveticin M, helveticin J, and enterolysin A, produced by *L. crispatus, L. helveticus,* and *E. faecalis*); and (iv) protein/lipid/carbohydrate complexes named bacteriolysins (leuconocin S and lactocin 27). Most of the sourdough bacteriocins considered good candidates to replace chemical preservatives belong to class II, as pediocin-like bacteriocins are the most widespread. Class IIa are active against *Listeria monocytogenes,* thanks to the specific structure including conserved tyrosine-glycine-asparagine-glycine-valine (YGNGV) motif, and disulfide bonds in the N-terminal region. Moreover, these Class IIa bacteriocins could be degraded by gastrointestinal proteases (Kumariya et al., 2019). Bacteriocin-encoding operons are under inductive expression control, thus requiring the presence of auto-inducer peptides, and the switching on of a two-component regulatory system. As the major reason for the antibacterial activity of LAB potentially originates from bacteriocin production, the studied probiotic sourdough strains are often tested for the presence of curvacin A, plantaricin A, pediocin, and sakacin P genes by polymerase chain reaction (PCR). Demirbaş et al. (2017) reported that *Leuc. mesenteroides, W. paramesenteroides, W. cibaria, Leuc. pseudomesenteroides, Lp. paraplantarum, L. curvatus, Fur. rossiae, Lev. brevis, Comp. paralimentarius,* and *Leuc. citreum* strains were positive for plantaricin A gene, as well as *L. plantarum* strain SC-9. The last strain, two strains of *Leuc. citreum* and *La. graminis* were also positive for the sakacin P gene. The best characterized bacteriocins produced by sourdough lactobacilli are bavaricin, plantaricin, and pentocin. Bavaricin A was one of the first bacteriocins isolated from sourdough LAB, from *L. bavaricus* MI401, which was able to synthesize it at temperatures from 4°C to 30°C. Bavaricin is composed of 41 amino acids and has a molecular weight of 3500–4000 Da. It displayed antimicrobial activity against closely related species, and this fact, coupled to the heat resistance and alkaline treatment sensitivity, proved its bacteriocin nature. Bavaricin was purified by ammonium sulfate precipitation, ion exchange, and reverse-phase chromatography. Interestingly, it had a bactericidal mode of action on 90% of the tested *Listeria monocytogenes* strains. The mechanism of action was clarified by studying another bacteriocin of the same class, bavaricin MN produced by *Latilactobacillus sakei*. Kaiser and Montville (1996) showed that bavaricin reduces the membrane potential (Δp) of energized *Listeria monocytogenes* cells in a concentration-dependent fashion, as a concentration of 9.0 μg/ml decreases Δp by 85%. However, bavaricin activity is inhibited at high concentrations of sodium chloride.

Plantaricin is another promising natural bio-preservative produced by *Lp. plantarum* strains. Isolated, purified, and characterized are plantaricin ST31 (isolated from sourdough), plantaricin C19 (3.8 kDa), plantaricin ASM1 (5045.7 Da), plantaricin Y (4.2 kDa), plantaricin 163 (3.5 kDa), bacteriocin B391, bacteriocin JLA-9, plantaricin ZJ2008, and plantaricin LPL-1 (Wang et al., 2018a). The molecular weight of the newly studied plantaricin, produced by *Lp. plantarum* strain LPL-1 is

4.3 kDa. This bacteriocin is composed of 41 amino acids and contains a conserved YGNGV motif. Plantaricin LPL-1, as the other bacteriocins of Class IIa, is highly thermostable (121°C, 20 min), surfactant-stable, and possesses bactericidal activity against foodborne spoilage and pathogenic bacteria. Its mode of action is membrane permeabilization.

The BLIS of *Fr. sanfranciscensis* strain C57 is active against *Listeria monocytogenes*. It is produced in pH diapason 4.0 to 5.0 (in the stationary phase of microbial growth) and is chromosomally encoded. Pentocin is a fungistatic BLIS formed by *Lp. pentosus* TV35b (Okkers et al., 1999). Besides preventing the growth of *Candida albicans*, it inhibits *Clostridium sporogenes, Cl. Tyrobutyricum, Listeria innocua*, and *Propionibacterium acidipropionici*. The molecular size of pentocin TV35b is 3929.63, and it consists of 33 amino acids. The antagonistic properties of pentocin 31-1 produced by *Lp. Pentosus* strain 31-1 (5592.2 Da) against *Listeria* and those of pentocin JL-1 (2987.23 Da) against multidrug-resistant *Staphylococcus aureus* have also been discussed recently. The production of such substances, however, seems to be quite rare. In fact, Corsetti et al. (2004) verified that, after screening of 437 strains from 70 tests, only 5 BLIS producers were detected (*Lp. pentosus* 2MF8 and 8CF, *Lp. plantarum* 4DE and 3DM, and *Lactobacillus* sp. CS1). Other bacteriocin-producing sourdough LAB are *L. amylovorus* DCE 471 and *Lactococcus lactis* M30 that were found to produce BLIS effective against *Bacillus, Staphylococcus*, and *Listeria* spp. (Messens et al., 2002). Other strains can modulate the microbial counts of selected sourdough starters. For example, *Li. reuteri* strain LTH2584 increases the microbial counts of *Li. reuteri* in sourdough by the formation of reutericyclin, while BLIS synthesis of *Lactococcus lactis* M30 contributes to the stable persistence of *Fr. sanfranciscensis* and reduces the growth of other LAB (Gänzle and Vogel, 2003).

10.4.3 LAB AND EPS PRODUCTION FOR SOURDOUGH BREAD WITH IMPROVED QUALITY

EPS are a heterogeneous group of polymers that are secreted by LAB and perform diverse functions in more than 30 sourdough species (*Lc. casei, L. acidophilus, Lev. brevis, Lat. curvatus, Li. reuteri, La. sakei, Li. fermentum, Lentilactobacillus parabuchneri, Fr. sanfranciscensis, Lentilactobacillus diolivorans, L. delbrueckii, Leuc. mesenteroides* subsp. *mesenteroides, Leuc. citreum, Leuc. mesenteroides* subsp. *dextrtanicum, Pediococcus damnosus, P. parvulus, Lactococcus lactis* subsp. *cremoris, Streptococcus mutans, Str. downei, Str. sobrinus, Str. salivarius, Str. gordonii, W. cibaria*, etc. (Zeidan et al., 2017; Petrova et al., 2021). LAB-derived EPS play a major role in sourdough fermentation, rheology, and bread quality. They have been the subject of increasing scientific interest in recent years as a new generation of bread improvers replace the currently employed hydrocolloids (Galle and Arendt, 2014; Gobbetti et al., 2020). EPS have a positive effect on bread quality as confirmed by a good positive correlation between sensory analysis and instrumental measurements (Di Monaco et al., 2015). Besides, EPS prolong the shelf life of sourdough bread, an issue of economic importance, as the losses from bread staling worldwide are considerable (Taglieri et al., 2021). A microbial starter culture composed of *Lactococcus lactis* 95A and *Lat. curvatus* 69B2, both potent EPS producers, applied to wheat sourdough resulted in some very durable bread indeed: its springiness remained unchanged over 21 days, quite unlike the samples with EPS non-producers (*Leuconostoc lactis* 68A and *Lat. curvatus* 68A2) in which decay was pronounced (Di Monaco et al., 2015).

LAB microbiota of sourdough is extremely complex and so is the EPS profile associated with it. Most EPS formed by sourdough LAB are homopolysaccharides, glucans (reuteran, alternan, dextran, and mutan), and fructans (levan and inulin). The enzymes participating in EPS synthesis are glucansucrases of the GH family 70 (forming α-Glucans) and fructansucrases of the GH family 68 (forming β-fructans). The last enzymes, catalyzing the synthesis of inulin and levan, are divided into inulosucrase and levansucrase, as most enzymes are characterized to date, and produce EPS of the levan-type. Very few strains of *Lactobacillus*, *Leuconostoc*, and *Streptococcus* spp. were reported to produce inulin or to possess inulosucrase-encoding genes (Zannini et al., 2016). Both types of enzymes use the energy from hydrolysis of the glycosidic bonds in sucrose to catalyze glycosyl residue transfer. Fructansucrase may also use raffinose as a substrate in addition to sucrose (Gänzle et al., 2007). One of the most active fructans producers is *Fr. sanfranciscensis*. This species forms prebiotic fructooligosaccharides (FOS), levan, and kestose during its growth in wheat and rye sourdoughs, as well as the rare heterooligosaccharides, erlose and arabsucrose (Tieking et al., 2005). The cereal isolate *La. curvatus* produces a heteropolysaccharide composed of galactosamine, galactose, and glucose in a ratio of 2:3:1, respectively (Van der Meulen et al., 2007).

β-Glucans are essential EPS in sourdough because they are highly fermentable by the microbiota of the gastrointestinal tract (GIT) and can enhance both the growth rate and lactic acid production. *L. acidophilus* LA5, *Lp. plantarum* WCFS1, *Lp. plantarum* CETC 8328, and *Li. fermentum* CECT 8448 grow better in the presence of barley β-glucans. Duenas-Chasco et al. (1997) isolated *Pediococcus damnosus* strain 2.6 producing (1–3)-β-D-glucan. Several studies discussed the therapeutic uses of these EPS, but since the biological effects of β-D-glucans are influenced by the branching degree, chain length, and tertiary structure, the subject needs further investigation in the future.

Among 41 LAB strains isolated from durum wheat sourdoughs for *cornetto*, a type of traditional bread from Southern Italy, six species from three genera were identified: *Lp. plantarum* (49%), *Leuc. mesenteroides* (17%), *La. curvatus* (15%), *Lp. paraplantarum* (12%), *Weissella cibaria* (5%), and *Lp. pentosus* (2%). Their ability to produce EPS was found to depend on the substrate and the medium (higher in liquid medium). All strains of *Leuc. mesenteroides* and *W. cibaria* produced dextran from sucrose (but not from glucose or maltose). The highest EPS production (140–297 mg/l) was observed by eight strains of *Lp. plantarum* and *Lp. paraplantarum* in liquid medium with maltose as a carbon source. Interestingly, there was no correlation between EPS concentration and viscosity (Zotta et al., 2008).

Dextran is the most widely used EPS in bakery, generally regarded as safe (GRAS), and considered to have nutritional properties similar to starch. Industrially, it is produced from sucrose by *Leuc. mesenteroides* subsp. *mesenteroides* and is applied as a sourdough ingredient in concentrations of up to 5%, providing an improvement of the softness, crumb texture, and loaf volume of bread (Zannini et al., 2016). LAB-derived dextran has been proposed as a promising alternative to current hydrocolloids to improve the quality of wheat bread enriched with faba bean flour. However, the effect appears to be species-specific. When wheat flour was replaced with 43% faba bean sourdough fermented by the dextran-producing *Weissella confusa* VTT E–143403 (E3403), the result was a bread with increased specific volume and reduced crumb hardness. The same positive effects were not reproduced with *Leuc. pseudomesenteroides* DSM 20193, probably because of its lower dextran level (3.6 *versus* 5.2% for *W. confusa*) and higher acidity (Wang et al., 2018b).

Consumption of EPS-rich sourdough provides the opportunity to improve human gut health *via* prebiotic fibers. Besides FOS, *Fr. sanfranciscensis* synthesizes levan, while *Leuc. mesenteroides* synthesizes isomaltooligosaccharides (IMO). IMO have well-described prebiotic properties *in vitro*, are not fermented by yeast, and remain stable in bread (Galle and Arendt, 2014). Besides prebiotic functionality, EPS from LAB exhibit anti-tumor, cholesterol-lowering, and immunomodulatory effects. EPS serve also as a barrier inhibiting pathogens or bacterial toxin adhesion to epithelial surfaces, thus preventing the initial stages of an infection, for instance, with *Helicobacter pylori* or *E. coli* (Korakli and Vogel, 2006). As it was reviewed by Zannini et al. (2016), the biological role of EPS is to facilitate cellular recognition, control the quorum-sensing, help in genetic information exchange, and protect microbial cells in harsh environments. Hydrophilic EPS have a high water-retention capacity, maintaining a hydrated microenvironment and enabling survival. Neutral and charged EPS form a hydrated polymer network, mediating the mechanical stability of biofilms, and may serve as a source of carbon, nitrogen, and phosphorus-containing compounds for utilization by the biofilm ecosystem. Charged and hydrophobic EPS promote polysaccharide gel formation.

10.5 POTENTIAL APPLICATIONS OF SOURDOUGH MICROORGANISMS AND THEIR METABOLITES

The statistics of the Food and Agriculture Organization (FAO) of the United Nations stated that around 1.3 billion tons of food are lost or discarded each year due to microbial contaminations, 30% of which are cereals and bread (Leyva Salas et al., 2017). These losses in industrialized countries are observed at retail or consumer levels, and the most frequent reason for such a huge global food loss is microbial spoilage, which affects the organoleptic quality of products and threatens hygiene, food safety, and public health. Besides, fungal genera such as *Aspergillus*, *Penicillium*, *Alternaria*, and *Fusarium* produce mycotoxins as secondary metabolites, with possible fatal effects on humans and animals. Therefore, bio-preservation of food products is the primary application of sourdough microorganisms. Other uses include the development of technologies to produce gluten-reduced foods, fortified pasta and cereals, fibers synthesis, and, in some "white" biotechnologies, for microbial

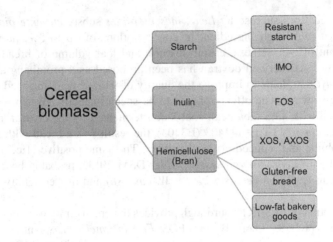

FIGURE 10.4 Possible routes for valorization of cereals and pseudocereals.

synthesis of valuable metabolites from biomass. Figure 10.4 schematizes some of the main routes toward valorization of cereals, pseudocereals, and their by-products.

10.5.1 Development of Sourdough Starters with Antifungal Activity

Currently, almost all preservatives used to avoid fungal spoilage in foods are chemical substances: benzoate, propionate, sorbate, nitrate, nitrite, and sulfites. There is only one exception, natamycin (E235), a microbial compound produced by *Streptomyces nataliensis* which belongs to the group of polyene macrolide antimycotics (Silva et al., 2016). A good alternative for bread protection, in terms of hygiene and food safety, may be the implementation of sourdough LAB with antifungal properties, such as the most studied species *Lp. plantarum* (Crowley et al., 2013). Other potential antifungal strains are *Fur. Rossiae* (Garofalo et al., 2012), *L. amylovorus* (Arendt et al., 2009), *L. harbinensis* (Delavenne et al., 2013), *Li. Reuteri* (Axel et al., 2016), *Lev. Brevis*, and *L. spicheri* (Le Lay et al., 2016). Besides other "conventional" antimicrobial substances, these LAB also produce rare and very active cyclic dipeptides. Their key role as antifungal metabolites was clarified for the first time by Niku-Paavola et al. (1999), who characterized cyclo (Gly-L-Leu) generated by *Lp. plantarum*. Then, Ström et al. (2002) reported two fungistatic bacteriocin-like dipeptides, cyclo(L-Phe–L-Pro) and cyclo(L-Phe–trans-4-OH-L-Pro), produced by *Lp. plantarum* strain MiLAB 393. Dal Bello et al. (2007) isolated *Lp. plantarum* FST1.7 from wheat sourdough and showed that its antifungal activity was due to the presence of cyclo(L-Leu-L-Pro) and cyclo(L-Phe-L-Pro). The cyclic dipeptides act in synergy with 3-PLA (Ryan et al., 2009), and their concentrations gradually increase upon sourdough fermentation (Axel et al., 2017). Other antifungal cyclopeptides produced by sourdough LAB are cyclo(L-Pro-L-Pro), cyclo(L-Tyr-L-Pro), cyclo(L-Met-L-Pro), cyclo(L-His-L-Pro), and cyclo(Leu-Leu) (Axelsson et al., 1989; Broberg et al., 2007; Black et al., 2013; Valerio et al., 2016; Mieszkin et al., 2017).

10.5.2 DEVELOPMENT OF FUNCTIONAL CEREAL PRODUCTS

Sourdoughs, especially those made from wholemeal flours, are known to be a rich source of macronutrients (proteins, lipids, and carbohydrates), but also dietary fiber, antioxidants, vitamins, minerals, phenolic acids, lignans, and phytosterols. Epidemiological studies show that consumption of whole grains reduces the risk of some socially significant diseases such as cardiovascular, diabetes, and obesity. A diet rich in functional foods is a response to these problems with increasing scientific evidence and clinical data, revealing that the consumption of whole grain sourdough products normalizes blood glucose, lowers blood pressure and cholesterol, and stops progressive atherosclerosis (Novotni et al., 2020; Skendi et al., 2020; Rocha, 2011). The regulations concerning functional cereal foods are included in the Codex Alimentarius of the FAO to develop food standards. Dietary fibers contained in them can be (i) water-insoluble (cellulose, chitin, hemicellulose, pentoses, lignin, xanthan gum, and resistant starch) and (ii) water-soluble, such as β-glucans and arabinoxylans (AXOS), and galactosaccharides (GOS) and fructooligosaccharides (FOS). All of them are considered prebiotics – fermentable fibers that are not amenable to digestion by the human GIT enzymes and benefit the host by selectively stimulating the growth and/or activity of probiotic bacteria in the colon (Goh and Klaenhammer, 2015).

The capability of probiotic LAB to metabolize cereal prebiotic fibers such as FOS, xylooligosaccharides (XOS), AXOS, and resistant starch (RS) further expands the substrate spectrum and provides beneficial effects on human health through prebiotic/probiotic synergistic action (Petrova and Petrov, 2020). Other emerging prebiotic substances are isomaltooligosaccharides, lactosucrose, gluco- and soybean oligosaccharides, polydextrose, gluconic acid, melibiose-, mannan-, pectin- and N-acetyl-chitin derived oligosaccharides, lactose-, glutamine- and hemicellulose-rich substrates, as well as germinated barley foodstuffs, oligo-dextrans, and lactoferrin-derived peptides (Petrova and Petrov, 2017).

Most prebiotic fibers that are widespread in cereal flours and sourdough are FOS (up to 0.4% of the dry weight of wheat, rye, and barley), XOS, and AXOS. Xylooligosaccharides are composed of β-(1,4)-linked xylose units, but xylan consists of a backbone of β-1,4-linked xylose units with diverse substituent including side chain carbohydrates, uronic acids, phenolic acids, and acetyl groups. In AXOS, these residues are arabinofuranosyl and provide diverse biological properties.

The genetic basis of the probiotic/prebiotic interaction lies in gene-encoding glycoside-hydrolase enzymes in the GIT and sourdough microflora. Among *Lactobacillus* species that metabolize FOS, this occurs through one of two metabolic pathways. With the first FOS, they are transported intact and hydrolyzed by cytoplasmic GH32 β-fructosidases. In *L. acidophilus*, *L. crispatus*, and *L. jensenii*, FOS transport is mediated by adenosine triphosphate (ATP)-dependent binding cassette; while in *Lp. Plantarum,* this process is carried out by the sucrose-dependent phosphoenolpyruvate phosphotransferase system (PTS). Cytoplasmic β-fructosidase is encoded by the *msm* operon containing *LacI* regulator (*msmR*), ATP-binding cassette (ABC) transporter, the β-fructosidase gene (*bfrA*), and the phosphorylase gene.

The operon expression is regulated by catabolic repression and is specifically induced by FOS and sucrose. This mode of using FOS gives lactobacilli a competitive advantage in the GIT because the substrate is transported in the cell before hydrolysis. However, in probiotics of the species *Lc. Paracasei*, another way to convert FOS is available. This is performed by an enzyme of the GH32 family of β-fructosidases, which is cell-wall anchored. In this way, FOS are hydrolyzed extracellularly, and fructose residues are transported into the cell. XOS and AXOS are chemically heterogenous, and several enzymes are engaged in their degradation: endo-xylanase, β-d-xylosidase, α-glucuronidase, α-arabinofuranosidase, and acetyl-xylan esterase. Endo-xylanases and β-D-xylosidases hydrolyze xylan, degrading β-1,4-linked xylan chain, whereas xylosidases perform xylooligomer hydrolysis and release xylose. The key enzyme for arabinoxylan utilization is α-l-arabinofuranosidase, which can cleave arabinose and acts synergistically with endoxylanases. IMO hydrolysis in sourdough lactobacilli occurs thanks to the presence of gene clusters *malEFG* and *msmK* (*L. acidophilus*) encoding four-component ABC transport system that imports maltodextrins into the cytosol. The gene *malP* in *L. acidophilus* (named *mapA* in *Fr. sanfranciscensis*) encodes maltose phosphorylase capable of phosphorylating the maltose, whereas *pgmB* (*L. acidophilus*), named *pgmA* in *Fr. sanfranciscensis* encodes β-phosphoglucomutase that converts β-D-glucose-1-phosphate to β-D-glucose-6-phosphate (Gänzle and Follador, 2012). Other genes engaged in maltose metabolism are *malH* (in *L. acidophilus*) and *simA* (in *Lc. casei*) that hydrolyze the phosphorylated sucrose isomers trehalulose, turanose, maltulose, leucrose, and palatinose (isomaltulose).

Several enzymes of the glycoside-hydrolase family are engaged in RS degradation by lactobacilli: amylase, α-glucosidase, glycogen phosphorylase, maltose phosphorylase, neopullulanase, and olygo-1,6-glucosidase (Petrova et al., 2013). Lactic acid generated before starch gelatinization increases the share of resistant starch much more efficiently than acetic acid (De Angelis et al., 2007; Maioli et al., 2008; Fois et al., 2018). Prebiotic fibers enhance the growth of lactobacilli and bifidobacteria in the GIT, thus providing the popular bifidogenic effect of fiber consumption. In turn, amylolytic probiotics, such as *B. pseudolongum* ATCC 25526, *B. adolescentis* VTT E-001561, *B. adolescentis* ATCC 15703, *B. animalis* pentos. *lactis*, *B. longum* pentos. *longum* ATCC 15697, *B. breve* DSM 20213, *L. plantarum*, *L. pentoses*, and *L. paracasei* display an impressive activity against pathogenic and food-spoiling bacteria like *E. coli*, *Vibrio cholerae*, *Klebsiella pneumoniae*, *Bacillus subtilis* (Fijan, 2016; Petrova and Petrov, 2011), and *Clostridium perfringens*, *Bacteroides fragilis*, *Staphylococcus aureus*, and *Salmonella typhimurium* (Petrova and Petrov, 2017). Therefore, functional cereal products could be a source of probiotic bacteria alleviating health issues.

Fifteen new species of lactobacilli were isolated from sourdough during the last 20 years. These are *Lev. Acidifarinae*, *Lev. Zymae*, *Lev. Hammesii*, *Lev. Namurensis*, *Lev. Spicheri*, *Companilactobacillus crustorum*, *Comp. mindensis*, *Comp. nantensis*, *Comp. nodensis*, *Li. Frumenti*, *Li. Panis*, *Li. Pontis*, *Li. Secaliphilus*, *Furfurilactobacillus siliginis*, and *Fur. rossiae*. They all produce, as is the case with sourdough species in general, many low-molecular antimicrobial compounds, vitamins, and BLIS (De Vuyst et al., 2009; Demirbaş et al., 2017).

Notably, the sourdough microbiota share several species with that of the human intestine (Gobbetti et al., 2016). As mentioned above, LAB probiotics can perform gluten detoxification in celiac patients, thus producing an anti-inflammatory effect on the intestine. This knowledge inspired the production of pasta with decreased gluten content by the introduction of biotechnological steps based on LAB fermentation (Di Cagno et al., 2005). Capozzi et al. (2011) used a similar fermentation step of wheat semolina with *Lp. plantarum* to produce riboflavin-enriched pasta with the same organoleptic properties as the one obtained by traditional methods. An advanced approach to producing gluten-free pasta is the substitution of durum wheat semolina with pseudocereal flours of buckwheat, amaranth, or quinoa. Caperuto et al. (2000) mixed quinoa and corn flours for the manufacture of spaghetti with satisfactory appearance and sensory properties.

The current demand for healthy and low-calorie foods has stimulated the development of innovative ways to introduce dietary fiber into sourdough foods and meet contemporary consumer expectations. In the production of bread, the inclusion of a large amount of fiber to provide beneficial physiological effects destroys the continuous sourdough matrix, causing unfavorable technological effects and, therefore, adversely affects the viscoelasticity of the sourdough and the functional quality of the obtained bio-fortified bread. The effects of a combination of different fibers could be an appropriate strategy to overcome the shortcomings of the individual fiber and the counteraction to their harmful effect and, possibly, would improve the dough-handling properties.

10.5.3 REDUCING THE AMOUNT OF ANTI-NUTRIENTS IN SOURDOUGH

As mentioned previously, anti-nutrient compounds in cereals include not only phytic acid but also tannins, lectins, and protease inhibitors, with phytate being the most studied among them. LAB strains can diminish the phytate content of sourdough. According to Sharma et al. (2020), dozens of LAB species in sourdough display phytase activity, but the highest is that of *L. amylovorus* (125–146 U/mL) and *L. plantarum* (60–74.4 U/mL).

Another anti-nutrient in sourdough is the protein complex of amylase/trypsin inhibitors (ATI), a mixture of cysteine-rich proteins with Mw ~12 kDa acting as dimers or tetramers. Their presence in sourdough induces an immunological disorder known as non-celiac wheat sensitivity with similar symptoms to those of celiac disease and irritable bowel syndrome. ATIs correspond to about 4% of the whole protein content of wheat and are resistant to gastrointestinal proteases. *Lactobacillus* strains with high ATI-destroying capability include *L. salivarius* H32.1, *L. mucosae* D5a1, and *L. rhamnosus* LE3, while *L. fermentum* R39.3 and *L. reuteri* R12.22 are weaker (Caminero et al., 2019). Huang et al. (2020) used a mice model to show that strains of *Fr. sanfranciscensis*, *Li. Reuteri*, or *La. sakei* digest ATI and thus can alleviate NCWS *in vivo*.

10.5.4 FERMENTATION OF PHENOLIC COMPOUNDS FOR INCREASED ANTIOXIDANT ACTIVITY OF CEREAL PRODUCTS

The role of phenolic compounds in the development of functional foods is controversial. They have many positive effects as antioxidants with antimicrobial activity but, on the other hand, are also considered undesirable as dietary supplements because some of them inhibit digestive enzymes and, thus, the proper absorption of nutrients. Polyphenols are common organic compounds present in cereals, but not synthesized by humans or animals. They have several aromatic rings and hydroxyl groups, which determine their various antioxidant properties: to chelate metal ions, to neutralize free radicals by donation of an electron or hydrogen atom, to modify some cellular signaling processes, or to increase the levels of antioxidant enzymes. The main polyphenol classes include phenolic acids, stilbenes, lignans, flavonoids (catechins, flavones, isoflavones, flavanones, flavanols, and anthocyanins) and tannins, which are high-molecular esters of the ellagic or gallic acids.

Wheat, barley, oats, millet, sorghum, and rice contain up to 1 g/kg of low-molecular-weight phenolic acids. Ferulic, coumaric, and vanillic acids comprise more than 50% of this content, but cereals also contain high amounts of flavonoids and tannins. Phenolic acids in wheat and rye include also hydroxycinnamic and hydroxybenzoic acids (Adebo and Medina-Meza, 2020; Skendi et al., 2020). Ferulic acid is a powerful antioxidant, slowing down the aging process. Furthermore, it is known to prevent Alzheimer's disease, atherosclerosis, cervical cancer, high cholesterol, hypertension, diabetes, and osteoporosis. It has also anti-inflammatory, antimicrobial, anti-allergic, antiviral, antithrombotic, and hepatoprotective effects (Kumar and Goel, 2019). Polyphenols have antibacterial activity against *Staphylococcus aureus*, *Salmonella enteritidis*, *Serratia marcescens*, *Cronobacter sakazakii*, *Escherichia coli*, *Pseudomonas aeruginosa*, *Bacillus subtilis*, and *Enterococcus faecalis*, as well as antifungal properties inhibiting the growth of *Candida albicans*. Due to their hydrophobic character, they act on the phospholipid layer of the bacterial plasma membrane, disturbing the synthesis of proteins and ATP (Piekarska-Radzik and Klewicka, 2021).

Remarkably, phenolic acids have quite the opposite effect on LAB bacteria. Many bacteria from the genus *Lactobacillus* can ferment them and, thus, significantly decrease the toxic level of the primary compounds. That is why polyphenols are evaluated as substances with prebiotic features, which can increase the number of probiotic bacteria in the gastrointestinal tract. The growth of LAB such as *Lp. Rhamnosus*, *L. delbrueckii* subsp. *bulgaricus*, *Lp. Paracasei* subsp. *paracasei*, and *Lactococcus lactis* subsp. *cremoris* is stimulated by phenolic compounds in cereals and sourdough. The feruloylated arabinoxylan (containing 0.43% of ferulic acid) increases the growth of *Bifidobacterium* and *Lactobacillus* genera. Liu et al. (2021) managed to produce up to 6.60 mg/g ferulic acid from the natural substrate wheat arabinoxylan after cloning, and overexpression of feruloyl esterase from *L. acidophilus*. *Lc. paracasei*, *Lp. Plantarum*, and *L. acidophilus* can even increase polyphenol amount by 50–60% during fermentation of raw mulberry juice, reaching final concentrations in the range 1.5–1.7 mg/l (Kwaw et al., 2018).

Among sourdough LAB, the most studied converters of phenolic acids are *Lp. plantarum, Fur. rossiae, Li. reuteri, Li. fermentum, Lc. pentosus, Lev. brevis, Secundilactobacillus collinoides*, and *Lp. plantarum*. They all possess genes encoding at least three different phenolic reductases, feruloyl esterases, and decarboxylases. *Lp. plantarum* is the LAB species with the greatest capacity for enzymatic transformation of polyphenols. The complete genome sequence of *Lp. plantarum* WCFS1 shows the presence of genes encoding benzyl alcohol dehydrogenase (lp_3054), phenolic acid decarboxylase, and tannin acyl hydrolase. The first enzyme reversibly oxidizes some aromatic alcohols to aldehydes coupled with nicotinamide adenine dinucleotide (NAD^+) reduction, the second reduces the amounts of p-coumaric, caffeic, and ferulic acids, while the third can hydrolyze the ester bonds in tannins and the esters of gallic acid. Svensson et al. (2010) showed that the fermentation of sorghum sourdough by a microbial starter culture containing *Lp. plantarum, Lc. casei, Li. Fermentum*, and *Li. reuteri* results in complex metabolic transformations of: caffeic acid to dihydrocaffeic acid, ethylcatechol, and vinyl catechol; ferulic acid to dihydroferulic acid; and naringenin-7-O-glucoside to naringenin – thus benefiting the nutritional value and antimicrobial activity of sorghum sourdough.

Gaur et al. (2020) recently clarified the enzymatic pathway in *Lp. plantarum*, by which hydroxycinnamic acid is transformed into phenylpropionic acid, both compounds with notable antioxidant, anti-inflammatory, and antimicrobial activity. Ripari et al. (2019) investigated the polyphenolic metabolism of 114 *Lactobacillus* strains obtained from 35 sourdough samples. In wheat and rye malt sourdoughs prepared in the laboratory: *Lp. plantarum* metabolized free ferulic acid; and *L. hammesii* DSM 16381 converted syringic and vanillic acids and reduced the levels of bound ferulic acid. When both strains were combined, the process was enhanced and the free ferulic acid was transformed into dihydroferulic acid and volatile metabolites. In another study, *Lev. brevis* was also shown to metabolize ferulic (and caffeic) acids in a synthetic sourdough medium. Aiming to observe whether the sourdough microbiome can convert ferulic acid into antioxidants, Boudaoud et al. (2021) co-cultured in laboratory sorghum sourdough the yeast *Kazachstania* and *Saccharomyces cerevisiae* with LAB strains derived from French natural sourdoughs. *Fr. sanfranciscensis, Lp. plantarum, Lp. xiangfangensis, Lev. hammesii, La. Curvatus*, and *La. sakei* metabolized ferulic acid into dihydroferulic acid and 4-vinylguaiacol, a metabolite with significant activity against drug-resistant human colorectal cancer (Luo et al., 2021).

10.5.5 SYNTHESIS OF PHYTOESTROGENS

Phytoestrogens (PE) are organic molecules of the polyphenol class with a structure like estradiol. That is why their presence in food is known to alleviate menopause symptoms. The Asiatic diet, containing more soy than others, is associated with decreased risk of breast and prostate cancer, cardiovascular problems, and type-2 diabetes (Rietjens et al., 2017). Phytoestrogens cover five different types: isoflavones, lignans, ellagitannins, stilbenes, and coumestans. The highest PE content in wheat, oats, barley, and rice sourdough is that of lignans (Peirotén et al., 2020). Interestingly,

PE functions are mediated by the intestinal LAB microbiota, which transforms the cereal phenolic compounds into their bioactive PE form, as LAB in the GIT can convert isoflavones, ellagitannins, and lignans to the bioactive equol, urolithins, and enterolignans. Bravo et al. (2017) analyzed the production of equol, 5-hydroxy-equol, enterodiol, and enterolactone by 70 strains of *Lactobacillus*, *Lactococcus*, and *Enterococcus*. The authors found that *L. salivarius* strains INIA P448 and INIA P183 and *L. gasseri* INIA P508 strain metabolize lignans into enterolactone and enterodiol, and the obtained amounts were significant (6–46 µM). Recently, Peirotén et al. (2019) revealed that intestinal isolates of *Bifidobacterium bifidum* INIA P466, *B. catenulatum* INIA P732, and *B. pseudolongum* INIA P2 synthesize enterodiol (2–11 µM) from lignan extracts, while *B. pseudocatenulatum* INIA P946 produced secoisolariciresinol.

10.6 AN INTEGRATED INDUSTRIAL OUTLOOK

In the last decade, food production and consumption have experienced rapid changes related to the growing demand for natural and healthy foods. The food industry needs to keep up with new scientific knowledge and continuously monitor the progress and potential of using functional food ingredients. Their goal is to produce food products that will meet the demands of modern consumers and support them in maintaining a healthy lifestyle. In this context microorganisms and their metabolites from sourdough play an important role (Fernández-Peláez et al., 2020). Their role is not only to meet the nutritional needs but also to meet the specific requirements of certain consumer groups in terms of beneficial effects on the mental and physical condition of the body and prevention of some relevant chronic diseases (Vitali Čepo et al., 2020).

Fermentation of sourdough is one of the oldest methods of processing grain matrices, which has been mainly studied for its contribution to the sensory, structural, nutritious, and functional properties of bakery products (Michel et al., 2016). Researchers are mainly focused on the large microbial diversity of sourdough and their effects on human intestinal health (Won et al., 2008; Minervini, et al., 2014; Fernández-Peláez et al., 2020). The detection and use of microorganisms that provide probiotic properties and can survive food production processes are primary objectives of the food industry. Accordingly, isolation, identification, and characterization of native foodborne strains are considered the most important issues for the food industry due to their socioeconomic impacts (Kalui et al., 2010; Palla et al., 2019).

The term "probiotic" is commonly used for living microorganisms that can provide people with health benefits when consumed (FAO/WHO, 2006). Native lactobacilli strains isolated from traditional sourdough have shown great potential for use in the food industry (Denkova et al., 2012; Doğan and Tekiner, 2019). Some of them have good resistance to antibiotics, resistance to high content of table salt and low temperatures, which is why they have a wide range of applications (Zangeneh et al., 2019). The most common application is in bakery products (e.g. bread, biscuits, cookies, pizza, and pasta), dairy products (e.g. cheeses, yogurt, and baby foods), alcoholic and non-alcoholic beverages (e.g. wine, beer, juices, and non-fermented and fermented cereal drinks), nutraceuticals (dietary supplements and food additives), products of other industrial sectors such as feed and pet food, cosmetics, and pharmaceuticals (Bartkiene et al., 2020).

For the food industry it is very important that certain probiotics and their metabolites are presented to consumers through health claims on product labels, especially those that are not known to a wide range of consumers. Acceptance of a particular functional food depends on the consumer's knowledge of the health effects of certain ingredients (Rijkers et al., 2011). Previous research efforts have shown that consumers recognize foods to which probiotics and prebiotics have been added because of the positive health effects, and they are generally willing to pay a higher price for these types of products (Hieke et al., 2015).

Furthermore, all functional foods, including those containing microorganisms from sourdough and their metabolites, must comply with existing regulations, which slows the progress of placing new and innovative products on the market. Particular attention must be paid to regulations governing the field of food for special nutritional and medical needs, food supplements, novel foods, food additives, and nutritional and health claims.

Regulation (European Commission, EC) no. 1924/2006 with amendments, regulates the use of nutrition and health claims on foods and food additives throughout the European Union (EU). According to this regulation, all claims that state, suggest, or imply a relationship between food and health may be considered health claims (European Parliament and the Council, 2006). All health claims that can be used within the EU and have passed the strictly prescribed approval procedure, are included in the Register of approved health claims available on the official website of the European Commission. The disadvantage of the approval procedure is that it can be a lengthy process and an obstacle to the development of new products in useful time for the industry and their usage as functional food ingredients.

The biological effects of probiotics largely depend on the type of involved microorganisms, but also on the strain level (Azaïs-Braesco et al., 2010). For purposes of approving some novel health claim, long-term and usually expensive clinical trials should be conducted for each individual strain or their combinations (Agarbati, et al., 2020). It means that industry should invest in research that is not a guarantee for obtaining a certain health claim. This will not motivate the food industry to include new probiotic strains and their products in food production, and consortia between academics and industry will become fundamental.

According to existing regulations, the term "probiotic" is considered a health claim because it implies that the product contains a substance that may be beneficial to health. The European Food Safety Authority (EFSA) has rejected all health claims relating to probiotics in 2010, citing an unsubstantiated link between the claim and the health effects of each probiotic. It is only allowed to indicate the name of the microorganisms in the list of ingredients in accordance with Regulation (EU) no. 1169/2011 on food labeling. Manufacturers may provide this information on a voluntary basis, but the health benefits arising from their usage cannot be specifically highlighted (European Parliament and of the Council, 2011).

The health claim related to living microorganisms approved in the EU is only that for yogurt: "Live cultures in yogurt improve lactose digestion in people who have difficulty digesting lactose." This claim can be used where the specific cultures listed in the EFSA opinion are present, and yogurt must contain at least 108 living microorganisms per gram (*Streptococcus thermophilus* and *Lactobacillus delbrueckii* subsp. *Bulgaricus*) (EFSA, 2011).

In approving health claims relating to living microorganisms and their metabolites, it is important to have high-quality, standardized clinical studies, which will be acceptable to the regulatory authorities. The assessment of individual requirements by regulators is a very complex process, subject to case-by-case assessment because there are no clear criteria for defining immune system health or bowel health (Riedl et al., 2017). During the characterization of products with probiotics and their metabolites, the food matrix, background diet, microbiota, age, and health status of the examined population must be taken into account (De Roos et al., 2019).

Considering all the above, a close cooperation between research institutions, food producers, and regulators will be crucial for the application of microorganisms from sourdough and their metabolites in the food industry, creating an environment that will promote faster development of wider range of new food products and innovation in the food industry.

10.7 CONCLUSION

The use of sourdough has a very long history in human nutrition and is currently receiving more scientific attention. This strong interest in food fermentation can also be shown by the in-depth study of lactic acid bacteria and the recently derived extensive taxonomic changes for this important group of food-related microorganisms. Among other reasons, the improving scientific understanding of the complex microbiota in sourdough and its detailed biochemistry offers an enormous potential for future applications tailored to human nutrition. This starts with increasing knowledge about the microbial composition and dynamics of the microbiota in flours from cereals and pseudocereals and sourdoughs, and enabling tailored optimization of bread quality (leavening, acidification, flavor, and aroma formation) and safety (*e.g.* antifungal and antibacterial activity). Further, the deepened knowledge about the formed microbial metabolites and their function and properties allows the investigation and development of tailored functional foods. This coincides with an increased interest of the consumers in balanced and natural foods with regard to health aspects in their personal diet. In more detail, the reduction of anti-nutrients and the increased antioxidative activity via sourdough fermentation sourdough are of specific interest. In summary, sourdough-related microorganisms and their application show great potential application in the agro-food chain in general and in personalized diets as well. However, robust cooperation between the food industry and in-depth scientific research will be critical in reaching this goal and in overcoming the associated legal hurdles.

ACKNOWLEDGMENTS

This work is based upon the work from COST Action 18101 SOURDOMICS – Sourdough biotechnology network toward novel, healthier and sustainable food and bioprocesses (https://sourdomics.com/; https://www.cost.eu/actions/CA18101/, where the author João Miguel F. Rocha is the Chair and Grant Holder Scientific Representative, Vera Fraberger and Domig Konrad Johan are committee members from Austria, Görkem Özülkü is non-committee member from Turkey, Penka Petrova and Kaloyan Petrov are committee and non-committee members from Bulgaria, respectively, and Knežević Nada is committee member from Croatia. Cost Action SOURDOMICS is supported by COST (European Cooperation in Science and Technology) (https://www.cost.eu/). COST is a funding agency for research and innovation networks. Penka Petrova and Kaloyan Petrov are thankful to the Bulgarian Ministry of Education and Science for the financial support by the "*Healthy Foods for a Strong Bio-Economy and Quality of Life*" National Research Programme approved by DCM # 577/17.08.2018 and Contract KP-06-COST7 with national co-financing to COST Action 18101. Regarding the author João Miguel F. Rocha, this work was also financially supported by: (i) Base Funding – UIDB/00511/2020 of the Laboratory for Process Engineering, Environment, Biotechnology and Energy – LEPABE – funded by national funds through the FCT/MCTES (PIDDAC); (ii) Project PTDC/EQU-EQU/28101/2017 – SAFEGOAL – Safer Synthetic Turf Pitches with Infill of Rubber Crumb from Recycled Tires, funded by FEDER funds through COMPETE2020 – Programa Operacional Competitividade e Internacionalização (POCI) and by national funds (PIDDAC) through FCT/MCTES.

REFERENCES

Adebo, O.A., and Medina-Meza, I.G. 2020. Impact of fermentation on the phenolic compounds and antioxidant activity of whole cereal grains: A mini review. *Molecules* 25:927.

Agarbati, A., Canonico, L., Marini, E., Zannini, E., Ciani, M., and Comitini, F. 2020. Potential probiotic yeasts sourced from natural environmental and spontaneous processed foods. *Foods* 9:287. doi: 10.3390/foods9030287

Arena, M.P., Capozzi, V., Russo, P., Drider, D., Spano, G., and Fiocco, D. 2018. Immunobiosis and probiosis: Antimicrobial activity of lactic acid bacteria with a focus on their antiviral and antifungal properties. *Applied Microbiology and Biotechnology* 102:9949–58.

Arendt, E.K., Dal Bello, F., and Ryan, L. 2009. Increasing the shelf life of bakery and Patisserie products by using the antifungal *Lactobacillus amylovorus* DSM 19280. US Patent WO200,9141,427 A2.

Arora, K., Ameur, H., Polo, A., Di Cagno, R., Rizzello, C.G., and Gobbetti, M. 2020. Thirty years of knowledge on sourdough fermentation: A systematic review. *Trends in Food Science & Technology* 108:71–83. doi:10.1016/j.tifs.2020.12.008

Axel, C., Brosnan, B., Zannini, E., Peyer, L.C., Furey, A., Coffey, A., and Arendt, E.K. 2016. Antifungal activities of three different *Lactobacillus* species and their production of antifungal carboxylic acids in wheat sourdough. *Applied Microbiology and Biotechnology* 100:1701–11.

Axel, C., Zannini, E., and Arendt, E.K. 2017. Mold spoilage of bread and its biopreservation: A review of current strategies for bread shelf life extension. *Critical Reviews in Food Science and Nutrition* 57:3528–42.

Axelsson, L.T., Chung, T.C., Dobrogosz, W.J., and Lindgren, S.E. 1989. Production of a broad spectrum antimicrobial substance by *Lactobacillus reuteri*. *Microbial Ecology in Health and Disease* 2:131–6.

Azaïs-Braesco, V., Bresson, J.L., Guarner, F., and Corthier, G. 2010. Not all lactic acid bacteria are probiotics but some are. *British Journal of Nutrition* 103 (7):1079–81. doi: https://doi.org/10.1017/S0007114510000723 103

Bartkiene, E., Bartkevics, V., Pugajeva, I., Krungleviciute, V., Mayrhofer, S., and Domig, K. 2017. The contribution of P. acidilactici, L. plantarum, and L. curvatus starters and L-(+)-lactic acid to the acrylamide content and quality parameters of mixed rye: Wheat bread. *LWT* 80:43–50. doi:10.1016/j.lwt.2017.02.005

Bartkiene, E., Zavistanaviciute, P., Lele, V., Ruzauskas, M., Bartkevics, V., Bernatoniene, J., and Santini, A. 2018. Lactobacillus plantarum LUHS135 and paracasei LUHS244 as functional starter cultures for the food fermentation industry: Characterisation, mycotoxin-reducing properties, optimisation of biomass growth and sustainable encapsulation by using dairy by-products. *LWT: Food Science and Technology* 93:649–58. doi:10.1016/j.lwt.2018.04.017

Bartkiene, E., Lele, V., Ruzauskas, M., Domig, K.J., Starkute, V., Zavistanaviciute, P., and Rocha, J.M. 2020. Lactic acid bacteria isolation from spontaneous sourdough and their characterization including antimicrobial and antifungal properties evaluation. *Microorganisms* 8 (1):20. Retrieved from https://www.ncbi.nlm.nih.gov/pubmed/31905993. doi: 10.3390/microorganisms8010064

Ben Omar, N., and Ampe, F. 2000. Microbial community dynamics during production of the Mexican fermented maize dough pozol. *Applied and Environmental Microbiology* 66(9):3664–73.

Bendheim, P.E., Poeggeler, B., Neria, E., Ziv, V., Pappolla, M.A., and Chain, D.G. 2002. Development of indole-3-propionic acid (OXIGON) for Alzheimer's disease. *Journal of Molecular Neuroscience* 19:213–7.

Birt, D.F., Boylston, T., Hendrich, S., Jane, J.L., Hollis, J., Li, L., McClelland, J., Moore, S., Phillips, G.J., Rowling, M., Schalinske, K., Scott, M.P., and Whitley, E.M. 2013. Resistant starch: Promise for improving human health. *Advances in Nutrition* 4:587–601.

Black, B.A., Zannini, E., Curtis, J.M., and Gänzle, M.G. 2013. Antifungal hydroxy fatty acids produced during sourdough fermentation: Microbial and enzymatic pathways, and antifungal activity in bread. *Applied and Environmental Microbiology* 79:1866–73.

Böswald, C., Engelhardt, G., Vogel, H., and Wallnöfer, P.R. 1995. Metabolism of the Fusarium mycotoxins zearalenone and deoxynivalenol by yeast strains of technological relevance. *Natural Toxins* 3(3):138–44. https://doi.org/10.1002/nt.2620030304

Boudaoud, S., Aouf, C., Devillers, H., Sicard, D., and Segond, D. 2021. Sourdough yeast-bacteria interactions can change ferulic acid metabolism during fermentation. *Food Microbiol* 98:103790. doi:10.1016/j.fm.2021.103790

Bravo, D., Peirotén, A., Alvarez, I., and Landete, J.M. 2017. Phytoestrogen metabolism by lactic acid bacteria: Enterolignan production by *Lactobacillus salivarius* and *Lactobacillus gasseri* strains. *Journal of Functional Foods* 37:373–8.

Broberg, A., Jacobsson, K., Strom, K., and Schnurer, J. 2007. Metabolite profiles of lactic acid bacteria in grass Silage. *Applied and Environmental Microbiology* 73:5547–52.

Caminero, A., McCarville, J.L., Zevallos, V.F., Pigrau, M., Yu, X.B., Jury, J., Galipeau, H.J., Clarizio, A.V., Casqueiro, J., Murray, J.A., Collins, S.M., Alaedini, A., Bercik, P., Schuppan, D., and Verdu, E.F. 2019. Lactobacilli degrade wheat amylase trypsin inhibitors to reduce intestinal dysfunction induced by immunogenic wheat proteins. *Gastroenterology* 156:2266–80.

Caperuto, L., Amaya-Farfan, J., and Camargo, C. 2000. Performance of quinoa (*Chenopodium quinoa* Willd. flour in the manufacture of gluten-free spaghetti. *Journal of the Science of Food and Agriculture* 81:95–101.

Capozzi, V., Menga, V., Digesu, A.M., De Vita, P., van Sinderen, D., Cattivelli, L., Fares, C., and Spano, G. 2011. Biotechnological production of vitamin B2-enriched bread and pasta. *Journal of Agricultural and Food Chemistry* 59:8013–20.

Chaves-Lopez, C., Serio, A., Delgado-Ospina, J., Rossi, C., Grande-Tovar, C.D., and Paparella, A. 2016. Exploring the bacterial microbiota of Colombian fermented maize dough "Masa Agria"(Maiz Añejo). *Frontiers in Microbiology* 7:1168.
Chaves-López, C., Rossi, C., Maggio, F., Paparella, A., and Serio, A. 2020. Changes occurring in spontaneous maize fermentation: An overview. *Fermentation* 6:36.
Chen, C.C., Lai, C.C., Huang, H.L., Huang, W.Y., Toh, H.S., Weng, T.C., and Tang, H.J. 2019. Antimicrobial activity of Lactobacillus species against Carbapenem-resistant Enterobacteriaceae. *Frontiers in Microbiology* 10:789. doi:10.3389/fmicb.2019.00789
Chen, X.Y., Levy, C., and Ganzle, M.G. 2016. Structure-function relationships of bacterial and enzymatically produced reuterans and dextran in sourdough bread baking application. *International Journal of Food Microbiology* 239:95–102. doi:10.1016/j.ijfoodmicro.2016.06.010
Coda, R., Di Cagno, R., Gobbetti, M., and Rizzello, C.G. 2014. Sourdough lactic acid bacteria: Exploration of non-wheat cereal-based fermentation. *Food Microbiology* 37:51–8.
Corsetti, A., Gobbetti, M., Rossi, J., and Damiani, P. 1998. Antimould activity of sourdough lactic acid bacteria: Identification of a mixture of organic acids produced by *Lactobacillus sanfrancisco* CB1. *Applied Microbiology and Biotechnology* 50:253–6.
Corsetti, A., Gobbetti, M., De Marco, B., Balestrieri, F., Paoletti, F., Russi, L., and Rossi, J. 2000. Combined effect of sourdough lactic acid bacteria and additives on bread firmness and staling. *Journal of Agricultural and Food Chemistry* 48(7):3044–51. doi:10.1021/jf990853e
Corsetti, A., Settanni, L., and Van Sinderen, D. 2004. Characterization of bacteriocin-like inhibitory substances (BLIS. from sourdough lactic acid bacteria and evaluation of their in vitro and in situ actvity. *Journal of Applied Microbiology* 96:521–34.
Crowley, P., Schober, T., Clarke, C., and Arendt, E. 2002. The effect of storage time on textural and crumb grain characteristics of sourdough wheat bread. *European Food Research and Technology* 214(6):489–96. doi:10.1007/s00217-002-0500-7
Crowley, S., Mahony, J., and van Sinderen, D. 2013. Current perspectives on antifungal lactic acid bacteria as natural bio-preservatives. *Trends in Food Science and Technology* 33:93–109.
Dal Bello, F., Clarke, C.I., Ryan, L.A.M., Ulmer, H., Schober, T.J., Strom, K., Sjogren, J., van Sinderen, D., Schnurer, J., Arendt, E.K., Strom, K., Sjogren, J., and Schnurer, J. 2007. Improvement of the quality and shelf life of wheat bread by fermentation with the antifungal strain *Lactobacillus plantarum* FST 1.7. *Journal of Cereal Science* 45:309–318.
De Angelis, M., Mariotti, L., Rossi, J., Servili, M., Fox, P.F., Rollán, G., and Gobbetti, M. 2002. Arginine catabolism by sourdough lactic acid bacteria: Purification and characterization of the arginine deiminase pathway enzymes from *Lactobacillus sanfranciscensis* CB1. *Applied and Environmental Microbiology* 68:6193–201.
De Angelis, M., Rizzello, C.G., Alfonsi, G., Arnault, P., Cappelle, S., Di Cagno, R., and Gobbetti, M. 2007. Use of sourdough lactobacilli and oat fibre to decrease the glycaemic index of white wheat bread. *British Journal of Nutrition* 98:1196–205.
De Mello, V.D, Paananen, J., Lindström, J., Lankinen, M.A., Shi, L., Kuusisto, J. et al. 2017. Indolepropionic acid and novel lipid metabolites are associated with a lower risk of type 2 diabetes in the Finnish diabetes prevention study. *Scientific Reports* 7:46337.
De Roos, B., Bronze, M., Aura, A.M., Cassidy, A., Garcia Conesa, M.-T., Gibney, E.R., Greyling, A., Kaput, J., Kerem, Z., Knežević, N., Kroon, P., Landberg, R., Manach, C., Milenkovic, D., Rodriguez-Mateos, A., and Tomás-Barberán, F.A. 2019. Targeting the delivery of dietary plant bioactives to those who would benefit most: from science to practical applications. *European Journal of Nutrition* 58(Supplement 2):65–73.
De Vuyst, L., Schrijvers, V., Paramithiotis, S., Hoste, B., Vancanneyt, M., Swings, J., and Messens, W. 2002. The biodiversity of lactic acid bacteria in Greek traditional wheat sourdoughs is reflected in both composition and metabolite formation. *Applied and Environmental Microbiology* 68(12):6059–69. doi:10.1128/aem.68.12.6059-6069.2002

De Vuyst, L., Vrancken, G., Ravyts, F., Rimaux, T., and Weckx, S. 2009. Biodiversity, ecological determinants, and metabolic exploitation of sourdough microbiota. *Food Microbiology* 26:666–75.

De Vuyst, L., Harth, H., Van Kerrebroeck, S., and Leroy, F. 2016. Yeast diversity of sourdoughs and associated metabolic properties and functionalities. *International Journal of Food and Microbiology* 239:26–34. doi:10.1016/j.ijfoodmicro.2016.07.018

De Vuyst, L., Van Kerrebroeck, S., and Leroy, F. 2017. Chapter Two - Microbial ecology and process technology of sourdough fermentation. In S. Sariaslani and G.M. Gadd (Eds.), *Advances in Applied Microbiology* (Vol. 100, pp. 49–160): Academic Press.

Delavenne, E., Ismail, R., Pawtowski, A., Mounier, J., Barbier, G., and Le Blay, G. 2013. Assessment of lactobacilli strains as yogurt bioprotective cultures. *Food Control* 30:206–13.

Demirbaş, F., İspirli, H., Kurnaz, A.A., Tahsin Yilmaz, M.T., and Dertli, E. 2017. Antimicrobial and functional properties of lactic acid bacteria isolated from sourdoughs. *LWT: Food Science and Technology* 79:361–6.

Denkova, R., Ilieva, S., DImbareva, D., and Denkova, Z. 2012. Probiotic properties of lactobacillus acidophilus z10, isolated from naturally fermented sourdough, food and environment safety. *Journal of Faculty of Food Engineering* 11:3.

Dertli, E., Mercan, E., Arıcı, M., Yılmaz, M.T., and Sağdıç, O. 2016. Characterisation of lactic acid bacteria from Turkish sourdough and determination of their exopolysaccharide (EPS) production characteristics. *LWT: Food Science and Technology* 71:116–24. doi:10.1016/j.lwt.2016.03.030

Di Cagno, R., de Angelis, M., Alfonsi, G., de Vincenzi, M., Silano, M., Vincentini, O., and Gobbetti, M. 2005. Pasta made from durum wheat semolina fermented with selected lactobacilli as a tool for a potential decrease of the gluten intolerance. *Journal of Agricultural and Food Chemistry* 53:4393–402.

Di Monaco, R., Torrieri, E., Pepe, O.R., Masi, P., and Cavella, S. 2015. Effect of sourdough with exopolysaccharide (EPS)-producing lactic acid bacteria (LAB) on sensory quality of bread during shelf life. *Food and Bioprocess Technology* 8:691–701.

Diana, M., Quílez, J., and Rafecas, M. 2014a. Gamma-aminobutyric acid as a bioactive compound in foods: A review. *Journal of Functional Foods* 10:407–20. doi:10.1016/j.jff.2014.07.004

Diana, M., Rafecas, M., and Quílez, J. 2014b. Free amino acids, acrylamide and biogenic amines in gamma-aminobutyric acid enriched sourdough and commercial breads. *Journal of Cereal Science* 60:639–44.

Dinardo, F.R., Minervini, F., De Angelis, M., Gobbetti, M., and Gänzle, M.G. 2019. Dynamics of Enterobacteriaceae and lactobacilli in model sourdoughs are driven by pH and concentrations of sucrose and ferulic acid. *LWT* 114:108394.

Diowksz, A., Malik, A., Jaśniewska, A., and Leszczyńska, J. 2020. The inhibition of amylase and ACE enzyme and the reduction of immunoreactivity of sourdough bread. *Foods* 9:656.

Doğan, M., and Tekiner, İ.H 2019. Assessment of probiotic properties of lactic acid bacteria from traditional sourdoughs for bread-making in turkey against some gut conditions. *Journal of Applied Food Technology* 6(2):34–40.

Dordevic, T.M., Siler-Marinkovic, S.S., and Dimitrijevic-Brankovic, S.I. 2010. Effect of fermentation on antioxidant properties of some cereals and pseudo cereals. *Food Chemistry* 119:957–63.

Duenas-Chasco, M.T., Rodriguez-Carvajal, M.A., Mateo, P.T., Franco-Rodriguez, G., Espartero, J.L., Irastorza-Iribas, A., and Gil-Serrano, A.M. 1997. Structural analysis of the exopolysaccharide produced by *Pediococcus damnosus* 2.6. *Carbohydrate Research* 303:453–8.

Ercolini, D., Pontonio, E., De Filippis, F., Minervini, F., La Storia, A., Gobbetti, M., and Di Cagno, R. 2013. Microbial ecology dynamics during rye and wheat sourdough preparation. *Applied and Environmental Microbiology* 79(24):7827–36.

European Food Safety Authority, EFSA. 2011. Scientific opinion on the substantiation of health claims related to Lactobacillus fermentum ME-3 and decreasing potentially pathogenic gastro-intestinalmicroorganisms (ID 3025) pursuant to Article 13(1) of Regulation (EC) No 1924/2006. *EFSA Journal* 9(4):2025. Retrieved from https://efsa.onlinelibrary.wiley.com/doi/pdf/10.2903/j.efsa.2011.2025

European Parliament, and the Council of the European Union. 2006. Regulation (EC) No. 1924/2006 of 20 December 2006 on nutrition and health claims made on foods (consolidated text). Retrieved from https://eur-lex.europa.eu/legal-content/EN/TXT/?uri=CELEX%3A02006R1924-20141213

European Parliament, and the Council of the European Union. 2011. Regulation (EU) 1169/2011 of 25 October 2011 on the provision of food information to consumers, amending Regulations (EC) No 1924/2006 and (EC) No 1925/2006 of the European Parliament and of the Council, and repealing Commission Directive 87/250/EEC, Council Directive 90/496/EEC, Commission Directive 1999/10/EC, Directive 2000/13/EC of the European Parliament and of the Council, Commission Directives 2002/67/EC and 2008/5/EC and Commission Regulation (EC) No 608/2004 (consolidated text). Retrieved from https://eur-lex.europa.eu/legal-content/EN/TXT/?uri=CELEX%3A02011R1169-20180101

Fazeli, M.R., Shahverdi, A.R., Sedaghat, B., Jamalifar, H., and Samadi, N. 2004. Sourdough-isolated *Lactobacillus fermentum* as a potent anti-mould preservative of a traditional Iranian bread. *European Food Research and Technology* 218:554–6.

Fernández-Peláez, J., Paesani, C., and Gómez, M. 2020. Sourdough technology as a tool for the development of healthier grain-based products: An update. *Agronomy* 10:1962. https://doi.org/10.3390/agronomy10121962

Fijan, S. 2016. Antimicrobial effect of probiotics against common pathogens. In: *Probiotics and Prebiotics in Human Nutrition and Health*, V. Rao and L.G. Rao (Eds.). Intech Open. doi: 10.5772/63141.

Fois, S., Piu, P.P., Sanna, M., Roggio, T., and Catzeddu, P. 2018. Starch digestibility and properties of fresh pasta made with semolina-based liquid sourdough. *LWT* 89:496–502.

Food and Agricultural Organization of the United Nations, and World Health Organization, FAO/WHO. 2006. Probiotics in food Health and nutritional properties and guidelines for evaluation. Retrieved from: http://www.fao.org/3/a0512e/a0512e.pdf

Fraberger, V., Call, L.M., Domig, K.J., and D'Amico, S. 2018. Applicability of yeast fermentation to reduce fructans and other FODMAPs. *Nutrients* 10(9):1247. doi:10.3390/nu10091247

Fraberger, V., Ammer, C., and Domig, K.J. 2020a. Functional properties and sustainability improvement of sourdough bread by lactic acid bacteria. *Microorganisms* 8(12). doi:10.3390/microorganisms8121895

Fraberger, V., Ladurner, M., Nemec, A., Grunwald-Gruber, C., Call, L.M., Hochegger, R., and D'Amico, S. 2020b. Insights into the potential of sourdough-related lactic acid bacteria to degrade proteins in wheat. *Microorganisms* 8(11):1689. doi:doi:10.3390/microorganisms8111689

Fraberger, V., Unger, C., Kummer, C., and Domig, K.J. 2020c. Insights into microbial diversity of traditional Austrian sourdough. *LWT: Food Science and Technology* 127. doi:10.1016/j.lwt.2020.109358

Franco, W., Pérez-Díaz, I.M., Connelly, L., and Diaz, J.T. 2020. Isolation of exopolysaccharide-producing yeast and lactic acid bacteria from quinoa (Chenopodium Quinoa) sourdough fermentation. *Foods* 9(3):337.

Fujimoto, A., Ito, K., Narushima, N., and Miyamoto, T. 2019. Identification of lactic acid bacteria and yeasts, and characterization of food components of sourdoughs used in Japanese bakeries. *Journal of Bioscience and Bioengineering* 127(5):575–81. doi:10.1016/j.jbiosc.2018.10.014

Galle, S., and Arendt, E.K. 2014. Exopolysaccharides from sourdough lactic acid bacteria. *Critical Reviews in Food Science and Nutrition* 54:891–901.

Gänzle, M.G. 2015. Lactic metabolism revisited: Metabolism of lactic acid bacteria in food fermentations and food spoilage. *Current Opinion in Food Science* 2:106–17. doi:10.1016/j.cofs.2015.03.001

Gänzle, M.G., and Follador, R. 2012. Metabolism of oligosaccharides and starch in lactobacilli: A review. *Frontiers in Microbiology* 3:1–15.

Gänzle, M.G., and Vogel, R.F. 2003. Contribution of reutericyclin production to the stable persistence of *Lactobacillus reuteri* in an industrial sourdough fermentation. *International Journal of Food Microbiology* 80:31–45.

Gänzle, M.G., Vermeulen, N., and Vogel, R.F. 2007. Carbohydrate, peptide and lipid metabolism of lactobacilli in sourdough. *Food Microbiology* 24:128–38.

Gänzle, M.G., Loponen, J., and Gobbetti, M. 2008. Proteolysis in sourdough fermentations: Mechanisms and potential for improved bread quality. *Trends in Food Science and Technology* 19:513–21.

Garofalo, C., Zannini, E., Aquilanti, L., Silvestri, G., Fierro, O., Picariello, G., and Clementi, F. 2012. Selection of sourdough lactobacilli with antifungal activity for use as biopreservatives in bakery products. *Journal of Agricultural and Food Chemistry* 60:7719–28.

Gaur, G., Oh, J.-H., Filannino, P., Gobbetti, M., van Pijkeren, J.-P., and Gänzle, M. 2020. Genetic determinants of hydroxycinnamic acid metabolism in heterofermentative lactobacilli. *Applied and Environmental Microbiology* 86:e02461-19.

Gezginc, Y., and Kara, Ü. 2019. The effect of exopolysaccharide producing Lactobacillus plantarum strain addition on sourdough and wheat bread quality. *Quality Assurance and Safety of Crops & Foods* 11(1):95–106. doi:10.3920/qas2018.1361

Gobbetti, M., De Angelis, M., Corsetti, A., and Di Cagno, R. 2005. Biochemistry and physiology of sourdough lactic acid bacteria. *Trends in Food Science & Technology* 16:57–69.

Gobbetti, M., Minervini, F., Pontonio, E., Di Cagno, R., and De Angelis, M. 2016. Drivers for the establishment and composition of the sourdough lactic acid bacteria biota, *International Journal of Food Microbiology* 239:3–18.

Gobbetti, M., De Angelis, M., Di Cagno, R., Polo, A., and Rizzello, C.G. 2020. The sourdough fermentation is the powerful process to exploit the potential of legumes, pseudocereals and milling by-products in baking industry. *Critical Reviews in Food Science and Nutrition* 60:2158–73.

Goh, Y.J., and Klaenhammer, T.R. 2015. Genetic mechanisms of prebiotic oligosaccharide metabolism in probiotic microbes. *Annual Reviews in Food Science and Technology* 6:137–56.

Guerzoni, M.E., Gianotti, A., and Serrazanetti, D.I. 2011. Fermentation as a tool to improve healthy properties of bread. In: *Flour and Breads and Their Fortification in Health and Disease Prevention*, V.R. Preedy, R.R. Watson, and V.B. Patel (Eds.). Academic Press, Elsevier, pp. 385–93.

Guimarães, A., Venancio, A., and Abrunhosa, L. 2018. Antifungal effect of organic acids from lactic acid bacteria on *Penicillium nordicum*. *Food Additives & Contaminants: Part A* 35:1803–18.

Hamad, S.H., Böcker, G., Vogel, R.F., and Hammes, W.P. 1992. Microbiological and chemical analysis of fermented sorghum dough for Kisra production. *Applied Microbiology and Biotechnology* 37(6):728–31.

Hansen, A., Lund, B., and Lewis, M.J. 1989. Flavor production and acidification of sourdoughs in relation to starter culture and fermentation temperature. *Lebensmittel-Wissenschaft & Technologie* 22(4):145–49. Retrieved from <Go to ISI>://WOS:A1989AR63500002

Harth, H., Van Kerrebroeck, S., and De Vuyst, L. 2016. Community dynamics and metabolite target analysis of spontaneous, backslopped barley sourdough fermentations under laboratory and bakery conditions. *International Journal of Food Microbiology* 228:22–32.

Hieke, S. Kuljanic, N. Wills, J.M. Pravst, I. Kaur, A., Raats, M.M., van Trijp, H.C.M., Verbeke, W., and Grunert, K.G. 2015. The role of health-related claims and health-related symbols in consumer behaviour: Design and conceptual framework of the CLYMBOL project and initial results. *Nutrition Bulletin* 40(1):66–72.

Huang, J., Mei, L.H., Wu, H., and Lin, D.Q. 2007. Biosynthesis of γ-aminobutyric acid (GABA) using immobilized whole cells of *Lactobacillus brevis*. *World Journal of Microbiology and Biotechnology* 23:865–71.

Huang, X., Schuppan, D., Rojas Tovar, L.E., Zevallos, V.F., Loponen, J., and Ganzle, M. 2020. Sourdough fermentation degrades wheat alpha-amylase/trypsin inhibitor (ATI) and reduces pro-inflammatory activity. *Foods* 9(7):943. doi:10.3390/foods9070943

Hüttner, E.K., Dal Bello, F., and Arendt, E.K. 2010. Identification of lactic acid bacteria isolated from oat sourdoughs and investigation into their potential for the improvement of oat bread quality. *European Food Research and Technology* 230(6):849–57.

Huys, G., Daniel, H.M., and De Vuyst, L. 2013. Taxonomy and biodiversity of sourdough yeasts and lactic acid bacteria. In *Handbook on Sourdough Biotechnology* (pp. 105–54). Springer.

Kaiser, A.L., and Montville, T.J. 1996. Purification of the bacteriocin bavaricin MN and characterization of its mode of action against *Listeria monocytogenes* Scott A cells and lipid vesicles. *Applied and Environmental Microbiology* 62:4529–35.

Kalui, C.M., Mathara, J.M., and Kutima, P.M. 2010. Probiotic potential of spontaneously fermented cereal based foods: A review. *African Journal of Biotechnology* 9(17):2490–98.

Karaman, K., Sagdic, O., and Durak, M.Z. 2018. Use of phytase active yeasts and lactic acid bacteria isolated from sourdough in the production of whole wheat bread. *LWT* 91:557–67. doi:10.1016/j.lwt.2018.01.055

Katina, K., Arendt, E., Liukkonen, K.H., Autio, K., Flander, L., and Poutanen, K. 2005. Potential of sourdough for healthier cereal products. *Trends in Food Science & Technology* 16(1–3):104–12. doi:10.1016/j.tifs.2004.03.008

Kaufmann, S.H.E. 2018. Indole propionic acid: A small molecule links between gut microbiota and tuberculosis. *Antimicrobial Agents and Chemotherapy* 62:e00389-18.

Koistinen, V.M., Mattila, O., Katina, K., Poutanen, K., Aura, A.M., and Hanhineva, K. 2018. Metabolic profiling of sourdough fermented wheat and rye bread. *Scientific Reports* 8(1):1–11.

Komatsuzaki, N., Shima, J., Kawamoto, S., Momose, H., and Kimura, T. 2005. Production of γ-aminobutyric acid (GABA) by *Lactobacillus paracasei* isolated from traditional fermented foods. *Food Microbiology* 22:497–504.

Korakli, M., and Vogel, R. 2006. Structure/function relationship of homopolysaccharide producing glycansucrases and therapeutic potential of their synthesized glycans. *Applied Microbiology and Biotechnology* 71:790–803.

Küley, E., Özyurt, G., Özogul, I., Boga, M., Akyol, I., Rocha, J.M., and Özogul, F. 2020. The role of selected lactic acid bacteria on organic acid accumulation during wet and spray-dried fish-based silages. Contributions to the winning combination of microbial food safety and environmental sustainability. *Microorganisms (MDPI)* 8(2):172; Special Issue: *Microbial Safety of Fermented Products*; https://doi.org/10.3390/microorganisms8020172.

Kumar, N., and Goel, N. 2019. Phenolic acids: Natural versatile molecules with promising therapeutic applications. *Biotechnology Reports* 24:e00370.

Kumariya, R., Garsa, A.K., Rajput, Y.S., Sood, S.K., Akhtar, N., and Patel, S. 2019. Bacteriocins: Classification, synthesis, mechanism of action and resistance development in food spoilage causing bacteria. *Microbial Pathogenesis* 128:171–7.

Kwaw, E., Ma, Y., Tchabo, W., Apaliya, M.T., Wu, M., Sackey, A.S., Xiao, L., and Tahir, H.E. 2018. Effect of *Lactobacillus* strains on phenolic profile, color attributes and antioxidant activities of lactic-acid fermented mulberry juice. *Food Chemistry* 250:148–54.

Lanciotti, R., Patrignani, F., Bagnolini, F., Guerzoni, M.E., and Gardini, F. 2003. Evaluation of diacetyl antimicrobial activity against *Escherichia coli*, *Listeria monocytogenes* and *Staphylococcus aureus*. *Food Microbiology* 20:537–43.

Lavermicocca, P., Valerio, F., Evidente, A., Lazzaroni, S., Corsetti, A., and Gobbetti, M. 2000. Purification and characterization of novel antifungal compounds from the sourdough *Lactobacillus plantarum* strain 21B. *Applied and Environmental Microbiology* 66:4084–90.

Lavermicocca, P., Valerio, F., and Visconti, A. 2003. Antifungal activity of phenyllactic acid against moulds isolated from bakery products. *Applied and Environmental Microbiology* 69:634–40.

Le Lay, C., Mounier, J., Vasseur, V., Weill, A., Le Blay, G., Barbier, G., and Coton, E. 2016. In vitro and in situ screening of lactic acid bacteria and propionibacteria antifungal activities against bakery product spoilage molds. *Food Control* 60:247–55.

Leyva Salas, M., Mounier, J., Valence, F., Coton, M., Thierry, A., and Coton, E. 2017. Antifungal microbial agents for food biopreservation: A review. *Microorganisms* 5:37.

Liu, A., Jia, Y., Zhao, L., Gao, Y., Liu, G., Chen, Y., and Liu, S. 2018. Diversity of isolated lactic acid bacteria in Ya'an sourdoughs and evaluation of their exopolysaccharide production characteristics. *LWT* 95:17–22. doi:10.1016/j.lwt.2018.04.061

Liu, S., Soomro, L., Wei, X., Yuan, X., Gu, T., Li, Z., Wang, Y., Bao, Y., Wang, F., Wen, B., and Xin, F. 2021. Directed evolution of feruloyl esterase from *Lactobacillus acidophilus* and its application for ferulic acid production. *Bioresource Technology*:124967.

Lotong, V., Iv, E.C., and Chambers, D.H. 2000. Determination of the sensory attributes of wheat sourdough bread. *Journal of Sensory Studies* 15(3):309–26. doi:10.1111/j.1745-459X.2000.tb00273.x

Luo, Y., Wang, C.Z., Sawadogo, R., Yuan, J., Zeng, J., Xu, M., Tan, T., and Yuan, C.S. 2021. 4-Vinylguaiacol, an active metabolite of ferulic acid by enteric microbiota and probiotics, possesses significant activities against drug-resistant human colorectal cancer cells. *ACS Omega* 6:4551–61.

Ma, S., Wang, Z., Guo, X., Wang, F., Huang, J., Sun, B., and Wang, X. 2021. Sourdough improves the quality of whole-wheat flour products: Mechanisms and challenges: A review. *Food Chemistry* 360:130038. doi:10.1016/j.foodchem.2021.130038

Maioli, M., Pes, G.M., Sanna, M., Cherchi, S., Dettori, M., Manca, E., and Farris, G.A. 2008. Sourdough-leavened bread improves postprandial glucose and insulin plasma levels in subjects with impaired glucose tolerance. *Acta Diabetologica* 45:91–6.

Menezes, L.A.A., Molognoni, L., de Sá Ploêncio, L.A., Costa, F.B.M., Daguer, H., and Dea Lindner, J.D. (2019). Use of sourdough fermentation to reducing FODMAPs in breads. *European Food Research and Technology* 245(6):1183–95. doi:10.1007/s00217-019-03239-7

Meroth, C.B., Hammes, W.P., and Hertel, C. 2003. Identification and population dynamics of yeasts in sourdough fermentation processes by PCR-denaturing gradient gel electrophoresis. *Applied and Environmental Microbiology* 69(12):7453–61.

Messens, W., Neysens, P., Vansieleghem, W., Vanderhoeven, J., and De Vuyst, L. 2002. Modelling growth and bacteriocin production by *Lactobacillus amylovorus* DCE 471 in response to temperature and pH values used for sourdough fermentations. *Applied and Environmental Microbiology* 68:1431–5.

Michel, E., Monfort, C., Deffrasnes, M., Guezenec, S., Lhomme, E., Barret, M., Sicard, D., Dousset, X., and Onno, B. 2016. Characterization of relative abundance of lactic acid bacteria species in French organic sourdough by cultural, qPCR and MiSeq high-throughput sequencing methods. *International Journal of Food Microbiology* Elsevier 239:35–43. Retrieved from https://hal.archives-ouvertes.fr/hal-01527569/document. https://doi.org/10.1016/j.ijfoodmicro.2016.07.034

Mieszkin, S., Hymery, N., Debaets, S., Coton, E., Le Blay, G., Valence, F., and Mounier, J. 2017. Action mechanisms involved in the bioprotective effect of *Lactobacillus harbinensis* K.V9.3.1.Np against *Yarrowia lipolytica* in fermented milk. *International Journal of Food Microbiology* 248:47–55.

Mimori, S., Kawada, K., Saito, R., Takahashi, M., Mizoi, K., Okuma, Y., Hosokawa, M., and Kanzaki, T. 2019. Indole-3-propionic acid has chemical chaperone activity and suppresses endoplasmic reticulum stress-induced neuronal cell death. *Biochemical and Biophysical Research Communications* 517:623–8.

Minervini, F., De Angelis, M., Di Cagno, R., and Gobbetti, M. 2014. Ecological parameters influencing microbial diversity and stability of traditional sourdough. *International Journal of Food Microbiology* 171:136–46. doi: 10.1016/j.ijfoodmicro.2013.11.021

Minervini, F., Celano, G., Lattanzi, A., De Angelis, M., and Gobbetti, M. 2016. Added ingredients affect the microbiota and biochemical characteristics of durum wheat type-I sourdough. *Food Microbiology* 60, 112–23. doi:10.1016/j.fm.2016.05.016

Minervini, F., Di Cagno, R., Lattanzi, A., De Angelis, M., Antonielli, L., Cardinali, G., Cappelle, S., and Gobbetti, M. 2012. Lactic acid bacterium and yeast microbiotas of 19 sourdoughs used for traditional/typical Italian breads: Interactions between ingredients and microbial species diversity. *Applied and Environmental Microbiology* 78(4):1251–64.

Minervini, F., Dinardo, F.R., Celano, G., De Angelis, M., and Gobbetti, M. 2018. Lactic acid bacterium population dynamics in artisan sourdoughs over one year of daily propagations is mainly driven by flour microbiota and nutrients. *Frontiers in Microbiology* 9:1984. doi:10.3389/fmicb.2018.01984

Minervini, F., Dinardo, F.R., De Angelis, M., and Gobbetti, M. 2019. Tap water is one of the drivers that establish and assembly the lactic acid bacterium biota during sourdough preparation. *Scientific Report* 9(1):570. doi:10.1038/s41598-018-36786-2

Mohammed, S.I., Steenson, L.R., and Kirleis, A.W. 1991. Isolation and characterization of microorganisms associated with the traditional sorghum fermentation for production of Sudanese kisra. *Applied and Environmental Microbiology* 57(9):2529–33.

Moroni, A.V., Dal Bello, F., and Arendt, E.K. 2009. Sourdough in gluten-free bread-making: An ancient technology to solve a novel issue? *Food Microbiology* 26(7):676–84.

Moroni, A.V., Arendt, E.K., and Dal Bello, F. 2011. Biodiversity of lactic acid bacteria and yeasts in spontaneously-fermented buckwheat and teff sourdoughs. *Food Microbiology* 28(3):497–502.

Mu, W., Yu, S., Zhu, L., Zhang, T., and Jiang, B. 2012. Recent research on 3-phenyllactic acid, a broad-spectrum antimicrobial compound. *Applied Microbiology and Biotechnology* 95:1155–63.

Nachi, I., Fhoula, I., Smida, I., Ben Taher, I., Chouaibi, M., Jaunbergs, J., and Hassouna, M. 2018. Assessment of lactic acid bacteria application for the reduction of acrylamide formation in bread. *LWT* 92:435–41. doi:10.1016/j.lwt.2018.02.061

Niku-Paavola, M.L., Laitila, A., Mattila-Sandholm, T., and Haikara, A. 1999. New types of antimicrobial compounds produced by *Lactobacillus plantarum*. *Journal of Applied Microbiology* 86:29–35.

Ninety.com, and Brandt, M.J. 2019. Industrial production of sourdoughs for the baking branch: An overview. *International Journal of Food and Microbiology* 302:3–7. doi:10.1016/j.ijfoodmicro.2018.09.008

Novotni, D., Gänzle, M., and Rocha, J.M. 2020. Chapter 5. Composition and activity of microbiota in sourdough and their effect on bread quality and safety. In: *Trends in Wheat and Bread Making*, C.M. Galanakis (Ed.). Elsevier-Academic Press, 469 pp.; Galanaksis, Charis M. (Ed), https://doi.org/10.1016/B978-0-12-821048-2.00005-2, *Galanaksis-TWBM-1632435*, ISBN 978-0-12-821048-2.

Nuraida, L., Wacher, M.C., and Owens, J.D. 1995. Microbiology of pozol, a Mexican fermented maize dough. *World Journal of Microbiology and Biotechnology* 11(5):567–71.

Okada, T., Sugishita, T., Murakami, T., Murai, H., Saikusa, T., Horino, T., Onoda, A., Kajimoto, O., Takahashi, R., and Takahashi, T. 2000. Effect of the defatted rice germ enriched with GABA for sleeplessness, depression, autonomic disorder by oral administration. *Journal of the Japanese Society for Food Science and Technology* 47:596–603.

Okkers, D.J., Dicks, L.M., Silvester, M., Joubert, J.J., and Odendaal, H.J. 1999. Characterization of pentocin TV35b, a bacteriocin-like peptide isolated from *Lactobacillus pentosus* with a fungistatic effect on *Candida albicans*. *Journal of Applied Microbiology* 87:726–34.

Păcularu-Burada, B., Georgescu, L.A., Vasile, M.A., Rocha, J.M., and Bahrim, G.-E. 2020. Selection of wild lactic acid bacteria strains as promoters of postbiotics in gluten-free sourdoughs. *Microorganisms (MDPI)* 8 (5):643, Special Issue: *Microbial Safety of Fermented Products*; doi: https://doi.org/10.3390/microorganisms8050643.

Păcularu-Burada, B., Turturică, M., Rocha, J.M., and Bahrim, G.-E. 2021. Statistical approach to potentially enhance the postbiotication of gluten-free sourdough. *Applied Sciences (MDPI)* 11(11) 5306. Special Issue: *Advances of Lactic Fermentation for Functional Food Production*. doi: https://doi.org/10.3390/app11115306

Palla, M., Cristani, C., Giovannetti, M., and Agnolucci, M. 2017. Identification and characterization of lactic acid bacteria and yeasts of PDO Tuscan bread sourdough by culture dependent and independent methods. *International Journal of Food and Microbiology* 250:19–26. doi:10.1016/j.ijfoodmicro.2017.03.015

Palla, M., Agnoluccia, M., Calzonea, A., Giovannettia, M., Di Cagnoc, R., Gobbettic, M., Rizzellod, C.G., and Pontoniod, E. 2019. Exploitation of autochthonous Tuscan sourdough yeasts as potential starters. *International Journal of Food Microbiology* 302:59–68.

Peirotén, A., Gaya, P., Alvarez, I., Bravo, D., and Landete, J.M. 2019. Influence of different lignan compounds on enterolignan production by *Bifidobacterium* and *Lactobacillus* strains. *International Journal of Food Microbiology* 289:17–23.

Peirotén, A., Bravo, D., and Landete, J.M. 2020. Bacterial metabolism as responsible of beneficial effects of phytoestrogens on human health. *Critical Reviews in Food Science and Nutrition* 60:1922–37.

Pepe, O., Villani, F., Oliviero, D., Greco, T., and Coppola, S. 2003. Effect of proteolytic starter cultures as leavening agents of pizza dough. *International Journal of Food Microbiology* 84:319–26.

Pétel, C., Onno, B., and Prost, C. 2017. Sourdough volatile compounds and their contribution to bread: A review. *Trends in Food Science & Technology* 59:105–23.

Petrova, P., and Petrov, K. 2011. Antimicrobial activity of starch-degrading *Lactobacillus* strains isolated from boza. *Biotechnology & Biotechnological Equipment* 25(sup1):114–6.

Petrova, P., and Petrov, K. 2017. Prebiotic–probiotic relationship: The genetic fundamentals of polysaccharides conversion by *Bifidobacterium* and *Lactobacillus* genera. In: *Handbook of Food Bioengineering*. 1st ed., Elsevier Inc., Volume 2, pp. 237–78.
Petrova, P., and Petrov, K. 2020. Lactic acid fermentation of cereals and pseudocereals: Ancient nutritional biotechnologies with modern applications. *Nutrients* 12:1118.
Petrova, P., Petrov, K., and Stoyancheva, G. 2013. Starch–modifying enzymes of lactic acid bacteria: Structures, properties, and applications. *Starch/Stärke* 65:34–47.
Petrova, P., Ivanov, I., Tsigoriyna, L., Valcheva, N., Vasileva, E., Parvanova-Mancheva, T., Arsov, A., and Petrov, K. 2021. Traditional Bulgarian dairy products: Ethnic foods with health benefits. *Microorganisms* 9:480.
Pfliegler, W.P., Pusztahelyi, T., and Pocsi, I. 2015. Mycotoxins: Prevention and decontamination by yeasts. *J Basic Microbiol*, 55(7), 805–18. doi:10.1002/jobm.201400833
PiekarskaRadzik, L., and Klewicka, E. 2021. Mutual influence of polyphenols and Lactobacillus spp. bacteria in food: A review. *European Food Research and Technology* 247:9–24.
Pizarro, F., and Franco, F. 2017. Volatile organic compounds at early stages of sourdough preparation via static headspace and GC/MS analysis. *Current Research in Nutrition and Food Science*, 5:2.
Reid, G. 2008. Probiotic lactobacilli for urogenital health in women. *Journal of Clinical Gastroenterology* 3(42 Suppl. 3 Pt 2):234–6.
Riedl, A., Gieger, C., Hauner, H., Daniel, H., and Linseisen, J. 2017. Metabotyping and its application in targeted nutrition: An overview. *British Journal of Nutrition* 117:1631–44.
Rietjens, I.M.C.M., Louisse, J., and Beekmann, K. 2017. The potential health effects of dietary phytoestrogens. *British Journal of Pharmacology* 174:1263–80.
Rijkers, G.T. de Vos, W.M. Brummer, R.J. Morelli, L. Corthier, G., and Marteau, P. 2011. Health benefits and health claims of probiotics: Bridging science and marketing. *British Journal of Nutrition* 106(9):1291–6. doi:10.1017/S000711451100287X
Ripari, V., Gänzle, M.G., and Berardi, E. 2016. Evolution of sourdough microbiota in spontaneous sourdoughs started with different plant materials. *International Journal of Food Microbiology* 232:35–42.
Ripari, V., Bai, Y., and Gänzle, M.G. 2019. Metabolism of phenolic acids in whole wheat and rye malt sourdoughs. *Food Microbiology* 77:43–51.
Rizzello, C.G., Lorusso, A., Montemurro, M., and Gobbetti, M. 2016. Use of sourdough made with quinoa (Chenopodium quinoa) flour and autochthonous selected lactic acid bacteria for enhancing the nutritional, textural and sensory features of white bread. *Food Microbiology* 56:1–13.
Rocha, J.M. 2011. *Microbiological and lipid profiles of broa: Contributions for the characterization of a traditional portuguese bread*. PhD thesis dissertation. Instituto Superior de Agronomia, Universidade de Lisboa (ISA-UL), Lisbon, Portugal, 705 pages. Thesis available at http://hdl.handle.net/10400.5/3876
Rocha, J.M., and Malcata, F.X. 1999. On the microbiological profile of traditional Portuguese sourdough. *Journal of Food Protection* 62 (12):1416–29. ISSN: 0362-028X. http://www.scopus.com/inward/record.url?eid=2-s2.0-0345201643&partnerID=MN8TOARS.
Rocha, J.M., and Malcata, F.X. 2012. Microbiological profile of maize and rye flours, and sourdough used for the manufacture of traditional Portuguese bread. *Food Microbiology* 31:72–88. doi: 10.1016/j.fm.2012.01.008.
Rocha, J.M., and Malcata, F.X. 2016a. Microbial ecology dynamics in Portuguese broa sourdough. *Journal of Food Quality* 39(6):634–48.
Rocha, J.M., and Malcata, F.X. 2016b. Behavior of the complex micro-ecology in maize and rye flour and mother-dough for broa throughout storage. *Journal of Food Quality* 39:218–33. doi: 10.1111/jfq.12183.

Röcken, W. 1996. Applied aspects of sourdough fermentation. *Advances in Food Sciences* 18:212–6.
Ryan, L.A.M., Dal Bello, F., Arendt, E.K., and Koehler, P. 2009. Detection and quantitation of 2,5-diketopiperazines in wheat sourdough and bread. *Journal of Agricultural and Food Chemistry* 57:9563–8.
Ryan, L.A., Zannini, E., Dal Bello, F., Pawlowska, A., Koehler, P., and Arendt, E.K. 2011. Lactobacillus amylovorus DSM 19280 as a novel food-grade antifungal agent for bakery products. *International Journal of Food Microbiology* 146(3):276–83. doi:10.1016/j.ijfoodmicro.2011.02.036
Scherf, K.A. 2019. Immunoreactive cereal proteins in wheat allergy, non-celiac gluten/wheat sensitivity (NCGS) and celiac disease. *Current Opinion in Food Science* 25:35–41. doi:10.1016/j.cofs.2019.02.003
Schwenninger, S.M., Lacroix, C., Truttmann, S., Jans, C., Sporndli, C., Bigler, L., and Meile, L. 2008. Characterization of low-molecular-weight anti yeast metabolites produced by a food-protective *Lactobacillus-Propionibacterium* coculture. *Journal of Food Protection* 71:2481–7.
Sekwati-Monang, B., Valcheva, R., and Gänzle, M.G. 2012. Microbial ecology of sorghum sourdoughs: Effect of substrate supply and phenolic compounds on composition of fermentation microbiota. *International Journal of Food Microbiology* 159(3):240–6.
Sharma, N., Angural, S., Rana, M., Puri, N., Kondepudi, K.K., and Gupta, N. 2020. Phytase producing lactic acid bacteria: Cell factories for enhancing micronutrient bioavailability of phytate rich foods. *Trends in Food Science & Technology* 96:1–12.
Silva, A.R., Grosso, A.C., Deleure-Matos, C., and Rocha, J.M. 2019. Comprehensive review on the interaction between natural compounds and brain receptors: Benefits and toxicity. *European Journal of Medicinal Chemistry* 174:87–115. Elsevier. ISSN: 0223-5234. 17 April 2019. doi: https://doi.org/10.1016/j.ejmech.2019.04.028.
Silva, M., and Lidon, F. 2016. Food preservatives: An overview on applications and side effects. *Emirates Journal of Food and Agriculture* 28:366–73.
Skendi, A., Zinoviadou, K.G., Papageorgiou, M., and Rocha, J.M. 2020. Advances on the valorisation and functionalization of by-products and wastes from cereal-based processing industry. *Foods (MDPI)* 9 (9) 1243. doi:10.3390/foods9091243.
Spicher, G., and Stephan, H. 1966. Microflora of sourdough. 3. Studies on lactobacilli occurring in "spontaneous sourdoughs" and their significance for baking techniques. *Zentralblatt fur Bakteriologie, Parasitenkunde, Infektionskrankheiten und Hygiene. Zweite naturwissenschaftliche Abt.: Allgemeine, landwirtschaftliche und technische Mikrobiologie*, 120(7), 685–702. Retrieved from https://www.scopus.com/inward/record.uri?eid=2-s2.0-0013987961&partnerID=40&md5=0a8ef0234c2c03d6eda68b9647b4fce5
Sterr, Y., Weiss, A., and Schmidt, H. 2009. Evaluation of lactic acid bacteria for sourdough fermentation of amaranth. *International Journal of Food Microbiology* 136(1):75–82.
Ström, K., Sjogren, J., Broberg, A., and Schnurer, J. 2002. *Lactobacillus plantarum* MiLAB 393 produces the antifungal cyclic dipeptides cyclo(L-Phe-L-Pro) and cyclo(L-Phe-trans-4-OH-L-Pro) and 3-phenyllactic acid. *Applied and Environmental Microbiology* 68:4322–7.
Suo, B., Chen, X., and Wang, Y. 2021. Recent research advances of lactic acid bacteria in sourdough: Origin, diversity, and function. *Current Opinion in Food Science* 37:66–75.
Šušković, J., Kos, B., Beganović, J., Pavunc, A.L., Habjanič, K., and Matošić, S. 2010. Antimicrobial activity-the most important property of probiotic and starter lactic acid bacteria. *Food Technology and Biotechnology* 48:296–307.

Svensson, L., Sekwati-Monang, B., Lutz, D.L., Schieber, A., and Gänzle, M.G. 2010. Phenolic acids and flavonoids in nonfermented and fermented red sorghum (Sorghum bicolor (L.) Moench). *Journal of Agricultural and Food Chemistry* 58:9214–20.

Taglieri, I., Macaluso, M., Bianchi, A., Sanmartin, C., Quartacci, M.F., Zinnai, A., and Venturi, F. 2021. Overcoming bread quality decay concerns: Main issues for bread shelf life as a function of biological leavening agents and different extra ingredients used in formulation. A review. *Journal of the Science of Food and Agriculture* 101:1732–43.

Tejero-Sariñena, S., Barlow, J., Costabile, A., Gibson, G.R., and Rowland, I. 2012. In vitro evaluation of the antimicrobial activity of a range of probiotics against pathogens: Evidence for the effects of organic acids. *Anaerobe* 18:530–8.

Tieking, M., Kuhnl, W., and Gänzle, M.G. 2005. Evidence for formation of heterooligosaccharides by *Lactobacillus sanfranciscensis* during growth in wheat sourdough. *Journal of Agricultural and Food Chemistry* 53:2456–61.

Trakselyte-Rupsiene, K., Juodeikiene, G., Alzbergaite, G., Zadeike, D., Bartkiene, E., Özogul, F., Ruller, L., Robert, J., and Rocha, J.M. 2021. Bio-refinery of plant drinks press cake permeate using ultrafiltration and Lactobacillus fermentation into antimicrobials and its effect on the growth of wheatgrass in vivo. *Food Bioscience* 46:101427. In press. https://doi.org/10.1016/j.fbio.2021.101427

Turpin, W., Humblot, C., and Guyot, J.-P. 2011. Genetic screening of functional properties of lactic acid bacteria in a fermented pearl millet slurry and in the metagenome of fermented starchy foods. *Applied and Environmental Microbiology* 77:8722–34.

Urien, C., Legrand, J., Montalent, P., Casaregola, S., and Sicard, D. 2019. Fungal species diversity in french bread sourdoughs made of organic wheat flour. *Frontiers in Microbiology* 10:201. doi:10.3389/fmicb.2019.00201

Valerio, F., Di Biase, M., Lattanzio, V.M.T., and Lavermicocca, P. 2016. Improvement of the antifungal activity of lactic acid bacteria by addition to the growth medium of phenylpyruvic acid, a precursor of phenyllactic acid. *International Journal of Food Microbiology* 222:1–7.

Van der Meulen, R., Scheirlinck, I., Van Schoor, A., Huys, G., Vancanneyt, M., Vandamme, P., and De Vuyst, L. 2007. Population dynamics and metabolite target analysis of lactic acid bacteria during laboratory fermentations of wheat and spelt sourdoughs. *Applied and Environmental Microbiology* 73:4741–50.

Van Kerrebroeck, S., Maes, D., and De Vuyst, L. 2017. Sourdoughs as a function of their species diversity and process conditions, a meta-analysis. *Trends in Food Science & Technology* 68:152–9. doi:10.1016/j.tifs.2017.08.016

Velikova, P., Stoyanov, A., Blagoeva, G., Popova, L., Petrov, K., Gotcheva, V., Angelov, A., and Petrova, P. 2016. Starch utilization routes in lactic acid bacteria: New insight by gene expression assay. *Starch–Stärke* 68:953–60.

Venturi, F., Sanmartin, C., Taglieri, I., Nari, A., Andrich, G., and Zinnai, A. 2016. Effect of the baking process on artisanal sourdough bread-making: A technological and sensory evaluation. *Agrochimica* 60(3):222–34. doi:10.12871/00021857201635

Venturi, M., Galli, V., Pini, N., Guerrini, S., and Granchi, L. 2019. Use of selected lactobacilli to increase gamma-aminobutyric acid (GABA) content in sourdough bread enriched with amaranth flour. *Foods* 8(6):1–13. doi:10.3390/foods8060218

Villegas, J.M., Brown, L., Savoy de Giori, G., and Hebert, E.M. 2016. Optimization of batch culture conditions for GABA production by Lactobacillus brevis CRL 1942, isolated from quinoa sourdough. *LWT: Food Science and Technology* 67:22–6. doi:10.1016/j.lwt.2015.11.027

Vitali Čepo, D., Prusac, M., Velkovski Škopić, O., and Tatarević, A. 2020. Pharmacist Recommendations on the Use of Probiotics. *Medicus* 29(1):115–34.

Vogelmann, S.A., Seitter, M., Singer, U., Brandt, M.J., and Hertel, C. 2009. Adaptability of lactic acid bacteria and yeasts to sourdoughs prepared from cereals, pseudocereals and cassava and use of competitive strains as starters. *International Journal of Food Microbiology* 130(3):205–12.

Vrancken, G., Rimaux, T., Weckx, S., Leroy, F., and De Vuyst, L. 2011. Influence of temperature and backslopping time on the microbiota of a type I propagated laboratory wheat sourdough fermentation. *Applied and Environmental Microbiology* 77(8):2716–26.

Wang, Y., Qin, Y., Xie, Q., Zhang, Y., Hu, J., and Li, P. 2018a. Purification and characterization of Plantaricin LPL-1, a novel class IIa Bacteriocin produced by *Lactobacillus plantarum* LPL-1 isolated from fermented fish. *Frontiers in Microbiology* 9:2276.

Wang, Y., Sorvali, P., Laitila, A., Maina, N.H., Coda, R., and Katina, K. 2018b. Dextran produced in situ as a tool to improve the quality of wheat-faba bean composite bread. *Food Hydrocolloids* 84:396–405.

Weckx, S., Van der Meulen, R., Maes, D., Scheirlinck, I., Huys, G., Vandamme, P., and De Vuyst, L. 2010. Lactic acid bacteria community dynamics and metabolite production of rye sourdough fermentations share characteristics of wheat and spelt sourdough fermentations. *Food Microbiology* 27(8):1000–8.

Won, J.S., Kim, W.J., Geun Lee, K., Woo Kim, C., and Seob Noh, W. 2008. Fermentation characteristics of exopolysaccharide-producing lactic acid bacteria from sourdough and assessment of the isolates for industrial potential. *Journal of Microbiology and Biotechnology* 18(7):1266–73.

Wu, Q., Tun, H.M., Law, Y.S., Khafipour, E., and Shah, N.P. 2017. Common distribution of gad operon in Lactobacillus brevis and its GadA contributes to efficient GABA synthesis toward cytosolic near-neutral pH. *Frontiers in Microbiology* 8:206. doi:10.3389/fmicb.2017.00206

Xu, D., Zhang, Y., Tang, K., Hu, Y., Xu, X., and Ganzle, M.G. 2019. Effect of mixed cultures of yeast and lactobacilli on the quality of wheat sourdough bread. *Frontiers in Microbiology* 10:2113. doi:10.3389/fmicb.2019.02113

Yagmur, G., Tanguler, H., Leventdurur, S., Elmaci, S., Turhan, E., Francesca, N., and Erten, H. 2016. Identification of predominant lactic acid bacteria and yeasts of Turkish sourdoughs and selection of starter cultures for liquid sourdough production using different flours and dough yields. *Polish Journal of Food and Nutrition Sciences* 66(2):99–107. doi:10.1515/pjfns-2015-0041

Yildirim, R.M., and Arici, M. 2019. Effect of the fermentation temperature on the degradation of phytic acid in whole-wheat sourdough bread. *LWT: Food Science and Technology* 112:1–9. doi:10.1016/j.lwt.2019.05.122

Zadeike, D., Vaitkeviciene, R., Bartkevics, V., Bogdanova, E., Bartkiene, E., Lele, V., and Valatkeviciene, Z. 2021. The expedient application of microbial fermentation after whole-wheat milling and fractionation to mitigate mycotoxins in wheat-based products. *LWT: Food Science and Technology* 137. doi:10.1016/j.lwt.2020.110440

Zangeneh, M., Khaleghi, M., and Khorrami, S. 2019. Isolation of Lactobacillus plantarum strains with robust antagonistic activity, qualified probiotic properties, and without antibiotic-resistance from traditional sourdough. *Avicenna Journal of Clinical Microbiology and Infection* 6(2):66–74. doi:10.34172/ajcmi.2019.13

Zannini, E., Garofalo, C., Aquilanti, L., Santarelli, S., Silvestri, G., and Clementi, F. 2009. Microbiological and technological characterization of sourdoughs destined for breadmaking with barley flour. *Food Microbiology* 26(7):744–53.

Zannini, E., Waters, D.M., Coffey, A., and Arendt, E.K. 2016. Production, properties, and industrial food application of lactic acid bacteria-derived exopolysaccharides. *Applied Microbiology and Biotechnology* 100:1121–35.

Zaunmüller, T., Eichert, M., Richter, H., and Unden, G. 2006. Variations in the energy metabolism of biotechnologically relevant heterofermentative lactic acid bacteria during growth on sugars and organic acids. *Applied Microbiology and Biotechnology* 72:421–9.

Zeidan, A.A., Poulsen, V.K., Janzen, T., Buldo, P., Derkx, P.M.F., Øregaard, G., and Neves, A.R. 2017. Polysaccharide production by lactic acid bacteria: from genes to industrial applications. *FEMS Microbiological Reviews* 41(Supplement 1):S168–S200.

Zhang, L.S., and Davies, S.S. 2016. Microbial metabolism of dietary components to bioactive metabolites: Opportunities for new therapeutic interventions. *Genome Medicine* 8:46.

Zheng, J., Wittouck, S., Salvetti, E., Franz, C.M.A.P., Harris, H.M.B., Mattarelli, P., and Lebeer, S. 2020. A taxonomic note on the genus Lactobacillus: Description of 23 novel genera, emended description of the genus Lactobacillus Beijerinck 1901, and union of Lactobacillaceae and Leuconostocaceae. *International Journal of Systematic and Evolutionary Microbiology* 70(4):2782–858. doi:10.1099/ijsem.0.004107

Zokaityte, E., Cernauskas, D., Klupsaite, D., Lele, V., Starkute, V., Zavistanaviciute, P., Ruzauskas, M., Gruzauskas, R., Juodeikiene, G., Rocha, J.M., Bliznikas, S., Viskelis, P., Ruibys, R., and Bartkiene, E. 2020. Bioconversion of milk permeate with selected lactic acid bacteria strains and apple by-products into beverages with antimicrobial properties and enriched with galactooligosaccharides. *Microorganisms* 8(8):1182. doi: https://doi.org/10.3390/microorganisms8081182. ISSN 2076-2607, Publication date: 03/08/2020. Special issue *"Screening and characterization of the diversity of food microorganisms and their metabolites"*.

Zotta, T., Piraino, P., Parente, E., Salzano, G., and Ricciardi, A. 2008. Characterization of lactic acid bacteria isolated from sourdoughs for Cornetto, a traditional bread produced in Basilicata (Southern Italy). *World Journal of Microbiology and Biotechnology* 24:1785–95.

11 Isolation, Technological Functionalization, and Immobilization Techniques Applied to Cereals and Cereal-Based Products and Sourdough Microorganisms

Zlatina A. Genisheva,
Pedro Ferreira-Santos, and José A. Teixeira

CONTENTS

11.1 Introduction 311
11.2 Microorganisms Isolated from Sourdough 313
11.3 Functional Metabolites and Enzymes 315
    11.3.1 Enzymes of High Industrial Interest 316
11.4 Cell Immobilization 320
    11.4.1 Advantages and Disadvantages of Cell Immobilization 322
    11.4.2 Types of Supports 323
    11.4.3 Uses of Cell Immobilization in Bread Making 325
11.5 Conclusion 330
Acknowledgment 330
References 330

11.1 INTRODUCTION

Bread has been a staple food in western countries for centuries, and bread formulation and bread technology have undergone constant evolution over time. There are seven main ingredients in bread composition: flour, water, salt, sugars, leavening agent (microorganisms or chemicals), additives (stabilizer, emulsifier, oxidant, and gums), and enzymes (amylase, protease, hemicellulase, lipase, and lipoxygenase). Moreover, bread composition is multivariable since, for each bread ingredient, various products of different physical and chemical properties are available (Gugerli et al. 2004). Furthermore, bread undergoes four main process stages: mixing, leavening, baking, and storage. Each of these stages can be further optimized, thus increasing the complexity of the bread making process. Existing research in bread making is mainly focused on new technologies to achieve better mechanical properties of doughs, to extend preservation time, to improve the flavor, and to enhance the nutritional quality of bread (Plessas, Trantallidi, et al. 2007).

One way to achieve leavening of the dough is by a fermentation process using selected yeasts and sometimes lactic acid bacteria (LAB). The dough leavening includes biochemical, rheological, and thermodynamic phenomena. Carbon dioxide (CO_2) is produced inside the dough during leavening, which causes its rise and the creation of a porous structure. The use of sourdough as a starter in bread making was considered the objective of mimicking the traditional process of bread making and to satisfy the demand of consumers for natural technologies. Sourdough is a mixture of flour and water, yeasts, and lactic acid bacteria, where the diversity of LAB is larger than that of the yeast microbiota. The use of sourdough instead of baker's yeast itself, improves some important characteristics of bread, such as flavor and texture, nutritional value, and visual aspect. Moreover, the *in situ* production of antimicrobial compounds (organic acids, bacteriocins, etc.) extend the preservation time of the bread (Plessas, Trantallidi, et al. 2007). The use of sourdough in bread making decreases the glycemic index, the content of gluten, phytic acid, and fermentable oligosaccharides, disaccharides, monosaccharides, and polyols (FODMAPS). At the same time, it increases the release of bioactive peptides and phenolic compounds (Canesin and Cazarin 2021).

The cereal chemical composition used to make the flours for bread making has a great impact on the bread characteristics. The main cereal used in the sourdough bread is wheat. To improve the quality of the bread, different conventional or non-conventional flours (rye, barley, quinoa, triticale, sorghum, oat, and maize) can be used. Flours may also have anti-nutritional factors in their composition that decrease the nutritional quality of bread. However, during the fermentation process, these compounds may be reduced or even extinguished. The modification of the structure and composition of the flours during fermentation is very important to the quality of the final product (Canesin and Cazarin 2021).

Leavening of the bread is a solid-state fermentation process mostly carried out by free cells. The use of immobilization systems instead of free cells have been proven to have various advantages in the fermentation process (Genisheva, Teixeira, and Oliveira 2014). With the immobilization of microorganisms can be achieved higher cell density and biological activity, and also higher resistance to negative environmental conditions (high substrate concentrations, temperature, inhibitors, etc.). Moreover, the immobilized cells are easier to collect and reuse. In general, viability and function of the immobilized microorganisms are affected by the type of carriers used. The selection of a carrier is very important, especially when used in the food industry (Genisheva, Teixeira, and Oliveira 2014). Carriers should be easily accepted by the consumers and have the required food-grade purity. In this chapter, we will describe and evaluate the role of the immobilization technique in bread making, specifically in sourdough.

11.2 MICROORGANISMS ISOLATED FROM SOURDOUGH

Hundreds of different types of traditional sourdough breads exist, especially in Europe. Sourdough has a unique microbiota that is result of a synergy between yeasts and LAB. The diversity of sourdough microbial communities depends on process technologies, types of flour, and other ingredients traditionally associated with local culture and origin (Palla et al. 2017; De Vuyst et al. 2016). The diversity of microbial species identified in sourdoughs is wide, with more than 60 species of LAB and 30 species of yeasts. However, only one or two dominant species of yeasts and LAB are found per sourdough (Boudaoud et al. 2021; Carbonetto et al. 2018). Yeast species contribute to the leavening and aroma compounds of bread. The LAB species' primary contributions are acidification, flavor, and leavening of the dough (Carbonetto et al. 2018).

Saccharomyces cerevisiae is the most frequent yeast species isolated from sourdough; however, other subspecies of *Saccharomyces, Candida, Pichia,* and *Hansenula* have occasionally been isolated. Yeasts belonging to the genera *Kazachstania, Wickerhamomyces,* and *Torulaspora* were also identified in sourdough (Carbonetto et al. 2018). It was considered that the most geographically widespread species of yeasts in sourdoughs are *S. cerevisiae, Wickerhamomyces anomalus, Torulaspora delbrueckii, Pichia kudriavzevii, Kazachstania exigua,* and *Kazachstania humilis*. These six species were found in samples from Asia and Europe, and some from America, Africa, and Australia (Carbonetto et al. 2018). In Europe, yeast species' diversity in sourdough has been mainly studied in Italy, Belgium, and France. *S. cerevisiae* was present in almost 80% of the sourdough's yeast, but not necessarily the dominant yeast. *K. humilis* and *W. anomalus* were present in 16% and 8% of the samples, respectively. Moreover, the species *W. anomalus* was found in Belgium, while the species *Kazachstania bulderi* was detected only in France (Carbonetto et al. 2018). It was found that most of the commercial sourdoughs used in bakeries in Finland contained yeasts similar to *Candida milleri*. Other yeasts identified in Finish sourdough were *Torulopsis holmii, S. cerevisiae,* and *Torulopsis stellate*. Contaminants were also identified as *Endomycopsis fibuliger* and *Hansenula anomala* (Mäntynen et al. 1999). The most encountered yeast in a Portuguese sourdough were *S. cerevisiae* and *Candida pelliculosa* (Rocha and Malcata 1999). Even though sourdough bread has only a short history in Japan, study samples from four bakeries in the Kansai region demonstrated that the main present yeast is *S. cerevisiae*, with *Candida humilis* (*Candida milleri*) detected in some samples (Fujimoto et al. 2019).

LAB represent heterogeneous species with the common feature of lactic acid production as a result of sugar metabolism which leads to an acidification of the environment. LAB present in sourdough can be homofermentative or heterofermentative.

The homofermentative metabolize hexoses into lactic acid only, while the heterofermentative convert hexoses into lactic acid, acetic acid, ethanol, and CO_2 (Boudaoud et al. 2021). LAB belonging to the genera of *Lactobacillus, Leuconostoc, Pediococcus, Weissella*, and *Enterococcus* have been detected in sourdoughs. In general, heterofermentative *Lactobacillus* species dominate the sourdough microbiota (De Vuyst and Vancanneyt 2007). The main LAB found in Japanese bakeries were *Levilactobacillus brevis* (former *Lactobacillus brevis*), *Companilactobacillus alimentarius* (former *Lactobacillus alimentarius*), *Lactiplantibacillus pentosus* (former *Lactobacillus pentosus*), *Lactobacillus vaccinostercus, Fructilactobacillus sanfranciscensis* (former *Lactobacillus sanfranciscensis*), and *Lactilactubacillus sakei* (former *Lactobacillus sakei*). The most frequently isolated LAB in Portuguese sourdough were the genera *Leuconostoc* (heterofermentative) and *Lactobacillus* (homofermentative) (Rocha and Malcata 1999). In Chinese samples of sourdough, the identified LAB belong to the genera *Lactobacillus* and *Pediococcus* (Suo, Chen, and Wang 2021). A study on the LAB diversity of spontaneously-fermented hull-less barley sourdough from Turkey concluded that the dominant LAB was *Pediococcus*. From the 80 isolates obtained in this study was revealed the presence of 32 different strains belonging to 9 different species: *Pediococcus, Levilactobacillus brevis, Latillactobacillus curvatus* (former *Lactobacillus curvatus*), *Limosilactobacillus fermentum* (former *Lactobacillus fermentum*), *Lactobacillus musae, Lactiplantibacillus plantarum* (former *Lactobacillus plantarum*), *Lactobacillus paralimentarius, Leuconostoc mesenteroides*, and *Lactobacillus equigenerosi* (Çakır, Arıcı, and Durak 2020). The most commonly detected bacterial species, in sourdough, worldwide are *Levilactobacillus brevis, Limosilactobacillus pontis* (former *Lactobacillus pontis*), *Limosilactobacillus reuteri* (former *Lactobacillus reuteri*), *Lactiplantibacillus plantarum*, and *Fructilactobacillus sanfranciscensis*. From these, *Fructilactobacillus sanfranciscensis*, was the dominant LAB of more than 75% of the sourdoughs (Gänzle and Ripari 2016). In accordance, 99% of LAB isolates from PDO (protected designation of origin) Tuscan sourdough bread showed identity with *Fructilactobacillus sanfranciscensis* (Palla et al. 2017).

Sourdough can have a simple (few species) or very complex microbiota consortia. There is controversy between studies, some saying that the microbiota of sourdough has little or almost no variation over time, while others demonstrated changes during proliferation at the laboratory and bakery levels (Minervini et al. 2015). It is expected that, in laboratory conditions, the microbiota are more stable, as the conditions are more controllable, and the level of contamination is supposed to be lower compared to an artisan bakery. Actually, the stability of the sourdough microbiota at the artisanal bakery and laboratory levels is different and depends on various factors, such as flour microbial composition, microbial interactions, flour composition (carbohydrates and free amino acids), endogenous enzymatic activities, specific technology parameters (e.g., leavening and storage temperature, pH and redox potential, dough hydration, number of sourdough refreshment steps, fermentation time between refreshments, and the use of starters and/or baker's yeast), and bakery environment (Minervini et al. 2015).

11.3 FUNCTIONAL METABOLITES AND ENZYMES

The consumption of whole wheat bread is linked with certain health benefits because of its high content of fiber, vitamins, bioactive compounds, and phytic acid. However, there is a negative point related to the phytate in whole grain flours, as they inhibit the absorption of minerals (Fekri et al. 2020). The traditional sourdough is a source of different types of microorganisms some species of which are known as probiotics. According to the Food and Agriculture Organization/World Health Organization (FAO/WHO), probiotics are live, safe, and non-toxic microorganisms, which, if ingested in an adequate amount, bring health benefits to the host. Probiotics in sourdough have not only a positive effect on the gastrointestinal tract, but also help in the biotransformation and liberation of some bioactive compounds, as is the case of phenolic compounds. Several genera and strains of probiotic bacteria (*Lactobacillus, Bifidobacteri-um, Lactococcus, Leuconostoc, Pediococcus, S. salivarius subsp. thermophilus, E. faecium, E. faecalis, E. coli, B. cereus, B. subtilis, B. clausii, B. coagulans, B. licheniformis,* and *B. polyfermenticus*) and yeast *Saccharomyces boulardii* have been identified in sourdough (Liptáková, Matejčeková, and Valík 1989).

The volatile profile of breads turned out to be richer when fermentation occurs with a mixture of yeast and LAB, as is the case in sourdough bread. During sourdough fermentation of wheat flour, the yeasts and LAB functions promote flavor formation. In addition to ethanol and carbon dioxide, yeasts produce several metabolites during dough fermentation, such as higher alcohols, organic acids, diacetyl, and esters that specifically affect the flavor. LAB are responsible for the formation of various types of volatile aroma compounds, including alcohols, aldehydes, ketones, esters, and organic acids. The volatile compound profile of the bread crumb is determined by the yeast strains and their interaction with LAB (De Vuyst et al. 2016). It was concluded that yeast strains with higher growth rates produced a higher concentration of aroma compounds (Carbonetto et al. 2018).

Lactic acid bacteria may produce antimicrobial substances that protect the food against undesirable yeast, bacteria and molds, thus increasing its shelf life. The production of antimicrobial agents depends mostly on the temperature and LAB growth phase. These antimicrobials include organic acids, reuterin, hydrogen peroxide, hydroxyl fatty acids, and phenolic and proteinaceous compounds. Bacteriocins are heterogeneous peptides with antimicrobial effects that kill or inhibit the growth of other bacterial strains. Typically, LAB bacteriocins have a narrow antibacterial spectrum, but some strains may also produce bacteriocins with a broad antibacterial spectrum (Liptáková, Matejčeková, and Valík 1989). The main antimicrobial activity of the LAB is due to the production of lactic acid and consequent pH decrease. Hydrogen peroxide is produced by most LAB and has antimicrobial activity (Liptáková, Matejčeková, and Valík 1989). Reuterin is a product of glycerol fermentation, under anaerobic conditions, formed by some LAB species. Reuterin has inhibitory activity against Gram-negative and Gram-positive bacteria, yeasts, fungi and protozoa (Liptáková, Matejčeková, and Valík 1989).

11.3.1 ENZYMES OF HIGH INDUSTRIAL INTEREST

Enzymes are the natural catalysts of biological systems. Similar to any other catalyst, an enzyme brings the catalyzed reaction to its equilibrium point faster than it would occur otherwise. The power of enzymes has been recognized for thousands of years, and its proteinaceous nature was identified by the early 1800s. The industrial use of enzymes, as we know it today, started in 1913 when Otto Röhm discovered the efficacy of pancreatic trypsin for the removal of proteinaceous stains from clothes (Aehle 2007). Thereafter, the knowledge of the chemistry of proteins drew heavily on the improving techniques and concepts of organic chemistry (isolation, separation, purification, and the synthesis of peptides) culminated in the protein structure of enzymes.

Today, enzymes are used frequently in the biotech and food industries, including pulp and paper, detergents, textiles, pharmaceuticals, chemicals, biofuels, food and beverages, personal care, among others. In the specific case of the bread making industry, enzymes are added in bread formulations to enhance the production and quality of the final product. Usually, enzymatic processes can replace conventional chemical processes to make the foods free of chemicals (e.g., ascorbic acid, mono- and diglycerides, propionates, etc.), making enzymes preferred over chemical ones (Dahiya et al. 2020). On the other hand, unlike chemical-dough improvers, the use of many enzymes requires stable process conditions, such as pH, time, and temperature control.

Additionally, in food manufacturing, enzymes are generally recognized as safe (GRAS) by the US Food and Drug Administration (FDA); and in the final baked products, the enzymes do not show any activity ("Enzyme Preparations Used in Food (Partial List) | FDA," n.d.). Enzymes contribute positively as chemical-dough improvers do; however, enzymes being of biological origin can provide a clean option to make the bread natural and, in some ways, "healthier."

Microorganisms have been shown to be one of the largest and most useful sources of many enzymes (Adrio and Demain 2014). Nature provides a vast amount of enzyme resources, and different environments have been used as sources of microbial enzymes with potential for biocatalytic applications, for example, from animal digestive tracts, volcanic zones, marine environments, soil, agri-food wastes, caves, etc. (Hess et al. 2011; Kennedy, Marchesi, and Dobson 2008; Zhang, Wang, and Jiang 2021).

The enzymes used in bread making processes are generally obtained by fermentation using bacteria or fungi (Dahiya et al. 2020) (see Table 11.1). Various enzymes employed in bread making include xylanase, phytase, lipase, cellulase, amylase, protease, and glucose oxidase (Miguel et al. 2013; Dahiya et al. 2020), classified according to the type of reaction catalyzed, and generally act by breaking down the dough components such as starch, lipids, proteins, and fibers (Hans et al. 2009).

Enzyme supplementation (individually or in complex mixtures) of flour and dough are frequently used to change dough rheology, viscosity, texture, gas retention, and softness of bread (controlling the browning from the Maillard reaction, increasing the crumb properties); reduce acrylamide formation in bakery products;

TABLE 11.1
The Different Microorganisms Producing Various Enzymes, Mechanisms of Action, and Effects of Enzymes in Bread Making

Microorganism	Enzyme	Mechanism of action	Bread making effects	References
Sporotrichum thermophile *Streptomyces* strain *Trichoderma reesei*	Cellulase	Hydrolysis of cellulose complex unit into a simpler form, i.e. glucose, oligosaccharides, and cellobiose	• Decrease water absorption capacity that reduces baking time • Increase the loaf volume, producing softer crumb • Improve the bread qualities • Reduce the use of emulsifiers in bread preparation • Increase the phenolic content • Increase the formation of organic acids	(Bala and Singh 2017; Yassien, Jiman-Fatani, and Asfour 2014; Dahiya et al. 2020; Verma, Kumar, and Bansal 2018; Jayasekara and Ratnayake 2019; Barkiene et al. 2017; De Vuyst et al. 2016)
Pseudomonas fluorescens *Bacillus subtilis* *Aspergillus niger* *Penicillium* sp. *Fusarium* sp.	Lipase	Catalyze the hydrolysis of triacylglycerols and form monoacylglycerols, diacylglycerols, glycerol, and free fatty acids during bread making process	• Increase bread volume and stability of dough • Form emulsifiers • Reduce or retard staling • Develop flavors • Enhancing the dough rheology	(Nema et al. 2019; Tanyol, Uslu, and Yönten 2015; Iqbal and Rehman 2015; Chandra et al. 2020)

(*Continued*)

TABLE 11.1 (CONTINUED)
The Different Microorganisms Producing Various Enzymes, Mechanisms of Action, and Effects of Enzymes in Bread Making

Microorganism	Enzyme	Mechanism of action	Bread making effects	References
Rhizopus oryzae *Bacillus subtilis* *Bacillus pumilus* *Bacillus clausii*	Protease	Hydrolysis of peptide bonds	• Control dough rheology and viscoelastic qualities of gluten • Improve dough extensibility and increase bread or loaf volume • Form flavors and amino acids • Give crispness property on bread crust • Formation of gluten-free products • Cross-linking between gluten and other peptides • Enhance gas retention • Increase bread crumb strength and height and dough stability • Protect frozen dough from damage	(Bhange, Chaturvedi, and Bhatt 2016; Razzaq et al. 2019; Benabda et al. 2019)
Rhizopus oryzae *Bacillus subtilis* *Streptomyces badiun*	Amylase	Catalyze the breakdown of α-1,4-glycosidic bonds in the inner part of the amylopectin or amylose	• Reduce viscosity of dough • Improve volume and texture of the bread • Anti-staling property • Increases the softness and shelf life of the bread • Reduce fermentation time • Production of prebiotics	(Trabelsi et al. 2019; Rana et al. 2017; L and T 2017; Benabda et al. 2019; Bhange, Chaturvedi, and Bhatt 2016)
Streptomyces sp. *Sporotrichum thermophile* *Pichia pastoris* *Bacillus subtilis* *Myceliophthora thermophila* *Aspergillus foetidus*	Xylanase	Hydrolysis of xylan	• Increases volume and shelf life, reduces stickiness, and improves crumb shape • Degrade the unextractable arabinoxylan and form water extractable xylan followed by redistribution of water in the dough • Increase softness, extensibility and elasticity of the dry dough • Provide more soluble dietary fibers in bread	(Cunha et al. 2018; Adigüzel and Tunçer 2016; Guo et al. 2018; Bala and Singh 2017)

(Continued)

TABLE 11.1 (CONTINUED)
The Different Microorganisms Producing Various Enzymes, Mechanisms of Action, and Effects of Enzymes in Bread Making

Microorganism	Enzyme	Mechanism of action	Bread making effects	References
Lacticaseibacillus casei *Enterobacter* sp. ACSS *Sporotrichum thermophile* *Aspergillus niger*	Phytase	Catalyze the hydrolysis of phytate, an anti-nutritional component present in the flours of cereals (wheat, maize, and rye) used in bread making	• Better shape of bread • Increase bread volume, soft crumb • Decrease phytate content • Improve mineral and protein content • Increase the aroma, taste, and chewability of the bread	(Chanderman et al. 2016; Ranjan and Satyanarayana 2016; Buddhiwant et al. 2016)
Aspergillus niger *Penicillium notatum*	Glucose Oxidase	Catalyzes the glucose oxidation into hydrogen peroxide and gluconic acid	• Control browning from Maillard reaction • Increase in properties of crumb • Oxidizing agent in bread preparation, replacing potassium bromate (a carcinogen) • Improve the final texture of bread • Increase the loaf volume	(Kiesenhofer, Mach, and Mach-Aigner 2017; Hassan, Jebor, and Ali 2018)

reduce fermentation time; improve the dough stability and increase prebiotic production (Vidal et al. 2016; Renzetti, Dal Bello, and Arendt 2008; Cappelli, Oliva, and Cini 2020; Almeida and Chang 2012; Hans et al. 2009; Adrio and Demain 2014).

11.4 CELL IMMOBILIZATION

Cell immobilization is the process of confining cells in a restricted area while retaining their catalytic activity (Karel, Libicki, and Robertson 1985). It is one of the great tools for the development of ecologically and economically available biocatalysts (Hameed and Ismail 2018). Different immobilization methods are available, and the nature of the application often commands its choice. There are four main immobilization techniques for microbial cells (Figure 11.1): attachment to a surface, entrapment within a porous matrix, cell aggregation (flocculation), and containment behind barriers (Genisheva, Teixeira, and Oliveira 2014). This division is based on the physical mechanism of cell localization and the nature of the support mechanisms (Verbelen et al. 2006). Moreover, these techniques can be applied to essentially all of the whole cell systems of potential interest: microorganisms and animal and plant cells. This form of heterogeneous catalysis may impact on cell physiology and cell mobility, physical interactions of immobilized cells with the support, and the creation of a microenvironment (Karel, Libicki, and Robertson 1985).

Attachment to a surface. The attachment to a surface (Figure 11.1 A) can be done by natural adsorption, electrostatic forces, or covalent binding, with cross-linking agents. As a spontaneous and natural process, this type of immobilization technique is used often. It is an easy immobilization technique, which in some cases

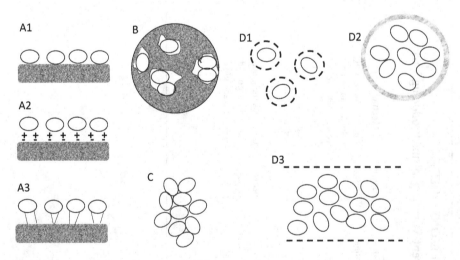

FIGURE 11.1 Immobilization techniques. A – attachment to a surface; A1 – natural adsorption; A2 – attachment by electrostatic forces; A3 – attachment by covalent bindings. B – entrapment within a porous surface. C – cell flocculation. D – containment behind a barrier; D1 – encapsulation; D2 – interfacial encapsulation; D3 – containment between microporous membrane.

can be achieved by simple contact of a cell suspension with the carrier for a small time period, and weak binding forces are established. It is the most used technique for yeast cell immobilization to be further applied in winemaking (Genisheva et al. 2014). The adsorption is affected first by the physiological conditions of the cells, like age, cell membrane charges, and hydrophobicity; second, by the medium characteristics, its composition, and pH (Mehrotra et al. 2021); and third, by the surface of the immobilization support, as rough surfaces allow higher cell retention into the support's cavities (Genisheva et al. 2011). Natural adhesion can be conferred also by electrostatic, ionic (Lewis acid/base), and hydrophobic (Lifshitz–van der Waals) interactions. Cellular attachment can be induced using linking agents – such as metal oxides, glutaraldehyde or aminosilanes – creating chemical bonds (covalent) between cells and the support (Verbelen et al. 2006; Genisheva, Teixeira, and Oliveira 2014). However, this immobilization procedure is generally incompatible with cell viability, since the cross-linking agents are highly toxic for the microbial cells and decrease their activity. As a consequence, this method of immobilization is no longer used for microbial cells but still remains suitable for the immobilization of enzymes (Genisheva, Teixeira, and Oliveira 2014). The main drawback of attachment of cells to a surface is that there is no barrier between the liquid and the immobilized cell, and the cells can be easily detached from the support. At some point of cell growth, an equilibrium is established between free and immobilized cells. Moreover, the detachment of cells depends on the age of the cell, cellular wall composition, and pH and ionic composition of the medium (Genisheva, Teixeira, and Oliveira 2014). Also, the desorption of cell is compensated as new cells grow on the support. Compared to other types of immobilization, the attachment to a surface is advantageous as the oxygen transfer is good and no scale-up drawback exists (Genisheva, Teixeira, and Oliveira 2014).

Entrapment. The entrapment of cells in a porous matrix (Figure 11.1 B) can be achieved in two ways. In the first one, cells are allowed to diffuse into a preformed porous material, where cells begin to grow, and their mobility is delimited by the presence of other cells and by the matrix itself. Sponge, sintered glass, ceramics, silicon carbide, polyurethane foam, chitosan, and stainless steel fibers are commonly used materials (Verbelen et al. 2006). In the second type of entrapment, a solid matrix is synthesized *in situ* around the cells. For this type of entrapment, the most common materials are of natural origin as proteins (gelatin and collagens) and polysaccharides (cellulose, alginate, agar, and carrageenan), or synthetic, such as polyacrylamide (Genisheva, Teixeira, and Oliveira 2014). Though high biomass concentration can be achieved with this technique, there are various drawbacks, such as diffusion limitations of nutrients, metabolites and oxygen; instability of the gel beads and detachment of cells; and limited cell division (Genisheva, Teixeira, and Oliveira 2014; Kregiel, Berlowska, and Ambroziak 2013). Calcium alginate gel is the most commonly used material for cell entrapment in the food industry. The cells inside of the alginate beads can be released, and because of this fact, it was proposed in the 1980s to make an external layer of sterile alginate and produce double layer alginate beads (Genisheva, Teixeira, and Oliveira 2014).

Cell aggregation. Cell aggregation or flocculation is a type of immobilization that does not need an external carrier (Figure 11.1 C). Flocculation is a naturally occurring process, but it can be also artificially induced using cross-linking agents. Cell aggregation is a simple and inexpensive immobilization technique, however, it is difficult to predict and control. It is a complex process connected with the expression of flocculation genes such as FLO1, FLO5, FLO8, and FLO11 (Genisheva, Teixeira, and Oliveira 2014). The flocculation depends not only on the genetic characteristics of the cell strain, but also on parameters, such as medium pH and composition, dissolved oxygen, fermentation conditions (temperature and agitation), structure and surface charges of the cell wall, as well as the cell growth phase (Bekatorou, Plessas, and Mallouchos 2016). An important issue for the success of this system is the selection of a proper yeast strain and fermentation system. The main applications of flocculation in the food industry are for alcohol production, some kinds of beers, and sparkling wines. It is of great importance in the brewing industry as an environmentally friendly, easy, cost-effective method of immobilization that facilitates the separation of the yeast cells from the green beer at the end of the fermentation. Moreover, flocculation affects the beer quality and fermentation productivity (Bekatorou, Plessas, and Mallouchos 2016; Genisheva, Teixeira, and Oliveira 2014). Flocculation is a very important yeast characteristic in sparkling wine production by the traditional method, facilitating the process of disgorging (Torresi, Frangipane, and Anelli 2011).

Containment behind barriers. The containment of cells behind a barrier (Figure 11.1 D) can be attained by entrapment of the cells in microcapsules, and using microporous membrane filters, or by cell immobilization onto an interaction surface of two immiscible liquids (Genisheva, Teixeira, and Oliveira 2014). The entrapment of cells in microcapsules is more laborious. Firstly, the cells are encapsulated in a spherical gel and a second coating is made with a polymer such as polyethylenimine. Then, the gel is dissolved, but the cells are left in suspension, contained behind the polymer barrier. The microporous membrane filters are normally made of polymers, e.g. polyvinylchloride or polypropylene. The containment of cells behind a barrier allows very high cell concentrations, and at the same time the product is cell free. However, this implies some major disadvantages of membrane immobilization techniques as mass transfer limitations (supply of oxygen and nutrients to the cells and the removal of carbon dioxide), and possible membrane plugging caused by cell growth (Bekatorou, Plessas, and Mallouchos 2016). Entrapment of cells in microcapsules (size range between 1 and 1000 μm) overcomes these problems, as they have larger specific surface area that allows good diffusion of nutrients and metabolites. Microencapsulation is used in various biotechnological processes, such as encapsulation of probiotics, additives and bioactives for food production, development of encapsulated biocatalysts for fermentation processes, and environmental bioremediation (Bekatorou, Plessas, and Mallouchos 2016).

11.4.1 Advantages and Disadvantages of Cell Immobilization

The technique of cell immobilization is used mainly in submerged fermentations, and to a lesser extent in solid state fermentations. Immobilization can be performed in all cells, yeast, or bacteria, as well as on enzymes. The main advantages of cell immobilization in comparison to free cells in fermentation processes are: short fermentation times, increased fermentation rates, increased tolerance to inhibitors, reusability of the immobilized cells, better conduction and control of the process and feasibility of continuous processes, possibility to use smaller fermentation facilities, easy product recovery, reusability of the biocatalyst, etc. Immobilized cells are used in bioethanol production (Santos et al. 2018), brewing (Kopsahelis, Kanellaki, and Bekatorou 2007), wine production (Genisheva et al. 2013, 2012), and different other biotechnological processes. In wine production, cell immobilization is used for alcohol and malolactic fermentations (Genisheva et al. 2013, 2014, 2012). Immobilized yeast cells were able to carry out the complete alcoholic fermentation of white wine in 4 days compared to the 7 days needed for the traditional free cell system (Genisheva et al. 2014). Immobilized cells are more protected against the negative effect of inhibitors. For example, immobilized cells were able to conduct wine fermentation with high concentrations of free SO_2, while fermentations with free cells did not start at all (Genisheva et al. 2013).

Recently, more attention has been focused on cell immobilization techniques for the treatment of a variety of waste. It is often used in dye wastewater decolorization, petroleum wastewater treatment, and biohydrogen production. Biodegradation is one of the most effective and environmentally friendly methods for removal of organic pollutants from wastewater. However, biodegradation is often limited by cell growth, cell separation, and substrate inhibition. The immobilization technique showed a superior activity not only in overcoming the toxicity of the inhibitor compounds but also by making the process cheaper with the possibility to reuse the immobilized complex (Hameed and Ismail 2018). Immobilized bacteria were able to decolorize wastewater even in high concentrations of dye, while the free cells' function was inhibited (Hameed and Ismail 2018).

The disadvantages that must be pondered when using immobilized cell systems are the mechanical stability of the support used to immobilize the microbial cells and the loss of cell activity on prolonged operation.

11.4.2 Types of Supports

An important aspect in the immobilization technology are the support materials used, as they play a crucial role in the function of microorganisms. Many different carrier materials are used. A suitable support material must correspond to some prerequisites, like high mechanical strength and chemical stability, low toxicity to cells, and good mass transfer performance (Lou et al. 2019). The selection of the support will depend on the processes where it will be applied. Immobilized cells have been widely used in applications, such as pharmaceutical, environmental engineering, food industry, and biosensor applications (Mehrotra et al. 2021). And there would be different prerequisites for supports to be used, for example, in environmental engineering and in the food industry. Supports that will be used in biological treatment processes must correspond to criteria, like light-weight, flexible, cost effective, mechanically and chemically stable, inert, non-biodegradable, non-polluting, non-toxic, insoluble in aqueous medium, amenable to a simple method for bacterial immobilization, and possess high diffusivity with high biomass retention (Mehrotra et al. 2021). Immobilization supports for implementation in the food industry must have additional prerequisites, such as being easily accepted by the consumer, of food-grade purity, low cost, abundance, and physically and biologically stable. Moreover, its selection depends on the process in which it will be applied as well as the process conditions, for example, immobilization supports suitable for the wine industry should be also suitable for low-temperature fermentation (Torresi, Frangipane, and Anelli 2011).

The supports can be used in their natural form or submitted to some treatment to modify the surface in contact with the biomass (Genisheva, Teixeira, and Oliveira 2014). The previous acid/basic pretreatment of the support make it "cleaner" but at the same time brings a loss of support material, sometimes up to 75% (Genisheva et al. 2011). In some cases, the pretreatment of the support did not improve the ethanol production nor cell adhesion, thus making it an unnecessary step in the cell immobilization (Genisheva et al. 2011). In other cases, the pretreatment provides higher immobilization cell loads on the support material (Brányik et al. 2001).

The carriers used for immobilization can be classified as organic and inorganic and further in synthetic and natural (Figure 11.2). The synthetic materials feature a high mechanical and chemical stability, while the advantages of the natural carriers are biocompatibility, biodegradability, and safety for the environment.

For whole cell immobilization, organic carriers are preferred over inorganic support materials (Mehrotra et al. 2021). However, inorganic supports are thought to

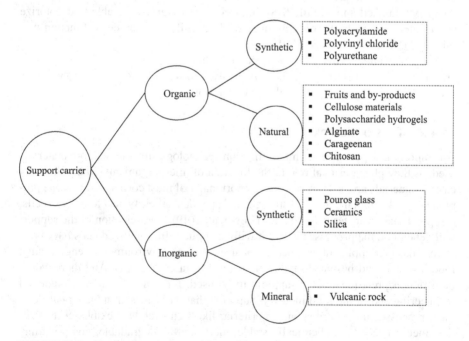

FIGURE 11.2 Types of supports used for cell immobilization.

be more attractive than organic supports due to their low cost, abundance in nature, reusability, and environmental friendliness (Genisheva, Teixeira, and Oliveira 2014). Organic matrices are disposed to bacterial contamination and also have less mechanical and operational stability. Among the organic carriers, the synthetic polymeric material provides a more stable matrix formation in solutions of varying pH. The natural organic carriers are available, inexpensive, biodegradable, non-toxic, and usually have higher cell load capacity compared to the inorganics (due to existing functional groups on the organic matrices' surfaces providing additional improvement in the absorption capacity) (Mehrotra et al. 2021). Natural organic polymers are carrageenan, pectate, agar, agarose, chitosan, charcoal, cellulose, gelatin, collagen, bacterial cellulose, and alginate (derived from algal polysaccharides). Disadvantages associated with the use of natural organic polymers are particle disruption due to high cell load concentration, low mechanical and chemical stability, low reproducibility, low adaptability, and mass transfer problems. In comparison, the synthetic organic polymers have higher mechanical stability but lower diffusion. Synthetic organic gels include polyvinyl alcohol, polyacrylamide, polycarbamoyl sulfonate, and polyethylene glycol. Synthetic plastics include polyethylene, polyurethane, polypropylene, polyacrylonitrile, and polyvinyl chloride (Mehrotra et al. 2021). Synthetic carriers like polyvinyl alcohol (PVA) and acrylamide may have the problems of agglomeration or cytotoxicity as carriers (Wu et al. 2021).

Inorganic carriers that can be used for immobilization are activated charcoal, clay, porous glass, zeolite, anthracite, ceramics, silica, and others. Inorganic carriers may have low adsorption capacity and affinity for cells, resulting in high microbial loss (Wu et al. 2021). Carbonaceous materials (activated charcoal and biochar) have been frequently used as immobilized carriers for the degradation of organic pollutants (Lou et al. 2019). Biochar is an environmentally friendly material that has attracted much attention due to its cost-effectiveness, various sources, and good physical and chemical properties, like large specific surface area and high porosity, providing a valuable habitat for bacteria, and increased cell concentration (Lou et al. 2019).

Lately, nanomaterials have gained popularity as an immobilizing matrix because of their unique physicochemical properties. Various nanomaterials have been exploited for immobilization, however silica-based materials are considered the most advantageous because of their ease of functionalization, biocompatibility, microbial resistance, chemical inertness, higher thermal and mechanical stability, etc. (Mukundan et al. 2020).

11.4.3 Uses of Cell Immobilization in Bread Making

Cell immobilization is mostly used in submerged fermentations. It is not usual in solid-state fermentations, such as dough making. In the literature, there are few examples in the use of immobilization in bread making. Immobilization was applied in straight dough and in sourdough breads. The found published data on this matter are summarized in the following lines, as well as in Table 11.2.

Straight dough and sourdough bread were made with three different leavening agents. The first one was baker's yeast immobilized on a starch–gluten–milk matrix

TABLE 11.2
Immobilization Technology Used in Bread Making

Process	Immobilization method	Microorganism/molecule	Support	Reference
sourdough bread straight dough	attachment	kefir baker's yeast	*trahanas* orange peel	(Plessas, Bekatorou, et al. 2007)
barley sourdough bread	attachment	bacteria (*P. acidilactici*)	apple pomace	(Bartkiene et al. 2017)
frozen dough	encapsulation	yeast (*S. cerevisiae*)	calcium alginate calcium alginate–starch	(Cozmuta et al. 2021)
refrigerated dough	encapsulation	enzyme (amylase) yeast	alginate/gelatin	(Gugerli et al. 2004)
sourdough bread straight dough	attachment	kefir baker's yeast bacteria (*L. casei*)	brewer's spent grains	(Plessas, Trantallidi, et al. 2007)
straight dough	entrapment	enzyme glucose oxidase	chitosan sodium tripolyphosphate	(Tang et al. 2014)
sourdough bread	attachment	bacteria *P. pentosaceus* SP2	wheat bran	(Plessas, Mantzourani, and Bekatorou 2020)
sourdough bread	attachment	bacteria *L. paracasei* K5 *L. bulgaricus* ATCC 11842	delignified wheat bran	(Mantzourani et al. 2019)
sourdough bread	attachment	kefir	corn grains	(Plessas et al. 2012)

(*trahanas*, a type of fermented cereal food) and on orange peel; the second used kefir immobilized on orange peel; and the third used free baker's yeast cells. The resulting bread was evaluated for its shelf life and overall quality, including the volatile composition. The results showed that the cells immobilized on *trahanas* led to an increase of shelf life of 30%, compared to the bread made with free cell (Plessas, Bekatorou, et al. 2007). When these immobilized cells were used to make sourdough bread, the shelf life increased even more (~130%). Similar results were obtained when cells immobilized on orange peel were used.

The shelf life of sourdough bread made with kefir immobilized on orange peel increased by more than 50%. In terms of increased shelf life of bread, the best results were obtained with sourdough bread made with cells immobilized on *trahanas*. The authors concluded that these differences are due to the LAB microbiota and the different composition of the used supports (Plessas, Bekatorou, et al. 2007). The breads made with immobilized cells, especially sourdough bread made with cells immobilized on *trahanas*, had richer aroma composition. Breads with immobilized cells were found to have improved flavor quality. Sourdough breads had compact texture that resulted in higher retention of moisture levels after baking and reduced moisture loss rates. Authors attributed this variation in texture as the effect of the biocatalyst particles on leavening, as well as the different amounts and distribution of carbon dioxide produced in the dough mass compared to free baker's yeast cells. The distribution of the yeasts was higher on the surface of the immobilization supports, and this resulted in fewer holes of higher size. In the conventional baker's yeast bread, the cells are homogenously distributed within the dough mass, and the bread leavening is homogenous (Plessas, Bekatorou, et al. 2007). The final conclusions of the authors were that the use of immobilized cells is a good alternative to increase shelf life, delay staling, and improve the overall quality of bread compared to the conventional baker's yeast method.

In another study, *Pediococcus acidilactici* LUHS29 were immobilized in apple pomace to apply to barley sourdough fermentation for functional bread production (Bartkiene et al. 2017). The resulting bread was compared to bread resulting from spontaneous sourdough fermentation. Samples were analyzed for total phenolic content, lactic acid content, and antioxidant activity. Immobilized cells were able to produce 15.3% more lactic acid compared to spontaneous fermentations. Moreover, the production of L-lactic acid increased up to 92.7% from the total lactic acid content. This is important, as the D (-) isomer cannot be metabolized by humans, and excessive intake can result in acidosis, which is a disturbance in the acid–alkali balance in the blood. The highest total phenolic compound contents (up to 142.02 ± 0.16 mg_{GA}/100 g) were determined in sourdoughs fermented with immobilized cells. However, there was not found significant correlation between total phenolic content and antioxidant activity of the samples. In conclusion, the application of immobilized bacterial cells on apple pomace was considered a good alternative for the food industry due to its bioactive potential (Bartkiene et al. 2017).

Storage of frozen dough is a common practice in the bakery industry. Yeast encapsulation was used as a tool to improve the sensorial characteristics of frozen dough bread (Cozmuta et al. 2021). *S. cerevisiae* were encapsulated in calcium alginate or calcium alginate–starch gels. Fresh or frozen doughs containing free or encapsulated yeast were used for bread making, and their physical and sensorial characteristics were examined. Encapsulated yeast was stored frozen for one and three months. After that period, the morphology, structure, and cryotolerance of the encapsulated yeast cells were characterized and compared with those of free yeast cells. The use of starch in the alginate mixture increased the entrapment efficiency of the yeast cells. Moreover, calcium alginate–starch beads had stronger cryopreservation properties, indicating the structural support provided by the starch. However, the final results demonstrated that the starch, at the same time, inhibits the expansion of the bread volume due to the low leakage of yeast cells into the dough. Breads made with yeasts immobilized on calcium alginate had higher specific volumes compared with those containing alginate–starch–yeast beads. From all frozen dough, the highest scores for individual attributes were accounted by breads made with immobilized cells. Between all samples, bread made with frozen dough (3 months) with cells immobilized on calcium alginate received the best score evaluation. Large, irregular, and unevenly distributed pores in the fresh dough breads with immobilized cells were also noticed in this study. However, the final conclusions were that alginate-encapsulated yeast is a good approach to overcoming the cryosensitivity of baker's yeast. Moreover, the resulting breads has higher scores in sensorial characteristics, than those made with free yeast cells (Cozmuta et al. 2021).

This immobilization technique was also used to extend the shelf life of refrigerated dough (Gugerli et al. 2004). The main issue when dealing with refrigerated dough is the control of the fermentation. Yeasts continue to metabolize even at low temperatures, as production of CO_2 continues over storage time. This uncontrolled yeast activity can negatively affect the organoleptic and rheological properties of the dough. The flour fermentations have two steps: the first one is controlled by enzymes (amylase) and the second by yeasts. Regarding this, to improve the refrigerated storage of dough, two different immobilization methods were studied: enzyme (amylase) immobilization and yeast immobilization in alginate/gelatin micro beads. The enzyme immobilization was considered unsuccessful, as a high leaching of enzyme was observed. In contrast, the use of immobilized yeast completely stopped the flour fermentation in ambient and low temperatures (Gugerli et al. 2004).

Straight dough and sourdough breads were produced using baker's yeast, kefir, or *Lacticaseibacillus casei* (former *Lactobacillus casei*) immobilized on brewer's spent grains (Plessas, Trantalidi, et al. 2007). The quality of the resulting breads was compared to commercial type baker's yeast bread. The breads produced with sourdough had lower pH and higher total titratable acidity compared to the other breads produced in the study, a fact that had positive effect on the sourdough bread preservation by doubling its shelf life. The breads were also evaluated for their aroma volatile composition. Sourdough breads had the highest number and concentrations of aroma compounds, followed by breads made with immobilized kefir. Breads produced using baker's yeast cells (free or immobilized) by the straight-dough method had the lowest number of aroma compounds. Consumer preference, in terms of aroma, taste, and overall quality, was for sourdough bread. It was concluded that the biocatalysts made by immobilization of baker's yeast, kefir, or *L. casei* cells on brewer's spent grains were found efficient for bread making either by the straight-dough or the sourdough method and have great potential for replacing the conventional pressed baker's yeast (Plessas, Trantallidi, et al. 2007).

Kefir was also immobilized on boiled corn grains to be used in sourdough bread making. The produced sourdough breads had good loaf volumes and good sensory characteristics, as well as higher resistance to mold spoilage compared to the control bread (Plessas et al. 2012).

Enzymes like glucose oxidase are used in bread making to improve different aspects like reinforcing gluten structure, strengthening dough stability, increasing specific volume, texture, elasticity, and water-holding capacity. Glucose oxidase catalyzes the oxidation of β-$_D$-glucose to $_D$-gluconic acid with the release of hydrogen peroxide (Tang et al. 2014). Hydrogen peroxide improves dough handling properties, gas retention capability, and bread quality, especially for commercial wheat flour. However, the use of enzyme has some weaknesses, like fast oxidation and low stability in flour. In an attempt to improve these drawbacks, glucose oxidase was immobilized on chitosan sodium tripolyphosphate and further implied in bread making (Tang et al. 2014). The results demonstrated that immobilized enzyme retained high activity. Moreover, it slowed down the release of hydrogen peroxide, compared to the free enzyme, and increased gas retention properties in the dough. The method was considered promising for the improvement of bread making quality of commercial wheat flour (Tang et al. 2014).

The recently isolated *Pediococcus pentosaceus* SP2 strain was immobilized on wheat bran and used for sourdough bread making (Plessas, Mantzourani, and Bekatorou 2020). Moreover, the bacteria were used in fresh, freeze-dried, and immobilized forms. Physicochemical characteristics, resistance to spoilage, flavor-related compounds, and consumer acceptance of the resulting breads were assessed. Immobilization enhanced the *P. pentosaceus* viability. Bread made with immobilized bacteria was more resistant to mold spoilage compared to the rest of the samples. Moreover, it had higher total titratable acidity and higher content of acids (lactic acid and propionic acid). Volatile composition of sourdough breads, independent of the type of bacteria used (fresh, freeze-dried, or immobilized), were similar. The sensory evaluation of the produced sourdough breads also did not show differences between samples. The immobilization of the novel strain *P. pentosaceus* SP2 was considered to have potential to be used for industrial purposes (Plessas, Mantzourani, and Bekatorou 2020).

Two lactic acid bacteria, *Lacticaseibacillus paracasei* (former *Lactobacillus paracasei*) K5 and *Lactobacillus bulgaricus* ATCC 11842, were immobilized on delignified wheat bran (in single and mixed form) and freeze-dried to be used in sourdough bread making (Mantzourani et al. 2019). The resulting breads were analyzed for their physicochemical characteristics, mold and rope spoilage appearance, volatile composition, and organoleptic characteristics and compared to a control sample. Although both immobilized bacteria demonstrated good fermentative activity, the best results were achieved when freeze-dried, immobilized *L. paracasei* K5 was used. This sourdough bread was more resistant against mold and rope spoilage and had higher acidity. All produced sourdough breads were accepted by the consumers. Only the control sample had lower score evaluation after 2 days of storage. Both lactic acid bacteria were produced using cheese whey, that according to the authors, decreases the costs of the proposed process (Mantzourani et al. 2019).

From the previous examples, it is seen that in sourdough bread making, the most used immobilization method is the natural adhesion, attachment to a surface. Moreover, the immobilization supports are mainly from natural food, for easier acceptance by the consumer.

11.5 CONCLUSION

The use of sourdough in bread making improves its physical and chemical characteristics and extends its shelf life compared to the breads made with baker's yeast. Furthermore, the use of cell immobilization in bread making has a big potential. It has been proven that the use of immobilization technology improves the sensory and physical characteristics of the resulting breads. Sourdough breads with immobilization microorganisms were easily accepted by the consumers with positive score evaluations. However, more investigation is needed in order for this methodology to be implemented in bakeries and industry.

ACKNOWLEDGMENT

This study was supported by the Portuguese Foundation for Science and Technology (FCT) under the scope of the strategic funding of UIDB/04469/2020 unit.

REFERENCES

Adigüzel, Ali Osman, and Münir Tunçer. 2016. "Production, Characterization and Application of a Xylanase from Streptomyces Sp. AOA40 in Fruit Juice and Bakery Industries." *Food Biotechnology* 30 (3): 189–218. https://doi.org/10.1080/08905436.2016.1199383.
Adrio, Jose L., and Arnold L. Demain. 2014. "Microbial Enzymes: Tools for Biotechnological Processes." *Biomolecules* 4 (1): 117–39. https://doi.org/10.3390/biom4010117.
Aehle, Wolfgang, ed. 2007. *Enzymes in Indurstry: Production and Applications*, 3rd Edition. Hoboken: Wiley.
Almeida, Eveline Lopes, and Yoon Kil Chang. 2012. "Effect of the Addition of Enzymes on the Quality of Frozen Pre-Baked French Bread Substituted with Whole Wheat Flour." *LWT: Food Science and Technology* 49 (1): 64–72. https://doi.org/10.1016/j.lwt.2012.04.019.
Bala, Anju, and Bijender Singh. 2017. "Concomitant Production of Cellulase and Xylanase by Thermophilic Mould Sporotrichum Thermophile in Solid State Fermentation and Their Applicability in Bread Making." *World Journal of Microbiology and Biotechnology* 33 (6): 1–10. https://doi.org/10.1007/S11274-017-2278-6.
Bartkiene, Elena, Donata Vizbickiene, Vadims Bartkevics, Iveta Pugajeva, Vita Krungleviciute, Daiva Zadeike, Paulina Zavistanaviciute, and Grazina Juodeikiene. 2017. "Application of Pediococcus Acidilactici LUHS29 Immobilized in Apple Pomace Matrix for High Value Wheat-Barley Sourdough Bread." *LWT - Food Science and Technology* 83: 157–64. https://doi.org/10.1016/j.lwt.2017.05.010.
Bekatorou, Argyro, Stavros Plessas, and Athanasios Mallouchos. 2016. "Cell Immobilization Technologies for Applications in Alcoholic Beverages." In *Applications of Encapsulation and Controlled Release*, 441–63. https://doi.org/10.1201/9780429299520-20.

Benabda, Olfa, Sana M'Hir, Mariam Kasmi, Wissem Mnif, and Moktar Hamdi. 2019. "Optimization of Protease and Amylase Production by Rhizopus Oryzae Cultivated on Bread Waste Using Solid-State Fermentation." *Journal of Chemistry* 2019. https://doi.org/10.1155/2019/3738181.

Bhange, Khushboo, Venkatesh Chaturvedi, and Renu Bhatt. 2016. "Simultaneous Production of Detergent Stable Keratinolytic Protease, Amylase and Biosurfactant by Bacillus Subtilis PF1 Using Agro Industrial Waste." *Biotechnology Reports* 10 (June): 104. https://doi.org/10.1016/J.BTRE.2016.03.007.

Boudaoud, Sonia, Chahinez Aouf, Hugo Devillers, Delphine Sicard, and Diego Segond. 2021. "Sourdough Yeast-Bacteria Interactions Can Change Ferulic Acid Metabolism during Fermentation." *Food Microbiology* 98 (October 2020): 103790. https://doi.org/10.1016/j.fm.2021.103790.

Brányik, T., A.A. Vicente, J.M. Machado Cruz, and J.A. Teixeira. 2001. "Spent Grains - A New Support for Brewing Yeast Immobilisation." *Biotechnology Letters* 23 (13): 1073–78. https://doi.org/10.1023/A:1010558407475.

Buddhiwant, Priyanka, Kavita Bhavsar, V. Ravi Kumar, and Jayant M. Khire. 2016. "Phytase Production by Solid-State Fermentation of Groundnut Oil Cake by Aspergillus Niger: A Bioprocess Optimization Study for Animal Feedstock Applications." *Preparative Biochemistry & Biotechnology* 46 (6): 531–38. https://doi.org/10.1080/10826068.2015.1045606.

Çakır, Elif, Muhammet Arıcı, and Muhammed Zeki Durak. 2020. "Biodiversity and Techno-Functional Properties of Lactic Acid Bacteria in Fermented Hull-Less Barley Sourdough." *Journal of Bioscience and Bioengineering* 130 (5): 450–56. https://doi.org/10.1016/j.jbiosc.2020.05.002.

Canesin, Míriam Regina, and Cínthia Baú Betim Cazarin. 2021. "Nutritional Quality and Nutrient Bioaccessibility in Sourdough Bread." *Current Opinion in Food Science* 40: 81–86. https://doi.org/10.1016/j.cofs.2021.02.007.

Cappelli, Alessio, Noemi Oliva, and Enrico Cini. 2020. "A Systematic Review of Gluten-Free Dough and Bread: Dough Rheology, Bread Characteristics, and Improvement Strategies." *Applied Sciences* 10 (18): 6559. https://doi.org/10.3390/APP10186559.

Carbonetto, Belén, Johan Ramsayer, Thibault Nidelet, Judith Legrand, and Delphine Sicard. 2018. "Bakery Yeasts, a New Model for Studies in Ecology and Evolution." *Yeast* 35 (11): 591–603. https://doi.org/10.1002/yea.3350.

Chanderman, Ashira, Adarsh Kumar Puri, Kugen Permaul, and Suren Singh. 2016. "Production, Characteristics and Applications of Phytase from a Rhizosphere Isolated Enterobacter Sp. ACSS." *Bioprocess and Biosystems Engineering* 39 (10): 1577–87. https://doi.org/10.1007/s00449-016-1632-7.

Chandra Prem, Enespa, Ranjan Singh, and Pankaj Kumar Arora. 2020. "Microbial Lipases and Their Industrial Applications: A Comprehensive Review." *Microbial Cell Factories* 19 (1): 1–42. https://doi.org/10.1186/S12934-020-01428-8.

Cozmuta, A. A. Mihaly, R. Jastrzębska, M.P. Apjok, L.M. Cozmuta, A. Peter, and C. Nicula. 2021. "Immobilization of Baker's Yeast in the Alginate-Based Hydrogels to Impart Sensorial Characteristics to Frozen Dough Bread." *Food Bioscience*. https://doi.org/10.1016/j.buildenv.2020.107229.

Cunha, Luana, Raquel Martarello, Paula Monteiro De Souza, Marcela Medeiros De Freitas, Kleber Vanio Gomes Barros, Edivaldo Ximenes Ferreira Filho, Mauricio Homem-De-Mello, and Pérola Oliveira Magalhães. 2018. "Optimization of Xylanase Production from Aspergillus Foetidus in Soybean Residue." *Enzyme Research* 2018. https://doi.org/10.1155/2018/6597017.

Dahiya, Seema, Bijender Kumar Bajaj, Anil Kumar, Santosh Kumar Tiwari, and Bijender Singh. 2020. "A Review on Biotechnological Potential of Multifarious Enzymes in Bread Making." *Process Biochemistry* 99 (May): 290–306. https://doi.org/10.1016/j.procbio.2020.09.002.

"Enzyme Preparations Used in Food (Partial List) | FDA." n.d. Generally Recognized as Safe (GRAS). https://www.fda.gov/food/generally-recognized-safe-gras/enzyme-preparations-used-food-partial-list.

Fekri, Arezoo, Mohammadali Torbati, Ahmad Yari Khosrowshahi, Hasan Bagherpour Shamloo, and Sodeif Azadmard-Damirchi. 2020. "Functional Effects of Phytate-Degrading, Probiotic Lactic Acid Bacteria and Yeast Strains Isolated from Iranian Traditional Sourdough on the Technological and Nutritional Properties of Whole Wheat Bread." *Food Chemistry* 306 (September 2019): 125620. https://doi.org/10.1016/j.foodchem.2019.125620.

Fujimoto, Akihito, Keisuke Ito, Noriko Narushima, and Takahisa Miyamoto. 2019. "Identification of Lactic Acid Bacteria and Yeasts, and Characterization of Food Components of Sourdoughs Used in Japanese Bakeries." *Journal of Bioscience and Bioengineering* 127 (5): 575–81. https://doi.org/10.1016/j.jbiosc.2018.10.014.

Gänzle, Michael, and Valery Ripari. 2016. "Composition and Function of Sourdough Microbiota: From Ecological Theory to Bread Quality." *International Journal of Food Microbiology* 239: 19–25. https://doi.org/10.1016/j.ijfoodmicro.2016.05.004.

Genisheva, Z., S. Macedo, S.I. Mussatto, J.A. Teixeira, and J.M. Oliveira. 2012. "Production of White Wine by Saccharomyces Cerevisiae Immobilized on Grape Pomace." *Journal of the Institute of Brewing* 118 (2). https://doi.org/10.1002/jib.29.

Genisheva, Z., S.I. Mussatto, J.M. Oliveira, and J.A. Teixeira. 2011. "Evaluating the Potential of Wine-Making Residues and Corn Cobs as Support Materials for Cell Immobilization for Ethanol Production." *Industrial Crops and Products* 34 (1). https://doi.org/10.1016/j.indcrop.2011.03.006.

Genisheva, Z., S.I. Mussatto, J.M. Oliveira, and J.A. Teixeira. 2013. "Malolactic Fermentation of Wines with Immobilised Lactic Acid Bacteria: Influence of Concentration, Type of Support Material and Storage Conditions." *Food Chemistry* 138 (2–3): 1510–14. https://doi.org/10.1016/j.foodchem.2012.11.058.

Genisheva, Z., J.A. Teixeira, and J.M. Oliveira. 2014. "Immobilized Cell Systems for Batch and Continuous Winemaking." *Trends in Food Science and Technology* 40 (1). https://doi.org/10.1016/j.tifs.2014.07.009.

Genisheva, Z., M. Vilanova, S.I. Mussatto, J.A. Teixeira, and J.M. Oliveira. 2014. "Consecutive Alcoholic Fermentations of White Grape Musts with Yeasts Immobilized on Grape Skins - Effect of Biocatalyst Storage and SO$_2$ Concentration on Wine Characteristics." *LWT - Food Science and Technology* 59 (2P1). https://doi.org/10.1016/j.lwt.2014.06.046.

Gugerli, R., V. Breguet, U. Von Stockar, and I.W. Marison. 2004. "Immobilization as a Tool to Control Fermentation in Yeast-Leavened Refrigerated Dough." *Food Hydrocolloids* 18 (5): 703–15. https://doi.org/10.1016/j.foodhyd.2003.11.008.

Guo, Yalan, Zhen Gao, Jiaxing Xu, Siyuan Chang, Bin Wu, and Bingfang He. 2018. "A Family 30 Glucurono-Xylanase from Bacillus Subtilis LC9: Expression, Characterization and Its Application in Chinese Bread Making." *International Journal of Biological Macromolecules* 117 (October): 377–84. https://doi.org/10.1016/j.ijbiomac.2018.05.143.

Hameed, Basma B., and Zainab Z. Ismail. 2018. "Decolorization, Biodegradation and Detoxification of Reactive Red Azo Dye Using Non-Adapted Immobilized Mixed Cells." *Biochemical Engineering Journal* 137: 71–77. https://doi.org/10.1016/j.bej.2018.05.018.

Hans, Goesaert, Slade Louise, Levine Harry, and Delcour JanA. 2009. "Amylases and Bread Firming – An Integrated View." *Journal of Cereal Science*. 50 (3): 345–52. https://doi.org/10.1016/J.JCS.2009.04.010.

Hassan, Baydaa A., Mohammed A. Jebor, and Zahra M. Ali. 2018. "Purification and Characterization of the Glucose Oxidase from Penicillium Notatum." *International Journal of Pharmaceutical Quality Assurance* 9 (01): 273–81. https://doi.org/10.25258/IJPQA.V9I01.11360.

Hess, M., A. Sczyrba, R. Egan, T.-W. Kim, H. Chokhawala, G. Schroth, S. Luo, et al. 2011. "Metagenomic Discovery of Biomass-Degrading Genes and Genomes from Cow Rumen." *Science* 331 (6016): 463–67. https://doi.org/10.1126/science.1200387.

Iqbal, Syeda Abeer, and Abdul Rehman. 2015. "Characterization of Lipase from **Bacillus Subtilis** I-4 and Its Potential Use in Oil Contaminated Wastewater." *Brazilian Archives of Biology and Technology* 58 (5): 789–97. https://doi.org/10.1590/S1516-89132015050318.

Jayasekara, Sandhya, and Renuka Ratnayake. 2019. "Microbial Cellulases: An Overview and Applications." In *Cellulose*, edited by Alejandro Rodríguez Pascual and María E. Eugenio Martín. London: IntechOpen. https://doi.org/10.5772/INTECHOPEN.84531.

Karel, Steven, Shari B. Libicki, and Channing R. Robertson. 1985. "The Immobilization of Whole Cells: Engineering Principles." *Chemical Engineering Science* 40 (8): 1321–54. https://doi.org/10.1016/0009-2509(85)80074-9.

Kennedy, Jonathan, Julian R Marchesi, and Alan DW Dobson. 2008. "Marine Metagenomics: Strategies for the Discovery of Novel Enzymes with Biotechnological Applications from Marine Environments." *Microbial Cell Factories* 7 (1): 27. https://doi.org/10.1186/1475-2859-7-27.

Kiesenhofer, Daniel P., Robert L. Mach, and Astrid R. Mach-Aigner. 2017. "Glucose Oxidase Production from Sustainable Substrates." *Current Biotechnology* 6 (3): 238–44. https://doi.org/10.2174/2211550105666160712225517.

Kopsahelis, Nikolaos, Maria Kanellaki, and Argyro Bekatorou. 2007. "Low Temperature Brewing Using Cells Immobilized on Brewer's Spent Grains." *Food Chemistry* 104 (2): 480–88. https://doi.org/10.1016/j.foodchem.2006.11.058.

Kregiel, Dorota, Joanna Berlowska, and Wojciech Ambroziak. 2013. "Growth and Metabolic Activity of Conventional and Non-Conventional Yeasts Immobilized in Foamed Alginate." *Enzyme and Microbial Technology* 53 (4): 229–34. https://doi.org/10.1016/j.enzmictec.2013.05.010.

Liptáková, Denisa, Zuzana Matejčeková, and Ľubomír Valík. 1989. "Lactic Acid Bacteria and Fermentation of Cereals and Pseudocereals." *Intech* 32 (tourism): 137–44. https://www.intechopen.com/books/advanced-biometric-technologies/liveness-detection-in-biometrics.

Lou, Liping, Qian Huang, Yiling Lou, Jingrang Lu, Baolan Hu, and Qi Lin. 2019. "Adsorption and Degradation in the Removal of Nonylphenol from Water by Cells Immobilized on Biochar." *Chemosphere* 228: 676–84. https://doi.org/10.1016/j.chemosphere.2019.04.151.

Mäntynen, V.H., M. Korhola, H. Gudmundsson, H. Turakainen, G.A. Alfredsson, H. Salovaara, and K. Lindström. 1999. "A Polyphasic Study on the Taxonomic Position of Industrial Sour Dough Yeasts." *Systematic and Applied Microbiology* 22 (1): 87–96. https://doi.org/10.1016/S0723-2020(99)80031-9.

Mantzourani, Ioanna, Antonia Terpou, Athanasios Alexopoulos, Eugenia Bezirtzoglou, and Stavros Plessas. 2019. "Assessment of Ready-to-Use Freeze-Dried Immobilized Biocatalysts as Innovative Starter Cultures in Sourdough Bread Making." *Foods* 8 (1). https://doi.org/10.3390/foods8010040.

Mehrotra, Tithi, Subhabrata Dev, Aditi Banerjee, Abhijit Chatterjee, Rachana Singh, and Srijan Aggarwal. 2021. "Use of Immobilized Bacteria for Environmental Bioremediation: A Review." *Environmental Chemical Engineering* 2 (2): 1–61. https://doi.org/10.1016/j.jece.2021.105920.

Miguel, Ângelo Samir Melim, Tathiana Souza Martins-Meyer, Érika Veríssimo da Costa Figueiredo, Bianca Waruar Paulo Lobo, and Gisela Maria Dellamora-Ortiz. 2013. "Enzymes in Bakery: Current and Future Trends." In *Food Industry*, edited by Innocenzo Muzzalupo. London: IntechOpen. https://doi.org/10.5772/53168.

Minervini, Fabio, Anna Lattanzi, Maria De Angelis, Giuseppe Celano, and Marco Gobbetti. 2015. "House Microbiotas as Sources of Lactic Acid Bacteria and Yeasts in Traditional Italian Sourdoughs." *Food Microbiology* 52: 66–76. https://doi.org/10.1016/j.fm.2015.06.009.

Mukundan, Soumya, Jose Savio Melo, Debasis Sen, and Jitendra Bahadur. 2020. "Enhancement in β-Galactosidase Activity of Streptococcus Lactis Cells by Entrapping in Microcapsules Comprising of Correlated Silica Nanoparticles." *Colloids and Surfaces B: Biointerfaces* 195 (June): 111245. https://doi.org/10.1016/j.colsurfb.2020.111245.

Nema, Ashutosh, Sai Haritha Patnala, Venkatesh Mandari, Sobha Kota, and Santhosh Kumar Devarai. 2019. "Production and Optimization of Lipase Using Aspergillus Niger MTCC 872 by Solid-State Fermentation." *Bulletin of the National Research Centre* 43 (1): 1–8. https://doi.org/10.1186/S42269-019-0125-7.

Palla, Michela, Caterina Cristani, Manuela Giovannetti, and Monica Agnolucci. 2017. "Identification and Characterization of Lactic Acid Bacteria and Yeasts of PDO Tuscan Bread Sourdough by Culture Dependent and Independent Methods." *International Journal of Food Microbiology* 250: 19–26. https://doi.org/10.1016/j.ijfoodmicro.2017.03.015.

Plessas, Stavros, Athanasios Alexopoulos, Argyro Bekatorou, and Eugenia Bezirtzoglou. 2012. "Kefir Immobilized on Corn Grains as Biocatalyst for Lactic Acid Fermentation and Sourdough Bread Making." *Journal of Food Science* 77 (12). https://doi.org/10.1111/j.1750-3841.2012.02985.x.

Plessas, Stavros, Argyro Bekatorou, Maria Kanellaki, Athanasios A. Koutinas, Roger Marchant, and Ibrahim M. Banat. 2007. "Use of Immobilized Cell Biocatalysts in Baking." *Process Biochemistry* 42 (8): 1244–49. https://doi.org/10.1016/j.procbio.2007.05.023.

Plessas, Stavros, Ioanna Mantzourani, and Argyro Bekatorou. 2020. "Evaluation of Pediococcus Pentosaceus SP2 as Starter Culture on Sourdough Bread Making." *Foods* 9 (1): 1–11. https://doi.org/10.3390/foods9010077.

Plessas, Stavros, Marillena Trantallidi, Argyro Bekatorou, Maria Kanellaki, Poonam Nigam, and Athanasios A. Koutinas. 2007. "Immobilization of Kefir and Lactobacillus Casei on Brewery Spent Grains for Use in Sourdough Wheat Bread Making." *Food Chemistry* 105 (1): 187–94. https://doi.org/10.1016/j.foodchem.2007.03.065.

Rana, Neerja, Neha Verma, Devina Vaidya, Y S Parmar, Bhawna Dipta, and Correspondence Bhawna Dipta. 2017. "Application of Bacterial Amylase in Clarification of Juices and Bun Making." *Journal of Pharmacognosy and Phytochemistry* 6 (5): 859–64.

Ranjan, Bibhuti, and T. Satyanarayana. 2016. "Recombinant HAP Phytase of the Thermophilic Mold Sporotrichum Thermophile: Expression of the Codon-Optimized Phytase Gene in Pichia Pastoris and Applications." *Molecular Biotechnology* 58 (2): 137–47. https://doi.org/10.1007/s12033-015-9909-7.

Razzaq, Abdul, Sadia Shamsi, Arfan Ali, Qurban Ali, Muhammad Sajjad, Arif Malik, and Muhammad Ashraf. 2019. "Microbial Proteases Applications." *Frontiers in Bioengineering and Biotechnology* 0 (JUN): 110. https://doi.org/10.3389/FBIOE.2019.00110.

Renzetti, Stefano, Fabio Dal Bello, and Elke K. Arendt. 2008. "Microstructure, Fundamental Rheology and Baking Characteristics of Batters and Breads from Different Gluten-Free Flours Treated with a Microbial Transglutaminase." *Journal of Cereal Science* 48 (1): 33–45. https://doi.org/10.1016/J.JCS.2007.07.011.

Rocha, J. Miguel, and F. Xavier Malcata. 1999. "On the Microbiological Profile of Traditional Portuguese Sourdough." *Journal of Food Protection* 62 (12): 1416–29. https://doi.org/10.4315/0362-028X-62.12.1416.

Santos, Eduardo Leal Isla, Magdalena Rostro-Alanís, Roberto Parra-Saldívar, and Alejandro J. Alvarez. 2018. "A Novel Method for Bioethanol Production Using Immobilized Yeast Cells in Calcium-Alginate Films and Hybrid Composite Pervaporation Membrane." *Bioresource Technology* 247 (September 2017): 165–73. https://doi.org/10.1016/j.biortech.2017.09.091.

Shivlata, L., and Satyanarayana, T. 2017. "Characteristics of Raw Starch-Digesting α-Amylase of Streptomyces Badius DB-1 with Transglycosylation Activity and Its Applications." *Applied Biochemistry and Biotechnology* 181 (4): 1283–1303. https://doi.org/10.1007/S12010-016-2284-4.

Suo, Biao, Xinyi Chen, and Yuexia Wang. 2021. "Recent Research Advances of Lactic Acid Bacteria in Sourdough: Origin, Diversity, and Function." *Current Opinion in Food Science* 37: 66–75. https://doi.org/10.1016/j.cofs.2020.09.007.

Tang, Lele, Ruijin Yang, Xiao Hua, Chaohua Yu, Wenbin Zhang, and Wei Zhao. 2014. "Preparation of Immobilized Glucose Oxidase and Its Application in Improving Breadmaking Quality of Commercial Wheat Flour." *Food Chemistry* 161: 1–7. https://doi.org/10.1016/j.foodchem.2014.03.104.

Tanyol, Mehtap, Gülşad Uslu, and Vahap Yönten. 2015. "Optimization of Lipase Production on Agro-Industrial Residue Medium by Pseudomonas Fluorescens (NRLL B-2641) Using Response Surface Methodology." *Biotechnology & Biotechnological Equipment* 29 (1): 64–71. https://doi.org/10.1080/13102818.2014.991635.

Torresi, Sara, Maria Teresa Frangipane, and Gabriele Anelli. 2011. "Biotechnologies in Sparkling Wine Production. Interesting Approaches for Quality Improvement: A Review." *Food Chemistry* 129 (3): 1232–41. https://doi.org/10.1016/j.foodchem.2011.05.006.

Trabelsi, Sahar, Sameh Ben Mabrouk, Mouna Kriaa, Rihab Ameri, Mouna Sahnoun, Monia Mezghani, and Samir Bejar. 2019. "The Optimized Production, Purification, Characterization, and Application in the Bread Making Industry of Three Acid-Stable Alpha-Amylases Isoforms from a New Isolated Bacillus Subtilis Strain US586." *Journal of Food Biochemistry* 43 (5): e12826. https://doi.org/10.1111/JFBC.12826.

Verbelen, Pieter J., David P. De Schutter, Filip Delvaux, Kevin J. Verstrepen, and Freddy R. Delvaux. 2006. "Immobilized Yeast Cell Systems for Continuous Fermentation Applications." *Biotechnology Letters* 28 (19): 1515–25. https://doi.org/10.1007/s10529-006-9132-5.

Verma, Nitin, Vivek Kumar, and Mukesh C. Bansal. 2018. "Utility of Luffa Cylindrica and Litchi Chinensis Peel, an Agricultural Waste Biomass in Cellulase Production by Trichoderma Reesei under Solid State Cultivation." *Biocatalysis and Agricultural Biotechnology* 16 (October): 483–92. https://doi.org/10.1016/J.BCAB.2018.09.021.

Vidal, Arnau, Asier Ambrosio, Vicente Sanchis, Antonio J. Ramos, and Sonia Marín. 2016. "Enzyme Bread Improvers Affect the Stability of Deoxynivalenol and Deoxynivalenol-3-Glucoside during Breadmaking." *Food Chemistry* 208: 288–296. https://doi.org/10.1016/j.foodchem.2016.04.003.

Vuyst, Luc De, and Marc Vancanneyt. 2007. "Biodiversity and Identification of Sourdough Lactic Acid Bacteria." *Food Microbiology* 24 (2): 120–27. https://doi.org/10.1016/j.fm.2006.07.005.

Vuyst, Luc De, Henning Harth, Simon Van Kerrebroeck, and Frédéric Leroy. 2016. "Yeast Diversity of Sourdoughs and Associated Metabolic Properties and Functionalities." *International Journal of Food Microbiology* 239: 26–34. https://doi.org/10.1016/j.ijfoodmicro.2016.07.018.

Wu, Ping, Zeyu Wang, Amit Bhatnagar, Paramsothy Jeyakumar, Hailong Wang, Yujun Wang, and Xiaofang Li. 2021. "Microorganisms-Carbonaceous Materials Immobilized Complexes: Synthesis, Adaptability and Environmental Applications." *Journal of Hazardous Materials* 416 (January): 125915. https://doi.org/10.1016/j.jhazmat.2021.125915.

Yassien, Mahmoud Abdul-Megead, Asif Ahmad Mohammad Jiman-Fatani, and Hani Zakaria Asfour. 2014. "Production, Purification and Characterization of Cellulase from Streptomyces Sp." *African Journal of Microbiology Research* 8 (4): 348–54. https://doi.org/10.5897/AJMR2013.6500.

Zhang, Shuhang, Jingjing Wang, and Hong Jiang. 2021. "Microbial Production of Value-Added Bioproducts and Enzymes from Molasses, a by-Product of Sugar Industry." *Food Chemistry* 346 (June): 128860. https://doi.org/10.1016/j.foodchem.2020.128860.

12 Sourdough Microorganisms in Food Applications

Mensure Elvan and Sebnem Harsa

CONTENTS

12.1 Introduction 338
12.2 Importance of Sourdough Microorganisms 341
12.3 Effect of Sourdough Microorganisms on Nutrition and Health 341
    12.3.1 Enzymatic Activities 341
        12.3.1.1 Phytase-Active LAB and Mineral Availability 342
    12.3.2 Vitamin Stability 343
    12.3.3 Gamma-Aminobutyric Acid 343
    12.3.4 Salt Reduction and Production of Hypotensive Compounds 344
    12.3.5 Gluten Degradation 344
    12.3.6 Fermentable Oligo-, Di-, Monosaccharides, and Polyols 345
    12.3.7 Antioxidant Activity 345
    12.3.8 Probiotic Properties 346
    12.3.9 Glycemic Index 346
12.4 Effect of Sourdough Microorganisms on the Shelf Life of Products 347
    12.4.1 Antifungal Activity 347
    12.4.2 Anti-bacillus Activity 348
    12.4.3 Production of EPS 348
12.5 Effect of Sourdough Microorganisms on Sensorial Properties 349
12.6 Food Applications of Sourdough Microorganisms 352
    12.6.1 Whole Grain Bread 353
    12.6.2 Gluten-Free Bread 353
    12.6.3 Pasta 354
    12.6.4 Steamed Bread 355
    12.6.5 Burger Buns 355
    12.6.6 Pizza 355
12.7 Future Applications of Sourdough in the Food Industry 356
References 356

12.1 INTRODUCTION

Fermentation is one of the earliest technologies applied to food, used in biological preservation, improvement of nutritional and sensory properties, and increasing the shelf life and microbiological safety of foods (Ricci 2019). Over the last years, sourdough has gained special interest due to its beneficial effects in bread and bakery products. Sourdough is not only applied for sourdough bread production, but also for conventional bread, pizza, snacks, and bakery food products. Sourdough fermentation improves several features of dough, such as volume, texture, flavor, and nutritional quality, while also retarding staling and preventing microbial spoilage (De Vuyst and Vancanneyt 2007).

Sourdough consists of over 60 lactic acid bacteria (LAB) (Labat, Morel, and Rouau 2000) species belonging to *Pediococcus, Leuconostoc, Weissella, Enterococcus,* and *Lactobacillus* genera (De Vuyst et al. 2014) and 6 yeast species: *Kazachstania exigua, Pichia kudriavzevii, Wickerhamomyces anomalus, Saccharomyces cerevisiae, Candida humilis,* and *Torulaspora delbrueckii* (Huys, Daniel, and De Vuyst 2013). These microorganisms mainly stem from flour, dough ingredients, and/or the environment (De Vuyst and Vancanneyt 2007), and their composition is affected by endogenous factors – such as enzymes and chemical composition of the flour – and by exogenous factors, such as redox potential and temperature (Hammes and Gänzle 1998). In addition, geographical and climatological conditions increase the complexity of these fermented environments. Recently, the microflora of sourdough from plain, basin, and mountain regions of China were evaluated, and it was revealed that Firmicutes, Proteobacteria, and Cyanobacteria were the predominant phyla found across these regions. Sourdough microflora from different environments were different from each other, *Lactobacillus* being dominant in sourdoughs from plain and mountain regions, while *Pediococcus pentosaceus* was dominant in sourdoughs from basin regions, and Acetobacter represented the microflora of mountain sourdough samples (Xing et al. 2020).

There is also different microflora in a variety of sourdough treated using different techno-bio-functional processes (De Vuyst et al. 2014, Fu et al. 2020). For example, *Lacticaseibacillus paracasei, Limosilactobacillus fermentum, Lactiplantibacillus plantarum, Latilactobacillus curvatus, Companilactobacillus paralimentarius, Lactobacillus gallinarum, Companilactobacillus kimchii, Limosilactobacillus pontis, Levilactobacillus spicheri,* and *Schleiferilactobacillus perolens* have been isolated from industrially produced rice sourdoughs (Meroth, Hammes, and Hertel 2004); while mainly *Leuconostoc mesenteroides, Lacticaseibacillus casei, L. plantarum, Lentilactobacillus farraginis, L. curvatus, L. paracasei Levilactobacillus brevis, Loigolactobacillus coryniformis, Lactobacillus uvarum, Enteroccocus pseudoavium, P. pentosaceus,* and *Pediococcus acidilactici* species have been isolated from spontaneously fermented rye sourdoughs (Bartkiene et al. 2020); and *Fructilactobacillus sanfranciscensis, C. paralimentarius, L. plantarum, L. brevis, Lactiplantibacillus paraplantarum, L. curvatus, Furfurilactobacillus rossiae, Leuconostoc pseudomesenteroides, Leuconostoc mesenteroides, Weissella paramesenteroides*, and *Weissella cibaria* strains have been isolated from Turkish sourdoughs (Dertli et al. 2016). LAB strains, such as *Companilactobacillus alimentarius, L. brevis, Lactiplantibacillus pentoses, F. sanfranciscensis, Latilactobacillus sakei*, and *Paucilactobacillus vaccinostercus*, and yeasts e.g. *S. cerevisiae* and *C. humilis* were isolated from sourdough in Japanese bakeries (Fujimoto et al. 2019). In wheat sourdoughs of Ya'an (China), the microflora consists of *Lactococcus raffinolactis, Lactococcus lactis, Lacticaseibacillus pantheris, L. plantarum, Weissella viridescens, L. pseudomesenteroides, L. mesenteroides*, and *Leuconostoc citreum* (Liu et al. 2018), whereas the microflora of spontaneously fermented chia sourdoughs contains mainly *Enterococcus* species along with *Lactobacillus, Lactococcus*, and *Weissella* species (Maidana et al. 2020). In quinoa sourdough fermentation, LAB was found to be predominant in the microflora, similar to the others stated for cereal and gluten-free cereal sourdoughs (Ruiz Rodríguez et al. 2016).

Kluyveromyces yeast strains, such as *Kluyveromyces aestuarii, Kluyveromyces lactis*, and *Kluyveromyces marxianus* were first isolated from Iranian traditional sourdough (Fekri et al. 2020). In addition, *Gluconobacter oxydans* was first isolated from Chinese traditional sourdoughs (Xing et al. 2020). Spontaneously fermented sourdough is an outstanding source of LAB with high potential for overcoming technical limitations encountered in the food industry and responding to a variety of needs, as shown in the previous studies. Antifungal and antimicrobial activities, along with metabolic ability to ferment multiple carbohydrate sources, show the high techno-economic industrial potential of sourdough LAB or their starter cultures (Bartkiene et al. 2020). The main microorganisms present in sourdough have been summarized in Table 12.1.

Sourdough fermentation depends on the microflora composition and fermentation conditions in two basic stages, namely lactic acid and alcohol biosynthesis. During lactic acid fermentation, fermentable carbohydrates are utilized by LAB, while during alcoholic fermentation, carbohydrates are converted to ethanol and carbon dioxide by yeasts anaerobically through the Embden-Meyerhof-Parnas pathway. LAB mainly accounts for acidification of the products, while the main role of yeasts is the leavening of the dough through releasing CO_2 (Hammes and Gänzle 1998). In general, a sourdough includes a varying number of LAB and yeasts, with a ratio of approximately 100:1 (Gobbetti 1998).

The microorganisms isolated from sourdough differ according to the type of sourdough. Three types of sourdough have been described, depending on the type of technology used for their production (De Vuyst and Neysens 2005): type I sourdough, or traditional sourdough, is produced by the traditional process in which the starter cultures are not added but rather continuous daily refreshments are carried out; type II sourdough, or accelerated sourdough, is produced by fermentation in which a starter culture is added to a flour/water mixture; type III sourdough, or dried sourdough production, is initiated by starter cultures applied as aroma carriers and acidity developers during the bread making process. The main characteristics of each sourdough type and their specific microflora are summarized in Figure 12.1.

TABLE 12.1
Microorganisms Most Frequently Found in Sourdough

Microbiota	Heterofermentative lactic acid bacteria	Homofermentative lactic acid bacteria	Yeasts
	Fructilactobacillus sanfranciscensis	Lactobacillus amylovorus	Saccharomyces cerevisiae
	Fructilactobacillus fructivorans	Lactobacillus acidophilus	Candida humilis
	Levilactobacillus brevis	Lactobacillus delbrueckii	Kazachstania exigua
	Limosilactobacillus fermentum	Companilactobacillus farciminis	Pichia kudriavzevii
	Limosilactobacillus reuteri	Companilactobacillus paralimentarius	Wickerhamomyces anomalus
	Limosilactobacillus pontis	Pediococcus pentosaceus	Torulaspora delbrueckii
	Limosilactobacillus panis		
	Furfurilactobacillus rossiae		
	Lactiplantibacillus plantarum subsp. plantarum		
	Lacticaseibacillus casei		
	Lacticaseibacillus rhamnosus		
	Companilactobacillus alimentarius		
	Leuconostoc mesenteroides		
	Leuconostoc citreum		
	Weissella confusa		
	Weissella cibaria		

Sourdough Microorganisms in Food Applications

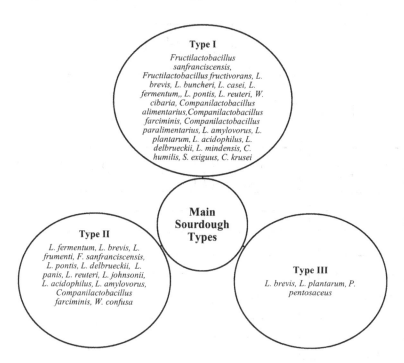

FIGURE 12.1 Summary of the main sourdough types and their predominant microflora.

12.2 IMPORTANCE OF SOURDOUGH MICROORGANISMS

The nutritional, sensorial, and technological properties of bread or bakery products are mainly affected by starter cultures of sourdoughs. Desirable nutritional properties include the degradation of phytic acid, which can act as an anti-nutritional factor and an increased production of functional compounds, such as bioactive peptides, free phenolic acids, soluble arabinoxylans, and soluble fiber (Katina and Poutanen 2013). The sensorial properties influenced by the starter cultures are mainly related to enhanced textural properties of foods (Montemurro, Coda, and Rizzello 2019) and other technological improvements of the products, including acidification, microbial growth (Coda et al. 2010), production of exopolysaccharides (EPS) (Galle and Arendt 2014), antifungal activity (Coda et al. 2013), and biosynthesis of antimicrobial components. All these beneficial properties are represented in Figure 12.2.

12.3 EFFECT OF SOURDOUGH MICROORGANISMS ON NUTRITION AND HEALTH

12.3.1 Enzymatic Activities

The sourdough indigenous microflora is an important source of genetic diversity able to produce several enzymes suitable for biotechnological applications. In this

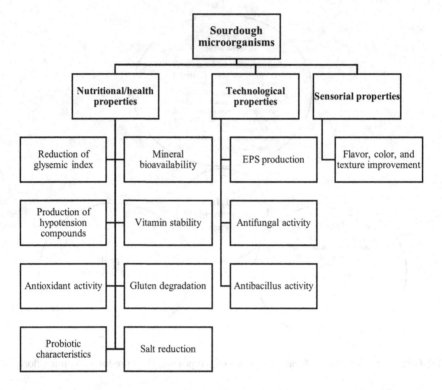

FIGURE 12.2 Nutritional/health, technological, and sensorial properties of sourdough microorganisms.

respect, phytase and xylanase enzymes can be useful improving the nutritional and technological potential of sourdough. Xylanases are mainly used to improve the viscosity of dough, increase the bread volume, and extend shelf life in the bakery industries (Poutanen, Flander, and Katina 2009). Manini et al. (2016) isolated new strains of LAB species including *L. brevis*, *L. sakei*, *L. mesenteroides*, *L. curvatus*, *L. citreum*, *P. pentosaceus*, and *L. casei* from spontaneously fermented wheat bran sourdough. Moreover, LAB species with phytate breakdown ability can be used as starter cultures in the fermentation of food products rich in phytic acid to improve their nutritional quality by increasing mineral bioavailability.

12.3.1.1 Phytase-Active LAB and Mineral Availability

Whole grain food products are important to the human diet, as they contain high levels of vitamin E, vitamin B and minerals, such as P, Cu, Mg, Zn, and Fe. However, these minerals are bounded to phytic acid and thus, their bioavailability is decreased (Kumar et al. 2010). In order to eliminate this anti-nutritional effect, it is necessary to reduce the levels of phytate in the products. The addition of sourdough has shown promise in breaking down phytate in commercial rye bread (Nielsen et al. 2007). Phytase-active species have been isolated from rye and wheat sourdough, such as

L. panis, *L. reuteri*, *L. fermentum*, and *P. pentosaceus*, with *L. panis* having the highest activity (Nuobariene et al. 2015). Phytase activity was also found in water/salt-soluble sourdough extracts when *L. plantarum* and *P. pentosaceus* were used as starter cultures (Coda et al. 2011). Lopez et al. (2000) used *L. mesenteroides*, *L. acidophilus*, and *L. plantarum* isolated from rye sourdough, wheat sourdough, and rye/wheat sourdoughs, respectively, to hydrolyze phytate in pasteurized wheat flour. LAB strains are currently used in fermentations to improve the nutritional quality of cereals and cereal-based foods by reducing certain anti-nutritional factors, such as tannins, phytate, and enzyme inhibitors. LAB species are involved in the production of many metabolites including EPS, organic acids, antimicrobial compounds, and beneficial enzymes during cereal fermentation. Moreover, LAB strains encoding phytases can be suitable starter cultures for cereal and legume fermentations (Rollán, Gerez, and LeBlanc 2019, Sumengen, Dincer, and Kaya 2013).

12.3.2 VITAMIN STABILITY

Cereal-based foods are important sources of vitamin E, thiamine, riboflavin, and folates. In addition, whole grains are also an excellent source of tocols, lignans, phenolic acids, and phytosterols (Liukkonen et al. 2003). Processing may change the levels of these compounds and, in some cases, limit their bioavailability (Slavin, Jacobs, and Marquart 2000). Yeast fermentation can increase the amount of folate during wheat and rye baking compared to LAB fermentation (Kariluoto et al. 2006, Liukkonen et al. 2003). Additionally, yeast fermentation increased the content of thiamine that normally decreases during the baking process (Martinez-Villaluenga et al. 2009). Moreover, two sourdough isolates of *L. plantarum* have riboflavin synthesis capacity and riboflavin-overproducing stability (Capozzi et al. 2011). Folate, riboflavin, and phytase LAB strains may have positive effects during fermentation. The presence of *L. mesenteroides*, *Lactobacillus* species, and yeast during the fermentation of Indian local fermented breakfast food has resulted in increased folate and riboflavin concentrations (Ghosh and Chattopadhyay 2011). In addition, the amount of these two vitamins in the breakfast food increased due to the presence of *Saccharomyces boulardii* and *L. lactis* during fermentation (Chandrasekar Rajendran et al. 2017). In a recent study, sourdough pasta produced with quinoa flour by the addition of *L. plantarum* strains (*L. plantarum* CRL 1964 and *L. plantarum* CRL 2107) showed increased levels of folates and riboflavin, enhancing the bioavailability of vitamins and minerals in cereal- or pseudocereal-based food products (Carrizo et al. 2020).

12.3.3 GAMMA-AMINOBUTYRIC ACID

LAB have a positive effect on the sensory and nutritional quality of bakery products due to the release of biologically active metabolites. Gamma-aminobutyric acid (GABA) is a non-protein amino acid commonly found in fermented foods that acts as an inhibitory neurotransmitter of the central nervous system (Krnjević 1974). GABA is synthesized by the enzyme glutamate decarboxylase that catalyzes the decarboxylation of L-glutamate (Siragusa et al. 2007). LAB with glutamate decarboxylase enzyme enhanced the concentrations of GABA in sourdough (Venturi et al. 2019). However, potential GABA-producing LAB isolated from sourdough is strain-dependent. During sourdough fermentation, a high level of GABA content was found using micronized by-products from wheat germ (Rizzello et al. 2010) and debranned durum wheat (Rizzello, Coda, et al. 2012). The findings from the Wu et al. (2017) study revealed that *L. brevis* has the GAD gene. More recently, sourdough isolates *L. plantarum, L. paracasei, Levilactobacillus hammesii, L. brevis, L. senmaizukei,* and *L. coryniformis* have exhibited the GAD gene (Fraberger, Ammer, and Domig 2020). Although only some part of sourdough isolates demonstrates the presence of the GAD gene, they have potential for GABA production (Villegas et al. 2016). GABA exerts beneficial effects on health, such as reduction of anxiety and depression, demonstration of antidiabetic and hypotensive effects, and neurotransmission (Diana, Quílez, and Rafecas 2014, Yılmaz and Gökmen 2020).

12.3.4 SALT REDUCTION AND PRODUCTION OF HYPOTENSIVE COMPOUNDS

Sourdough fermentation has also demonstrated beneficial effects on lowering salt content and enriching bread and bakery products with antihypertensive substances such as free amino acids and peptides, depending on the LAB strain used during fermentation (Gobbetti et al. 2019). GABA has an important function in lowering blood pressure in hypertensive patients. However, clinical studies are needed to prove the efficacy of GABA-enriched sourdough bread or other baked goods when inducing hypotension (Gobbetti et al. 2019). LAB may also synthesize angiotensin I-converting enzyme (ACE) and inhibitory peptides during sourdough fermentation that also can reduce blood pressure (Gobbetti et al. 2014).

12.3.5 GLUTEN DEGRADATION

Sourdough-containing foods have high nutritional indexes and good protein digestibility due to the enzymatic activity of LAB (Kopeć et al. 2011). Sourdough LAB has proteolytic activity due to the presence of intracellular peptidases and cell-wall proteinases, causing the hydrolysis of native proteins to peptides and free amino acids during sourdough LAB fermentation. Proteolysis can be applied to reduce gluten content (Gobbetti et al. 2014) and improve the digestibility of protein-rich flours used in the fortification of cereal-based foods (Curiel et al. 2015). Providing fully hydrolyzed or lowered amounts of gluten in cereal-based products is important for celiac patients (Gobbetti et al. 2014). In one study, semolina pasta was produced by fermenting semolina with selected LAB and demonstrated increased gluten degradation in the product without any changes in the sensory profiles of the pasta (Di Cagno et al. 2005). Gluten hydrolyzing microflora, such as *Wickerhamomyces anomalus, Bacillus cereus, Bacillus megaterium, Enterococcus faecalis,* and *Enterococcus mundtii* have been isolated from Pakistani fermented sourdough. Among the isolated microflora, *W. anomalus* and *E. mundtii* showed the highest gluten degradation ability. In addition, the yeast strains and *Enterococcus* species demonstrated more proteolytic activity in comparison with the control strain used (*L. plantarum* ATCC 14917) and, thus, these isolates can be applied to produce gluten-free products for celiac patients (Sakandar, Usman, and Imran 2018). Overall, LAB can be thought of as evolving cell factories that provide functional biological molecules and ingredients to produce value-added gluten-free functional foods (Zannini et al. 2012).

12.3.6 FERMENTABLE OLIGO-, DI-, MONOSACCHARIDES, AND POLYOLS

The prevalence of wheat-related disorders, such as gluten intolerance, celiac disease, non-celiac wheat sensitivity (NCWS), and inflammatory bowel syndrome (IBS) have increased all over the world (Casella et al. 2010). Fructose is a metabolic product of fructan, and it is included as part of the fermentable oligo-, di-, monosaccharides, and polyols (FODMAPs) group that plays an important role in the NCWS. *Lactobacillus* species secretes a wide variety of enzymes that break down oligosaccharides (Zhao and Gänzle 2018). Carbohydrate active enzymes are applied industrially to prepare/ produce low-FODMAP breads and bakery products by degrading fructans, mannitol, and raffinose (Loponen and Gänzle 2018). According to the findings of a recent study, more than half of sourdough's LAB have the potential to degrade fructose. Therefore, well-tolerated bakery products can be produced, creating a synergistic effect with certain fructan-degrading yeasts (Fraberger et al. 2018).

Sourdough biotechnology and its microorganisms also have the potential to be used in reducing protein allergy. Sourdough bacteria reduced soy and dairy protein immunoreactivity (Rui et al. 2019). Moreover, according to a new study, novel strains of *P. pentosaceus, P. acidilactici*, and *L. sakei subsp. sakei* isolated from Chinese sourdough effectively reduced the allergenicity of wheat proteins (Fu et al. 2020). Thus, traditional fermentation methods, such as sourdough, can be used to produce or prepare high-quality, gluten-free, and hypoallergenic food products (Zhang and He 2013).

12.3.7 ANTIOXIDANT ACTIVITY

Antioxidant activity has been associated with the hydrolyzation of complex phenolic compounds to phenolic acids and the release of antioxidant peptides from native proteins during sourdough fermentation (Gobbetti et al. 2019, Rizzello, Nionelli, et al. 2012). Previous studies have demonstrated that sourdough biotechnology is responsible for the high phenolic compounds and antioxidant activity of sourdough bread (Banu, Vasilean, and Aprodu 2010, Bustos et al. 2017). Sourdough-based fermentations can produce bioactive peptides with anti-inflammatory, hypocholesterolemic, immunomodulatory, and antiproliferative/ anticancer properties (Rizzello et al. 2016). The ability of selected LAB to synthesize antioxidants was demonstrated through sourdough fermentation of several cereal flours (Coda et al. 2012, Rizzello et al. 2008, Rizzello, Nionelli, et al. 2012). More recently, new sourdough isolates of *C. milleri, L. citreum*, and *F. sanfranciscensis* species (Martorana et al. 2018) were used to produce sourdough bread and resulted in breads with high content of phenolic compounds and DPPH (2,2-diphenyl-1-picrylhydrazyl) inhibition (Sidari et al.

2020). This study demonstrated that using a versatile starter isolated from sourdough in industrial and artisanal bread production can improve the antioxidant properties of breads.

12.3.8 Probiotic Properties

Probiotics can produce bioactive molecules, phenolic compounds, and EPS, increasing the quality and nutritional value of fermented foods. Sourdough as a spontaneously fermented food, can be a source of potential probiotic microorganisms, such as *P. pentosaceus* CE65 (Manini et al. 2016). Microorganisms isolated from Pakistani local sourdough have shown resistance to high acidity and bile salts and have antipathogenic activity, as well as being effective in gluten degradation. Among the isolates, new yeast strains of *W. anomalus* and *E. mundtii* have strong probiotic potential (Sakandar, Usman, and Imran 2018). Potential probiotic yeasts (*S. cerevisiae* strain-2 and *S. cerevisiae* strain-4) have been isolated from Altamura (Italy) sourdough with promising functionality when used as starter cultures for preparing cereal-based products (Perricone et al. 2014). In a recent study, probiotic microflora having phytate degradation ability was isolated from Iranian traditional sourdough to produce whole wheat breads, thereby producing products with desirable functional and nutritional characteristics (Fekri et al. 2020).

12.3.9 Glycemic Index

When the amount of rapidly digestible carbohydrates increases in the diet, the blood sugar levels (glycemic index) rises fast, causing an excessive demand for insulin in the postprandial period. Hyperglycemia is a widely recognized risk factor for diseases mainly associated with metabolic syndrome (Barclay et al. 2008). Bakery products are one of the main sources of digestible carbohydrates (Katina and Poutanen 2013), and, although some are made with wholemeal flour, they often have high glycemic indexes (Foster-Powell, Holt, and Brand-Miller 2002). Macro- and micronutrients such as amylose, amylopectin, and proteins, found in various cereals, interact with each other, and even a small amount of proteins can significantly influence the digestibility of starch. Gelatinized starch formation in cereal foods leads to high digestibility values and glycemic responses upon consumption (Singh, Dartois, and Kaur 2010).

Sourdough fermentation causes chemical changes that may reduce the level of starch gelatinization, resulting in a decreased digestibility of starch (Östman 2003). These results were not appreciated when using baker's yeast for fermentation of the same type of bread (Scazzina et al. 2009). Moreover, the glycemic index of bread is reduced due to the production of organic acids by LAB that lowers the digestibility of starch (Poutanen, Flander, and Katina 2009). Additionally, peptides, amino acids, and free phenolic compounds regulated carbohydrate metabolism throughout sourdough fermentation (Katina et al. 2007, Nilsson, Holst, and Björck 2007), and reduced the glycemic and insulin indexes (Novotni et al. 2011, Solomon and Blannin 2007). Furthermore, LAB enzymes can contribute to fiber solubilization in foods.

For instance, xylanases act on arabinoxylan, increasing the content of water-extractable arabinoxylan, which is related to delayed digestion of carbohydrates, decreased rate of absorption, and low insulinemic and glycemic indexes (Katina et al. 2006). According to previous studies, the fermentation of rye and wheat flours with LAB from sourdough reduced the glycemic index of wholemeal barley (Östman 2003) and wheat breads (De Angelis et al. 2009 Lappi et al. 2010) and the insulin index of rye breads (Juntunen et al. 2003).

12.4 EFFECT OF SOURDOUGH MICROORGANISMS ON THE SHELF LIFE OF PRODUCTS

Shelf life of bread is limited due to microbiological spoilage, including mold growth, that induces ropiness and other physicochemical changes such as firming and staling. Bread firmness is affected by gluten and gluten/starch ratio, which change the elasticity of bread during storage. Sourdough LAB have capability to acidify the dough, hydrolyze starch, and proteolyzed gluten, which affect the physicochemical changes of bread and thus the shelf life of the products (Corsetti et al. 1998). Moreover, some sourdough LAB have been shown to produce EPS that can exert great potential as anti-staling agents (Korakli et al. 2003). In addition, bread staling can be reduced by pentosans formed by the solubilization of arabinoxylans during sourdough fermentation, and the gluten–starch interactions responsible for staling can be prevented (Gray and Bemiller 2003). Sourdough biotechnology and its microbiota can also exert antibacterial and antifungal activities resulting in an increased shelf life of bread. According to the studies, adding 30% of wheat sourdough protected the products against staling and extended the shelf life of the bread (Rinaldi et al. 2015, Torrieri et al. 2014). The use of sourdough has also been effective in slowing mold disease and preventing ropy-bread disease (Savkina et al. 2019). Control of bread staling and preserving the quality of bread for a longer time can provide economic benefits (Gamel, Abdel-Aal, and Tosh 2015).

12.4.1 ANTIFUNGAL ACTIVITY

Penicillium, Aspergillus (Garcia, Bernardi, and Copetti 2019), *Fusarium, Mucor,* and *Rhizopus* (Legan 1993) are the genera of molds that most commonly lead to bread spoilage. Bakery industries need to apply chemical preservatives, such as sorbic, benzoic, and propionic acids and their salts to prevent mold growth, and sourdough could be used as a natural preservative alternative to these chemicals (Axel, Brosnan, Zannini, Peyer, et al. 2016, Belz et al. 2012, Gerez et al. 2009). Gerez et al. (2009) and Fraberger, Ammer, and Domig (2020) revealed that the inhibition of mold species in bread was influenced by the type of LAB strains present. *L. reuteri, L. brevis,* and *L. plantarum* had antifungal activities, due to their ability to produce antifungal compounds, such as phenyllactic and acetic acids (Gerez et al. 2009). *L. brevis* S4.5, *L. plantarum* S4.2, *L. coryniformis* S4.4.2, *L. paracasei* S9.11, and *L. parabuchneri* S2.9, isolated from sourdough, strongly inhibited the growth of *Penicillium roqueforti* DSM1079 (Fraberger, Ammer, and Domig 2020). In a study performed by Manini et al. (2016), *L. curvatus* CE83; *L. plantarum* CE42, CE60, and CE84; and *P. pentosaceus* CE65 and CE23 (isolated from spontaneously fermented wheat bran sourdough-like) had antifungal activity against *A. oryzae* and *A. niger* (Smith et al. 2004). The ability of some *L. plantarum* species to form acetic acid, lactic acid, and hydrogen peroxide, as well as to produce bacteriocins and compete for nutrients, are the main mechanisms of antifungal activity of these species (Gupta and Srivastava 2014). *F. sanfranciscensis* CB1 demonstrated inhibitory activity against molds such as *Aspergillus*, *Fusarium*, *Monilia*, and *Penicillium*, due to producing different organic acids, particularly caproic acid (Corsetti et al. 1998). In another study, *L. brevis* isolated from rice bran sourdough has demonstrated antibacterial, antifungal, and anti-aflatoxigenic effects (against aflatoxin B1). LAB isolated from quinoa and rice sourdoughs have also demonstrated antifungal properties (Axel, Brosnan, Zannini, Furey, et al. 2016). *L. brevis*, *L. paraplantarum*, and *Companilactobacillus crustorum* species, isolated from spontaneously fermented Turkish einkorn sourdough, have also shown antifungal activity against *A. niger*, *A. flavus*, and *Penicillium carneum* (Çakır et al. 2020). While most LAB isolated from rye sourdough demonstrated antifungal properties against selected molds, only *L. curvatus* No.51, *L. coryniformis* No.71, *L. plantarum* No.122, *L. casei* No.210, and *L. paracasei* No.244 effectively inhibited selected pathogenic–opportunistic bacteria (Bartkiene et al. 2020). In addition, in another study, all sourdough-isolated LAB have shown antibacterial and antifungal properties (Demirbaş et al. 2017). The study confirms that sourdough microorganisms, such as LAB and probiotics, can be used as biological preservatives (Sadeghi et al. 2019).

12.4.2 ANTI-BACILLUS ACTIVITY

Bacillus subtilis, *Bacillus pseudomesenteroides*, *Bacillus licheniformis*, and *B. cereus* species cause rope spoilage in bread (Menteş, Ercan, and Akçelik 2007). Certain LAB strains showed fungistatic, fungicidal, and antibacterial activities against foodborne pathogens by producing inhibitory substances (Cizeikiene et al. 2013). Numerous studies focused on finding potential sourdough LAB to control these pathogens. Among isolated LAB strains from sourdough, *L. plantarum* and *L. paracasei* demonstrated strong inhibitory activity against *Bacillus* spp. (Fraberger, Ammer, and Domig 2020). Acetic and lactic acids are the main inhibitory compounds produced by LAB, whereas other antimicrobial compounds, such as diacetyl, hydrogen peroxide, and bacteriocins and bacteriocin-like inhibitory substances have shown effective inhibitory activity against *Bacillus* spp. (Digaitiene et al. 2012).

12.4.3 PRODUCTION OF EPS

Sourdough-based LAB can produce EPS, such as fructan and glucan (Galle and Arendt 2014). It was suggested that EPS isolated from LAB have beneficial effects on bakery technology, which are rheology, water absorption, machinability, storage stability of dough, and bread staling and loaf volume. Manini et al. (2016) determined that *L. plantarum* and *P. pentosaceus* were the major EPS producers in sourdough.

Previous studies have also shown that applying EPS-forming starter cultures during baking improves bread quality. I.e. *L. plantarum* and *W. cibaria* as sourdough starter cultures improved the sourdough viscosity and resulted in higher bread volume and lower firmness (Di Cagno et al. 2006). Lacaze, Wick, and Cappelle (2007) showed that using dextran-producer *L. mesenteroides* in preparing sourdough positively improved the mouthfeel, softness, crumb structure, and freshness of rye and wheat breads. In another study, EPS produced by *Propionibacterium freudenreichii* and *Weissella confusa* have demonstrated an impact on improving bread texture and loaf volume, staling retardation, and reducing crumb firming throughout the storage period compared to bread with added chemical organic acids (Tinzl-Malang et al. 2015). Furthermore, EPS provide functional properties such as hypocholesterolemic, antioxidant, immunomodulation, and prebiotic effects (Ryan et al. 2015). Some EPS-producer sourdough LAB have been summarized in Table 12.2.

Based on the findings from a study by Fekri et al. (2020), sourdough-isolated yeasts yielded higher levels of EPS when compared to the isolated bacteria. Therefore, it was stated that yeasts like *K. lactis* and *K. aestuarii* may have great potential to produce bread with high amounts of EPS. Moreover, ropy capsular polysaccharides, levan, and dextran can compensate for the adverse effects of antifungal chemical additives. Therefore, these ingredients demonstrate potential as anti-staling agents, and they improve the shelf life of bread (Tinzl-Malang et al. 2015).

Bacterial homoexopolysaccharides act as hydrocolloids in sourdoughs. Hydrocolloids alter gelatinization of starch and are used in producing gluten-free bread as fat replacers, gluten substituents, and fiber source (Rosell, Rojas, and De Barber 2001). EPS have recently gained relevance in the baking industry as hydrocolloids to improve sensory characteristics of bread (Galle and Arendt 2014).

12.5 EFFECT OF SOURDOUGH MICROORGANISMS ON SENSORIAL PROPERTIES

Sensory properties of food are important parameters influencing consumers' demand and/or acceptance. Sourdough and its microbiota have attracted attention due to not only their nutritional value but also to their improvements in the sensorial quality of products. Sourdough fermentation exerts characteristic smell, flavor, texture, and color profiles. EPS have beneficial effects on mouthfeel, texture, perception of taste, and stability of fermented food products. Furthermore, LAB synthesize lactic acid and acetic acid, that cause a sour taste and aroma of the foods, during fermentation. Moreover, protein degradation into amino acids allows browning reaction and leads to darker crumb and crust (Martins, Jongen, and Van Boekel 2000). Previously, it was reported that *F. sanfranciscensis* CD1 is the most proteolytic sourdough LAB (Gobbetti et al. 1996) and *L. plantarum* DC400, an amylolytic strain, can hydrolyze starch (Corsetti et al. 1998). These hydrolyzation processes affect the textural properties, such as viscosity and elasticity, of the bread and other bakery products, and the staling and starch retrogradation are delayed with the acid formation by LAB (Corsetti et al. 2000). Bio-acidified, gluten-free breads were found to be softer than their chemically acidified controls after a five-day storage period (Moore et al.

TABLE 12.2
EPS of LAB Isolated from Sourdoughs

LAB	EPS	Sourdough type	Reference
W. cibaria, L. mesenteroides, L. plantarum, L. paraplantarum	not structurally characterized	durum wheat sourdough	(Zotta et al. 2008)
L. citreum, L. mesenteroides, W. cibaria, W. confusa	glucan	sourdough	(Bounaix et al. 2009)
F. sanfranciscenis	fructan	sourdough	(Bounaix et al. 2009)
L. mesenteroides	glucan and fructan	sourdough	(Bounaix et al. 2009)
F. sanfranciscensis	fructan (levan)	sourdough	(Tieking et al. 2005)
L. mesenteroides	dextran	panettone	(Lacaze, Wick, and Cappelle 2007)
Weissella species	glucan and/or fructan	sourdough	(Tieking et al. 2003)
L. acidophilus		rice bran sourdough	(Abedfar et al. 2020)
F. sanfranciscensis	fructan	wheat/rye sourdough	(Korakli et al. 2001)
W. cibaria, L. plantarum	glucan and fructan	sourdough	(Di Cagno et al. 2006)
L. rhamnosus, W. cibaria		chia sourdough	(Maidana et al. 2020)
W. confusa	dextran	wheat sourdough	(Katina et al. 2009)
K. marxianus, K. lactis, K. aestuarii, E. faecium, P. pentosaceus, L. citreum	not structurally characterized	Iranian traditional sourdough	(Fekri et al. 2020)
F. sanfranciscensis, C. paralimentarius, L. plantarum, L. brevis, L. paraplantarum, L. curvatus, F. rossiae, L. pseudomesenteroides, L. mesenteroides, W. paramesenteroides, W. cibaria	not structurally characterized	Turkish sourdough	(Dertli et al. 2016)

2007). Moreover, acid production depends on the fermentable carbohydrates and LAB species present as well as on the dough yield, fermentation time, and temperature. A high fermentation temperature, high sourdough water content, and using wholemeal flour will increase the acid production in wheat sourdoughs (Lorenz and Brümmer 2003).

Flavor is one of the most popular sensory qualities. Numerous volatile and nonvolatile compounds from fermentation and ingredients importantly affect the flavor of bread. The release of amino acids, as well as the production of aldehydes, alcohols, acids, ketones, hydrocarbons, pyrazines, lactones, sulfur, furan derivatives, esters, ether derivatives, and pyrrol compounds contribute to the flavor of the final products. Other important flavor-active compounds include acetic acid, butanoic acid, phenylacetic acid, pentanoic acid, and 2- and 3-methylbutanoic acid (Czerny and Schieberle 2002). 2-phenylethanol produced by sourdough microorganisms remained high after baking and enhanced the flavor of bread (Fekri et al. 2020). The production of volatile compounds can be controlled by selecting the starter cultures and by changing the fermentation time and temperature, dough consistency, and flour extraction rate (Gobbetti et al. 1995). According to the results of a study investigating the effects of the use of different starter cultures, flour types, and ingredients on volatile substances in sourdough bread, the addition of sourdough positively affected the level of volatile compounds. After sourdough fermentation for up to 72 hours, the concentration of several volatiles increased, the main volatile compounds being alcohols, ethanol, aldehydes, terpenes, esters, and heterocyclic compounds. The presence of high levels of acetic acid, diacetyl and acetoin/ethyl acetate compounds contributing to the flavor of bread have also been identified in breads produced with sourdoughs, where *G. oxydans* IMDO A845 was used as a starter culture. Similarly, breads made with teff sourdoughs have been observed to have a different volatile compound profile than those of wheat-based sourdough breads (Van Kerrebroeck et al. 2018). In a study using mixed culture (*Pichia kudriavzevii* CGMCC 17607 and *Meyerozyma guilliermondii* CGMCC 17606 combined with *F. sanfranciscensis* DSM20451) as a starter, it was found that lowering the sourdough fermentation temperature to 10°C improved the formation of important volatiles and other flavor-active compounds such as 1-octen-3-ol, 3-methyl-1-butanol, isogeraniol, 3-methylbutanal, 2-methyl-1-propanol, benzene acetaldehyde, benzaldehyde, phenylethyl acetate furfural, 2-pentylfurane, ethyl octanoate, and ethyl decanoate (Xu et al. 2020).

The utilization of sourdough has also important influences on dough rheology in bakery products. Acid formation and enzymatic activities, in particular proteolytic activity, results in rheological changes in sourdough and sourdough-derived products (Katina 2005). The formation of a gluten network heavily affects the texture of bread as this network will trap the gas derived from the alcoholic fermentation (Cauvain 2012). The application of sourdough in bread making enhanced the volume of the loafs better when compared to chemically acidified breads (Clarke, Schober, and Arendt 2002). Furthermore, acidic sourdough is comparatively more effective than yeasted preferment in enhancing loaf volume (Corsetti et al. 2000). During sourdough fermentation, the gluten proteins are degraded, which caused less elasticity and softer texture of bread (Clarke, Schober, and Arendt 2002, Thiele, Grassl, and Gänzle 2004). Weaker gluten may allow the dough to expand higher, but this generally reduces gas retention. The increased volume in sourdough breads can be partially explained by arabinoxylan solubilization and EPS production (Korakli et al. 2001). Metabolites of sourdough microorganisms and their importance on bakery products are presented in Figure 12.3.

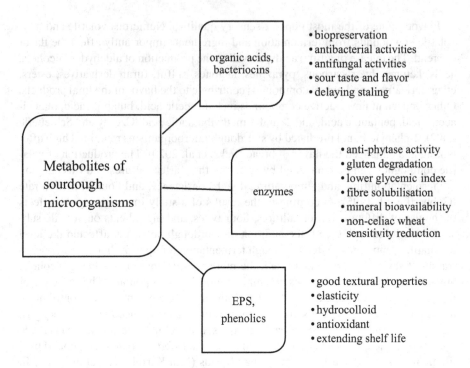

FIGURE 12.3 Relevance of metabolites produced by sourdough microorganisms on bakery products.

12.6 FOOD APPLICATIONS OF SOURDOUGH MICROORGANISMS

The growing popularity of sourdough consumption, along with its various health benefits, has led to the growth of this market (Plessas 2021). Moreover, sourdough microorganisms offer several advantages in pasta and bakery products mainly related to the reduction of pH during fermentation. These advantages include increased volume and softness, mineral availability, and shelf life, while preventing spoilage and decreasing glycemic index and phytate contents.

Sourdough technology results in bread with high-quality properties such as specific volume, shape coefficient, and porosity, but also low contents of acrylamide compared to control bread samples (Bartkiene et al. 2017). According to recent studies, it has been revealed that not only the use of sourdough microorganisms, but also their combination with microorganisms isolated from different fermented products has positive effects on bread and bakery products in many aspects of textural, nutritional, and sensory qualities (Comasio et al. 2020). Thereby, kombucha microorganisms enhanced the quality and shelf life of sourdough bread (Mohd Roby et al. 2020), the use of *P. pentosaceus* SP2 isolated from kefir grains resulted in a desirable acidity and organic acid content in sourdough breads (Plessas, Mantzourani, and

Bekatorou 2020), and the application of *Weissella koreensis* HO20 and *L. citreum* HO12, isolated from kimchi, delayed bread staling and prevented microbial spoilage (Choi et al. 2012). Moreover, recently a new sourdough bread has been produced using coconut water kefir starter during fermentation, resulting in breads with improved sensory properties, as well as prolonged shelf life (Limbad et al. 2020).

12.6.1 WHOLE GRAIN BREAD

The application of sourdough microorganisms for fermentation of whole grain bread can enhance the palatability and mouthfeel as well as improving flavor and texture of whole grain and bran-rich breads (Salmenkallio-Marttila, Katina, and Autio 2001). Isolates of *K. marxianus, K. aestuarii, K. lactis, P. pentosaceus, E. faecium, L. citreum*, and *S. cerevisiae* (positive control) were used during bread making to improve the physical characteristics of bread e.g. porosity and volume. According to these findings, LAB caused higher firmness in bread samples than yeasts. The use of *K. aestuarii* and *K. marxianus* resulted in high cohesiveness in breads. The use of *K. aestuarii* and *S. cerevisiae* caused the breads to reach high elasticity and low chewiness. On the one hand, the application of yeasts isolated from sourdough resulted in improved chewiness and hardness of bread samples. On the other hand, LAB possess proteolytic activity and thus they hydrolyze proline residues of gluten and cause low elastic, high extensive bread production. According to the sensory evaluation, breads produced with Iranian traditional sourdough microflora were more satisfying in terms of taste, texture, and flavor compared to those produced without this fermentation (Fekri et al. 2020).

Einkorn has specific qualities and is rich in nutrients, such as potassium, phosphorus, pyridoxine, beta carotene, lutein, lipid, and protein (Hidalgo and Brandolini 2014). Thirteen LAB species were isolated and identified from sourdough when einkorn sourdough was spontaneously fermented; among them *C. crustorum* was found as the dominant species in sourdough microflora. During einkorn sourdough bread making, using isolated *L. paraplantarum* 7285 and *L. brevis* R-1 as starter culture resulted in good sourdough structural properties and desirable bread quality (Çakır et al. 2020).

Rice bran, a by-product of milled rice, has been used for sourdough fermented bread production, such as Iranian rice bran bread (Farahmand et al. 2015), Iranian taftoon bread (Torkamani, Razavi, and Gharibzahedi 2015), and pan bread (Hanis-Syazwani et al. 2018). When *L. plantarum* was used as a starter culture in sourdough fermentation the rice bran sourdough bread provided a high specific volume, desirable sensorial quality, and longer shelf life when compared to wheat bran sourdough bread (Hanis-Syazwani et al. 2018).

12.6.2 GLUTEN-FREE BREAD

Patients with celiac disease react to the prolamins of barley, rye, and oats, and the gliadin fraction of wheat, with inflammation of the small intestine that causes malabsorption of various nutrients e.g. folate, iron, calcium, and vitamins (A, D, E, and

K). The healing of the intestinal mucosa and a healthy small intestine are achieved by following a strict gluten-free diet. Removing gluten has certain drawbacks for the bakery industry as gluten-free products exhibit poor flavor, texture, and mouthfeel (Arendt, O'brien, and Gormley 2002). Several studies have reported that application of sourdough can overcome some drawbacks as, during fermentation, gluten degradation occurs via the proteolytic activity of LAB and wheat enzymes (Kömen 2010). In one study, sourdough-containing, gluten-free breads were found as preferable compared to the control gluten-free ones (Katina et al. 2005). *L. plantarum* CF1 and *F. sanfranciscensis* (LS40, LS41) were used as sourdough starters for gluten-free bread production with enhanced nutritional, flavor, and textural characteristics (Di Cagno et al. 2008). Similarly, in another study, *C. paralimentarius* improved the crumb firmness in both millet and buckwheat breads, while *L. hammesii* positively affected the crumb firmness of millet breads. In addition, isolated *F. sanfranciscencis* strains enhanced the functional quality of both gluten-free breads (Bender et al. 2018).

Sorghum, a cheap replacement for wheat, has also gained attention for the production of gluten-free sourdough breads (Ogunsakin et al. 2017). The sorghum sourdough fermentation with functional starter cultures enhanced antioxidant properties, protein digestibility, and consumer acceptability of gluten-free breads (Olojede et al. 2020).

12.6.3 Pasta

Various studies have been performed to improve the functional and nutritional features of pasta and other bakery products using sourdough fermentation compounds. The attention of consumers to the nutritional and functional features of sourdough bakery food products increases day by day, therefore, pasta fortification with sourdough has been conducted in different ways, e.g. lowering GI and starch digestibility and the enrichment of B vitamins (Montemurro, Coda, and Rizzello 2019).

Sensorial attributes of pasta prepared with fermented flour differ from their conventional counterparts, mainly in their textural properties. LAB fermentation also provided acidic smell and flavor to the pasta and bakery products, and the solubilization of compounds in water occurs by cooking at high temperatures. Moreover, fermented flour content and fermentation parameters, such as proteolysis degree and fermentation time, led to differentiations between fermented and unfermented food products (Montemurro, Coda, and Rizzello 2019). On the one hand, in pasta produced by fermenting durum wheat semolina with LAB, a reduced firmness and increased cooking loss occurred due to the proteolysis that weakened the gluten network during fermentation (Di Cagno et al. 2005). On the other hand, in another study with pasta produced using fermented semolina by *S. cerevisiae* and *C. alimentarius*, the quality parameters of this pasta (optimal cooking time and cooking loss) decreased when compared to unfermented pasta products (Fois et al. 2018).

Curiel et al. (2014) produced gluten-free pasta using sourdough LAB species (*F. sanfranciscensis, L. brevis, C. alimentarius,* and *L. hilgardii*) and fungal proteases. The findings showed that nutritional index, amino acid profile, chemical scores, and biological value of experimental gluten-free pasta were higher compared to commercial durum wheat pasta. In addition, the sensorial properties of gluten-free pasta were found acceptable following a sensory evaluation.

12.6.4 STEAMED BREAD

Steamed bread, or steamed bun, is a staple food product in China. Sourdough is the main leavening agent used in the traditional steamed bread manufacturing (Li, Li, and Bian 2016). The application of sourdough biotechnology offered different level of aroma compounds compared to instant yeast, mainly due to the production of ethyl lactate, 2-pentylfuran, benzaldehyde, and 1-octen-3-ol (Xi et al. 2020). Potato steamed bread is becoming popular in the Chinese markets and is usually fermented with baker's yeast. However, this fermentation has some disadvantages from both processing and consumer points of view related to complications of working a highly viscous dough and the plain taste of the final products (Zhao, Mu, and Sun 2019a). To enhance the technological quality of potato steamed bread, microorganisms were isolated from five Chinese conventional sourdoughs, in which *Wickerhamomyces*, *Pediococcus*, and *Lactobacillus* were dominantly found. Among them, *Wickerhamomyces* showed great potential to improve resilience, hardness, and lightness of the breads (Zhao, Mu, and Sun 2019b). In one study, potato steamed bread fermentation was carried out by using different sourdoughs. However, sourdough biotechnology resulted in lower mineral content, dietary fiber, and protein content except for the total amount of amino acids in potato steamed bread. On the other hand, using sourdough fermentation caused a high resistant starch level and a low glycemic index (Zhao, Mu, and Sun 2019a).

12.6.5 BURGER BUNS

The increment of diabetes, cardiovascular diseases, and obesity in the population highlights the need to decrease the amount of sugar in sweet baked foods like burger buns. However, a reduction of sugar negatively affects the volume, taste, and texture of these buns (Sahin, Axel, and Arendt 2017). Sugar-alcohols can be a good sugar replacement. In a recent study, mannitol was reported as a promising agent for reducing sugar in burger buns (Sahin et al. 2018). Therefore, *L. citreum*, a mannitol-producer sourdough LAB, was used to lower sugar content in burger buns. The application of sourdough fermentation by *L. citreum* demonstrated high potential to improve volume, crumb structure, crust color, shelf life, and sweetness of these products (Sahin et al. 2019).

12.6.6 PIZZA

The use of sourdough in pizza is an old practice (Chavan and Chavan 2011). Pizza is still produced using sourdough biotechnology, offering higher flavor, sensory quality, and digestibility than that of baker's yeast (Corsetti and Settanni 2007). Obligatory heterofermentative *L. brevis*, *F. sanfranciscensis*, and *F. rossiae* species and facultative heterofermentative *L. plantarum*, *L. curvatus*, and *L. graminis* species were used for pizza production and exerted different effects on pizzas, such as volatile organic compounds, crust color, chewiness, crispness, morphology, weight loss, and presence of bubbles. As evaluated from the sensory panel, pizza samples prepared by using obligate facultative heterofermentative LAB had the highest crust color, taste persistency, crispness, and tearing resistance (Francesca et al. 2019).

12.7 FUTURE APPLICATIONS OF SOURDOUGH IN THE FOOD INDUSTRY

The positive effects of sourdough or its microorganisms in the food industry have been reported by multiple studies. However, alternative methods and new technologies are needed to preserve the viability and quality of sourdough microorganisms when used as industrial starter cultures. Cereal flour type, water ratio, fermentation temperature, and time are necessary to produce an effective sourdough, but the most important parameter is the appropriate selection of starter cultures (Gobbetti et al. 2019). The use of starter culture plays an important role in shortening the fermentation time, preventing fermentation failure, and standardizing the final product (Kavitake et al. 2018). A commercially viable sourdough starter culture should comprise of at least one homofermentative and one heterofermentative LAB to provide good aromatization and acidification (Chavan and Chavan 2011). The use of optimized starter cultures isolated from sourdough in other bakery products is also of great importance in terms of the quality of the foods including bakery products, such as whole wheat bread, gluten-free bread, pasta, steamed bread, hamburger buns, and pizza. The most important reason underlying this is sourdough microorganisms and their technological, nutritional, sensory, and even health effects. While internal and external factors differentiate the sourdough microbiota, they also increase the variety of beneficial effects and functionalities of the isolated microorganisms. From traditional to the future, sourdough microorganisms have a great potential to be explored in the food industry, especially in the bakery field.

REFERENCES

Abedfar, Abbas, Sepideh Abbaszadeh, Marzieh Hosseininezhad, and Maryam Taghdir. 2020. "Physicochemical and biological characterization of the EPS produced by L. acidophilus isolated from rice bran sourdough." *LWT* 127:109373.

Arendt, EK, CM O'brien, and Thomas Ronan Gormley. 2002. Development of Guten-Free Cereal Products. *Farm and Food* 12:21–27.

Axel, Claudia, Brid Brosnan, Emanuele Zannini, Ambrose Furey, Aidan Coffey, and Elke K Arendt. 2016a. "Antifungal sourdough lactic acid bacteria as biopreservation tool in quinoa and rice bread." *International Journal of Food Microbiology* 239:86–94.

Axel, Claudia, Brid Brosnan, Emanuele Zannini, Lorenzo C Peyer, Ambrose Furey, Aidan Coffey, and Elke K Arendt. 2016b. "Antifungal activities of three different Lactobacillus species and their production of antifungal carboxylic acids in wheat sourdough." *Applied Microbiology and Biotechnology* 100 (4):1701–1711.

Banu, Iuliana, Ina Vasilean, and Iuliana Aprodu. 2010. "Effect of lactic fermentation on antioxidant capacity of rye sourdough and bread." *Food Science and Technology Research* 16 (6):571–576.

Barclay, Alan W, Peter Petocz, Joanna McMillan-Price, Victoria M Flood, Tania Prvan, Paul Mitchell, and Jennie C Brand-Miller. 2008. "Glycemic index, glycemic load, and chronic disease risk: A meta-analysis of observational studies." *The American Journal of Clinical Nutrition* 87 (3):627–637.

Bartkiene, Elena, Vadims Bartkevics, Vita Krungleviciute, Iveta Pugajeva, Daiva Zadeike, and Grazina Juodeikiene. 2017. "Lactic acid bacteria combinations for wheat sourdough preparation and their influence on wheat bread quality and acrylamide formation." *Journal of Food Science* 82 (10):2371–2378.

Bartkiene, Elena, Vita Lele, Modestas Ruzauskas, Konrad J Domig, Vytaute Starkute, Paulina Zavistanaviciute, Vadims Bartkevics, Iveta Pugajeva, Dovile Klupsaite, and Grazina Juodeikiene. 2020. "Lactic acid bacteria isolation from spontaneous sourdough and their characterization including antimicrobial and antifungal properties evaluation." *Microorganisms* 8 (1):64.

Belz, Markus CE, Regina Mairinger, Emanuele Zannini, Liam AM Ryan, Kevin D Cashman, and Elke K Arendt. 2012. "The effect of sourdough and calcium propionate on the microbial shelf-life of salt reduced bread." *Applied Microbiology and Biotechnology* 96 (2):493–501.

Bender, Denisse, Vera Fraberger, Palma Szepasvári, Stefano D'Amico, S Tömösközi, G Cavazzi, H Jäger, Konrad J Domig, and Regine Schoenlechner. 2018. "Effects of selected lactobacilli on the functional properties and stability of gluten-free sourdough bread." *European Food Research and Technology* 244 (6):1037–1046.

Bounaix, Marie-Sophie, Valerie Gabriel, Sandrine Morel, Herve Robert, Philippe Rabier, Magali Remaud-Simeon, Bruno Gabriel, and Catherine Fontagne-Faucher. 2009. "Biodiversity of exopolysaccharides produced from sucrose by sourdough lactic acid bacteria." *Journal of Agricultural and Food Chemistry* 57 (22):10889–10897.

Bustos, Ana Yanina, Carla Luciana Gerez, Laura Beatriz Iturriaga, and María Pía Taranto. 2017. "Soybean Sourdough as Bio-ingredients to Enhances Physical and Functional Properties of Wheat Bakery Products." *Advance Journal of Food Science and Technology* 13 (4):161–169.

Çakır, Elif, Muhammet Arıcı, Muhammed Zeki Durak, and Salih Karasu. 2020. "The molecular and technological characterization of lactic acid bacteria in einkorn sourdough: Effect on bread quality." *Journal of Food Measurement and Characterization* 14:1646–1655.

Capozzi, Vittorio, Valeria Menga, Anna Maria Digesu, Pasquale De Vita, Douwe van Sinderen, Luigi Cattivelli, Clara Fares, and Giuseppe Spano. 2011. "Biotechnological production of vitamin B2-enriched bread and pasta." *Journal of Agricultural and Food Chemistry* 59 (14):8013–8020.

Carrizo, Silvana L, Alejandra de Moreno de LeBlanc, Jean Guy LeBlanc, and Graciela C Rollán. 2020. "Quinoa pasta fermented with lactic acid bacteria prevents nutritional deficiencies in mice." *Food Research International* 127:108735.

Casella, Giovanni, Renata D'Incà, Lydia Oliva, Marco Daperno, Valeria Saladino, Giorgio Zoli, Vito Annese, Walter Fries, and Claudio Cortellezzi. 2010. "Prevalence of celiac disease in inflammatory bowel diseases: An IG-IBD multicentre study." *Digestive and Liver Disease* 42 (3):175–178.

Cauvain, S. 2012. "Breadmaking: An overview." In *Breadmaking*, 9–31. Elsevier.

Chandrasekar Rajendran, SC, Bhawani Chamlagain, Susanna Kariluoto, Vieno Piironen, and Per EJ Saris. 2017. "Biofortification of riboflavin and folate in idli batter, based on fermented cereal and pulse, by Lactococcus lactis N8 and Saccharomyces boulardii SAA 655." *Journal of Applied Microbiology* 122 (6):1663–1671.

Chavan, Rupesh S, and Shraddha R Chavan. 2011. "Sourdough technology: A traditional way for wholesome foods: A review." *Comprehensive Reviews in Food Science and Food Safety* 10 (3):169–182.

Choi, Hyejung, Yeo-Won Kim, Inyoung Hwang, Jeongho Kim, and Sun Yoon. 2012. "Evaluation of Leuconostoc citreum HO12 and Weissella koreensis HO20 isolated from kimchi as a starter culture for whole wheat sourdough." *Food Chemistry* 134 (4):2208–2216.

Cizeikiene, Dalia, Grazina Juodeikiene, Algimantas Paskevicius, and Elena Bartkiene. 2013. "Antimicrobial activity of lactic acid bacteria against pathogenic and spoilage microorganism isolated from food and their control in wheat bread." *Food Control* 31 (2):539–545.

Clarke, CI, TJ Schober, and EK Arendt. 2002. "Effect of single strain and traditional mixed strain starter cultures on rheological properties of wheat dough and on bread quality." *Cereal Chemistry* 79 (5):640–647.

Coda, Rossana, Raffaella Di Cagno, Carlo G Rizzello, Luana Nionelli, Mojisola O Edema, and Marco Gobbetti. 2011. "Utilization of African grains for sourdough bread making." *Journal of Food Science* 76 (6):M329–M335.

Coda, Rossana, Luana Nionelli, Carlo G Rizzello, Maria De Angeles, Pierre Tossut, and Marco Gobbetti. 2010. "Spelt and emmer flours: Characterization of the lactic acid bacteria microbiota and selection of mixed starters for bread making." *Journal of Applied Microbiology* 108 (3):925–935.

Coda, Rossana, Carlo G Rizzello, Raffaella Di Cagno, Antonio Trani, Gianluigi Cardinali, and Marco Gobbetti. 2013. "Antifungal activity of Meyerozyma guilliermondii: Identification of active compounds synthesized during dough fermentation and their effect on long-term storage of wheat bread." *Food Microbiology* 33 (2):243–251.

Coda, Rossana, Carlo Giuseppe Rizzello, Daniela Pinto, and Marco Gobbetti. 2012. "Selected lactic acid bacteria synthesize antioxidant peptides during sourdough fermentation of cereal flours." *Applied and Environmental Microbiology* 78 (4):1087–1096.

Comasio, Andrea, Simon Van Kerrebroeck, Henning Harth, Fabienne Verté, and Luc De Vuyst. 2020. "Potential of Bacteria from Alternative Fermented Foods as Starter Cultures for the Production of Wheat Sourdoughs." *Microorganisms* 8 (10):1534.

Corsetti, Aldo, and Luca Settanni. 2007. "Lactobacilli in sourdough fermentation." *Food Research International* 40 (5):539–558.

Corsetti, Aldo, Marco Gobbetti, F Balestrieri, F Paoletti, L Russi, and J Rossi. 1998. "Sourdough lactic acid bacteria effects on bread firmness and stalin." *Journal of Food Science* 63 (2):347–351.

Corsetti, Aldo, Marco Gobbetti, B De Marco, F Balestrieri, F Paoletti, L Russi, and J Rossi. 2000. "Combined effect of sourdough lactic acid bacteria and additives on bread firmness and staling." *Journal of Agricultural and Food Chemistry* 48 (7):3044–3051.

Curiel, José Antonio, Rossana Coda, Isabella Centomani, Carmine Summo, Marco Gobbetti, and Carlo Giuseppe Rizzello. 2015. "Exploitation of the nutritional and functional characteristics of traditional Italian legumes: The potential of sourdough fermentation." *International Journal of Food Microbiology* 196:51–61.

Curiel, José Antonio, Rossana Coda, Antonio Limitone, Kati Katina, Mari Raulio, Giammaria Giuliani, Carlo Giuseppe Rizzello, and Marco Gobbetti. 2014. "Manufacture and characterization of pasta made with wheat flour rendered gluten-free using fungal proteases and selected sourdough lactic acid bacteria." *Journal of Cereal Science* 59 (1):79–87.

Czerny, Michael, and Peter Schieberle. 2002. "Important aroma compounds in freshly ground wholemeal and white wheat flour identification and quantitative changes during sourdough fermentation." *Journal of Agricultural and Food Chemistry* 50 (23):6835–6840.

De Angelis, Maria, Nicola Damiano, Carlo G Rizzello, Angela Cassone, Raffaella Di Cagno, and Marco Gobbetti. 2009. "Sourdough fermentation as a tool for the manufacture of low-glycemic index white wheat bread enriched in dietary fibre." *European Food Research and Technology* 229 (4):593–601.

De Vuyst, Luc, and Patricia Neysens. 2005. "The sourdough microflora: Biodiversity and metabolic interactions." *Trends in Food Science & Technology* 16 (1–3):43–56.

De Vuyst, Luc, and Marc Vancanneyt. 2007. "Biodiversity and identification of sourdough lactic acid bacteria." *Food Microbiology* 24 (2):120–127.

De Vuyst, Luc, Simon Van Kerrebroeck, Henning Harth, Geert Huys, H-M Daniel, and Stefan Weckx. 2014. "Microbial ecology of sourdough fermentations: Diverse or uniform?" *Food Microbiology* 37:11–29.

Demirbaş, Fatmanur, Hümeyra İspirli, Asena Ayşe Kurnaz, Mustafa Tahsin Yilmaz, and Enes Dertli. 2017. "Antimicrobial and functional properties of lactic acid bacteria isolated from sourdoughs." *LWT-Food Science and Technology* 79:361–366.

Dertli, Enes, Emin Mercan, Muhammet Arıcı, Mustafa Tahsin Yılmaz, and Osman Sağdıç. 2016. "Characterisation of lactic acid bacteria from Turkish sourdough and determination of their exopolysaccharide (EPS) production characteristics." *LWT: Food Science and Technology* 71:116–124.

Di Cagno, Raffaella, Maria De Angelis, Giuditta Alfonsi, Massimo de Vincenzi, Marco Silano, Olimpia Vincentini, and Marco Gobbetti. 2005. "Pasta made from durum wheat semolina fermented with selected lactobacilli as a tool for a potential decrease of the gluten intolerance." *Journal of Agricultural and Food Chemistry* 53 (11):4393–4402.

Di Cagno, Raffaella, Maria De Angelis, Antonio Limitone, Fabio Minervini, Paola Carnevali, Aldo Corsetti, Michael Gaenzle, Roberto Ciati, and Marco Gobbetti. 2006. "Glucan and fructan production by sourdough Weissella cibaria and Lactobacillus plantarum." *Journal of Agricultural and Food Chemistry* 54 (26):9873–9881.

Di Cagno, Raffaella, Carlo G Rizzello, Maria De Angelis, Angela Cassone, Giammaria Giuliani, Anna Benedusi, Antonio Limitone, Rosalinda F Surico, and Marco Gobbetti. 2008. "Use of selected sourdough strains of Lactobacillus for removing gluten and enhancing the nutritional properties of gluten-free bread." *Journal of Food Protection* 71 (7):1491–1495.

Diana, Marina, Joan Quílez, and Magdalena Rafecas. 2014. "Gamma-aminobutyric acid as a bioactive compound in foods: A review." *Journal of Functional Foods* 10:407–420.

Digaitiene, A, ÅS Hansen, G Juodeikiene, D Eidukonyte, and J Josephsen. 2012. "Lactic acid bacteria isolated from rye sourdoughs produce bacteriocin-like inhibitory substances active against Bacillus subtilis and fungi." *Journal of Applied Microbiology* 112 (4):732–742.

Farahmand, E, SH Razavi, MS Yarmand, and M Morovatpour. 2015. "Development of Iranian rice-bran sourdough breads: Physicochemical, microbiological and sensorial characterisation during the storage period." *Quality Assurance and Safety of Crops & Foods* 7 (3):295–303.

Fekri, Arezoo, Mohammadali Torbati, Ahmad Yari Khosrowshahi, Hasan Bagherpour Shamloo, and Sodeif Azadmard-Damirchi. 2020. "Functional effects of phytate-degrading, probiotic lactic acid bacteria and yeast strains isolated from Iranian traditional sourdough on the technological and nutritional properties of whole wheat bread." *Food Chemistry* 306:125620.

Fois, Simonetta, Piero Pasqualino Piu, Manuela Sanna, Tonina Roggio, and Pasquale Catzeddu. 2018. "Starch digestibility and properties of fresh pasta made with semolina-based liquid sourdough." *LWT* 89:496–502.

Foster-Powell, Kaye, Susanna HA Holt, and Janette C Brand-Miller. 2002. "International table of glycemic index and glycemic load values: 2002." *The American Journal of Clinical Nutrition* 76 (1):5–56.

Fraberger, Vera, Claudia Ammer, and Konrad J Domig. 2020. "Functional properties and sustainability improvement of sourdough bread by lactic acid bacteria." *Microorganisms* 8 (12):1895.

Fraberger, Vera, Lisa-Maria Call, Konrad J Domig, and Stefano D'Amico. 2018. "Applicability of yeast fermentation to reduce fructans and other FODMAPs." *Nutrients* 10 (9):1247.

Francesca, Nicola, Raimondo Gaglio, Antonio Alfonzo, Onofrio Corona, Giancarlo Moschetti, and Luca Settanni. 2019. "Characteristics of sourdoughs and baked pizzas as affected by starter culture inoculums." *International Journal of Food Microbiology* 293:114–123.

Fu, Wenhui, Huan Rao, Yang Tian, and Wentong Xue. 2020. "Bacterial composition in sourdoughs from different regions in China and the microbial potential to reduce wheat allergens." *LWT* 117:108669.

Fujimoto, Akihito, Keisuke Ito, Noriko Narushima, and Takahisa Miyamoto. 2019. "Identification of lactic acid bacteria and yeasts, and characterization of food components of sourdoughs used in Japanese bakeries." *Journal of Bioscience and Bioengineering* 127 (5):575–581.

Galle, Sandra, and Elke K Arendt. 2014. "Exopolysaccharides from sourdough lactic acid bacteria." *Critical Reviews in Food Science and Nutrition* 54 (7):891–901.

Gamel, Tamer H, El-Sayed M Abdel-Aal, and Susan M Tosh. 2015. "Effect of yeast-fermented and sour-dough making processes on physicochemical characteristics of β-glucan in whole wheat/oat bread." *LWT: Food Science and Technology* 60 (1):78–85.

Garcia, Marcelo Valle, Angélica Olivier Bernardi, and Marina Venturini Copetti. 2019. "The fungal problem in bread production: Insights of causes, consequences, and control methods." *Current Opinion in Food Science* 29:1–6.

Gerez, Carla Luciana, Maria Ines Torino, Graciela Rollán, and Graciela Font de Valdez. 2009. "Prevention of bread mould spoilage by using lactic acid bacteria with antifungal properties." *Food Control* 20 (2):144–148.

Ghosh, Debasree, and Parimal Chattopadhyay. 2011. "Preparation of idli batter, its properties and nutritional improvement during fermentation." *Journal of Food Science and Technology* 48 (5):610–615.

Gobbetti, M. 1998. "The sourdough microflora: Interactions of lactic acid bacteria and yeasts." *Trends in Food Science & Technology* 9 (7):267–274.

Gobbetti, Marco, Maria De Angelis, Raffaella Di Cagno, Maria Calasso, Gabriele Archetti, and Carlo Giuseppe Rizzello. 2019. "Novel insights on the functional/nutritional features of the sourdough fermentation." *International Journal of Food Microbiology* 302:103–113.

Gobbetti, Marco, Carlo G Rizzello, Raffaella Di Cagno, and Maria De Angelis. 2014. "How the sourdough may affect the functional features of leavened baked goods." *Food Microbiology* 37:30–40.

Gobbetti, Marco, MS Simonetti, Aldo Corsetti, F Santinelli, J Rossi, and P Damiani. 1995. "Volatile compound and organic acid productions by mixed wheat sour dough starters: Influence of fermentation parameters and dynamics during baking." *Food Microbiology* 12:497–507.

Gobbetti, Marco, Emanuele Smacchi, Patrick Fox, Leszek Stepaniak, and Aldo Corsetti. 1996. "The sourdough microflora. Cellular localization and characterization of proteolytic enzymes in lactic acid bacteria." *LWT-Food Science and Technology* 29 (5–6):561–569.

Gray, JA, and JN Bemiller. 2003. "Bread staling: Molecular basis and control." *Comprehensive Reviews in Food Science and Food Safety* 2 (1):1–21.

Gupta, Ruchi, and Sheela Srivastava. 2014. "Antifungal effect of antimicrobial peptides (AMPs LR14) derived from Lactobacillus plantarum strain LR/14 and their applications in prevention of grain spoilage." *Food Microbiology* 42:1–7.

Hammes, Walter Peter, and MG Gänzle. 1998. "Sourdough breads and related products." In *Microbiology of Fermented Foods*, 199–216. Springer.

Hanis-Syazwani, M, IF Bolarinwa, O Lasekan, and K Muhammad. 2018. "Influence of starter culture on the physicochemical properties of rice bran sourdough and physical quality of sourdough bread." *Food Research* 2 (4):340–349.

Hidalgo, Alyssa, and Andrea Brandolini. 2014. "Nutritional properties of einkorn wheat (Triticum monococcum L.)." *Journal of the Science of Food and Agriculture* 94 (4):601–612.

Huys, Geert, Heide-Marie Daniel, and Luc De Vuyst. 2013. "Taxonomy and biodiversity of sourdough yeasts and lactic acid bacteria." In *Handbook on Sourdough Biotechnology*, 105–154. Springer.

Juntunen, Katri S, David E Laaksonen, Karin Autio, Leo K Niskanen, Jens J Holst, Kari E Savolainen, Kirsi-Helena Liukkonen, Kaisa S Poutanen, and Hannu M Mykkänen. 2003. "Structural differences between rye and wheat breads but not total fiber content may explain the lower postprandial insulin response to rye bread." *The American Journal of Clinical Nutrition* 78 (5):957–964.

Kariluoto, Susanna, Marja Aittamaa, Matti Korhola, Hannu Salovaara, Liisa Vahteristo, and Vieno Piironen. 2006. "Effects of yeasts and bacteria on the levels of folates in rye sourdoughs." *International Journal of Food Microbiology* 106 (2):137–143.

Katina, Kati. 2005. "Sourdough: A tool for the improved flavour, texture and shelf-life of wheat bread." Doctoral dissertation, University of Helsinki, Helsinki, Finland

Katina, Kati, and Kaisa Poutanen. 2013. "Nutritional aspects of cereal fermentation with lactic acid bacteria and yeast." In *Handbook on Sourdough Biotechnology*, 229–244. Springer.

Katina, Kati, E Arendt, K-H Liukkonen, Karin Autio, Laura Flander, and Kaisa Poutanen. 2005. "Potential of sourdough for healthier cereal products." *Trends in Food Science & Technology* 16 (1–3):104–112.

Katina, Kati, Arja Laitila, Riikka Juvonen, K-H Liukkonen, S Kariluoto, V Piironen, R Landberg, P Åman, and Kaisa Poutanen. 2007. "Bran fermentation as a means to enhance technological properties and bioactivity of rye." *Food Microbiology* 24 (2):175–186.

Katina, Kati, Ndegwa Henry Maina, Riikka Juvonen, Laura Flander, Liisa Johansson, Liisa Virkki, Maija Tenkanen, and Arja Laitila. 2009. "In situ production and analysis of Weissella confusa dextran in wheat sourdough." *Food Microbiology* 26 (7):734–743.

Katina, Kati, Marjatta Salmenkallio-Marttila, Riitta Partanen, Pirkko Forssell, and Karin Autio. 2006. "Effects of sourdough and enzymes on staling of high-fibre wheat bread." *LWT: Food Science and Technology* 39 (5):479–491.

Kavitake, Digambar, Sujatha Kandasamy, Palanisamy Bruntha Devi, and Prathapkumar Halady Shetty. 2018. "Recent developments on encapsulation of lactic acid bacteria as potential starter culture in fermented foods: A review." *Food Bioscience* 21:34–44.

Kopeć, A, M Pysz, B Borczak, E Sikora, CM Rosell, C Collar, and M Sikora. 2011. "Effects of sourdough and dietary fibers on the nutritional quality of breads produced by bake-off technology." *Journal of Cereal Science* 54 (3):499–505.

Korakli, Maher, Melanie Pavlovic, Michael G Gänzle, and Rudi F Vogel. 2003. "Exopolysaccharide and kestose production by Lactobacillus sanfranciscensis LTH2590." *Applied and Environmental Microbiology* 69 (4):2073–2079.

Korakli, Maher, Andreas Rossmann, Michael G Gänzle, and Rudi F Vogel. 2001. "Sucrose metabolism and exopolysaccharide production in wheat and rye sourdoughs by Lactobacillus sanfranciscensis." *Journal of Agricultural and Food Chemistry* 49 (11):5194–5200.

Kömen, Gökçen. 2010. *Structural Changes of Gliadins During Sourdough Fermentation as a Promissing Approach to Gluten-free Diet.* İzmir Institute of Technology.

Krnjević, K. 1974. "Chemical nature of synaptic transmission in vertebrates." *Physiological Reviews* 54 (2):418–540.

Kumar, Vikas, Amit K Sinha, Harinder PS Makkar, and Klaus Becker. 2010. "Dietary roles of phytate and phytase in human nutrition: A review." *Food Chemistry* 120 (4):945–959.

Labat, E, Marie Helene Morel, and Xavier Rouau. 2000. "Effects of laccase and ferulic acid on wheat flour doughs." *Cereal Chemistry* 77 (6):823–828.

Lacaze, G, M Wick, and S Cappelle. 2007. "Emerging fermentation technologies: Development of novel sourdoughs." *Food Microbiology* 24 (2):155–160.

Lappi, Jenni, Emilia Selinheimo, Ursula Schwab, Kati Katina, Pekka Lehtinen, Hannu Mykkänen, Marjukka Kolehmainen, and Kaisa Poutanen. 2010. "Sourdough fermentation of wholemeal wheat bread increases solubility of arabinoxylan and protein and decreases postprandial glucose and insulin responses." *Journal of Cereal Science* 51 (1):152–158.

Legan, JD. 1993. "Mould spoilage of bread: The problem and some solutions." *International Biodeterioration & Biodegradation* 32 (1–3):33–53.

Li, Zhijian, Haifeng Li, and Ke Bian. 2016. "Microbiological characterization of traditional dough fermentation starter (Jiaozi) for steamed bread making by culture-dependent and culture-independent methods." *International Journal of Food Microbiology* 234:9–14.

Limbad, Mansi, Noemi Gutierrez Maddox, Nazimah Hamid, and Kevin Kantono. 2020. "Sensory and physicochemical characterization of sourdough bread prepared with a coconut water Kefir Starter." *Foods* 9 (9):1165.

Liu, Aiping, Yuhan Jia, Linzhi Zhao, Ya Gao, Guirong Liu, Yuran Chen, Guilin Zhao, Lizemin Xu, Li Shen, and Yuntao Liu. 2018. "Diversity of isolated lactic acid bacteria in Ya'an sourdoughs and evaluation of their exopolysaccharide production characteristics." *LWT* 95:17–22.

Liukkonen, Kirsi-Helena, Kati Katina, Annika Wilhelmsson, Olavi Myllymaki, Anna-Maija Lampi, Susanna Kariluoto, Vieno Piironen, Satu-Maarit Heinonen, Tarja Nurmi, and Herman Adlercreutz. 2003. "Process-induced changes on bioactive compounds in whole grain rye." *Proceedings of the Nutrition Society* 62 (1):117–122.

Lopez, Hubert W, Ariane Ouvry, Elisabeth Bervas, Christine Guy, Arnaud Messager, Christian Demigne, and Christian Remesy. 2000. "Strains of lactic acid bacteria isolated from sour doughs degrade phytic acid and improve calcium and magnesium solubility from whole wheat flour." *Journal of Agricultural and Food Chemistry* 48 (6):2281–2285.

Loponen, Jussi, and Michael G Gänzle. 2018. "Use of sourdough in low FODMAP baking." *Foods* 7 (7):96.

Lorenz, Klaus, and JM Brümmer. 2003. "Preferments and sourdoughs for German breads." In *Handbook of dough fermentations*, 275–298. CRC Press.

Maidana, Stefania Dentice, Cecilia Aristimuño Ficoseco, Daniela Bassi, Pier Sandro Cocconcelli, Edoardo Puglisi, Graciela Savoy, Graciela Vignolo, and Cecilia Fontana. 2020. "Biodiversity and technological-functional potential of lactic acid bacteria isolated from spontaneously fermented chia sourdough." *International Journal of Food Microbiology* 316:108425.

Manini, F, MC Casiraghi, K Poutanen, M Brasca, D Erba, and C Plumed-Ferrer. 2016. "Characterization of lactic acid bacteria isolated from wheat bran sourdough." *LWT: Food Science and Technology* 66:275–283.

Martinez-Villaluenga, Cristina, A Michalska, Juana Frías, Mariusz K Piskula, Concepción Vidal-Valverde, and Halina Zieliński. 2009. "Effect of flour extraction rate and baking on thiamine and riboflavin content and antioxidant capacity of traditional rye bread." *Journal of Food Science* 74 (1):C49–C55.

Martins, Sara IFS, Wim MF Jongen, and Martinus AJS Van Boekel. 2000. "A review of Maillard reaction in food and implications to kinetic modelling." *Trends in Food Science & Technology* 11 (9–10):364–373.

Martorana, Alessandra, Angelo Maria Giuffrè, Marco Capocasale, Clotilde Zappia, and Rossana Sidari. 2018. "Sourdoughs as a source of lactic acid bacteria and yeasts with technological characteristics useful for improved bakery products." *European Food Research and Technology* 244 (10):1873–1885.

Menteş, Özay, Recai Ercan, and Mustafa Akçelik. 2007. "Inhibitor activities of two Lactobacillus strains, isolated from sourdough, against rope-forming Bacillus strains." *Food Control* 18 (4):359–363.

Meroth, Christiane B, Walter P Hammes, and Christian Hertel. 2004. "Characterisation of the microbiota of rice sourdoughs and description of Lactobacillus spicheri sp. nov." *Systematic and Applied Microbiology* 27 (2):151–159.

Mohd Roby, Bizura Hasida, Belal J Muhialdin, Muna Mahmood Taleb Abadl, Nor Arifah Mat Nor, Anis Asyila Marzlan, Sarina Abdul Halim Lim, Nor Afizah Mustapha, and Anis Shobirin Meor Hussin. 2020. "Physical properties, storage stability, and consumer acceptability for sourdough bread produced using encapsulated kombucha sourdough starter culture." *Journal of Food Science* 85 (8):2286–2295.

Montemurro, Marco, Rossana Coda, and Carlo Giuseppe Rizzello. 2019. "Recent advances in the use of sourdough biotechnology in pasta making." *Foods* 8 (4):129.

Moore, MM, B Juga, TJ Schober, and EK Arendt. 2007. "Effect of lactic acid bacteria on properties of gluten-free sourdoughs, batters, and quality and ultrastructure of gluten-free bread." *Cereal Chemistry* 84 (4):357–364.

Nielsen, Merete Møller, Marianne Linde Damstrup, Agnete Dal Thomsen, Søren Kjærsgård Rasmussen, and Åse Hansen. 2007. "Phytase activity and degradation of phytic acid during rye bread making." *European Food Research and Technology* 225 (2):173–181.

Nilsson, Mikael, Jens J Holst, and Inger ME Björck. 2007. "Metabolic effects of amino acid mixtures and whey protein in healthy subjects: Studies using glucose-equivalent drinks." *The American Journal of Clinical Nutrition* 85 (4):996–1004.

Novotni, Dubravka, Duška Ćurić, Martina Bituh, Irena Colić Barić, Dubravka Škevin, and Nikolina Čukelj. 2011. "Glycemic index and phenolics of partially-baked frozen bread with sourdough." *International Journal of Food Sciences and Nutrition* 62 (1):26–33.

Nuobariene, Lina, Dalia Cizeikiene, Egle Gradzeviciute, Åse S Hansen, Søren K Rasmussen, Grazina Juodeikiene, and Finn K Vogensen. 2015. "Phytase-active lactic acid bacteria from sourdoughs: Isolation and identification." *LWT: Food Science and Technology* 63 (1):766–772.

Ogunsakin, AO, V Vanajakshi, KA Anu-Appaiah, SVN Vijayendra, SG Walde, K Banwo, AI Sanni, and P Prabhasankar. 2017. "Evaluation of functionally important lactic acid bacteria and yeasts from Nigerian sorghum as starter cultures for gluten-free sourdough preparation." *LWT: Food Science and Technology* 82:326–334.

Olojede, AO, AI Sanni, K Banwo, and AT Adesulu-Dahunsi. 2020. "Sensory and antioxidant properties and in-vitro digestibility of gluten-free sourdough made with selected starter cultures." *LWT* 129:109576.

Östman, Elin. 2003. *Fermentation as a Means of Optimizing the Glycaemic Index-food Mechanisms and Metabolic Merits with Emphasis on Lactic Acid in Cereal Products.* Lund University.

Perricone, Marianne, Antonio Bevilacqua, Maria Rosaria Corbo, and Milena Sinigaglia. 2014. "Technological characterization and probiotic traits of yeasts isolated from Altamura sourdough to select promising microorganisms as functional starter cultures for cereal-based products." *Food Microbiology* 38:26–35.

Plessas, Stavros. 2021. *Innovations in Sourdough Bread Making.* Multidisciplinary Digital Publishing Institute.

Plessas, Stavros, Ioanna Mantzourani, and Argyro Bekatorou. 2020. "Evaluation of Pediococcus pentosaceus SP2 as Starter Culture on Sourdough Bread Making." *Foods* 9 (1):77.

Poutanen, Kaisa, Laura Flander, and Kati Katina. 2009. "Sourdough and cereal fermentation in a nutritional perspective." *Food microbiology* 26 (7):693–699.

Ricci, A.; Bernini, V.; Maoloni, A.; Cirlini, M.; Galaverna, G.; Neviani, E.; Lazzi, C. . 2019. "Vegetable by-product lacto-fermentation as a new source of antimicrobial compounds." *Microorganisms* 7:607.

Rinaldi, Massimiliano, Maria Paciulli, Augusta Caligiani, Elisa Sgarbi, Martina Cirlini, Chiara Dall'Asta, and Emma Chiavaro. 2015. "Durum and soft wheat flours in sourdough and straight-dough bread-making." *Journal of Food Science and Technology* 52 (10):6254–6265.

Rizzello, CG, A Cassone, R Di Cagno, and M Gobbetti. 2008. "Synthesis of angiotensin I-converting enzyme (ACE)-inhibitory peptides and γ-aminobutyric acid (GABA) during sourdough fermentation by selected lactic acid bacteria." *Journal of Agricultural and Food Chemistry* 56 (16):6936–6943.

Rizzello, Carlo Giuseppe, Rossana Coda, Francesco Mazzacane, Davide Minervini, and Marco Gobbetti. 2012. "Micronized by-products from debranned durum wheat and sourdough fermentation enhanced the nutritional, textural and sensory features of bread." *Food Research International* 46 (1):304–313.

Rizzello, Carlo Giuseppe, Luana Nionelli, Rossana Coda, Maria De Angelis, and Marco Gobbetti. 2010. "Effect of sourdough fermentation on stabilisation, and chemical and nutritional characteristics of wheat germ." *Food Chemistry* 119 (3):1079–1089.

Rizzello, Carlo Giuseppe, Luana Nionelli, Rossana Coda, and Marco Gobbetti. 2012. "Synthesis of the cancer preventive peptide lunasin by lactic acid bacteria during sourdough fermentation." *Nutrition and cancer* 64 (1):111–120.

Rizzello, Carlo Giuseppe, Davide Tagliazucchi, Elena Babini, Giuseppina Sefora Rutella, Danielle L Taneyo Saa, and Andrea Gianotti. 2016. "Bioactive peptides from vegetable food matrices: Research trends and novel biotechnologies for synthesis and recovery." *Journal of Functional Foods* 27:549–569.

Rollán, Graciela C, Carla L Gerez, and Jean G LeBlanc. 2019. "Lactic fermentation as a strategy to improve the nutritional and functional values of pseudocereals." *Frontiers in Nutrition* 6:98.

Rosell, Cristima M, Jose A Rojas, and C Benedito De Barber. 2001. "Influence of hydrocolloids on dough rheology and bread quality." *Food Hydrocolloids* 15 (1):75–81.

Rui, Xin, Jin Huang, Guangliang Xing, Qiuqin Zhang, Wei Li, and Mingsheng Dong. 2019. "Changes in soy protein immunoglobulin E reactivity, protein degradation, and conformation through fermentation with Lactobacillus plantarum strains." *LWT* 99:156–165.

Ruiz Rodríguez, L, Esteban Vera Pingitore, G Rollan, Pier Sandro Cocconcelli, C Fontana, L Saavedra, G Vignolo, Elvira Maria Hebert. 2016. "Biodiversity and technological-functional potential of lactic acid bacteria isolated from spontaneously fermented quinoa sourdoughs." *Journal of Applied Microbiology* 120 (5):1289–1301.

Ryan, PM, RP Ross, GF Fitzgerald, NM Caplice, and C Stanton. 2015. "Sugar-coated: Exopolysaccharide producing lactic acid bacteria for food and human health applications." *Food & Function* 6 (3):679–693.
Sadeghi, Alireza, Maryam Ebrahimi, Mojtaba Raeisi, and Zohreh Nematollahi. 2019. "Biological control of foodborne pathogens and aflatoxins by selected probiotic LAB isolated from rice bran sourdough." *Biological Control* 130:70–79.
Sahin, Aylin W, Claudia Axel, and Elke K Arendt. 2017. "Understanding the function of sugar in burger buns: A fundamental study." *European Food Research and Technology* 243 (11):1905–1915.
Sahin, Aylin W, Claudia Axel, Emanuele Zannini, and Elke K Arendt. 2018. "Xylitol, mannitol and maltitol as potential sucrose replacers in burger buns." *Food & Function* 9 (4):2201–2212.
Sahin, Aylin W, Tom Rice, Emanuele Zannini, Claudia Axel, Aidan Coffey, Kieran M Lynch, and Elke K Arendt. 2019. "Leuconostoc citreum TR116: In-situ production of mannitol in sourdough and its application to reduce sugar in burger buns." *International Journal of Food Microbiology* 302:80–89.
Sakandar, Hafiz Arbab, Khadija Usman, and Muhammad Imran. 2018. "Isolation and characterization of gluten-degrading Enterococcus mundtii and Wickerhamomyces anomalus, potential probiotic strains from indigenously fermented sourdough (Khamir)." *LWT* 91:271–277.
Salmenkallio-Marttila, Marjatta, Kati Katina, and Karin Autio. 2001. "Effects of bran fermentation on quality and microstructure of high-fiber wheat bread." *Cereal Chemistry* 78 (4):429–435.
Savkina, O, L Kuznetsova, O Parakhina, M Lokachuk, and E Pavlovskaya. 2019. "Impact of using the developed starter culture on the quality of sourdough, dough and wheat bread." *Agronomy Research* 17 (S2):1435–1451.
Scazzina, Francesca, Daniele Del Rio, Nicoletta Pellegrini, and Furio Brighenti. 2009. "Sourdough bread: Starch digestibility and postprandial glycemic response." *Journal of Cereal Science* 49 (3):419–421.
Sidari, Rossana, Alessandra Martorana, Clotilde Zappia, Antonio Mincione, and Angelo Maria Giuffrè. 2020. "Persistence and effect of a multistrain starter culture on antioxidant and rheological properties of novel wheat sourdoughs and bread." *Foods* 9 (9):1258.
Singh, Jaspreet, Anne Dartois, and Lovedeep Kaur. 2010. "Starch digestibility in food matrix: A review." *Trends in Food Science & Technology* 21 (4):168–180.
Siragusa, S, Maria De Angelis, Raffaella Di Cagno, Carlo G Rizzello, Rossana Coda, and Marco Gobbetti. 2007. "Synthesis of γ-aminobutyric acid by lactic acid bacteria isolated from a variety of Italian cheeses." *Applied and Environmental Microbiology* 73 (22):7283–7290.
Slavin, Joanne L, David Jacobs, and Len Marquart. 2000. "Grain processing and nutrition." *Critical Reviews in Food Science and Nutrition*. 40 (4):309–326.
Smith, James P, Daphne Phillips Daifas, Wassim El-Khoury, John Koukoutsis, and Anis El-Khoury. 2004. "Shelf life and safety concerns of bakery products: A review." *Critical Reviews in Food Science and Nutrition* 44 (1):19–55.
Solomon, TPJ, and AK Blannin. 2007. "Effects of short-term cinnamon ingestion on in vivo glucose tolerance." *Diabetes, Obesity and Metabolism* 9 (6):895–901.
Sumengen, Melis, Sadik Dincer, and Aysenur Kaya. 2013. "Production and characterization of phytase from Lactobacillus plantarum." *Food Biotechnology* 27 (2):105–118.
Thiele, Claudia, Simone Grassl, and Michael Gänzle. 2004. "Gluten hydrolysis and depolymerization during sourdough fermentation." *Journal of Agricultural and Food Chemistry* 52 (5):1307–1314.

Tieking, Markus, Matthias A Ehrmann, Rudi F Vogel, and Michael G Gänzle. 2005. "Molecular and functional characterization of a levansucrase from the sourdough isolate Lactobacillus sanfranciscensis TMW 1.392." *Applied Microbiology and Biotechnology* 66 (6):655–663.

Tieking, Markus, Maher Korakli, Matthias A Ehrmann, Michael G Gänzle, and Rudi F Vogel. 2003. "In situ production of exopolysaccharides during sourdough fermentation by cereal and intestinal isolates of lactic acid bacteria." *Applied and Environmental Microbiology* 69 (2):945–952.

Tinzl-Malang, Saskia Katharina, Peter Rast, Franck Grattepanche, Janice Sych, and Christophe Lacroix. 2015. "Exopolysaccharides from co-cultures of Weissella confusa 11GU-1 and Propionibacterium freudenreichii JS15 act synergistically on wheat dough and bread texture." *International Journal of Food Microbiology* 214:91–101.

Torkamani, MG, SH Razavi, and SMT Gharibzahedi. 2015. "Critical quality attributes of Iranian 'Taftoon' breads as affected by the addition of rice bran sourdough with different lactobacilli." *Quality Assurance and Safety of Crops & Foods* 7 (3):305–311.

Torrieri, E, O Pepe, V Ventorino, P Masi, and S Cavella. 2014. "Effect of sourdough at different concentrations on quality and shelf life of bread." *LWT-Food Science and Technology* 56 (2):508–516.

Van Kerrebroeck, Simon, Andrea Comasio, Henning Harth, and Luc De Vuyst. 2018. "Impact of starter culture, ingredients, and flour type on sourdough bread volatiles as monitored by selected ion flow tube-mass spectrometry." *Food Research International* 106:254–262.

Venturi, Manuel, Viola Galli, Niccolò Pini, Simona Guerrini, and Lisa Granchi. 2019. "Use of selected lactobacilli to increase γ-Aminobutyric acid (GABA) content in sourdough bread enriched with amaranth flour." *Foods* 8 (6):218.

Villegas, Josefina M, Lucía Brown, Graciela Savoy de Giori, and Elvira M Hebert. 2016. "Optimization of batch culture conditions for GABA production by Lactobacillus brevis CRL 1942, isolated from quinoa sourdough." *LWT-Food Science and Technology* 67:22–26.

Wu, Qinglong, Hein Min Tun, Yee-Song Law, Ehsan Khafipour, and Nagendra P Shah. 2017. "Common distribution of gad operon in Lactobacillus brevis and its GadA contributes to efficient GABA synthesis toward cytosolic near-neutral pH." *Frontiers in Microbiology* 8:206.

Xi, Jinzhong, Dan Xu, Fengfeng Wu, Zhengyu Jin, Yun Yin, and Xueming Xu. 2020. "The aroma compounds of Chinese steamed bread fermented with sourdough and instant dry yeast." *Food Bioscience* 38:100775.

Xing, Xiaolong, Jingyi Ma, Zhongjun Fu, Yirui Zhao, Zhilu Ai, and Biao Suo. 2020. "Diversity of bacterial communities in traditional sourdough derived from three terrain conditions (mountain, plain and basin) in Henan Province, China." *Food Research International* 133:109139.

Xu, Dan, Huang Zhang, Jinzhong Xi, Yamei Jin, Yisheng Chen, Lunan Guo, Zhengyu Jin, and Xueming Xu. 2020. "Improving bread aroma using low-temperature sourdough fermentation." *Food Bioscience* 37:100704.

Yılmaz, Cemile, and Vural Gökmen. 2020. "Neuroactive compounds in foods: Occurrence, mechanism and potential health effects." *Food Research International* 128:108744.

Zannini, Emanuele, Erica Pontonio, Deborah M Waters, and Elke K Arendt. 2012. "Applications of microbial fermentations for production of gluten-free products and perspectives." *Applied Microbiology and Biotechnology* 93 (2):473–485.

Zhang, Guohua, and Guoqing He. 2013. "Predominant bacteria diversity in Chinese traditional sourdough." *Journal of Food Science* 78 (8):M1218–M1223.

Zhao, Xin, and Michael G Gänzle. 2018. "Genetic and phenotypic analysis of carbohydrate metabolism and transport in Lactobacillus reuteri." *International Journal of Food Microbiology* 272:12–21.

Zhao, Zheng, Taihua Mu, and Hongnan Sun. 2019a. "Comparative study of the nutritional quality of potato steamed bread fermented by different sourdoughs." *Journal of Food Processing and Preservation* 43 (9):e14080.

Zhao, Zheng, Taihua Mu, and Hongnan Sun. 2019b. "Microbial characterization of five Chinese traditional sourdoughs by high-throughput sequencing and their impact on the quality of potato steamed bread." *Food Chemistry* 274:710–717.

Zotta, Teresa, Paolo Piraino, Eugenio Parente, Giovanni Salzano, and Annamaria Ricciardi. 2008. "Characterization of lactic acid bacteria isolated from sourdoughs for Cornetto, a traditional bread produced in Basilicata (Southern Italy)." *World Journal of Microbiology and Biotechnology* 24 (9):1785–1795.

13 Production of Nutraceuticals or Functional Foods Using Sourdough Microorganisms

*Armin Mirzapour-Kouhdasht,
Samaneh Shaghaghian, and Marco Garcia-Vaquero*

CONTENTS

13.1 Introduction 370
13.2 Main Microorganisms of Sourdough.. 371
13.3 Enzymatic and Chemical Microbial Reactions Inducing the Production
of Bioactive Compounds from Foods 374
    13.3.1 Dietary Fiber... 374
    13.3.2 Proteins 376
    13.3.3 Lipids 377
    13.3.4 Mineral Bioavailability.. 377
    13.3.5 Vitamins 378
    13.3.6 Antimicrobial and Antifungal Compounds............................... 378
    13.3.7 Mycotoxin Elimination 379
    13.3.8 Bioactive Compounds and Associated Health Benefits........... 379
13.4 Uses of Sourdough Microorganisms in Food Products......................... 381
    13.4.1 Cereal and Pseudo-cereal Products 382
    13.4.2 Brewing Products... 383
    13.4.3 Dairy Products... 383
    13.4.4 Meat and Related Products 384
    13.4.5 Edible Coating 385
13.5 Conclusion and Future Perspectives 385
Acknowledgments.. 386
References.. 386

DOI: 10.1201/9781003141143-16

13.1 INTRODUCTION

Sourdough is produced through the fermentation of cereal, pseudo-cereal, or vegetable flours by lactic acid bacteria (LAB) and yeasts. The application of sourdough as a leavening agent in baking and brewing is one of the oldest biotechnological processes that dates back to ancient times [1, 2]. It is very important to note that the diversity and number of sourdough microbiota depend on the native microbial flora present in the environment (the place, surfaces, and hands of workers), as well as origin of flour, starter feeding frequency (when and how often), dough hydration level, type of pseudo/cereal, leavening conditions, fermentation period, and sourdough maintenance conditions [3–5].

Microbiological studies have detailed that sourdough includes over 50 types of LAB and over 20 species of yeasts, for the most part, types of the genus *Lactobacillus*, and genera *Saccharomyces* and *Candida*, and in some cases *Leuconostoc* spp., *Weissella* spp., *Pediococcus* spp., and *Enterococcus* spp. [6]. Moreover, the LAB and yeast species and strains can differ between sourdoughs due to the effects of different processing conditions and region-specific handling. The LAB that develops in the dough could be from natural flora in the flour or a starter culture that contains identified LAB species. In sour ferments, cell densities surpassing 10^8 colony forming units per g)CFU/g) are common [6].

Firstly, the focus of sourdough research was directed to improvements of sourdough technology on the rheology, shelf life, and flavor of the final bread as well as on the complex microbial biodiversity of these products [7–9]. Recently, attention has shifted toward understanding the functional aspects of sourdough fermentation [10–16] and the effect of these processes on the bioaccessibility of some nutritional compounds, such as phenols, vitamins, sterols, dietary fibers, and minerals [17]. Moreover, as the gluten and other proteins will be degraded during the fermentation process, there is an increased interest in studying the gluten allergenicity of these products as well as the release of bioactive peptides with especial health benefits, including antioxidant [18], anti-inflammatory [19], anticancer [9], and antihypertensive [20].

The composition and stability of sourdough's microbiota, which is regulated by a variety of exogenous and endogenous causes, is crucial to its applications in food products. Microorganisms present in sourdough are rich sources of proteins and micronutrients which can improve food or feed nutritional values by producing desirable compounds or degrading anti-nutritional factors during fermentation [21]. The growing consumer demand for less processed food, lower chemical preservatives, and minimum artificial additives in food systems brings attention to novel preservation techniques, such as bio-conservation and bacteriocins. In this regard, yeasts and LAB in sourdough are capable of releasing different multiple bioactive compounds playing a crucial role in food industries in this regard [21, 22].

This chapter briefly summarizes the main sourdough microorganisms and their enzymatic and chemical reactions that can induce production and release of bioactive compounds from foods as well as their applications in the production of nutraceuticals or functional foods.

13.2 MAIN MICROORGANISMS OF SOURDOUGH

The biodiversity and variability in the number of microorganisms of sourdough are affected by many factors, such as the chemical and microbial composition of the flour, dough hydration, fermentation temperature, and redox potential [23]. The initial microflora of the fermented dough reflects the indigenous microbial population of the flour or the commercially available sourdough starter. The greatest biodiversity of sourdough bacteria is identified in the genus *Lactobacillus* (more than 23 species), while over 20 species of yeasts, mostly of the genera *Saccharomyces* and *Candida*, are found. Generally, LAB to yeast associates in a ratio of 100: 1 for optimal activity and metabolic interactions [6, 24, 25]. The trophic and non-trophic interactions between sourdough LAB and yeasts result in lactic acid, acetic acid, ethanol, and carbon dioxide (CO_2) production through metabolizing fermentable sugar (mono- and disaccharides) [1, 6]. While LAB contribute to acidification, yeasts and LAB play a role in flavor formation, and yeasts and heterofermentative LAB species take part in leavening the sourdough [10]. There has been a continuous non-competitive association between bacteria and yeasts in sourdoughs. Throughout the fermentation, the redox potential of the dough decreases and provides the context for the growth of facultative anaerobes and LAB. The pH reduction, due to the lactic and acetic acid production of LAB, inhibits the *Enterobacteriaceae* growth. Following the fermentation process, LAB and yeast become the dominant microflora of the dough [8]. Figure 13.1 and Figure 13.2 indicate the microbial pathways and metabolites produced during sourdough fermentation.

According to the tolerance of microorganisms to acidic conditions and their abilities in carbohydrate and nitrogen metabolism, there has been recognized a typical three-phase evolution of the LAB population during the sourdough preparation. At the beginning of the fermentation, the genera *Enterococcus*, *Lactococcus*, and *Leuconostoc* are the dominant active microflora. Through the fermentation progress, the population of sourdough-specific LAB, such as *Lactobacillus*, *Pediococcus*, and *Weissella* increases. In the third phase, the adapted sourdough strains, including *Lactobacillus sanfranciscensis*, *Lactobacillus fermentum*, and *Lactobacillus plantarum* become the dominant bacterial populations [8]. Some antagonistic reactions and specific antimicrobial metabolites can also be produced by some sourdough LAB in a complex microbial consortium, leading to the dominancy of a single species or strain [26].

Sourdough can also be classified into three categories based on the applied technology and fermentation processes used.

- Type I sourdough, or backslopped, is the traditional sourdough fermented at 20–30° to reach pH ~4. The microorganisms are refreshed daily by a consecutive re-inoculation from a previously prepared sourdough to keep them in an active phase to produce high amounts of metabolites and gas.
- Type II sourdough, or semi-fluid accelerated sourdough, that was developed by industrializing the rye bread baking process. The fermentation is performed at > 30° for a long period (2–5 days) which leads to faster,

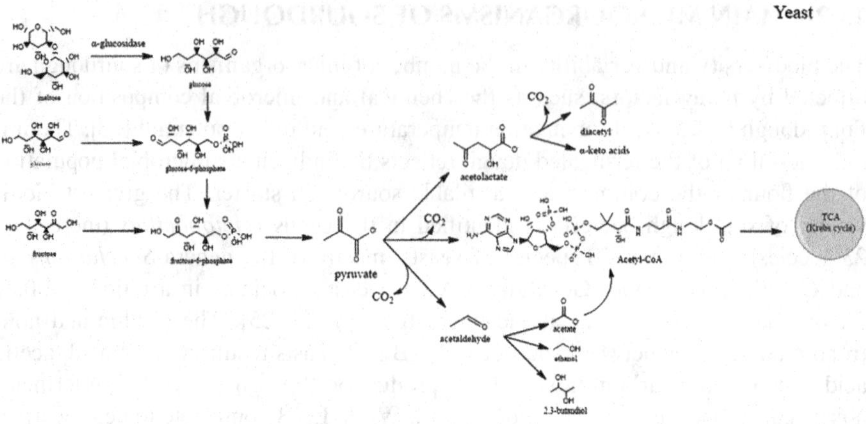

FIGURE 13.1 Overview of yeast metabolism pathways in a sourdough matrix.

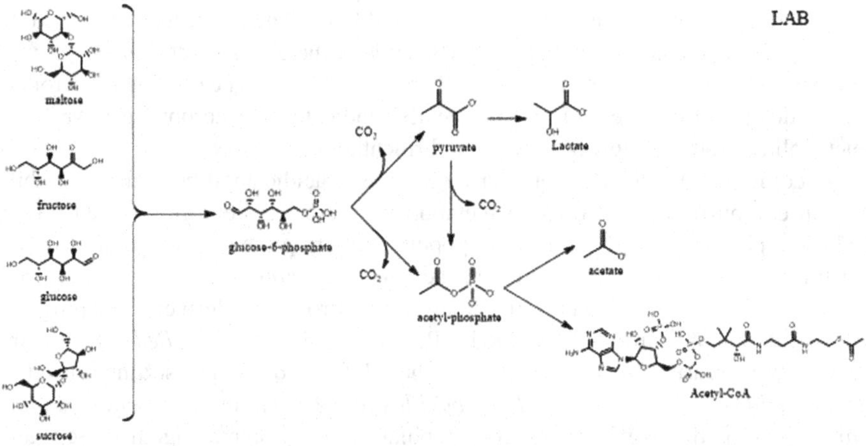

FIGURE 13.2 Overview of LAB metabolism pathways in a sourdough matrix.

controllable, and high yield of dough production. The pH of type II sourdoughs reaches < 3.5 after overnight fermentation. Due to either the presence or absence of yeasts and the low activity of microorganisms in the late stationary phase in type II sourdough, restricted metabolites are produced.
- Type III sourdough, or dried sourdough without baker's yeast, which mostly plays a role as an acidifier supplement and aroma carrier. This type of sourdough consists of resistant LAB to dry conditions, including *L. brevis* (heterofermentative), *P. pentosaceus*, and *L. plantarum* (facultative heterofermentative) strains. To initiate fermentation in a type III sourdough, applying starter cultures is essential [6, 23].

TABLE 13.1
Microflora Classification of Different Types of Sourdough (Sources: compiled from [3–7])

	Obligate heterofermentative	Facultative heterofermentative	Obligate homofermentative	Yeast
Type I	L. sanfranciscensis	L. alimentarius	L. acidophilus	C. humilis
	L. brevis	L. casei	L. delbrueckii	S. exiguus
	L. buchneri	L. (par)alimentarius	L. farciminis	C. krusei
	L. fermentum	L. plantarum	L. mindensis	S. cerevisiae
		P. acidilactici		
	L. fructivorans	L. spicheri	L. amylovorus	C. pelliculosa
	L. pontis		Leuc. Mesenteoides	C. colliculosa
	L. reuteri		Leuc. citreum	C. glabrata
	L. rossiae			
	W. cibaria			
Type II	L. brevis		L. acidophilus	No yeasts
	L. fermentum		L. delbrueckii	S. cerevisiae
	L. frumenti		L. amylovorus	may be
	L. pontis		L. farciminis	added
	L. panis		L. johnsonii	
	L. reuteri			
	L. sanfranciscensis			
	W. confusa			
Type III	L. brevis	L. plantarum		
		P. pentosaceus		
		P. acidilactici		

Each type of sourdough consists of specific microflora, regarding different intrinsic and extrinsic factors affecting sourdough fermentation, and is summarized in Table 13.1.

Obligate heterofermentative LAB are crucial in the sourdough process due to their high fermentation ability of carbohydrates (e.g., maltose) and their ability to assimilate amino acids selectively. That is, obligate heterofermentative LAB can convert arginine into ornithine, producing ammonia and ATP, and thereby coping with acid stress. Some LAB, such as *L. sanfranciscensis* and *L. (par) alimentarius*, are exclusively associated with sourdough fermentation; while other LAB, such as *L. brevis* and *L. plantarum*, can also be found in diverse fermented food systems [26]. There is a trophic mutualism between microorganisms of sourdough related to their preferred choice of carbohydrate to be used as a source of energy in their metabolism. While *S. cerevisiae* and *L. sanfranciscensis* (obligate heterofermentative) prefer to consume maltose, sourdough-specific yeasts such as *C. humilis* and *S. exiguous* are maltose-negative and hydrolyze sucrose and glucofructans instead [10].

Acquiring knowledge about sourdough microflora and the different microorganisms involved in the fermentation process is necessary for the development of effective methods to control of these processes aiming to produce high-quality products as well as for the identification of indicators that could potentially avail for the products' origin [6].

13.3 ENZYMATIC AND CHEMICAL MICROBIAL REACTIONS INDUCING THE PRODUCTION OF BIOACTIVE COMPOUNDS FROM FOODS

Sourdough is a non-aseptic biological ecosystem prepared through the fermentation of cereal or pseudo-cereal flour [27]. Fermentation can occur due to the original microflora of flours and processing equipment or by added LAB and yeasts. The synergic activity of LAB and yeast populations can lead to the secretion of metabolites that induce modifications of the macro- and micronutrients of food (Figure 13.3). Nevertheless, LAB are the main microorganisms related to the main nutritional modifications and production of functional compounds, while yeasts relate mainly to the production of aromatic and leavening agents [16, 28–30]. Recently, sourdough fermentation has gained public attention as a biotechnology process in producing nutraceuticals and value-added foods.

Acidification, synthesis of microbial metabolites, proteolysis and other enzymatic and chemical reactions induced by microorganisms under different fermentation conditions (pH, temperature, time, water content, and osmotic pressure) set the stage for biological fortification of food through increasing the protein digestibility and soluble fibers, as well as increments on phenolic compounds to enhance the antioxidant capacity and decrease the glycemic index, phytate content, and trypsin inhibitors present in foods [9, 16, 29].

13.3.1 Dietary Fiber

Dietary fiber is a functional component of food associated with relevant health benefits including lowering blood cholesterol and glucose levels, promoting health, and reducing the risk of chronic disease, obesity, diabetes, and colon cancer. The use of sourdough technology can improve the quality and bioavailability of several compounds including β-glucans, fructans, resistant starch, and arabinoxylans [27, 31, 32].

Enzymatic reactions during fermentation induce changes in insoluble to soluble fiber ratios through the effect of hydrolytic enzymes of flour, such as hemicellulase, amylase, xylosidase, arabinofuranosidase, β-glucans, endo-xylanase, cinnamoyl esterase, and glycolytic enzymes of lactic acid bacteria. Thereby, endo-xylanases activity of *Lactobacillus rhamnosus* degrade complicated arabinoxylans and make them digestible for intestinal microbiota besides improving dough properties under the hydrocolloid properties of water-extractable arabinoxylan [27, 29, 32].

Fermentable oligosaccharides, disaccharides, monosaccharides, and polyols (FODMAPs) are naturally present in various foods, from wheat and gluten-containing

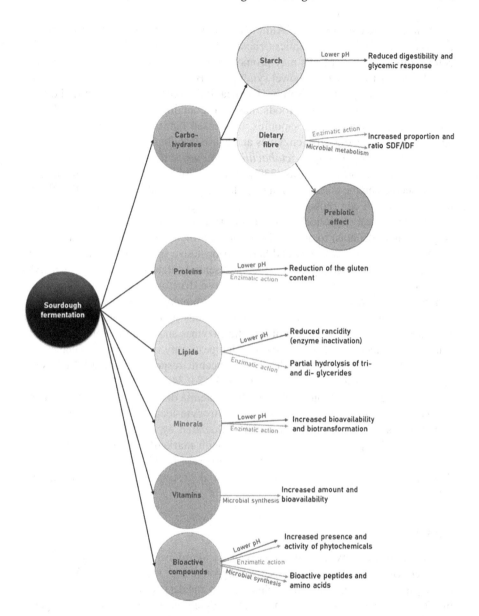

FIGURE 13.3 Effect of sourdough fermentation on macro- and micronutrients taken from [27], originally published by MDPI.

foods to corn, potato, quinoa, lentil, soy, and fruits such as pear, apple, and peach. Due to the lack of α-galactosidase (α-Gal) activity in the upper gastrointestinal tract to hydrolyze the galactooligosaccharides (α-GOS), the raffinose family of oligosaccharides (RFOs) (raffinose, stachyose, and verbascose) are fermented by the microflora of the large intestine. Although moderate doses of FODMAPs are beneficial as dietary fibers and prebiotics (fructans, RFOs, and α-GOS in particular), large concentrations of FODMAPs may promote gastrointestinal symptoms in individuals who suffer from irritable bowel syndrome (IBS) or non-celiac gluten sensitivity (NCGS), with symptoms of osmotic diarrhea, inflammation, and abdominal distension. In addition to raffinose, foods may also contain other anti-nutritional compounds, such as phytic acid, saponins, condensed tannins, and trypsin inhibitors, which can suppress digestive enzymes and cause intestinal disorders.

LAB in sourdough (e.g., *Lactobacilli* and *Leuconostoc*) are capable of decomposing RFOs (62–80%), tannins (23%), trypsin inhibitors (23–44%), and saponins (68%), decreasing the concentration of raffinose in the food matrix [16, 30–32]. The use of sourdough can reduce the content of fermentable carbohydrates of α-GOS, fructans and fructooligosaccharides (FOS), fructose, glucose, sucrose, lactose, and polyols (sugar-alcohols) to at least 30%. Homofermentative LAB in sourdough use FODMAPs as a carbon source and heterofermentative species use them as electron acceptors to produce energy and regenerate essential cofactors. Thus, the FODMAPs reduction with complete degradation of sucrose, fructose, and glucose due to the activity of lactic acid bacteria can alleviate the symptoms of IBS [27, 31]. Moreover, the production of lactic, acetic, and propionic acids by LAB induces acidic pH (< 3.5–4.0), and thus, the activation of specific enzymes and inhibition of α-amylase in sourdough, reducing FODMAPs and starch digestibility substantially, which results in controlling the blood glucose level and glycemic response to about 80% besides prolonging the gastric emptying rate [16, 27, 30, 33, 34].

Interestingly, the symptoms of hemolytic anemia disease, known as favism and caused by a genetic deficiency of the erythrocyte-located glucose-6-phosphate dehydrogenase in some humans, can be alleviated by *Lactobacillus plantarum* VTT E-133328, present in sourdough, through markedly decreasing the vicine and convicine (pyrimidine glycoside) derivatives of faba bean flour, limiting condensed tannins, trypsin inhibitor activity, and preventing their anti-nutritional effects [30].

13.3.2 PROTEINS

Proteases produced by LAB and yeasts in the sourdough, as well as those endogenous enzymes of the raw materials, lead to protein degradation, for example, gluten hydrolysis and gliadins/ glutenins solubilization and depolymerization in bread making. The special sequence of amino acids in gluten releases resistant immunogenic peptides in autoimmune disorders known as gluten sensitivity or celiac disease, whereas strain-specific LAB in sourdough (*Bacillus subtilis* GS 181 KX272352, *B. subtilis* GS 188 KX272353, *B. subtilis* GS 33 KX272356, *Bacillus cereus* GS 143 KX272357, *Lactobacillus alimentarius* 15M, *Lactobacillus brevis* 14G, and *Lactobacillus sanfranciscensis* 7A) can produce bioactive peptides and degrade toxic 33-mer peptide (considered the most immunogenic peptide responsible for triggering celiac disease) [16, 27, 31, 32]. De Angelis et al. claim that the peptide degradation is conducted during 14 h of fermentation through the activity of 5 different peptidases

(PepN-aminopeptidase type N, PepO-endopeptidase, PEP-prolyl endopeptidyl peptidase, PepX-prolyl dipeptidyl aminopeptidase, and PepQ-Prolinase) [35].

During primary proteolysis induced by flour endoproteases, oligopeptides are released, whereas small-chain bioactive polypeptides and free amino acids (FAAs), with antioxidant, antimicrobial, and anti-inflammatory properties, are revealed through enzymatic activity of LAB contributing also to the flavor of the products [16, 29, 32]. Not only does the acidified environment produced by LAB increase gluten solubility and provide ideal conditions for primary proteolysis, but the heterofermentative LAB (e.g., *Lactobacillus sanfranciscensis*) secretes glutathione reductase to produce reduced glutathione (GSH) through reduction of extracellular oxidized glutathione (GSSG). The GSH reacts with gluten proteins through thiol exchange reactions to decrease the guanosine monophosphate (GMP) molecular weight [16, 32].

Due to the proteolytic activity of sourdough microorganisms (e.g., *L. brevis* CECT 8183), the gluten concentration decreases, and the concentration of certain amino acids (leucine, isoleucine, histidine, lysine, arginine, and serine) increase. Short branched-chain peptides and amino acids produced by protein degradation of sourdough fermentation can regulate insulinemic response, providing protective effects against type II diabetes mellitus, and cardiovascular disease [27, 36, 37]. Bioactive peptides derived from food proteins can have a role in mineral-binding, immunomodulatory, hypocholesterolemic, antihypertensive, antimicrobial, and antioxidative activities leading to a decreased risk of degenerative aging diseases, such as cancer and arteriosclerosis [38, 39].

13.3.3 LIPIDS

The acidic pH of sourdough can also limit the lipase activity of food leading to a decrease in fat rancidity. The partial hydrolysis of the triglycerides and diglycerides occurs by the action of enzymes, following the increment of the monoglycerides and maintaining the sterol esters. Besides, certain strains of *Lactobacillus hammesii* can convert linoleic acid into monohydroxy octadecenoic acid with antifungal activity [27].

13.3.4 MINERAL BIOAVAILABILITY

Phytic acid can prevent diabetes through absorbing starch and sugar, interfering with cholesterol formation, and fat digestion [32]. On the other hand, phytic acid, as an abundant anti-nutritional factor, can prevent the absorption of Ca, Fe, K, Mg, Mn, and Zn by chelating and forming insoluble salts. Not only does the acidic condition of sourdough activate grain endogenous phytase, but also sourdough microflora activation promotes phytate degradation [16, 27, 30, 32]. Strains of *Lactobacillus* spp. (*L. brevis* HEB33 and *L. plantarum* ELB78) and yeasts (*Kluyveromyces marxianus*) in sourdough degrade phytic acid efficiently, with the help of their low pH resistant phytase activity, and increase mineral, FAA, and protein bioavailability [2, 3, 16]. Hydrolysis of phytic acid by phytase increases phosphate bioavailability through degrading phytic acid to myo-inositol and inorganic phosphate, with the help of inositol hexakisphosphatase. In this regard, oxidative stress can also be prevented due to the increased bioaccessibility of certain minerals, such as Se [27, 30, 32, 33].

13.3.5 VITAMINS

Vitamin B content, especially riboflavin (B_2) and nicotinamide (B_3) increase during sourdough LAB fermentation, whereas yeasts (*S. cerevisiae*) provide conditions for LAB to produce folates (B_9) and thiamine (B_1). The deficiency of folates, as cofactors of many enzymes, increases the mutation and cancer risk, while sufficient folate intake decreases coronary disease risk through lowering serum homocysteine levels [27, 34, 40]. Katina et al. concluded that the level of folates was not affected by the fermentation of rye flour by *L. plantarum* and *L. brevis*, while folates' concentration increased more than twofold by the addition of yeast (*S. cerevisiae*) to the microflora of sourdough [40].

13.3.6 ANTIMICROBIAL AND ANTIFUNGAL COMPOUNDS

Another benefit that can be exploited from sourdough LAB relates to the production of safer food/feed products. Bacteriocins (30–60 amino acids), bacteriocin-like inhibitory substances (BLIS), and the low-molecular-mass antimicrobial and antifungal metabolites, such as the reutericyclin produced by *L. reuteri* LTH2584, have the potential to be used against fungal infestations and mycotoxin contamination. Bacteriocins are proteinaceous toxins produced by different microorganisms to inhibit the growth of closely related species, while bioactive low-molecular-mass compounds are poorly known molecules with a wide activity spectrum against gram-positive and gram-negative bacteria as well as against some yeasts and molds [28, 29].

Many studies also focused on the antifungal activity of compounds produced during fermentation including acidic compounds (acetic, lactic, caproic/hexanoic, formic, propionic, butyric, phenyllactic, 4-hydroxy-phenyllactic, monohydroxy octadecenoic, and benzoic acids), fatty acids, volatile compounds (e.g., diacetyl and acetoin), cyclic dipeptides, hydrogen peroxide, reuterin, and peptidic compounds. Thereby, low amounts of two novel antifungal compounds, phenyllactic acid (PLA) and 4-hydroxyphenyllactic acid (OH-PLA), produced by *Lactobacillus plantarum*, showed high inhibitory activity toward the tested fungi [28, 29, 32]. The formation of monohydroxy octadecenoic acid (resulting from linoleic acid conversion) by *L. hammesii* strain, exerted an antifungal effect in the resulting sourdough breads [27, 28, 41]. *Lactobacillus amylovorus* DSM19280 showed the ability to produce carboxylic and fatty acids in addition to various cyclic dipeptides with antifungal properties [28, 32].

Antifungal sourdough starters can display synergistic effects in combination with some compounds (e.g., calcium propionate) against some resistant spores. During the fermentation phase by *L. plantarum* LB1, *L. plantarum* 1A7, *L. brevis* AM7, *Wickerhamomyces anomalus* LCF1695, and *L. rossiae* (LB2), peptides with fungi inhibitory effects are produced. The inhibition mechanism of most sourdough antifungal compounds is not clearly described, and more clear molecular interactions should be studied in order to use these compounds as future antimicrobials outside sourdough.

13.3.7 MYCOTOXIN ELIMINATION

While bacterial and fungal infections of food set the stage for mycotoxin contamination, strains of LAB may alleviate the issue through binding or bio-transforming mycotoxins enzymatically during long sourdough fermentations. Although the involved metabolic pathways and responsible enzymes are not determined precisely, sourdough LAB has proven to be effective when eliminating mycotoxins [28, 32, 42]. Previous studies of LAB from dairy and bakery products (*Lactobacillus*, *Lactococcus*, and *Bifidobacterium*) have also demonstrated the ability of these microorganisms to remove aflatoxin B1 by over 50% through cell-wall binding. For instance, *Lactobacillus helveticus* FAM22155 produced active enzymes against aflatoxin B1 during fermentation and degraded it into low-toxin organisms without a lactone ring structure. Hassan et al. also reported detoxification of deoxynivalenol with the help of *L. paracasei* subsp. tolerans, isolated from sourdough through the surface-binding mechanism [28, 32].

13.3.8 BIOACTIVE COMPOUNDS AND ASSOCIATED HEALTH BENEFITS

The microbial activity of sourdough following pH reduction leads to an increase in the release of different bioactive compounds and improves the antioxidant activity of the present phytochemicals in the food matrix, such as free phenolic acids, total phenolic compounds, free ferulic acid, alkylresorcinols, or quercetin. Not only the endogenous cereal enzymes but also phenolic acid reductase activities of lactobacilli take a role in converting phenolic compounds during sourdough fermentation. During sorghum fermentation, the pyrano-3-deoxyanthocyanidins is produced through hydroxycinnamic acid decarboxylation by decarboxylase-positive lactobacilli. The sourdough process favors hydrolysis of benzoxazinoids and degrades it into compounds with health benefits. On the other hand, the antioxidant activity of sourdough also relies on lactic acid bacterial secretion of enzymes, such as superoxide dismutase and synthesis of non-enzymatic antioxidants like glutathione and exopolysaccharides (EPS) [9, 27, 40, 43].

The phenolic compounds perform their antioxidant properties through 2 probable mechanisms: (1) activating Nrf2 (Nuclear factor-erythroid factor 2-related factor 2), a transcription factor of antioxidant enzymes; and (2) inhibiting NF-kB (Nuclear Factor kappa-light-chain-enhancer of activated B cells), an essential DNA transcriptional factor in inflammatory processes. Specific glycosyl hydrolase activity of *Lactobacillus plantarum* and *Lactobacillus rhamnosus* leads to a decomposition of the plants' cell-wall decomposition and a release of covalently bonded phenolic compounds. The phenolic compounds improve the antioxidant capacity of the products and contributes to an inhibition of α-amylase and α-glucosidases inhibiting starch hydrolysis [9, 16, 34, 38].

It has been investigated that an increase in EPS production and phenolic compounds concentration have occurred during the fermentation process conducted by *Pediococcus acidilactici* strains, *L. sanfranciscensis* and *L. brevis* strains, *L. acidophilus, Enterococcus faecium,* and *Kluyveromyces aestuarii* in sourdoughs. Enzymatic activity in the gastrointestinal tract or LAB fermented products provides a good environment for the release of bioactive peptides (usually composed of 3–20 residues), which scavenge free radicals and inhibit lipid peroxidation [9, 27, 38, 43, 44]. Some peptides and organic acids synthesized by LAB strains in sourdough have antioxidant and anti-inflammatory properties. The antioxidant activity of various peptides relates to their amino acid sequence, conformation and hydrophobicity. Overall, peptides with antioxidant properties have low molecular mass and specific amino acid composition, such as an increased relative abundance of Tyr, Trp, Met, Lys, Pro, Cys, His, Val, Leu, and Ala. Amino acids containing aromatic residues, or the presence of Ala, Leu, or His at the N or C terminus, improves the radical-scavenging activity of peptides, while Val, Leu, and Ile are normally present in those peptides promoting muscle synthesis [27, 33, 34, 38].

Along with fermentation of certain strains of *Lactobacillus* spp., a small-chain peptide with angiotensin-converting enzyme (ACE) inhibitory activity is formed through protein degradation at the C-terminal of angiotensin I and converting to the potent vasopressor angiotensin II. This leads to an increase in the aldosterone secretion and sodium reabsorption by the kidney to raise blood pressure. Additionally, vasodilator bradykinin (a peptide that promotes inflammation) is also inactivated by ACE [27, 31, 39].

The proteolytic activity of sourdough and decarboxylation of L-glutamic acid also results in the release of γ-amino butyric acid (GABA), a four-carbon nonprotein FAA, with inhibitory neurotransmitter properties in the nervous system. Certain LAB strains can increase GABA concentration, which is effective in controlling hyperglycemia, hypertension, inflammation, degeneration of target issue, and reducing oxidative stress. In a study by Coda et al., *Lactobacillus plantarum* C48 and *Lactobacillus lactis* subsp. Lactis PU1 (5×10^7 CFU/g) were introduced as well-characterized GABA-producing strains during fermentation for 24 h at 30°. In another study conducted by Diana et al., the level of produced GABA during 48 h fermentation by *L. brevis* CECT 8183 increased significantly. GABA concentration declines during fermentation, due to the yeast's consumption of amino acids, and further decreases during proofing and baking. In addition to its pharmaceutical application, GABA-enriched functional foods such as gammalone, gabaron tea, shochu, pickled vegetables, fermented meats and fishes, dairy and bakery products have recently become more desirable among consumers [9, 16, 27, 29, 30, 37, 38].

Some LAB strains can produce lunasin, a fragment of the larger 2S-albumins with anticancer properties, through peptidase activity. Among 40 strains of sourdough LAB tested by Rizzello et al., *L. curvatus* SAL33 and *L. brevis* AM7 showed the best lunasin production capacities [16, 27, 29, 31, 44]. Also *L. plantarum* and *L. rossiae* can synthesize bioactive benzoquinones with antitumor properties [27, 36]. Based on the high β-glucosidase activity of *L. plantarum* LB1 and *L. rossiae* LB5 in sourdough fermentation, non-glycosylated and physiologically active 2-methoxy benzoquinone (2-MBQ), and 2,6–dimethoxybenzoquinone (2,6-DMBQ) with antiproliferative activity toward human tumor cell lines (colon carcinoma and ovarian carcinoma) were released. The highest increment of 2-MBQ and 2,6-DMBQ were observed after 12–18 h of the fermentation process. Wheat germ is a good reservoir for the precursors of 2-MBQ and 2,6-DMBQ. Thus, sourdough fermented foods can also be counted as nutritional supplements with pharmaceutical and anticancer properties. The produced quinones can be used as anticancer chemotherapy drugs. A pharmaceutical product under the name Avemar®, uses the advantage of active compounds in fermented wheat germ against many cancer types, including leukemia, melanoma, and breast, colon testicular, head and neck, cervical, ovarian, gastric, thyroid, and brain carcinomas via disrupting the caspase-poly [ADP-ribose] polymerase pathway [36].

The level of acrylamide, a human carcinogen, can also be reduced in sourdough through the prevention of the Schiff base formation in the Maillard pathway. Some LAB strains (*L. brevis, L. plantarum, L. sakei, L. lactis, L. deuterium, L. rhamnosus, Pediococcus pentoseus,* and *Pediococcus acidilactici*) show the potential to reduce acrylamide content through pH reduction and glucose metabolism, whereas yeasts decrease acrylamide production through removal of its precursors (e.g., free asparagine) [31, 32].

Particular attention should be paid to anti-nutritive, lectin-related proteins in the food matrix comprising the lectin phytohemagglutinin (PHA), especially in legumes. Due to the high resistance of PHA to proteolysis and pH ranges, it leads to a coliform overgrowth in the gut lumen and results in toxic effects. However, genetically modified seeds are devoid of PHA, whereas the role of sourdough fermentation and microbial activity to conjugate polyphenols bioconversion in their free forms cannot be ignored [33].

In addition to all nutraceutical impacts of sourdough microorganisms, they have an unavoidable effect on foods' flavor through synthesizing volatile compounds produced by homo-fermentative LAB (diacetyl, acetaldehyde, and hexanal), heterofermentative LAB (ethyl acetate, alcohols, and aldehydes), and yeasts (aldehydes and ethyl acetate). In fact, Di Renzo et al. reported that *Lactobacillus paracasei* I1 in sourdough contributed greatly to the synthesis of aromatic ketones (diacetyl, acetoin, 2,6-dimethyl-4-heptanone, 5-methyl-3-hexanone, and 4-methyl-3-penten-2-one) [29, 45].

13.4 USES OF SOURDOUGH MICROORGANISMS IN FOOD PRODUCTS

Microorganisms present in sourdough are rich sources of proteins and micronutrients which can improve food or feed nutritional values by producing desirable compounds or degrading anti-nutritional factors during the fermentation of other foods [21].

Candida humilis, Kazachstania exigua, and *Saccharomyces cerevisiae* are the most common yeast species present in sourdough [26]. *Lactobacillus fermentum, Lactobacillus brevis, Lactobacillus plantarum, Lactobacillus sanfranciscensis,*

Lactobacillus panis, and *Weissella cibaria* are typical LAB strains that exist in sourdough, while *Lactobacillus plantarum* in sauerkraut, *Lactobacillus plantarum* and *Lactobacillus sanfranciscensis* in fermented dairy products, *Lactobacillus fermentum* in garri (a creamy, granular flour obtained by processing the starchy tuberous roots of freshly harvested cassava), *Weissella cibaria* in nham (an uncooked, fermented semi-dry Thai sausage), *Lactobacillus brevis* in Kefir, *Lactobacillus plantarum* and *Lactobacillus brevis* in Kimchi are found [26, 46].

13.4.1 Cereal and Pseudo-cereal Products

According to the probable food allergy caused by wheat products due to the presence of gluten, 7 strains of LAB and 5 strains of yeast were screened from sourdough starter cultures and studied to reduce allergenic proteins. Yeast activity during fermentation contributes to CO_2 production, which sets the stage for the thiol–disulfide reaction and thus gluten depolymerization. Sourdough yeasts (*Saccharomyces* or non-*Saccharomyces*) perform this hydrolytic activity by releasing cytoplasmic proteases. Although the interaction between various strains affects protein degradation, *Pediococccus acidilacticiXZ31* (XZ31) showed higher proteolytic activity against both casein and wheat protein as the substrates, and also high immunoglobulin (IgG)-binding reduction on human anti-serum [47].

EPS produced by LAB, can be applied in bakery products for rheological and sensory enhancement. The highest producers of EPS from sourdough are *Weissella* and *Leuconostoc* species. These species convert sucrose to dextran in high proportions [48]. With the growing interest of consumers for healthy and vegetarian non-dairy beverages, more studies have been conducted on novel dairy-like products. Different probiotic and EPS-producing LAB strains can improve nutritional and functional characteristics of yogurt-like beverages from quinoa flour through fermentation. *Lactobacillus rhamnosus* SP1, *Weissella confusa* DSM 20194, and *Lactobacillus plantarum* T6B10 generated around a 750% increase in the content of free amino acids, GABA, polyphenols, and also improved the antioxidant activity of the products and the nutritional quality of the products by starch hydrolysis and protein digestibility [49].

Also, some LAB-derived polyols can be used as alternatives to sugar in diet/ light products. *L. plantarum* and *L. lactis* can produce mannitol under proper fermentation conditions. Mannitol is a sugar alcohol that is 75% as sweet as true glucose but has a lower calorie count and is not absorbed by the body [46]. In a study conducted by Dat et al., there was a positive correlation between enzyme treatment of rice bran and GABA production through fermentation by *L. brevis* VTCC-B397. The hydrolysis activity of α-amylase, alcalase, and flavourzyme leads to higher reducing sugar and glutamic acid production as a substrate for LAB to synthesize the bioactive compound GABA [50].

Sourdough-isolated microorganisms can also help to extend food shelf life through releasing organic acids, antioxidant compounds, and bacteriocins. Higher antioxidant properties and shelf life of flavored non-dairy oat milk drink can be approached by using encapsulated *L. plantarum* isolated from the fermented grain flour in the matrix of starch and cellulose [51].

13.4.2 BREWING PRODUCTS

The mechanisms for producing functional biomolecules and food ingredients by LAB in cereal-based beverages have been reviewed in recent years. Alcoholic cereal beverages (malt fermenting drinks) and non-alcoholic beverages, such as thobwa and uji, have the advantage of EPS, mycotoxins, antioxidants, antifungal, and antibacterial compounds produced during the fermentation processes [52].

Fungi, esp. *Saccharomyces cerevisiae*, are another category of microorganisms that play a crucial role in sourdough fermentation and the brewing industry [21]. Yeast strains of sourdough cultures can be applied in low-alcohol beer brewing. Among the ascomycetes, maltose-negative non-Saccharomyces yeast species (*Kazachstania servazzii* and *Pichia fermentans*) were selected by Johansson et al. due to their bioflavoring potential and stress tolerance to produce low-alcohol wheat beers. Notably, maltose-negative yeasts are dependent on the heterofermentative activity of LAB in sourdough for energy production [53].

13.4.3 DAIRY PRODUCTS

EPS produced by LAB can be applied in low-fat cheeses and plant-based dairy beverages to improve their nutritional and sensory properties. EPS can be added as emulsifiers, stabilizers, thickeners, and gelling agents, to enhance moisture retention in these products. EPS can also improve the firmness of casein network in milk through interacting with proteins and micelles and can halt syneresis to increase shelf life in fermented dairy products, such as yogurt, kefir, cultured cream, cheese, and milk-based desserts [48].

Furthermore, LAB-derived EPS can protect the natural microflora of the food matrix against harsh production conditions (e.g., dehydration, osmotic stress, and pathogenic microbes) to preserve the probiotic properties of these food products. These biopolymers have protective effects on LAB passing through the human gut and induce beneficial antioxidant, antimicrobial, antiviral, anti-inflammatory, antitumor, immunomodulatory, and blood cholesterol-lowering activities [48]. Glucans produced by LAB can reduce the risk of cardiovascular diseases by absorbing the cholesterol and lowering its level in the blood, as well as enhancing rheological, textural, and mouthfeel characteristics of fermented foods. Glucans can be used as stabilizers when manufacturing low-fat products (e.g., salad dressings, ice cream, cheese, and yogurts) and it can also reduce glucose absorption rates and decrease glycaemic index against diabetes [46].

The growing consumer demand for less processed food, low use of chemical preservatives, and minimal use of artificial additives in food systems brings attention to novel preservation techniques using these microorganisms, including the use of bio-conservation processes and the use of these microorganisms for the production of bacteriocins. Yeasts and lactic acid bacteria in sourdough are capable of releasing bioactive compounds with promising aspects in these regards that could play a crucial role in food industries [21, 22]. *L. plantarum*, as one of the main LAB of sourdough and dairy products, shows strong antimicrobial activity against pathogenic bacteria. The antimicrobial activity of three *L. plantarum* strains (P1, S11, and M7) isolated from a fermented traditional Chinese dairy product against *Escherichia coli* and *Salmonella* was related to the production of organic acids (lactic and acetic acid), hydrogen peroxide, and antimicrobial peptides [54, 55].

During milk fermentation, antioxidant and ACE-inhibitory peptides are produced through proteolytic activities of LAB strains. In a study by Hutt et al., the potential effect of probiotic *L. plantarum* incorporated with cheese or yogurt on lowering diastolic and systolic blood pressure was investigated. The authors confirmed the impact of the LAB functional properties as well as on increasing the ACE-inhibitory peptide production in milk fermented by *L. plantarum* and the resulting lowered blood pressure to decrease the risk of cardiovascular disease [56, 57].

LAB are generally recognized as safe (GRAS) and can be used as probiotics that produce metabolites or postbiotics with health benefits and immunobiotics, which improve the immune system. GABA is a postbiotic with bioactive properties from controlling blood pressure to anxiety and depression treatment, which can be produced by LAB present in a sourdough culture. In another study by El-Fattah et al., a functional yogurt rich in bioactive peptides such as GABA, was produced by milk fermentation with a co-culture of LAB [55, 57].

13.4.4 MEAT AND RELATED PRODUCTS

LAB-derived EPS provide protection of *Lactobacilli* strains to survive the harsh conditions of food processing (e.g., low pH, low water activity, and osmotic stress) in fermented meat products. The texture and quality properties of various meat products, such as cooked ham, reconstructed ham, raw fermented sausages, and fat-reduced meat products were improved by the use of EPS-forming LAB [48].

Controlled LAB growth in meat and meat products can inhibit spoilage and pathogenic bacteria by different mechanisms, such as nutrient and oxygen competition with other microorganisms as well as synthesizing meat-borne bacteriocins. The application of LAB in meat production leads to the formation of bacteriocins and lactic and acetic acids to prevent indigenous bacterial growth and increase food safety. In this regard, bacteriocin-producing LAB found in sourdough and meat products (e.g., *L. sakei* and *L. plantarum*) can play a key role in extending the shelf life of meat products through preservative effects, as well as changing flavor and texture properties [22, 48].

Most meat-borne LAB bacteriocins are of small molecular weight (< 10 KDa), cationic and thermostable peptides. For instance, nisin secreted from *Lactococcus lactis* with strong anti-botulinic, anti-listerial, and anti-staphylococcal activities; and lactocin S, sakacin A, curvacin A, and sakacin K secreted from *L. sakei*, with inhibitory activities against other strains of *Lactobacillus, Pediococcus, Leuconostoc, Listeria, Enterococcus,* and *Clostridium* are identified bacteriocins in fermented sausages. Likewise, *L. bavaricus*, isolated from sourdough, produces the same type of bacteriocins.

The growth of *Leuconostoc monocytogenes* as a common contaminant of raw and processed meat can be controlled by LAB-produced bacteriocins. It is worthy to consider that the implementation of bacteriocin-producing culture in food systems depends significantly on the food formula, applied technology, and the cultural adaptation to the ecology of the food system [22].

13.4.5 EDIBLE COATING

The use of LAB-derived bacteriocins can reduce the application of chemical preservatives and physical treatments (e.g., heat). Nisin is a GRAS food bio-preservative produced by *Lactococcus lactis* strains. Bioactive food packaging can also contain immobilized bacteriocins. The effect of *L. plantarum* probiotics incorporated into carboxymethyl cellulose bioactive film was investigated on the shelf life of strawberries [58]. Due to the antifungal compounds (such as organic acids, low molecular weight proteins, and peptides) produced by *L. plantarum*, the growth of spoilage yeast and molds decreased [58]. The authors also did not appreciate any difference in the properties of the fruits using the coated package with *L. plantarum* compared to those fruits using a control package [58].

Kalantarmahdavi et al. [59] investigated a probiotic sourdough-based edible film covering on yogurt. The authors concluded that sourdough films can be effective bioactive edible films with probiotic properties. The prepared sourdough edible film, as a carrier for an isolated *L. plantarum subsp. plantarum* PTCC 1745 from pickled cabbage, not only can limit oxygen permeability and other spoilage factors penetration but also improve the product's quality and nutritional properties. In fact, synthesized exopolysaccharides provide a network to entrap the probiotics [59].

13.5 CONCLUSION AND FUTURE PERSPECTIVES

In conclusion, sourdough microflora, along with chemical and enzymatic reactions that occur during sourdough fermentation, can release compounds and by-products with enhanced biological activities with potential uses as antioxidants, anticancer, and antihypertension agents, among others. There is evidence that the complex flora found in sourdough, which is made by fermenting cereals or pseudocereals, affects the ratio of macro- and micronutrients in food. Additionally, fermentation can increase the digestibility, lower mycotoxin and anti-nutritional factors, and improve the sensory profile and taste of foods.

There are many commercial uses and many more promising benefits for using various sourdough microorganisms, particularly yeasts and LAB groups, in the manufacture of functional foods and nutraceuticals. In addition to sourdough fermentation approaches that use culture medium, bioinformatics in conjunction with DNA sequencing techniques can shed light on the composition of the fermentative microbial populations. One of the "OMICS" to be used in the examination of the microbiome of sourdough in relation to the organoleptic qualities of the finished products includes transcriptomic analysis with promising future in identifying these complex microbial populations and their future applications.

ACKNOWLEDGMENTS

Dr. Armin Mirzapour-Kouhdasht works within the project AMBROSIA (code: 2020HDHL102) funded within the Horizon 2020 Joint Programming Initiative (ERA HDHL PREVNUT 2020) project administered by the Department of Agriculture Food and the Marine (DAFM).

REFERENCES

1. R. S. Chavan and S. R. Chavan, "Sourdough technology: A traditional way for wholesome foods: A review," *Comprehensive Reviews in Food Science and Food Safety*, vol. 10, no. 3, pp. 169–182, 2011.
2. W. Hammes and M. Gänzle, *Microbiology of Fermented Foods* (B. J. Wood, ed.), Blackie, London, p. 199, 1998.
3. C. Garofalo, G. Silvestri, L. Aquilanti, and F. Clementi, "PCR-DGGE analysis of lactic acid bacteria and yeast dynamics during the production processes of three varieties of Panettone," *Journal of Applied Microbiology*, vol. 105, no. 1, pp. 243–254, 2008.
4. E. Lhomme, C. Urien, J. Legrand, X. Dousset, B. Onno, and D. Sicard, "Sourdough microbial community dynamics: An analysis during French organic bread-making processes," *Food Microbiology*, vol. 53, pp. 41–50, 2016.
5. M. Gobbetti, A. Corsetti, J. Rossi, F. l. Rosa, and S. d. Vincenzi, "Identification and clustering of lactic acid bacteria and yeasts from wheat sourdoughs of central Italy [for breadmaking, Umbria]," *Italian Journal of Food Science*, vol. 6, no 1, pp. 85–94, 1994.
6. L. De Vuyst and P. Neysens, "The sourdough microflora: Biodiversity and metabolic interactions," *Trends in Food Science & Technology*, vol. 16, no. 1–3, pp. 43–56, 2005.
7. L. De Vuyst, S. Van Kerrebroeck, and F. Leroy, "Microbial ecology and process technology of sourdough fermentation," *Advances in Applied Microbiology*, vol. 100, pp. 49–160, 2017.
8. F. Minervini, M. De Angelis, R. Di Cagno, and M. Gobbetti, "Ecological parameters influencing microbial diversity and stability of traditional sourdough," *International Journal of Food Microbiology*, vol. 171, pp. 136–146, 2014.
9. M. Gobbetti, M. De Angelis, R. Di Cagno, M. Calasso, G. Archetti, and C. G. Rizzello, "Novel insights on the functional/nutritional features of the sourdough fermentation," *International Journal of Food Microbiology*, vol. 302, pp. 103–113, 2019.
10. L. De Vuyst, H. Harth, S. Van Kerrebroeck, and F. Leroy, "Yeast diversity of sourdoughs and associated metabolic properties and functionalities," *International Journal of Food Microbiology*, vol. 239, pp. 26–34, 2016.
11. M. Gobbetti, C. G. Rizzello, R. Di Cagno, and M. De Angelis, "How the sourdough may affect the functional features of leavened baked goods," *Food Microbiology*, vol. 37, pp. 30–40, 2014.
12. C. J. Zhao, A. Schieber, and M. G. Gänzle, "Formation of taste-active amino acids, amino acid derivatives and peptides in food fermentations: A review," *Food Research International*, vol. 89, pp. 39–47, 2016.
13. C. Graça et al., "Yoghurt as a starter in sourdough fermentation to improve the technological and functional properties of sourdough-wheat bread," *Journal of Functional Foods*, vol. 88, p. 104877, 2022.
14. E. Çakır, M. Arıcı, and M. Z. Durak, "Biodiversity and techno-functional properties of lactic acid bacteria in fermented hull-less barley sourdough," *Journal of Bioscience and Bioengineering*, vol. 130, no. 5, pp. 450–456, 2020.

15. V. Fraberger, C. Ammer, and K. J. Domig, "Functional properties and sustainability improvement of sourdough bread by lactic acid bacteria," *Microorganisms*, vol. 8, no. 12, p. 1895, 2020.
16. C. Graça, A. Lima, A. Raymundo, and I. Sousa, "Sourdough fermentation as a tool to improve the nutritional and health-promoting properties of its derived-products," *Fermentation*, vol. 7, no. 4, p. 246, 2021.
17. K. Poutanen, L. Flander, and K. Katina, "Sourdough and cereal fermentation in a nutritional perspective," *Food Microbiology*, vol. 26, no. 7, pp. 693–699, 2009.
18. V. Galli et al., "Effect of selected strains of lactobacilli on the antioxidant and anti-inflammatory properties of sourdough," *International Journal of Food Microbiology*, vol. 286, pp. 55–65, 2018.
19. S. Luti et al., "Antioxidant and anti-inflammatory properties of sourdoughs containing selected Lactobacilli strains are retained in breads," *Food Chemistry*, vol. 322, p. 126710, 2020.
20. A. R. Girija, "Peptide nutraceuticals," in *Peptide Applications in Biomedicine, Biotechnology and Bioengineering*: Elsevier, 2018, pp. 157–181.
21. R. Singh, P. Gautam, M. Fatima, S. Dua, and J. Misri, "Microbes in foods and feed sector," in *Microbial Interventions in Agriculture and Environment*: Springer, 2019, pp. 329–352.
22. M. Hugas, "Bacteriocinogenic lactic acid bacteria for the biopreservation of meat and meat products," *Meat Science*, vol. 49, pp. S139–S150, 1998.
23. C. B. Meroth, J. Walter, C. Hertel, M. J. Brandt, and W. P. Hammes, "Monitoring the bacterial population dynamics in sourdough fermentation processes by using PCR-denaturing gradient gel electrophoresis," *Applied and Environmental Microbiology*, vol. 69, no. 1, pp. 475–482, 2003.
24. O. Ogunsakin, K. Banwo, O. Ogunremi, and A. Sanni, "Microbiological and physicochemical properties of sourdough bread from sorghum flour," *International Food Research Journal*, vol. 22, no. 6, pp. 2610–2618, 2015.
25. A. Paterson and J. R. Piggott, "Flavour in sourdough breads: A review," *Trends in Food Science & Technology*, vol. 17, no. 10, pp. 557–566, 2006.
26. L. De Vuyst, S. Van Kerrebroeck, H. Harth, G. Huys, H.-M. Daniel, and S. Weckx, "Microbial ecology of sourdough fermentations: Diverse or uniform?," *Food Microbiology*, vol. 37, pp. 11–29, 2014.
27. J. Fernández-Peláez, C. Paesani, and M. Gómez, "Sourdough technology as a tool for the development of healthier grain-based products: An update," *Agronomy*, vol. 10, no. 12, p. 1962, 2020.
28. Y. I. Hassan, T. Zhou, and L. B. Bullerman, "Sourdough lactic acid bacteria as antifungal and mycotoxin-controlling agents," *Food Science and Technology International*, vol. 22, no. 1, pp. 79–90, 2016.
29. L. Nionelli and C. G. Rizzello, "Sourdough-based biotechnologies for the production of gluten-free foods," *Foods*, vol. 5, no. 3, p. 65, 2016.
30. C. Rizzello, R. Coda, and M. Gobbetti, "Use of sourdough fermentation and non-wheat flours for enhancing nutritional and healthy properties of wheat-based foods," in *Fermented Foods in Health and Disease Prevention*: Elsevier, 2017, pp. 433–452.
31. M. R. Canesin and C. B. B. Cazarin, "Nutritional quality and nutrient bioaccessibility in sourdough bread," *Current Opinion in Food Science*, Vol. 40, pp. 81–86, 2021.
32. S. Ma et al., "Sourdough improves the quality of whole-wheat flour products: Mechanisms and challenges: A review," *Food Chemistry*, vol. 360, p. 130038, 2021.
33. M. Gabriele, F. Sparvoli, R. Bollini, V. Lubrano, V. Longo, and L. Pucci, "The impact of sourdough fermentation on non-nutritive compounds and antioxidant activities of flours from different Phaseolus Vulgaris L. genotypes," *Journal of Food Science*, vol. 84, no. 7, pp. 1929–1936, 2019.

34. R. Colosimo, M. Gabriele, M. Cifelli, V. Longo, V. Domenici, and L. Pucci, "The effect of sourdough fermentation on Triticum dicoccum from Garfagnana: 1H NMR characterization and analysis of the antioxidant activity," *Food Chemistry*, vol. 305, p. 125510, 2020.
35. M. De Angelis et al., "Mechanism of degradation of immunogenic gluten epitopes from Triticum turgidum L. var. durum by sourdough lactobacilli and fungal proteases," *Applied and Environmental Microbiology*, vol. 76, no. 2, pp. 508–518, 2010.
36. C. G. Rizzello et al., "Synthesis of 2-methoxy benzoquinone and 2, 6-dimethoxybenzoquinone by selected lactic acid bacteria during sourdough fermentation of wheat germ," *Microbial Cell Factories*, vol. 12, no. 1, pp. 1–9, 2013.
37. M. Diana, M. Rafecas, and J. Quílez, "Free amino acids, acrylamide and biogenic amines in gamma-aminobutyric acid enriched sourdough and commercial breads," *Journal of Cereal Science*, vol. 60, no. 3, pp. 639–644, 2014.
38. R. Coda, C. G. Rizzello, D. Pinto, and M. Gobbetti, "Selected lactic acid bacteria synthesize antioxidant peptides during sourdough fermentation of cereal flours," *Applied and Environmental Microbiology*, vol. 78, no. 4, pp. 1087–1096, 2012.
39. C. Rizzello, A. Cassone, R. Di Cagno, and M. Gobbetti, "Synthesis of angiotensin I-converting enzyme (ACE)-inhibitory peptides and γ-aminobutyric acid (GABA) during sourdough fermentation by selected lactic acid bacteria," *Journal of Agricultural and Food Chemistry*, vol. 56, no. 16, pp. 6936–6943, 2008.
40. K. Katina et al., "Fermentation-induced changes in the nutritional value of native or germinated rye," *Journal of Cereal Science*, vol. 46, no. 3, pp. 348–355, 2007.
41. M. Quattrini, N. Liang, M. G. Fortina, S. Xiang, J. M. Curtis, and M. Gänzle, "Exploiting synergies of sourdough and antifungal organic acids to delay fungal spoilage of bread," *International Journal of Food Microbiology*, vol. 302, pp. 8–14, 2019.
42. J. T. Oatley, M. D. Rarick, G. E. Ji, and J. E. Linz, "Binding of aflatoxin B1 to bifidobacteria in vitro," *Journal of Food Protection*, vol. 63, no. 8, pp. 1133–1136, 2000.
43. A. Abedfar, S. Abbaszadeh, M. Hosseininezhad, and M. Taghdir, "Physicochemical and biological characterization of the EPS produced by L. acidophilus isolated from rice bran sourdough," *LWT-Food Science and Technology*, vol. 127, p. 9, 2020.
44. C. G. Rizzello, L. Nionelli, R. Coda, and M. Gobbetti, "Synthesis of the cancer preventive peptide lunasin by lactic acid bacteria during sourdough fermentation," *Nutrition and Cancer*, vol. 64, no. 1, pp. 111–120, 2012.
45. T. Di Renzo, A. Reale, F. Boscaino, and M. C. Messia, "Flavoring production in Kamut®, quinoa and wheat doughs fermented by Lactobacillus paracasei, Lactobacillus plantarum, and Lactobacillus brevis: A SPME-GC/MS Study," *Frontiers in Microbiology*, vol. 9, p. 429, 2018.
46. D. Das and A. Goyal, "Lactic acid bacteria in food industry," in *Microorganisms in Sustainable Agriculture and Biotechnology*: Springer, 2012, pp. 757–772.
47. W. Fu, W. Xue, C. Liu, Y. Tian, K. Zhang, and Z. Zhu, "Screening of lactic acid bacteria and yeasts from sourdough as starter cultures for reduced allergenicity wheat products," *Foods*, vol. 9, no. 6, p. 751, 2020.
48. E. Korcz and L. Varga, "Exopolysaccharides from lactic acid bacteria: Technofunctional application in the food industry," *Trends in Food Science & Technology*, vol. 110, pp. 375–384, 2021.
49. A. Lorusso, R. Coda, M. Montemurro, and C. G. Rizzello, "Use of selected lactic acid bacteria and quinoa flour for manufacturing novel yogurt-like beverages," *Foods*, vol. 7, no. 4, p. 51, 2018.

50. Q. D. Lai, N. T. T. Doan, H. D. Nguyen, and T. X. N. Nguyen, "Influence of enzyme treatment of rice bran on gamma-aminobutyric acid synthesis by Lacto bacillus," *International Journal of Food Science & Technology*, vol. 56, no. 9, pp. 4722–4729, 2021.
51. S. Ravindran and S. RadhaiSri, "Probiotic oats milk drink with microencapsulated Lactobacillus plantarum–an alternative to dairy products," *Nutrition & Food Science*, vol. 51, no 3, pp. 471–482, 2020.
52. D. M. Waters, A. Mauch, A. Coffey, E. K. Arendt, and E. Zannini, "Lactic acid bacteria as a cell factory for the delivery of functional biomolecules and ingredients in cereal-based beverages: A review," *Critical Reviews in Food Science and Nutrition*, vol. 55, no. 4, pp. 503–520, 2015.
53. L. Johansson et al., "Sourdough cultures as reservoirs of maltose-negative yeasts for low-alcohol beer brewing," *Food Microbiology*, vol. 94, p. 103629, 2021.
54. C. H. Hu, L. Q. Ren, Y. Zhou, and B. C. Ye, "Characterization of antimicrobial activity of three Lactobacillus plantarum strains isolated from Chinese traditional dairy food," *Food Science & Nutrition*, vol. 7, no. 6, pp. 1997–2005, 2019.
55. L. Diez-Gutiérrez, L. San Vicente, L. J. R. Barron, M. del Carmen Villaran, and M. Chávarri, "Gamma-aminobutyric acid and probiotics: Multiple health benefits and their future in the global functional food and nutraceuticals market," *Journal of Functional Foods*, vol. 64, p. 103669, 2020.
56. P. Hütt, E. Songisepp, M. Rätsep, R. Mahlapuu, K. Kilk, and M. Mikelsaar, "Impact of probiotic Lactobacillus plantarum TENSIA in different dairy products on anthropometric and blood biochemical indices of healthy adults," *Beneficial Microbes*, vol. 6, no. 3, pp. 233–243, 2015.
57. A. Abd El-Fattah, S. Sakr, S. El-Dieb, and H. Elkashef, "Developing functional yogurt rich in bioactive peptides and gamma-aminobutyric acid related to cardiovascular health," *LWT*, vol. 98, pp. 390–397, 2018.
58. D. Khodaei and Z. Hamidi-Esfahani, "Influence of bioactive edible coatings loaded with Lactobacillus plantarum on physicochemical properties of fresh strawberries," *Postharvest Biology and Technology*, vol. 156, p. 110944, 2019.
59. M. Kalantarmahdavi, S. Khanzadi, and A. Salari, "Viability of Lactobacillus plantarum incorporated with sourdough powder-based edible film in set yogurt and subsequent changes during post fermentation storage," *Journal of Environmental Health and Sustainable Development*, vol 5, no 4, pp. 1117–11125, 2020.

14 Applications of Sourdough in Animal Feed

Kristina Kljak, Miona Belović, Marija Duvnjak, Vanja Jurišić, and Miloš Radosavljević

CONTENTS

14.1 Introduction 391
14.2 Fermentation with Lactic Acid Bacteria 392
    14.2.1 Ensiling 393
        14.2.1.1 Whole-Plant Silage 394
        14.2.1.2 Grain Silage 395
    14.2.2 Fermented Liquid Feed 396
    14.2.3 Cereal By-products from the Processing Industry 397
        14.2.3.1 By-products of Milling and Malting 397
        14.2.3.2 By-products of Distillation and Biofuel Production 398
    14.2.4 Nutritional and Health Effect of Fermented Feeds in Animal
        Nutrition 400
        14.2.4.1 Pigs and Poultry 400
        14.2.4.2 Cattle 405
14.3 Addition of Enzymes in Feed Mixtures 407
    14.3.1 Addition of Enzymes in Feed Mixtures for Monogastric
        Animals and Fish 408
    14.3.2 Enzymes in Feed for Ruminants 418
14.4 Future Trends in Animal Feed 422
14.5 Conclusion 423
Acknowledgments 425
References 425

14.1 INTRODUCTION

Feed costs account for the largest share of total production costs in livestock production, which can be as high as 70%. Therefore, improving feed efficiency is a constant endeavor in animal nutrition, not only to reduce production costs but also to reduce environmental impact. In this regard, feed processing, such as fermentation or the addition of various exogenous enzymes to diets are commonly used strategies to improve feed efficiency, having a positive effect on both nutrient digestibility and animal health. Thereby, fermentation can be a source of lactic acid bacteria (LAB) that can potentially reduce pathogenic bacteria, and the addition of exogenous enzymes can produce prebiotic oligosaccharides with potential as an alternative technology to replace antibiotic growth promotors in animal nutrition.

Cereal grains are widely used as an energy source in animal nutrition due to their high starch contents (Sauvant et al. 2004). Of the total world cereal production in 2020/21, 37.6% was used as animal feed, with coarse grains, such as maize, sorghum, and barley grown predominantly for animal feed, while wheat and rice were grown mainly for human consumption (FAO 2022). The proportion of cereal grains in the diet can be as high as 70% in pigs and poultry and 45% in cattle. These grains, because of their high proportion in the diet, are also an important source of non-essential amino acids, the limiting essential amino acids for monogastric animals being lysine, tryptophan in corn, threonine in sorghum and rice, and methionine for poultry (Kellems and Church 2010).

When considering cereals, whole plants and by-products from both the food and ethanol industries are often used as feed and are subjected to fermentation alongside the grain. Anti-nutritional factors (ANFs) in cereals include phytic acid and non-starch polysaccharides (NSP), and the addition of exogenous enzymes that degrade these compounds has a positive effect on feed efficiency. Phytic acid is a chelating agent that reduces the bioavailability of divalent cations, such as phosphorus (P), calcium (Ca), zinc (Zn), and iron (Fe), thereby decreasing their bioavailability (Lestienne et al. 2005). The content of NSP in cereals ranges from 10% to 20% and most of these compounds are arabinoxylans, β-glucans (comprising up to 3% of cereal fresh weight) and cellulose (Torbica et al. 2021). Arabinoxylans and β-glucans present in rye, wheat, barley, triticale, and oats are partially soluble and can form highly viscous digesta in the gastrointestinal tract of monogastric animals (Choct 2015), resulting in decreased nutrient digestibility.

This chapter focuses on the fermentation of cereal grains, whole plant, and by-products and on the effect of this technological process on nutrient digestibility. In addition, a brief overview of commonly used enzymes related to ANFs in cereals, their activities, and dosage are also provided. The effects of both fermentation and addition of exogenous enzymes in feed on animal health and performance are presented for both monogastric animals and ruminants.

14.2 FERMENTATION WITH LACTIC ACID BACTERIA

Cereal grains usually require minimal processing – simply cracking the kernel pericarp ensures efficient digestion by most domestic animals – but additional processing is required to substantially increase the utilization of nutrients. One such processing technique is fermentation, which is widely used in animal nutrition, not only to increase the nutritional value of the feed, but also as a preservation technique of feed available for short periods of time through the year. In addition, storage of cereal grains for extended periods of time can lead to nutrient deterioration unless at least 85% dry matter (DM) is achieved (Hackl et al. 2010). Fermentation of cereal grains at DM < 85% can be a cost-effective alternative to drying, providing more flexibility in harvest dates by not waiting for an adequate DM content of the grain for collection (Hackl et al. 2010).

Besides grains and dry by-products which are fermented to improve their nutritional value, feeds such as whole plants and wet by-products must be fermented for their storage and use throughout the year. Additionally, the complete diet or feed can also be fermented, although this strategy can result in loss of essential nutrients, such as vitamins and amino acids that are normally supplemented into the complete diet (Missotten et al. 2015). Fermentation could be carried out without the addition of water using plant material that has a low DM content, i.e. solid or semi-solid fermentation, or with the addition of water, known as submerged fermentation. The best-known example of semi-solid fermentation is silage, and when considering cereals, grains and whole plants can also be ensiled. On the other hand, liquid fermented feed, where cereal grains or complete diets are fermented after the addition of water, is the most common example of submerged fermentation.

14.2.1 ENSILING

Ensiling is a technique for the preservation of raw plant materials based on the effects of LAB on this biomass under anaerobic conditions. LAB ferment simple carbohydrates of the plant material into lactic acid, which lowers the pH, preventing silage spoilage due to the activity of undesirable microorganisms, such as yeasts and clostridia, allowing the year-round storage of this feed (Fabiszewska et al. 2019). Due to the presence of epiphytic LAB, fermentation of raw plant material can occur spontaneously (Pahlow et al. 2003), but the rate and efficiency of the process are variable. *Therefore*, to ensure a rapid drop in pH and consequently reduce nutrient losses, additives such as inoculants containing one or more LAB species and/ or strains are often used. Most commercial inoculants contain homofermentative LAB, which allow rapid and efficient fermentation and reduce the negative influence of other epiphytic microorganisms. The most commonly used homofermentative LAB are *Lactiplantibacillus plantarum* (*Lactobacillus plantarum*), *Lactobacillus acidophilus*, *Enterococcus faecium*, *Pediococcus acidilactici*, and *Pediococcus pentosaceus* (Fabiszewska et al. 2019). However, lactic acid alone is not sufficient to ensure aerobic stability of silage after exposure to air; heterofermentative LAB, such as *Lentilactobacillus buchneri* (*Lactobacillus buchneri*), produce acetic and propionic acids that inhibit the growth of yeasts and improve the aerobic stability of silage (Mari et al. 2009). Inoculants can be supplemented with enzymes, such as xylanases and β-glucanases, which can increase the content of simple sugars and provide more substrate for fermentation, particularly beneficial when ensiling raw material with low sugar content.

During ensiling, pH and content of water-soluble carbohydrates (WSC) decrease while the content of lactic and acetic acids and ethanol and buffering capacity increase (Filya 2003, Duvnjak et al. 2019). The content of other nutrients, such as crude protein (CP), fiber (more specifically, the insoluble fraction of the fiber consisting of cellulose, hemicellulose, and lignin, which is not soluble in neutral detergent, called NDF), and starch usually remain the same as in the raw plant material. The ensiling process is divided into four phases (Pahlow et al. 2003): the aerobic phase (lasts until trapped oxygen is depleted), the fermentation phase, the stable phase, and the feed-out phase. Compared to fresh plant material, silages have higher nutrient losses when the aerobic phase lasts longer due to the activity of plant enzymes and epiphytic aerobic microorganisms, such as molds, yeasts, and some bacteria (Pahlow et al. 2003). The fermentation phase usually lasts from a few days to more than a month, depending on the characteristics of the fresh material and the ensiling conditions. After the pH drops to between 3.8 and 5.0 (Oude Elferink et al. 2000), caused by the lactic and volatile fatty acids (VFAs) produced, the silage is stable if there is no contact with air. During this phase, only acid-tolerant enzymes continue to be active, and its length depends on the duration of time anaerobic conditions are maintained; in practice, it usually lasts until the next harvest season (Pahlow et al. 2003). The content of DM, WSC, and the buffer capacity of the materials are affected by the ensiling process, and the presence of butyric acid or high concentrations of ethanol (> 3–4%) and ammonia (> 10–15% of total nitrogen) may indicate the activity of clostridia, yeasts, and proteolytic enzymes (Kung Jr et al. 2018). Since numerous factors can affect the ensiling process and the resulting nutritional value of silage, efforts are made to achieve proper fermentation and maintain aerobic stability after exposure to air while minimizing nutrient losses.

As a range of livestock feeds, silages may contain mycotoxins or secondary metabolites produced by mycotoxin molds mainly from the genera *Aspergillus*, *Fusarium*, *Alternaria*, and *Penicillium*. Mycotoxins have a toxic effect, reducing diet intake, animal production, and growth, as well as increasing reproductive problems and death that can occur when feeding highly contaminated diets (Zain 2011). However, the presence of molds is not an indication of the presence of mycotoxins, as their production depends on the temperature, humidity, and presence of air during the ensiling process and storage. One strategy to prevent mycotoxin contamination in silages involves their inoculation with species/strains that exhibit mycotoxin detoxification effects. Cavallarin et al. (2011) showed that inoculation of corn plant material with *L. buchneri* during ensiling could reduce the production of aflatoxin after aerial exposure, while Martinez Tuppia et al. (2017) showed that fermentation of high moisture corn grain contaminated with fumonisin B_1 reduces toxin concentration up to 50% at day 141 after ensiling.

14.2.1.1 Whole-Plant Silage

Cereal whole-plant silages are a valuable feed for cattle due to their high energy content, however, the nutritional value of whole-plant silages varies, and some of the factors that influence this variation are plant species, stage of maturity of the plant, ensiling technique, harvesting practices, and addition of additives (Jacobs et al. 2009, Filya and Socu 2007). Nadeau (2007) showed that barley and triticale whole-plant silages are higher quality feeds than oats and wheat due to higher nutrient content, higher organic matter (OM) degradability, and better fermentation characteristics. Cereal whole plants can be harvested at different stages of maturity, from flowering to dough. In general, WSC content and NDF digestibility decrease while starch content increases with the maturity stage of the plants (Filya 2003, Nadeau 2007, Hargreaves et al. 2009). The DM degradability is generally higher at the dough stage compared to flowering maturity stage (Filya 2003, Nadeau 2007).

Of the cereal crops, corn silage production has increased significantly in many parts of the world in recent decades (Khan et al. 2015). The corn plant can be harvested at different kernel milk lines, but Filya (2004) showed that the highest DM and OM degradability, and yields of degradable fiber can be found at the two-thirds milk line stage. The recommended DM content of the corn plant for harvest is between 300 and 350 g/kg, and this content represents a compromise in relation to the content of starch (Khan et al. 2015). Regardless of the reduction in fiber digestibility due to the harvesting stage, the high content of starch compensates the digestibility of corn silage. Although corn silage is considered stable after fermentation, there are reports that some processes can still be active for prolonged periods of time. As a result, the contents of lactic acid, zein (storage protein in corn kernel), and hemicellulose decrease, while ammonia-nitrogen and starch digestibility increase during prolonged storage up to one year (Der Bedrosian et al. 2012, Duvnjak et al. 2016).

14.2.1.2 Grain Silage

In animal nutrition, fermented grain or grain silage include those made with high moisture grain (pre-mature grain ensiled around 70% DM, usually ground before ensiling) and the reconstituted grain (water is added to mature grain before ensiling to increase moisture content to about 70% DM). Grain silages have been effectively used to improve nutrient digestibility in diets of both ruminants (Wilkerson et al. 1997, Devant et al. 2015, Godoi et al. 2021) and monogastric animals (Niven et al. 2007, Pieper et al. 2011, Hackl et al. 2010, Humer et al. 2013, Puntigam et al. 2021). For example, Wilkerson et al. (1997), when comparing the apparent digestibility of non-fibrous carbohydrates, NDF, acid detergent fiber (ADF), hemicellulose, CP, OM, and DM of dry corn and high moisture corn (HMC) in diets of lactating cows, showed that the use of HMC had no effect on digestibility of fiber but significantly improved the digestibility of DM, OM, CP, and non-fiber carbohydrates (for 3, 5, 5, and 11%, respectively).

The ensiling of grain and whole plant differs in the moisture available for ensiling. For example, whole-plant corn is ensiled at 70% moisture, while HMC is ensiled at 30% moisture. Low moisture during fermentation and thus low LAB activity in HMC results in higher pH (4.0–4.5 vs. 3.7–4.0) and lower lactic acid (0.5–2.0% vs. 3–6%), acetic acid (<0.5% vs. 1–3%) and ethanol (0.2–2.0% vs. 1–3%) compared to whole-plant corn silages, while butyric and propionic acids are similar between both (Kung Jr et al. 2018).

Corn and sorghum are the most commonly ensiled grain silages, while other grains such as barley, wheat, rye, and triticale are used to a much lesser extent (Hackl 2010). Corn is a valuable feed ingredient in animal nutrition as a source of starch. However, starch in corn endosperm is surrounded by zein proteins (Holding and Larkins 2006), which are a barrier to starch digestion by both ruminal (Giuberti et al. 2014) and monogastric animals (Kljak et al. 2019). The fate of zein proteins during ensiling mainly influences starch digestibility in corn (Kung Jr et al. 2018).

Ensiling leads to zein degradation in both whole-plant corn silage (38% zein degradation during ensiling, Duvnjak et al. 2016) and HMC (Hoffman et al. 2011), resulting in higher starch digestibility in ensiled corn. Higher starch digestibility leads to a higher energy value of corn that must be taken into account when preparing animals' diets.

14.2.2 Fermented Liquid Feed

During fermentation of liquid feed, the feed is mixed with water in a ratio of 1:1.5 to 1:4, and the mixture is allowed to ferment spontaneously or by inoculation with LAB. Fermentation takes place for several days, during which lactic acid is produced, lowering the pH of the feed. Several parameters, such as time and temperature of the fermentation and the feed: water ratio, can affect the quality of the final product (Missotten et al. 2015). Desirable characteristics for fermented feed are pH 4.5, > 150 mM lactic acid, < 40 mM acetic acid, < 5 mM butyric acid, < 0.8 mM ethanol, and a total lactobacilli count of > 9 log10 CFU/m L (van Winsen et al. 2001). The faster the pH decreases, the faster pathogenic bacteria, such as *Salmonella* and *Escherichia coli*, are reduced, 75 mM being the lactic acid concentration needed to prevent the growth of *Salmonella* (Beal et al. 2002). The growth of LAB during fermentation of liquid feeds is accompanied by the growth of yeasts, which is associated with ethanol production and can affect the quality of the product both positively and negatively. Yeasts can bind enterobacteria, which blocks the binding of these bacteria to the intestinal epithelium (Mul and Perry 1994). On the other hand, high yeast concentration can lead to the production of "off-flavors," compounds that make the feed less palatable (Missotten et al. 2015).

In addition to spontaneous fermentation or inoculation with LAB, a technique known as "backslopping" can be used (Salovaara 1998). In this technique, part of the previous successful fermentation is used as inoculum for the new batch and mixed with a new mixture of feed and water. The advantage of this technique is that it allows gradual selection of LAB and accelerated fermentation, where fermented feed could be produced within a few hours (Missotten et al. 2015). When comparing different proportions (20, 33, and 42%), Moran et al. (2006) found that fermentation of a new batch of wheat occurred to the greatest extent when 20% of pre-fermented wheat was kept, and that there was no advantage to keeping more than this proportion.

Fermentation during preparation of fermented liquid feed has positive effects on the properties of the final product already mentioned. In addition to reducing mycotoxin content (Okeke et al. 2015), fermentation positively affects phytate degradation. Most of the P in cereals (54–85%) is bound to phytate (Humer and Zebeli 2015). Lowering pH during fermentation leads to phytate degradation, which can be partly explained by the optimal range for endogenous phytase activity (4.5–5.6), highest in rye, wheat, and barley (Humer and Zebeli 2015, Nkhata et al. 2018). However, the effect of fermentation on phytase activity depends on the concentration of lactic acid. Vötterl et al. (2019) reported that soaking for 48 h in solutions containing 10 g/kg lactic acid decreased phytase activity to a lesser extent compared to a solution containing 25 g/kg. Furthermore, in tannin-rich cereals such as sorghum, the increase in tannins during fermentation due to hydrolysis of condensed tannins counteracts the iron bioavailability (Nkhata et al. 2018). Moreover, fermentation also reduces trypsin inhibitors (Osman 2004) and increases available lysine (Hamad and Fields 1979).

14.2.3 CEREAL BY-PRODUCTS FROM THE PROCESSING INDUSTRY

Plant by-products are an important source of carbohydrates, lipids, proteins, minerals, and other phytochemicals (Jin et al. 2018); however, these resources are still underutilized, and recycling and reuse policies require not only technical knowledge but also a change in the global mindset (Chatzifragkou et al. 2015). On the other hand, the sustainability of livestock sectors should be shaped by the conversion of human-inedible inputs into human-edible animal proteins (Wilkinson 2011). The use of less food-competing feedstuffs in animal diets is a potential strategy that may reduce food-to-feed competition and mitigate the environmental impacts of livestock production (Salami et al. 2019).

Cereal by-products are generated during harvesting and processing of grain, including dry and wet milling, pearling, and malting that results in hulls, bran, germ, gluten feed and meal, brewer's spent grain (BSG), dried distillers' grains with solubles (DDGS), and distillers' grain with solubles (DGS) (Salazar-López et al. 2020). Currently, the most promising frontier for better utilization of plant by-products seems to be biorefineries. Biorefineries are industries that start from biomass feedstock and can recover the nutrients through extraction processes or chemical and biological reactions to produce value-added products and green energy (Carmona-Cabello et al. 2018). The spectrum of products from the wet milling process can substantially improve the overall economic performance of the biorefinery process (Jin et al. 2018).

14.2.3.1 By-products of Milling and Malting

Bran and hulls are concentrated sources of cellulose, lignin, and insoluble arabinoxylans, but β-glucan can also be present mainly in rye and wheat bran (Bach Knudsen 2014). Wheat and corn bran are the main agro-industrial by-products used in animal feeding. Wheat bran is a by-product of conventional dry milling of common wheat into flour and other wheat processing industries that include a bran removal step, such as semolina production from durum wheat. Corn bran is a by-product of several corn processing industries, including starch and ethanol production, and corn-based food manufacturing. Both wheat and corn also contain residual endosperm in addition to bran, resulting in a continuum of products with varying fiber: starch ratio (Verni et al. 2019).

These by-products can be subjected to further fermentation processes, as this allows for better digestibility of the by-product, making it more efficient to use as animal feed. Microbial fermentation, either with LAB, yeasts, or fungi, is a very efficient way to improve the nutritional and functional properties of cereal by-products to implement their use in food/feed production. Fermentation, alone or coupled with technological or biotechnological processing techniques, provides a variety of tools to modify cereal matrices. During fermentation, both endogenous and bacterial enzymes are capable of modifying grain constituents and affecting the structure, bioactivity, and bioavailability of nutrients (Verni et al. 2019).

Spontaneous fermentation of corn bran by co-dominance of *Weissella* spp., *Pediococcus* spp., and *Wickerhamomyces anomalus* at the initial stage and predominance of *La. plantarum*, *Levilactobacillus brevis* (*Lactobacillus brevis*), and *Kazachstania unispora* at the end of fermentation resulted in a significant reduction in fiber fraction, an increase in soluble dietary fiber, and a reduction in phytic acid content. There was also a significant increase in ferulic acid (Decimo et al. 2017). In addition, ensiling of peanut vine with the addition of *La. plantarum*, wheat bran, and corn stover resulted in improvement of fermentation characteristics and nutritional values in terms of increase in CP and reduction in fiber content of the ensiled mixture (Qin and Shen 2013).

Less common, but of interest, are plant by-products from various beverage production processes, namely BSG. During brewing, the soluble part of the grain is removed from the grain, which concentrates insoluble material in the BSG. BSG is an abundant, low-cost, and underutilized by-product rich in carbohydrates (arabinoxylans and cellulose), lignin, and protein (Radosavljević et al. 2018), which are useful when feeding cattle (Shen et al. 2019). The LAB biomass immobilized within BSG shows good potential as a probiotic additive or protein supplement in animal nutrition (Radosavljević et al. 2019).

14.2.3.2 By-products of Distillation and Biofuel Production

By-products of distillation and biofuel production represent the most common plant-derived by-products currently used as animal feed. The growth of biofuel production has been accompanied by increased output of animal feed by-products from common biofuel processes. Globally, these feed by-products are increasing in volume and importance (FAO 2012). Dry grind fermentation is the most popular process for converting corn to ethanol. The non-fermentable residues, including protein, fiber, and oil, are collected and dried as DDGS (see Figure 14.1). Typically, 0.37–0.4 l of ethanol and 0.31–0.32 kg of DDGS are produced from processing one kg of corn (Böttger and Südekum 2018). However, by-products of ethanol production may be high in sulfate due to the use of sulphuric acid in the manufacturing process, which can result in H_2S intoxication in cattle (Nietner et al. 2015). Additionally, mycotoxins are not degraded during ethanol production and can accumulate in distillers' grain (Zachariasova et al. 2014), while yeasts used for starch fermentation are estimated to contribute 200–500 g/kg of protein in the product (Böttger and Südekum 2018).

Distillers' grains (a by-product of ethanol distillation containing grains) and DGS (containing a mixture of wet grains and condensed distillers' solubles) have increasingly become an important source of energy, protein and fiber in ruminant rations due to the rapid expansion of the use of grains in the biofuel industry (Jin et al. 2018, FAO 2012). However, a negative effect of high dietary DDGS content on the growth performance of broiler chickens has been observed. DDGS generally has a significantly low digestibility of DM, gross energy and CP, and NSP-hydrolyzing enzymes might be the most efficient way to enable the use of increased amounts of DDGS (Swiatkiewicz et al. 2014). Lukasiewicz et al. (2012) found that broiler

Applications of Sourdough in Animal Feed

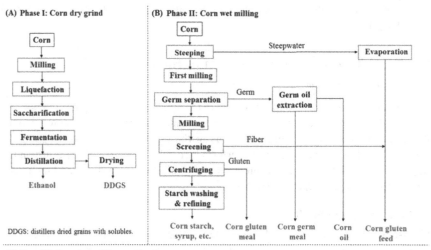

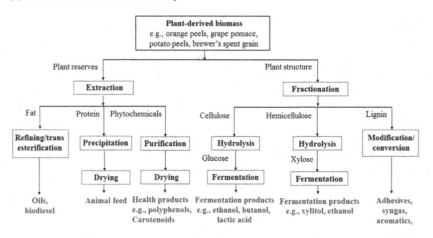

FIGURE 14.1 Corn dry grind biorefinery (phase I, A) and corn wet milling biorefinery (phase II, B) (Jin et al. 2018). During the dry grind fermentation, corn is milled, saccharified, and fermented to ethanol using yeast (Figure 14.1A). In phase II biorefinery, more products can be produced during the process. The corn wet milling process results in various products, such as starch, lactic acid, ethanol, corn syrup, and corn oil being produced (Figure 14.1B).

chickens fed a mixture with higher levels of DDGS had similar body weight and better FCE compared to birds fed a control diet. A properly balanced (fiber, energy, and amino acids) nutrient dose of the dried wheat decoction may be used as a good energy-protein component in broiler feed mixtures. It is a rational method of DDGS management, which is also a cheaper substitute for soybean meal. After a detailed study on the addition of wheat and corn DDGS in the diet of growing pigs, Ayoade et al. (2014) concluded that the inclusion of up to 30% DDGS in a corn-soybean meal diet reduced growth performance and carcass weight, but had no effect on visceral organ weights.

14.2.4 Nutritional and Health Effect of Fermented Feeds in Animal Nutrition

Fermented feeds are most commonly used in the nutrition of cattle, pigs, and poultry. As a result of properties arising from the fermentation process, fermented cereals affect nutrient digestibility and conditions in the digestive tract, and by doing so, affect animal health and production performance. The overview of literature on the effects of fermentation of cereals on nutrient digestibility in ruminant and monogastric animals is presented in Table 14.1.

14.2.4.1 Pigs and Poultry

The use of fermented cereals, whether as fermented grain or liquid, is widespread in pig farming and, to a much lesser extent, poultry nutrition. In poultry nutrition, complete diets are usually fermented rather than single feeds. Therefore, the number of studies on fermented liquid feeds or fermented grains for poultry is small compared to pigs, probably due to the assumption that wet feeds are less suitable for poultry and to the fact that temperatures in the poultry houses are high and not optimal for fermented feeds. However, poultry diets can contain up to 70% grains, and the greatest effect of fermented diets could occur with those high in NSP, such as those containing barley and oats, as poultry lack endogenous NSP enzymes (Sugiharto and Ranjitkar 2019).

Skrede et al. (2003) showed that the use of fermented barley and wheat in broiler chickens improved growth (1146.5 g in control diet vs 1223.4 g in fermented barley or wheat diets) and early feed conversion efficiency (FCE, up to 14 days). The same authors concluded that the use of β-glucan-degrading LAB reduced these ANFs and improved the nutritional value of barley and wheat, showing the aforementioned positive effects on early growth and FCE. Feeding high moisture stored oats resulted in an increased digestibility of DM, OM, and minerals compared to broilers fed dried oats (Svihus et al. 1997). In comparison to oats in the same study, the use of high moisture barley increased the digestibility of DM, OM, fat, and minerals, while the use of high moisture wheat increased only P and Ca digestibility in these supplemented groups compared to those receiving dried grains.

Fermentation, both liquid fermentation and fermentation of cereals, is widely used in pig nutrition and has an evident beneficial effect on pigs' performance, but the exact results are somewhat variable and seem to reflect differences in the type of cereals and fermentation conditions used. Sholly et al. (2011) showed that fermentation of barley and wheat increases ileal digestibility of carbohydrates by 2.3% and the digestibility of starch by 6% in barley and 3% in wheat-based diets, while the digestibility of cellulose was not affected. Starch digestibility is affected by granule swelling and gelatinization, which may also occur during fermentation, making starch more accessible to enzymes in the small intestine. Additionally, lactic acid

TABLE 14.1
Overview of Literature on the Effects of Fermented Cereals on Nutrient Digestibility in Ruminant and Monogastric Animal Nutrition

Animal	Type of fermentation	Cereals in feed	Main results	References
Monogastric animals				
Broiler chicken	Ensiled grain	Barley, oats, and wheat	Improved digestibility in fermented grain compared to dry grain, but to varying degrees. Digestibility coefficients improved significantly for DM, OM, and ash in fermented oats and barley (average increase for DM and OM of 22%, and for ash of 64%) with additional higher EE in fermented barley (average increase 56%). In fermented wheat, P and Ca digestibility were significantly improved (38% and 31%, respectively) compared to dry wheat.	Svihus et al. 1997
Broiler chicken	Ensiled grain, dried	Barley and wheat	The use of ensiled wheat or barley wholemeal flours improved ADG compared with the use of non-ensiled grain wholemeal flours (4% increase for barley and 14% increase for wheat, both at 40% level inclusion in the diet).	Skrede et al. 2003
Growing pigs	Fermented liquid feed	Barley and wheat	Fermentation improved ileal DM and OM digestibility and energy of barley by 6% and of wheat by 3%.	Jørgensen et al., 2010
Adult minipigs	High moisture grain silage / Low moisture grain silage	Triticale, barley, and wheat	Improved ATTD of OM, CF, CP, EE, starch, and total P when ensiled grain were fed to pigs. All digestibility coefficients were similar for high and low moisture silages except for wheat EE, where a significantly higher digestibility coefficient was obtained for high moisture silage (23% higher).	Pieper et al. 2011
Growing pigs	Fermented liquid feed	Barley, wheat	Fermentation increased ileal digestibility of carbohydrates and starch in barley diets (2.3% and 6%, respectively) and of starch in wheat diets (3%).	Sholly et al. 2011
Growing pigs	Grain silage – different DM level (70.1 – 80.9%)	Sorghum	Significant effect of silage DM on digestibility. ATTD of DM, OM, nitrogen-free extract, ash, and gross energy was significantly improved in low DM sorghum silages (70.1% DM), compared to high DM (80.9%) sorghum silages, but not to medium DM sorghum silages (73.8% DM). For starch low and medium DM silages showed significantly high starch digestibility.	Puntigam et al. 2021

(Continued)

TABLE 14.1 (CONTINUED)
Overview of Literature on the Effects of Fermented Cereals on Nutrient Digestibility in Ruminant and Monogastric Animal Nutrition

Animal	Type of fermentation	Cereals in feed	Main results	References
Ruminants				
Holstein cows	High-moisture grain silage	Corn	Cows fed diets containing rolled HMC (RHMC) and ground (GHMC) had improved ATTD compared to cows fed diets containing rolled dry corn (RC) and ground dry corn (GC). Mean ATTD improvements in fermented corn diets were 7.4%, 7.8%, 15%, 1.9%, and 4.7% for DM, OM, non-fiber carbohydrates, EE, and CP, respectively.	Wilkerson et al. 1997
Ruminants	Whole-plant silage	Corn, wheat	Plant maturity at ensiling affected digestibility of silage. In wheat silages, DM degradability increased from flowering to dough stage, while NDF degradability decreased. In corn silages, degradability of DM and OM increased from early dent to two-thirds milkline, with a decrease at blackline stage of growth, while NDF degradability decreased from early dent to blackline stage (early dent 32.6%, one-third milkline 31.2%, two-thirds milkline 30.7, blackline 18.1%).	Filya 2003, Filya 2004
Ruminants	Whole-plant silage	Corn	Improved *in situ* rumen starch degradation in fermented samples compared to dry corn irrelevant to the sample preparation (ground to 4 mm or 1 mm); on average 18.58% higher rate of degradation in silages.	Peyrat et al. 2014
Holstein heifers	High-moisture grain silage	Corn	Improved ATTD in diets with ground (0.4 cm screen) HMC, concentrate, and barley straw, but not improved in diets containing unprocessed HMC, concentrate, and barley straw compared with control diets containing corn meal and barley straw.	Devant et al., 2015
Ruminants	Whole-plant silage	Corn	Longer silage storage (3 vs. 8 weeks) results in increased *in vitro* starch and DM digestibility, and *in vitro* NDF digestibility, showing that silage storage time has an important effect on the nutritional value of silage.	Duvnjak et al., 2019

(Continued)

TABLE 14.1 (CONTINUED)
Overview of Literature on the Effects of Fermented Cereals on Nutrient Digestibility in Ruminant and Monogastric Animal Nutrition

Animal	Type of fermentation	Cereals in feed	Main results	References
Nellore bulls	High moisture grain silage Reconstituted grain silage	Corn	Improved ATTD and intestinal digestibility of DM, OM, and starch in HMC diets and reconstituted corn grain silage diets compared to dry corn diets (average 12% ATTD and average 111% intestinal digestibility improvement) with no effect on ruminal digestibility and with higher DM, OM, and starch intake in dry corn diets (20%, 20%, and 21%, respectively).	Godoi et al., 2021

Abbreviations: ADG (average daily gain), ADF (acid detergent fiber), ATTD (apparent total tract digestibility), CF (crude fiber), CP (crude protein), DC (dry corn), DM (dry matter), EE (ether extract), HMC (high moisture corn), NDF (neutral detergent fiber), OM (organic matter).

produced during fermentation could affect resistant starch (RS) content and thereby modify ileal digestibility of cereal starch (Giuberti et al. 2015). However, the direction of the effect on RS is dependent on the cereal species; while soaking for 48 h in a solution containing lactic acid decreased RS in corn by 53.7%, it increased starch content in wheat by 36.4% (Vötterl et al. 2019). When considering total-tract digestibility in pigs, fermentation reduced the fecal excretion of OM, CP, and fat (Sholly et al. 2011). Similarly, Jørgensen et al. (2010) reported that both ileal and total-tract digestibility of DM, OM, and energy increased after fermentation of the same cereals, but total-tract digestibility increased only by 1–3% in comparison to dried grain. An increase in protein digestibility probably contributed to an increase in DM and OM digestibility; intake of fermented feed causes decreases in stomach pH, after which may allow better proteolytic activity (Radecki et al. 1988). Humer et al. (2013) showed that there were no differences in DM and OM total-tract apparent digestibility when growing pigs were fed diets containing fermented corn and dry corn (DM digestibility in all treatments 89%, OM digestibility for dry corn and HMC 90% and for tightly closed fermented corn 89%) and concluded that since fermentation increases the solubility of fiber, the low fiber content in the study may have explained the lack of ameliorative effect of the fermentation process.

In cereal grains, phytic acid increases the endogenous losses of minerals in monogastric animals and thus, mineral and P excretion in manure (Selle and Ravindran 2008). In monogastric animals, this obstacle is usually solved by supplementation with exogenous phytase. However, since lactic acid bacteria are also capable of degrading phytate (Lopez et al. 2000), a decrease of insoluble P by up to 34% in HMC was found by Niven et al. (2007). Humer et al. (2013) showed that fermentation increased the apparent total-tract digestibility of P and Ca in growing pigs (42 and 71% in HMC vs. 28 and 61% in dry corn, respectively). Apparent total-tract digestibility of P also increased when triticale and wheat silages were fed to pigs, but only for silages with 75% DM and 22 g/kg lactic acid, while silages with 65% DM and 33 g/kg lactic acid had a digestibility value comparable to that of dry grains (Pieper et al. 2011). Furthermore, Puntigam et al. (2021) showed that sorghum silages had higher apparent total-tract digestibility of P (48 vs. 62.9%) with DM decrease (80.9 vs. 70.1%) and concluded that the addition of inorganic P could have been reduced by 0.39 g/kg using lower DM sorghum silages. Humer et al. (2013) discussed that phytate degradation activity in fermented grains is mainly due to the LAB phytase activity rather than endogenous phytase activity, with the exact degree of degradation depending on various fermentation conditions such as the type and metabolic activity of the fermented organisms, the solids/water ratio or the incubation temperature. Although the exact degree of P release varies, fermentation positively affects P digestibility in growing pigs and reduces the need for mineral P supplementation, thereby reducing P excretion in manure.

An important effect of fermentation of diet in poultry and pigs is on gut microflora (Missotten et al. 2015). In poultry, fermentation increases LAB while decreasing harmful bacteria such as *Salmonella* and *E. Coli* (Engberg et al. 2009). Liquid fermentation of the diet could reduce the transmission of *Salmonella* and *Campylobacter* (Heres et al. 2003a b). Heres et al. (2003b) explained the improved resistance to

Salmonella colonization by the acidifying effect of the fermented feed in the anterior part of the gastrointestinal tract and the reduced number of *Salmonella* bacteria reaching the intestine, where competition for colonization with *Enterobacteriaceae* and *Lactobacilli* takes place. In pigs, the most common changes reported in piglets are increased LAB and yeast in the stomach and small intestine (Canibe and Jensen 2007). This change is accompanied by a reduction in the incidence of various enteric diseases, such as Salmonellosis, post-weaning colibacillosis, porcine proliferative enteropathy, and swine dysentery (Canibe and Jensen 2012). The authors considered low pH and high concentration of lactic acid as the main factors responsible for the reduction of *Salmonella* spp. In addition to the changes in the microbial population of the gastrointestinal tract of pigs, the use of fermented liquid feed lowers the pH of the stomach, an important barrier against pathogens, especially in newly weaned pigs (Missotten et al. 2015). A similar beneficial effect was also observed when fermented corn was fed to growing pigs, with a significant reduction in biogenic amines in the feces of pigs fed fermented feed (1240.6 mol/kg DM) compared to dry corn (1791.8 mol/kg DM; Humer et al. 2014).

Based on the effects on nutrient digestibility and microbial population in the gastrointestinal tract, the inclusion of fermented liquid feed is expected to influence the production performance of pigs. A meta-analysis by Xu et al. (2020) showed that fermented feed ingredients boost the growth performance of weaner and grower pigs, however, when considering cereal-fermented liquid feed, the results of the studies were variable. Torres-Pitarch et al. (2020) reported that inclusion of fermented liquid feed containing barley (45%), wheat (42%), and wheat feed (12%) increased the body weight of grow-finisher pigs after the 28th and 55th day of age and average daily gain (ADG) without increase in diet intake. However, these changes did not improve FCE, although piglets fed diets containing fermented liquid feed had 2 kg heavier carcasses and 1% lower lean meat proportion at slaughter. On the other hand, O'Meara et al. (2020) concluded that the inclusion of cereal-fermented liquid feed had little additional benefit in terms of growth or FCE in grow-finisher pigs compared to dry cereals, while Scholten et al. (2002) showed that replacing 45% dry wheat in the diet with fermented feed improved FCE in weanling piglets. In the last study, the inclusion of cereal-fermented liquid feed resulted in higher villus height and villus: crypt ratio, and more favorable villus shape.

14.2.4.2 Cattle

Cattle are regularly fed diets high in carbohydrates in the form of fibers, WSC, and starch. WSC and starch are rapidly degraded in the rumen, whereas fiber degradation usually begins after a lag phase. End products of microbial fermentation in the rumen are VFAs (mainly acetate, propionate, and butyrate and, to a lesser extent, valerate, caproate, iso-butyrate, and iso-valerate), NH_3, gas (CO_2 and CH_4), and occasionally lactic acid (Nozière et al. 2010). Fermented cereal feeds are an important component of the diets of these animals with the inclusion of whole-plant silage as a forage i.e. source of fiber or grain silage as a starch source. In addition to fiber, whole-plant silages contain starch due to the presence of grain, and fermentation increases the effective degradability of starch due to the increase in the rapidly degradable fraction and degradation rate (Duvnjak et al. 2019, Peyrat et al. 2014). Although whole-plant silages may have lower DM and NDF digestibility than grass silages, their inclusion in the diet results in an increased dairy and beef cattle production performance, depending on the plant species (Keady 2005).

Forages affect feed intake, chewing behavior, ruminal pH, and NDF digestibility (Grant and Ferraretto 2018). Due to the ruminal effects of fiber, the term physically effective fiber (peNDF) has been introduced in ruminant nutrition to represent the proportion of fiber that promotes chewing activity (Zebeli et al. 2012). The peNDF are needed to maintain the pH in the rumen under optimal conditions and prevent disorders, such as sub-acute ruminal acidosis which can lead to milk fat depression, diarrhea, and laminitis. Since diets of high-performance animals usually contain a large amount of starch, its ruminal fermentation can lead to a drop in ruminal pH, an optimal balance must be found between peNDF and readily degradable carbohydrates in the diet. The peNDF content in the diet also affects fiber digestibility (Zebeli et al. 2006), which means that fiber digestibility is affected by both rumen conditions and the forage source. For example, a high starch content in the diet could negatively affect fiber digestibility by increasing the lag phase (Mertens and Loften 1980).

The proper functioning of the rumen is one of the essential factors in the production performance of ruminants. However, milk yield is negatively affected by the dietary NDF and positively affected by the ratio of non-structural carbohydrate and NDF contents due to the higher energy content of the diet (Zebeli et al. 2006). On the other hand, fiber content of the forage could influence milk composition by increasing milk fat contents (Firkins et al. 2001). When comparing different cereal species, feed intake of growing steers was more influenced by silage composition than plant species, but animal weight gain differed between corn, barley, and wheat silage (Oltjen and Bolsen 1980).

The use of HMC in ruminant feeding is cost-effective compared to dry corn as it offers a wider grain harvest window, higher yields, and reduced costs of drying, storage, and feed processing (Godoi et al. 2021). However, disadvantages of using HMC include low marketing potential, as these products are more difficult to sell than dry corn; possible product losses due to inadequate fermentation and storage, as these products may need a special bunker management different to that of dry corn; and possible need for additional storage and processing facilities on the farms (Lardy and Anderson 2016). The most important impact of using fermented corn is the higher availability and thus, higher rumen degradability, of starch compared to dry corn. Although cereals generally provide a substantial amount of nitrogen compounds for microbial synthesis because of their high content in the diet, increased starch degradation in the rumen must be synchronized with protein degradation to maximize the synthesis of microbial protein, the major source of amino acids for ruminants (Agle et al. 2010).

High moisture corn has been used as an adequate substitute for concentrates in fattening animals. In fattening Holstein heifers, the application of HMC resulted in similar performance and behavioral patterns as feeding concentrate and straw (Devant et al. 2015), while it was an adequate substitute for dry corn in finishing

Nellore bulls (Godoi et al. 2021). The DM and OM intake is generally lower in fermented corn compared to that of dry corn. Godoi et al. (2021) showed an average of 18.8% lower DM and OM intake of fermented corn compared to dry corn and discussed this reduction as a result of physical and chemical properties of feeds and DM intake of other feeds in diets, as well as animal physiological effects, such as energy limitation. However, the same authors also showed that fermented corn (regardless HMC or reconstituted corn grain silage) has higher starch digestibility (996 and 995 g/kg), DM digestibility (812 and 806 g/kg), and OM digestibility (849 and 841 g/kg) compared to dry corn (starch 846 g/kg, DM 746 g/kg, OM 774 g/kg). There is speculation that an increase in starch availability and rumen degradability of HMC can increase VFA production, leading to a pH below 6.2, and affecting NDF digestibility (Grant and Mertens 1992) or leading to rumen acidosis when pH falls below a value of 5.6 (Nagaraja and Titgemeyer 2007). This negative effect was not observed when HMC was given to cows in mid-lactation (pH 5.85; Krause et al. 2002), finishing Nellore bulls (pH 6.2; Godoi et al. 2021), or finishing Holstein heifers (average pH 5.8; Devant et al. 2015). Krause et al. (2002) showed that when dry corn was replaced with HMC, the high VFA production was nullified, as animals spent more time ruminating (427 min/day for HMC and 378 min/day for dry corn).

14.3 ADDITION OF ENZYMES IN FEED MIXTURES

Feed enzymes, mainly phytase and carbohydrate-degrading enzymes, have also shown to increase feed conversion by promoting the access of endogenous enzymes to their target substrates and decreasing their viscosity in the gut, thereby promoting feed digestibility. In addition, exogenous enzymes can produce prebiotic oligosaccharides that alter resident microbiota, promoting animal health (Castillo and Gatlin III 2015, Pestana et al. 2020). Therefore, exogenous enzymes have significant potential as an alternative technology to replace antibiotic growth promoters in pig and poultry feeding programs (Kiarie et al. 2013).

The first enzymes used in feed production were crude, and many decades passed by before the technology advanced far enough to produce enzymes in sufficient quantity and purity for their industrial application in feed mixtures. In recent years, there has been increased interest in producing enzymes using low-cost processes, such as solid-state fermentation (SSF) on abundant and practically free agro-residues and by-products. The fermentation process is easy to perform as SSF usually resemble the natural habitats of the microorganisms (Salim et al. 2017). Although this technique is still innovative for commercialization, it has numerous advantages including higher yields, productivity and extended stability of the products, and low production and energy costs, protein degradation, contamination risks, as well as lower or even absent catabolite repression in the presence of abundant substrates, such as glycerol, glucose, or other carbon sources compared to industrially used submerged fermentation techniques (Tanasković et al. 2021). The application of SSF allows the production of extracellular degrading enzymes of fungal or bacterial origin, including protease, amylase, cellulase, and lipase. An exhaustive number of microorganisms are capable of growing in a solid matrix and include, among others, *Aspergillus* spp.,

Penicillium spp., *Trichoderma* spp., *Rhizopus* spp. *Phanerochaete* spp., *Ganoderma* spp., *Saccharomyces* spp., *Bacillus* spp., *Escherichia* spp., and *Lactobacillus* spp. (Lizardi-Jiménez and Hernández-Martínez et al. 2017, Teng et al. 2017). Today, the desired target enzymes can be produced by selecting the producing microorganisms, fermentation media, and processing conditions.

For an enzyme to be effective, an appropriate enzyme-to-substrate ratio must be present in the diet. A complicating factor is that a particular substrate in one ingredient is not exactly the same as in another ingredient. Such differences arise from the location of the substrate in the ingredient matrix, the presence of other limiting factors, and differences in accessibility or solubility, which presents an additional challenge (Ravindran et al. 2013).

14.3.1 Addition of Enzymes in Feed Mixtures for Monogastric Animals and Fish

The application of enzymes in poultry diets has a long history, with the first product, named Protozyme, launched on the market in the 1920s. The first studies on the use of crude phytase to improve the P availability in plant material were reported in the 1970s; and 20 years later, the first commercial phytase and non-starch polysaccharide enzymes appeared on the market (Ravindran 2013). In addition to improving the growth performance of birds through enhanced feed digestion, enzymes can also improve the gastrointestinal health in these animals. In the last decade, NSP-degrading enzymes have been given the greatest attention as these enzymes are able to degrade the cell wall polysaccharides, cellulose, and hemicelluloses that cannot be digested by the gastrointestinal tract of poultry. Musigwa et al. (2021) proposed that the main mechanism of improvement of nutrient availability by enzymes is a reduction in intestinal digesta viscosity through the degradation of NSP. In the cereals, total content of NSP range from 8.3% to 9.8%, while the hemicellulose arabinoxylan represents the primary NSP, reaching up to 7.3% and 4.7% of wheat and corn dry matter, and 65% and 52% of the total NSP, respectively (Saleh et al. 2019). Besides arabinoxylan, cereals contain other NSPs, mainly cellulose and glucans (Ward 2021). Therefore, feed premixes for poultry usually contain a mixture of xylanase, cellulose, and β-glucosidase. However, in recent years, it has been proved that xylanase cannot be used to hydrolyze arabinoxylans since multiple arabinose substitutions in xylan backbone reduce the efficiency of xylanase, especially in corn and associated by-products. Arabinofuranosidase was successfully used to cleave the arabinose side chains and enable access of xylanase to the xylan backbone (Saleh et al. 2019, Ward 2021). Similarly, the addition of β-fructofuranoside fructohydrolase increased *in vitro* and *in vivo* galactoside hydrolysis by α-galactosidase in soybean and canola meals. It can be concluded that such supportive enzymes can improve the nutritional value of corn/soybean poultry diets (Ward 2021).

An overview of the literature regarding the addition of enzyme in feed mixtures for monogastric animals and fish is presented in **Table 14.2**. The application of commercial multi-carbohydrase Rovabio Advance (xylanase, β-glucanase, and arabinofuranosidase) in the broiler chicken diet showed beneficial effects, such as improved

TABLE 14.2
Overview of Literature Regarding the Addition of Enzyme in Feed Mixtures for Monogastric Animals and Fish

Animal	Cereals/legumes in feed	Enzyme/s added	Dose/activity of enzymes	Main results	References
Broiler chicken (*Gallus gallus domesticus*)	Wheat, soybean meal, wheat bran, and oat bran.	Multi-carbohydrase – Rovabio Advance (xylanase, β-glucanase, and arabinofuranosidase).	1,842; 2,745 and 7,412 VU/kg in standard crude protein diet and 1,854, 3,227, and 8,713 VU/kg in low protein diet for the recovered 200 β-glucanase, xylanase and arabinofuranosidase, respectively.	Increased feed energy and nitrogen efficiency, and lowered FCE, regardless of dietary CP content. If the released energy exceeds the optimal point of energy-to-CP ratio, feed and nitrogen intake will be lower.	Musigwa et al. 2021
Broiler chicken (*Gallus gallus domesticus*)	Corn and soybean meal.	β-mannanase, α-galactosidase, and a mixture of endoxylanases and β-glucanase.	0.02% β-mannanase (140 × 10⁶ U/kg); 0.01% α-galactosidase (750,000 U/kg); 0.05% endoxylanases and β-glucanase (3,000 U/kg and 400 U/kg, respectively).	Supplementation of 0.02% β-mannanase or 0.01% α-galactosidase or 0.05% xylanase + β-glucanase could improve feed conversion in corn-soybean meal diet by improving energy utilization and digestive physiology, and also supplementation of 0.05% xylanase + β-glucanase had a preferable efficacy in low energy diet.	Zou et al. 2013

(*Continued*)

TABLE 14.2 (CONTINUED)
Overview of Literature Regarding the Addition of Enzyme in Feed Mixtures for Monogastric Animals and Fish

Animal	Cereals/legumes in feed	Enzyme/s added	Dose/activity of enzymes	Main results	References
Broiler chicken (*Gallus gallus domesticus*)	Corn, DDGS, wheat bran and soybean meal.	Phytase alone, phytase and NSPase (separately or in combination).	Separately: Phytase (Axtra Phy T) 1,000 FTU/kg feed + NSPase (Rovabio Advance T) 1,250 VU/kg; Combo: Rovabio Advance Phy T (Phytase 1,000 FTU/kg + NSPase 1,250 VU/kg).	Phytase alone improved both mFCE ratio and bone mineralization, further supplementing with NSPase significantly improved overall performance including growth rate, mFCR, and performance index.	Poernama et al. 2021
Broiler chicken (*Gallus gallus domesticus*)	Corn and soybean meal in combination with 15% of *Spirulina* powder.	Carbohydrate-degrading enzymes (Rovabio Excel AP) or lysozyme.	0.005% of Rovabio Excel AP, predominantly β-xylanase and β-glucanase (Adisseo, Antony, France); 0.01% lysozyme powder obtained from chicken egg white (Merck KGaA, Darmstadt, Germany).	15% *Spirulina*, individually or combined with exogenous enzymes, reduced the birds' performance through a high digesta viscosity, which is likely associated with the gelation of microalga indigestible proteins. In addition, cell wall of *Spirulina* was successfully broken by the addition of lysozyme, but not by Rovabio Excel AP.	Pestana et al. 2020

(*Continued*)

TABLE 14.2 (CONTINUED)
Overview of Literature Regarding the Addition of Enzyme in Feed Mixtures for Monogastric Animals and Fish

Animal	Cereals/legumes in feed	Enzyme/s added	Dose/activity of enzymes	Main results	References
Broiler chicken (*Gallus gallus domesticus*)	Corn, soybean meal, and wheat middlings in combination with 10% or 15% flaxseed.	Omegazyme (Canadian Bio-Systems, Calgary, Canada) contains cellulase, xylanase, glucanase, mannanase, and galactanase.	Cellulase (5,600 U/g), xylanase (2,000 U/g), glucanase (1,200 U/g), mannanase (800 U/g), and galactanase (100 U/g).	There were no differences in BW, ADG, or diet intake during the starter or grower phase due to flax level or enzyme addition. Addition of enzyme led to large increases in the height and width of villi in the jejunum of birds fed 10% flaxseed and increases in crypt depth in the jejunum of birds fed 15% flaxseed.	Apperson and Cherian 2017
Broiler chicken (*Gallus gallus domesticus*)	Wheat in combination with copra or cassava leaf meal and soybean meal.	Challenzyme 1309A (Beijing, China).	8 enzymes with activities (U/g): β-glucanase, 800; xylanase, 15,000; β-mannanase, 100; α-galactosidase, 100; protease, 800; amylase, 500; pectinase, 500; and cellulose, 300 in 300 g/tonne of diet.	Diluting commercial feed with copra meal or cassava leaf meal at a concentration of 100 g/kg and 200 g/kg in the starter and finisher diets, respectively, and supplementation with enzyme products has no deleterious effects on the growth and carcass traits of broiler chickens.	Diarra and Anand 2020

(*Continued*)

TABLE 14.2 (CONTINUED)
Overview of Literature Regarding the Addition of Enzyme in Feed Mixtures for Monogastric Animals and Fish

Animal	Cereals/legumes in feed	Enzyme/s added	Dose/activity of enzymes	Main results	References
Broiler chicken (*Gallus gallus domesticus*)	Corn, soybean meal and DDGS.	Xylanase and phytase.	200 mg/kg of enzyme with endo-1,4-β-xylanase activity (1000 fungal xylanase units/g) and 200 mg/kg of enzyme with phytase activity (10,000 phytase units/g).	Enzymes (xylanase + phytase) can increase the nutritional efficacy of the diets with a high level of DDGS.	Swiatkiewicz et al. 2014
Pig (*Sus scrofa domesticus*)	Wheat, corn, barley and wheat bran, and soybean meal.	Xylanase and β-glucanase.	Rovabio Excel AP (3300 endo-β-1,4-xylanase visco-units and 300 endo-1,3(4)-β-glucanase units/kg of feed; 150 g/t of feed; Adisseo SAS, Antony, France).	In diets containing wheat bran, NSP increased soluble saccharides in the stomach and small intestine and increased VFA in the ileum without increasing fermentation in the hindgut, indicating that enzymes move partially from the large to the small intestine. Enzymes may also increase the fraction of other nutrients (starch and/or proteins) digested before the end of the ileum.	Cozannet et al. 2012

(*Continued*)

TABLE 14.2 (CONTINUED)
Overview of Literature Regarding the Addition of Enzyme in Feed Mixtures for Monogastric Animals and Fish

Animal	Cereals/legumes in feed	Enzyme/s added	Dose/activity of enzymes	Main results	References
Pig (*Sus scrofa domesticus*)	Corn, corn starch, soybean meal, 10% sugar beet pulp, and 15% DDGS.	Xylanase, β-glucanase and pectinase.	0.01% xylanase (Econase XT), 0.001% β-glucanase (Econase GT P), and 0.01% pectinase (Pectinase ABE), with activities of 190,000 XU/g xylanase, 2,320,000 BU/g β-glucanase, and 560 PE/g pectinase.	Inclusion of a soluble and highly fermentable fiber from sugar beet pulp with carbohydrase supplementation may help protect pigs against moderate enterotoxigenic *E. coli* infection.	Li et al. 2020
Nile tilapia (*Oreochromis niloticus*)	Corn, corn starch, soybean meal, wheat, wheat gluten meal, and wheat bran.	Phytase (*Buttiauxella* sp. phytase, DuPont Nutrition and Bioscience, Leiden, NL).	660 FTU/kg DM, diluted 1:50, added 5 ml/kg.	Improved the growth performance and nutrient digestibility independently of the quality of the diet.	Maas et al. 2020
		Phytase (the same as above) and xylanase (Danisco xylanase, DuPont Nutrition and Bioscience, Leiden, The Netherlands).	6000 U/kg DM, diluted 1:50, added 5 ml/kg.	Xylanase supplementation on top of phytase did not enhance growth performance, but enhanced the digestibility of the low quality diet.	

(*Continued*)

TABLE 14.2 (CONTINUED)
Overview of Literature Regarding the Addition of Enzyme in Feed Mixtures for Monogastric Animals and Fish

Animal	Cereals/legumes in feed	Enzyme/s added	Dose/activity of enzymes	Main results	References
Nile tilapia (*Oreochromis niloticus*)	Soybean protein, narrow-leafed lupin meal and corn starch.	RONOZYME®Hiphos (phytase), RONOZYME®ProAct (protease), and ROXAZYME®G2 (carbohydrase) from DSM Nutritional Products.	RONOZYME®Hiphos (10,000 FTU/g), RONOZYME®ProAct (75,000 protease units/g), ROXAZYME®G2 (2700 U/g xylanase, 700 U/g glucanase, 800 U/g cellulase); doses of phytase, protease, and carbohydrase were 300 mg/kg, 200 mg/kg, and 300 mg/kg, respectively.	Tilapia fed diet supplemented with phytase exhibited superior growth performance in contrast to fish fed the control diet. A significant difference was observed in the intestinal microbiota of tilapia fed the carbohydrase supplemented diet when compared to those fed the control diet.	Adeoye et al. 2016 http://dx.doi.org/10.1016/j.anifeedsci.2016.03.002

(*Continued*)

TABLE 14.2 (CONTINUED)
Overview of Literature Regarding the Addition of Enzyme in Feed Mixtures for Monogastric Animals and Fish

Animal	Cereals/legumes in feed	Enzyme/s added	Dose/activity of enzymes	Main results	References
Atlantic salmon (*Salmo salar*)	Soy protein concentrate, soy bean meal, pre-processed soy bean meal, corn gluten, wheat gluten, and wheat.	Enzyme complex containing phytase, protease, xylanase, and cellulase.	1253 FTU/g phytase, 0.196 U/g protease, 512 XU/g xylanase, 104 CMCU/g cellulase, added 0.6 g/kg.	Three different enzyme application strategies were evaluated: pre-processing soybean meal with the enzymes; addition of enzymes in the dry mix prior to extrusion; and enzyme coating post extrusion. Salmon fed the soybean meal diet with addition of enzymes in the dry mix prior to extrusion had significantly higher growth and diet intake than the fish fed the soybean protein concentrate diet without enzymes.	Jacobsen et al. 2018

(*Continued*)

TABLE 14.2 (CONTINUED)
Overview of Literature Regarding the Addition of Enzyme in Feed Mixtures for Monogastric Animals and Fish

Animal	Cereals/legumes in feed	Enzyme/s added	Dose/activity of enzymes	Main results	References
Grass carp (*Ctenopharyngodon idellus*)	Cereal waste from food processing plants and hotels (52 %).	Baker's yeast (*Saccharomyces cerevisiae*) + mixture of bromelain and papain (1:1).	Optimal dose was 25 g/kg yeast with enzymes bromelain (1%) and papain (1%).	Improved growth performance (FCE, protein efficiency ratio; and relative weight gain), stimulated fish immunity and enhanced resistance against *Aeromonas hydrophilia*	Mo et al. 2020

Abbreviations: ADG (average daily gain), BW (body weight), CP (crude protein), DDGS (dried distillers' grains with solubles), FCE (feed conversion efficiency), mFCE (mortality-corrected feed conversion efficiency), NSP (non-starch polysaccharides), VFA (volatile fatty acids), BU (β-glucanase units), CMCU (cellulase units), FTU (phytase units), PE (pectinase units), VU (visco-units), XU (xylanase units).

bird growth performance in several studies (Musigwa et al. 2021, Poernama et al. 2021). These effects were obtained when enzyme complexes were used combined with low energy feeds, including alternative feedstuffs. Even better results were obtained when multi-carbohydrase was used together with phytase. Zou et al. (2013) emphasized that a combination of xylanase and β-glucanase decreased the feed cost per kg of body weight gain and also the output of excrement, which led to an increased profit and a reduction of the environmental impacts of these farming systems. NSP-degrading enzymes were also efficient in poultry and pig diets to improve DDGS digestion (Swiatkiewicz et al. 2016).

Since pigs are also monogastric animals, they have similar digestion and gut microbiome to poultry. Therefore, it can be expected that the same combination of substrates and feed enzymes can be used in their diet (Kiarie et al. 2013). The DDGS was successfully used as an alternative ingredient for poultry and pig feed after fermentation with NSP-degrading enzymes (Swiatkiewicz et al. 2016, Jacobsen et al. 2018). Cozannet et al. (2012) researched the effect of Rovabio Excel AP commercial enzyme mixture (endo-β-1,4-xylanase and endo-1,3(4)-β-glucanase) in the diet of growing pigs containing mainly wheat, corn, barley, wheat bran, and soybean meal. The authors revealed that the addition of enzyme partially moved the digestion of dietary fiber from the large to the small intestine and improved digestion of starch and/or proteins in the ileum (Cozannet et al. 2012). Besides this effect, the addition of enzymes was shown to alter gut microbiota. The combination of soluble fiber from sugar beet pulp with multi-carbohydrase (xylanase, β-glucanase, and pectinase) may help protect pigs against moderate enterotoxigenic *E. coli* infection by maintaining or restoring microbial homeostasis (Li et al. 2020).

In general, contradictory results were reported in the literature regarding the supplementation of pigs' diets with exogenous enzymes. Torres-Pitarch et al. (2019) conducted a systematic review and meta-analysis on the effect of adding xylanase, xylanase+β-glucanase, mannanase, protease, or a multi-enzyme complex to grain meals (containing corn, wheat, barley, sorghum, rye, or co-product sources) on grow-finisher pigs. The overall conclusion was that ingredient composition of the diets is one of the most important sources of variation that may affect the efficacy of these enzymatic treatments. The meta-analysis results showed that dietary supplementation with mannanase and multi-enzyme complexes increased growth and feed efficiency in grow-finisher pigs. Despite the improvements found in nutrient digestibility in response to xylanase or xylanase+β-glucanase supplementation, they did not improve feed efficiency in grow-finisher pigs. The main cereal source influences response to enzyme supplementation in the diet formulation. Mannanase supplementation increased feed efficiency in corn-based diets, and multi-enzyme complex supplementation improved feed efficiency when corn-, wheat-, barley-, and co-product-based diets were fed to grow-finisher pigs.

A current trend in the aquaculture industry is replacing fishmeal with alternative feeds, including proteins of plant origin and by-products of the cereal industry (Castillo and Gatlin III 2015, Mo et al. 2020). However, plant-based feeds usually contain various ANFs such as phytin, NSP, and protease inhibitors, which can adversely affect fish performance and health. Therefore, enzymes with wide application in feeds for pigs and poultry, such as phytase, NSP-hydrolyzing enzymes, and protease, have been increasingly used in feed mixtures for fish. The addition of phytase to corn-soybean based fish feed mixtures improved growth performance of Nile tilapia (*Oreochromis niloticus*) (Adeoye et al. 2016, Maas et al. 2020). The addition of carbohydrase enzymes alone or in combination with phytase did not significantly affect growth performance; however, these enzymes improved the digestibility and influenced the intestinal microbiota of tilapia. On the other hand, Jacobsen et al. (2018) showed that salmon fed a soybean meal diet (which also contained wheat and corn gluten) with added enzymes (phytase, protease, xylanase, and cellulase) in the dry mixture before extrusion had significantly higher growth and feed intake than the fish fed the soybean protein concentrate diet without enzymes. An alternative diet consisting of 52% cereal waste, 25% meat waste, 10% fruit and vegetables, supplemented with baker's yeast (*Saccharomyces cerevisiae*) and a mixture of two proteases (bromelain and papain in a 1:1 ratio) improved growth performance (FCE, protein efficiency ratio and relative weight gain) and stimulated immunity of grass carp (*Ctenopharyngodon idellus*) (Mo et al. 2020).

14.3.2 ENZYMES IN FEED FOR RUMINANTS

An overview of the literature regarding the addition of enzyme in ruminant feed mixtures is presented in Table 14.3. A large proportion of diets for growing beef cattle consists of forage whose low digestibility in ruminant diets may limit the supply of energy and nutrients (Kondratovich et al. 2019). Exogenous enzymes can be used in ruminants to improve their diets. These enzymes derive from fungi (mainly *Trichoderma longibrachiatum*, *Aspergillus niger*, and *A. oryzae*) and bacteria (*Bacillus* spp, *Penicillium funiculosum*) with high hydrolytic activity on cellulose and hemicellulose, which are incorporated in liquid or granular form into the total mixed ration (hay, silages, concentrates, supplementary feeds, or premix) to increase the availability of nutrients present in the cell walls (Mendoza et al. 2014). The use of fibrolytic enzymes (*Trichoderma ressie* extract) in the diets of growing beef cattle stimulated diet intake and generated positive aspects for ruminal fermentation, regardless of the quality of growing diets offered. Also, the *in vitro* ADF digestion of several roughage substrates was also positively affected by exogenous fibrolytic enzymes, with additional benefits observed in intact sorghum grain and corn stalks (Kondratovich et al. 2019).

Increasing scientific and public health concerns about the use of subtherapeutic doses of antimicrobials for animal production purposes have led researchers to explore safe alternative strategies to enhance performance in ruminants. Alternative strategies to replace antibiotics include the use of exogenous enzymes, yeasts (*Saccharomyces cerevisiae*), phytogenic extracts, and essential oils to improve nutrient utilization and animal productivity (Kholif et al. 2017). The administration of exogenous enzymes and yeast increased the concentration of ruminal propionate, which is considered beneficial in dairy production. Degradation and solubilization of NSP by exogenous enzymes increased the available substrates for microbial

TABLE 14.3
Overview of Literature Regarding the Addition of Enzyme in Feed Mixtures for Ruminants

Animals	Cereals/legumes in feed	Enzyme/s added	Dose/activity of enzymes	Main results	References
Ruminally cannulated beef steers (*Bos taurus*)	Corn silage, alfalfa hay, steam-flaked corn, wet corn gluten feed (high quality diet); Corn silage, sorghum stalks hay, steam-flaked corn and wet corn gluten feed (low quality diet).	Fibrolytic enzymes produced by *Trichoderma reesei* (exogenous xylanase (EC 3.2.1.8) and cellulase (EC 3.2.1.4)).	Xylanase (21,000 µmol/min of reduced xylose equivalents) and cellulase (600 µmol/min of reduced glucose), 0.75 mL/kg of DM.	Fibrolytic enzymes positively affected digestion and might have additional benefit when used on unprocessed sorghum grain. Additionally, fibrolytic enzymes in beef cattle growing diets stimulated intake and generated positive impacts on ruminal fermentation.	Kondratovich et al. 2019
Indigenous Thailand cows (*Bos taurus*)	Rice straw silage and concentrated diet (cassava chip, palm kernel meal, rice bran, and soybean meal).	*Lacticaseibacillus casei* (*Lactobacillus casei*) TH14 and cellulase enzymes for ensiling of rice straw with molasses.	Cellulase 500,000 unit/g; 10,000 unit/kg fresh matter of rice straw.	Rice straw ensiled by *L. casei* TH14 with molasses or *L. casei* TH14 with molasses and cellulase provides greater results by preventing fermentation of CP, whereas DM and OM digestibility, rumen bacterial population, and propionic acid increased. In addition, there was no negative effect on nitrogen utilization and energy metabolism.	Cherdthong et al. 2021

(*Continued*)

TABLE 14.3 (CONTINUED)
Overview of Literature Regarding the Addition of Enzyme in Feed Mixtures for Ruminants

Animals	Cereals/legumes in feed	Enzyme/s added	Dose/activity of enzymes	Main results	References
Holstein dairy cows (*Bos taurus*)	Barley silage + alfalfa hay, barley grain, corn grain, canola meal, soybean meal, peas, corn distillers, and wheat and corn gluten meal.	Fibrolytic enzymes derived from *Trichoderma reesei* (mixture of xylanase and cellulase; AB Vista, Wiltshire, UK).	Xylanase (EC 3.2.1.8; xylanase activity = 350,000 BXU/g) and cellulase (EC 3.2.1.8; endoglucanase activity = 10,000 ECU/g) added in amount of 0.00, 0.25, 0.50, 0.75, 1.00, and 1.25 mL.	Pre-treating dairy cow barley silage-based diet with 0.75 mL of fibrolytic enzymes/kg of total mixed ration increased the milk production efficiency of dairy cows fed diet containing 34% barley silage on DM basis.	Refat et al. 2018
Dairy cows (*Bos taurus*)	Alfalfa hay, several corn and wheat products, and dried distiller's grains.	Fibrolytic enzymes derived from *Trichoderma reesei* (mixture of xylanase and cellulase).	Xylanase (EC 3.2.1.8; xylanase activity = 350,000 XU/g) and cellulase (EC 3.2.1.8; endoglucanase activity = 10,000 ECU/g).	Enzyme treatment to pens of cows in transition period and enzyme treatment during the lactation period increased yields of milk, fat, and protein and reduced the time to first breeding but did not alter time to pregnancy.	Golder et al. 2019
Dairy cows (*Bos taurus*)	Alfalfa hay, several corn and wheat products, and dried distiller's grains.	Fibrolytic enzymes derived from *Trichoderma reesei* (mixture of xylanase and cellulase).	Xylanase (EC 3.2.1.8; xylanase activity = 350,000 XU/g) and cellulase (EC 3.2.1.8; endoglucanase activity = 10,000 ECU/g).	Addition of fibrolytic enzymes improved the digestibility of feed, particularly protein and fiber fractions, with increase of milk fat production and milk protein percentage and yield.	Rossow et al. 2020

(*Continued*)

Applications of Sourdough in Animal Feed 421

TABLE 14.3 (CONTINUED)
Overview of Literature Regarding the Addition of Enzyme in Feed Mixtures for Ruminants

Animals	Cereals/legumes in feed	Enzyme/s added	Dose/activity of enzymes	Main results	References
Goat (*Capra aegagrus hircus*)	Egyptian berseem clover (*Trifolium alexandrinum*), crushed yellow corn, soybean meal, and wheat bran.	*Saccharomyces cerevisiae* and exogenous enzymes.	Yea-Sacc1026 (Alltech Inc., Nicholasville, KY, USA), containing a minimum of 5×10⁹ CFU of *S. cerevisiae*/g DM live yeast and Veta-Zyme Plus® (Vetagri Consulting Inc., Brampton, Canada), containing *L. acidophilus* at 2 × 10⁸ colony-forming units (CFU), protease at 2000 U/g, α-amylase at 550 U/g, and cellulase at 400 U/g.	Increased diet intake, enhanced nutrient digestibility, and enhanced ruminal fermentation, milk production, and composition. No synergism was observed between the exogenous enzymes and yeast.	Kholif et al. 2017
Lamb (*Ovis aries*)	Enzymatically treated olive cake added to oat hay + concentrate.	Acid cellulase + xylanase (50:50).	Xylanase (2267 μmol xylose released/min per ml), endoglucanase (1161 μmol glucose released/min per ml), and exoglucanase (113 μmol glucose released/min per ml) added in amounts of 20 or 80 ml (4 ml or 16 ml in the final mixture).	Supplementation of olive cake-based lamb diets with enzymes improved lamb growth performance as a result of increased feed intake and enhanced fiber digestibility, with no adverse effects on animal health.	Abid et al. 2020

Abbreviations: CP (crude protein), DM (dry matter), OM (organic matter), ECU (endo-glucanase units), XU (xylanase units).

fermentation in the cecum and hence the total VFA production. However, mixing the exogenous enzymes with yeast showed no synergistic effect.

Cherdthong et al. (2020, 2021) investigated the influence of adding *Lactobacillus casei* TH14, molasses, and cellulase to ensiled rice straw in Thai-indigenous beef cattle diet. The combination of *L. casei* TH14, molasses, and cellulose gave better results by preventing CP loss during fermentation while increasing DM and OM digestibility, rumen bacterial population, and propionic acid. In addition, there were no adverse effects on nitrogen utilization and energy metabolism in Thai-indigenous beef cattle. Inoculation of *L. casei* TH14 in combination with cellulase improved the quality and *in vitro* digestibility of silage. However, the addition of molasses is recommended to improve the fermentation quality as rice straw has low WSC content.

Fibrolytic enzymes have been added to feed to increase milk production. Several recent studies investigated the effect of adding fibrolytic enzyme mixtures (FEM) extracted from *Trichoderma reesei* (mixture of xylanase and cellulase) to the diet of dairy cows. Refat et al. (2018) showed that the addition of FEM to a barley silage-based diet increased *in vitro* DM digestibility, milk protein percentage, milk fat yield, fat-corrected milk, energy-corrected milk, and FCE. However, there were no effects on feeding behavior. Golder et al. (2019) observed that the introduction of diet FEM treatment in cows during the transition period and enzyme treatment during the lactation period increased milk, fat, and protein yield and shortened time to first breeding, but did not change time to pregnancy. While DM intake did not increase sufficiently, the observed increase in BW indicated that enzyme supplementation increased feed digestibility. Rossow et al. (2020) observed a similar trend and concluded that the addition of FEM improved the digestibility of the feed, especially the protein and fiber fractions, resulting in an increase in milk fat production and milk protein percentage and yield. In non-lactating cows, fibrolytic enzymes had little effect on the bacterial population in the rumen fluid and the transition layer between the rumen fluid and fiber mat as monitored with six major rumen bacterial species.

The addition of enzymes could also influence lamb growth. Abid et al. (2020) found that supplementation of olive cake-based lamb diets with exogenous fibrolytic enzymes (cellulase and xylanase), even at relatively low concentrations, improved lamb growth performance. The authors emphasize that the method of application and the amount of enzyme added may influence the magnitude of the effects, and it appears that pre-treatment of the feed with exogenous fibrolytic enzymes is necessary to create a stable enzyme-feed complex.

14.4 FUTURE TRENDS IN ANIMAL FEED

The demand for animal products has been driven primarily by an increase in human population that also led to the recent scientific and technological responses in different animal systems to adapt to these increased production needs. Growth in consumption and production of animal products is expected and the projections indicate that the total consumption of meat and dairy products will increase by 76% and 62%, respectively, between the period 2005/2007 to 2050 (Alexandratos and Bruinsma 2012). Developments in breeding, nutrition, and animal health will contribute to sustain this increased production potential by means of providing further efficiency in animal production as well as genetic gains. Also, future demand for animal products will be significantly affected by socioeconomic factors, such as human health concerns and changing socio-cultural values. All of these future developments will be bound to the availability of land and water, competition between food and feed resources, and all together focused on the need to operate in a highly carbon-constrained economy. Moreover, it is highly likely that animal production will be significantly affected by carbon constraints and environmental and animal welfare legislation.

Advances in animal nutrition are focused on better understanding the nutrient requirements of animals, determining the supply and availability of nutrients in feeds, and formulating cost-effective diets that efficiently match nutrient requirements and supply (Tona 2018). In this regard, biotechnological processing of feeds or complete diets will continue to be an important factor in improving nutrient utilization in animals. Future demands for ensiling and liquid feed fermentation will lead to changes in crops and by-products used, the methods employed in the fermentation process, and the means of accessing the composition and the quality of fermented product (Wilkinson and Muck 2019). Additionally, future research will continue to focus on finding better and more profitable approaches to improve the aerobic stability of fermented products. Due to the recognized effects on the gastrointestinal tract and the resulting impact on animal health and production, the use of fermented cereals as an alternative to antibiotic growth promotors is recognized as a promising alternative, and this aspect will continue to be of great importance in the future, particularly in monogastric animals. Although cereal by-products are already included in animal diets, food–feed competition and the rising cost of grain production will increase the focus on the use of by-products in animal nutrition. Fermentation has the potential to improve the nutritional value of cereal by-products, especially those with high fiber content, such as bran or DDGS, which will allow their inclusion in higher proportion in complete diets than currently in practice.

Since circular economy and sustainability principles dictate the use of a wide range of feeds, including poorly digestible feeds, the application of enzymes in feed mixtures for poultry and pigs presents a necessity. In order to enable the utilization of nutrients from alternative feeds, the supplementation of feed mixtures with multi-enzyme complexes will be a common practice. The results from studies conducted on fish suggest that the combination of enzymes with proven positive effect on fish growth performance (phytase and protease) can be used in feed mixtures from renewable resources, such as plant-based protein sources and cereal wastes in the future. Enzymes with hydrolytic activity on cellulose and hemicellulose (cellulase and xylanase) have great potential to replace antibiotics in complete diets for ruminants in the future. Accordingly, future research of prebiotics in animals should focus on immunological aspects, changes at the gut epithelium, and on quality of animal-derived products.

14.5 CONCLUSION

Improving the feed efficiency of cereals in animals can be achieved by fermenting whole plants, grains, or by-products prior to their inclusion in the diet or by adding endogenous enzymes to feed mixtures. Both of these approaches result in improved nutrient digestibility and animal performance compared to untreated feeds/feed mixtures. LAB have been used for centuries to produce fermented feeds and foods, and nowadays their use is widespread in animal nutrition for conservation of forage feeds and high moisture grain, i.e. in ensiling. The fermentation characteristics of an ensiled product will depend on a number of factors, and efforts are constantly being made to produce aerobically stable silage with low nutrient loss compared to fresh plant material. Ensiled whole-plant cereal silages are important sources of fiber and energy for ruminants, which influence the proper functioning of the rumen. Due to their health-promoting effects, feeds fermented with LAB have added value as probiotics in both ruminants and monogastric animals. In addition to the effect on nutrient digestibility and thus, improved availability, LAB fermentation has other beneficial effects, such as prevention of mycotoxin contamination and phytate degradation. However, in ruminants, the inclusion of ensiled grain in a complete diet has to be optimized with fiber to avoid the condition of sub-acute ruminal acidosis leading to negative effects on animals' health and production.

Without substantial processing of cereals and their by-products, significant improvement in animal growth performance could be achieved only by supplementing the complete diet with a combination of enzymes. Knowledge of the chemical structure and composition of NSP is essential for the development of enzyme combinations to improve the efficiency of energy recovery from feeds. ANFs, such as phytic acids and viscous components, depress digestion, allowing significant amounts of starch and/or protein to enter the large intestine, stimulating putrefactive bacteria activity and predisposing the animal to intestinal disease. Combinations of enzymes that attack only the backbone of NSP, such as xylanase and β-glucosidase, were only marginally effective in hydrolyzing hemicelullosic components present in cereal grains, due to their high level of branching and high content of bound phenolic compounds. Therefore, enzymes with high de-branching activity, such as arabinofuranosidases, should be used in combination with enzymes that attack the backbone of NSP to improve the digestibility of cereal-based feeds for poultry, especially corn-based feeds. Mannanase is another less-applied enzyme that has been shown to improve digestibility of cereal-based NSP and feed efficiency in grow-finisher pigs. In ruminant feeds, fibrolytic enzymes derived from various fungi have been successfully used to improve milk production and growth performance.

The application of exogenous enzymes has great potential to replace antibiotics in feeds for ruminants, pigs and poultry; to reduce environmental impact by enabling the utilization of by-products from various industries (milling industry, sugar industry, biofuel production, catering, etc.); to increase the nutritional value of low-value feeds and to reduce manure excretion and methane production. Increasing scientific evidence suggests that enzymes can be part of an integrated approach to contain enteric pathogens of economic importance. This value is reflected not only in reduced medication costs, but also in reduced variability in animal performance and reduced mortality by promoting gut health. The application of exogenous enzymes increases diet digestibility and alters selection pressure on the microbiota, which in turn moderates the efficiency of feed utilization by the host and immune interactions with the gut microbiome.

ACKNOWLEDGMENTS

The authors are grateful to Professor Darko Grbeša for constructive comments in preparation of the chapter. This work was supported by the Ministry of Science and Education of the Republic of Croatia, grants No. 178-1780496-0368 and 0178001, and the Ministry of Education, Science and Technological Development of the Republic of Serbia, grants No. 451-03-9/2021-14/200222 and 451-03-9/2021-14/200134.

REFERENCES

Abid, K., J. Jabri, H. Ammar, et al. 2020. Effect of treating olive cake with fibrolytic enzymes on feed intake, digestibility and performance in growing lambs. *Animal Feed Science and Technology* 261: 114405.

Adeoye, A. A., A. Jaramillo-Torres, S. W. Fox, D. L. Merrifield, S. J. Davies. 2016. Supplementation of formulated diets for tilapia (*Oreochromis niloticus*) with selected exogenous enzymes: Overall performance and effects on intestinal histology and microbiota. *Animal Feed Science and Technology* 215: 133–143.

Agle, M., A. N. Hristov, S. Zaman, C. Schneider, P. M. Ndegwa, V. K. Vaddella. 2010. Effect of dietary concentrate on rumen fermentation, digestibility, and nitrogen losses in dairy cows. *Journal of Dairy Science* 93: 4211–4222.

Alexandratos, N., J. Bruinsma. 2012. *World Agriculture Towards 2030/2050: The 2012 Revision*. Rome: FAO.

Apperson, K. D., G. Cherian. 2017. Effect of whole flax seed and carbohydrase enzymes on gastrointestinal morphology, muscle fatty acids, and production performance in broiler chickens. *Poultry Science* 96: 1228–1234.

Ayoade, D.I., Kiarie,E., Slominski, B.A., Nyachoti, C.M. 2014. Growth and physiological responses of growing pigs to wheat – corn distillers dried grains with solubles. *Journal of Animal Physiology and Animal Nutrition* 98: 569–577.

Bach Knudsen, K. E. 2014. Fiber and nonstarch polysaccharide content and variation in common crops used in broiler diets. *Poultry Science* 93: 2380–2393.

Beal, J. D., S. J. Niven, A. Campbell, P. H. Brooks. 2002. The effect of temperature on the growth and persistence of Salmonella in fermented liquid pig feed. *International Journal of Food Microbiology* 79: 99–104.

Böttger, C., K-H. Südekum. 2018. Protein value of distillers dried grains with solubles (DDGS) in animal nutrition as affected by the ethanol production process. *Animal Feed Science and Technology* 244: 11–17.

Canibe, N., B. B. Jensen. 2007. Fermented liquid feed and fermented grain to piglets-effect on gastrointestinal ecology and growth performance. *Livestock Science* 108: 198–201.

Canibe, N., B. B. Jensen. 2012. Fermented liquid feed: Microbial and nutritional aspects and impact on enteric diseases in pigs. *Animal Feed Science and Technology* 173: 17–40.

Carmona-Cabello, M., I. L. Garcia, D. Leiva-Candia, M. P. Dorado. 2018. Valorization of food waste based on its composition through the concept of biorefinery. *Current Opinion in Green and Sustainable Chemistry* 14: 67–79.

Castillo, S., D. M. Gatlin III. 2015. Dietary supplementation of exogenous carbohydrase enzymes in fish nutrition: A review. *Aquaculture* 435: 286–292.

Cavallarin, L., E. Tabacco, S. Antoniazzi, G. Borreani. 2011. Aflatoxin accumulation in whole crop maize silage as a result of aerobic exposure. *Journal of the Science of Food and Agriculture* 91: 2419–2425.

Chatzifragkou, A., O. Kosik, P. C. Prabhakumari, et al. 2015. Biorefinery strategies for upgrading Distillers' Dried Grains with Solubles (DDGS). *Process Biochemistry* 50: 2194–2207.

Cherdthong, A., C. Suntara, W. Khota. 2020. *Lactobacillus casei* TH14 and additives could modulate the quality, gas kinetics and the in vitro digestibility of ensilaged rice straw. *Journal of Animal Physiology and Animal Nutrition* 104: 1690–1703.

Cherdthong, A., C. Suntara, W. Khota, M. Wanapat. 2021. Feed utilization and rumen fermentation characteristics of Thai-indigenous beef cattle fed ensiled rice straw with *Lactobacillus casei* TH14, molasses, and cellulase enzymes. *Livestock Science* 245; 104405.

Choct, M. 2015. Feed non-starch polysaccharides for monogastric animals: Classification and function. *Animal Production Science* 55: 1360–1366.

Cozannet, P., A. Preynat, J. Noblet. 2012. Digestible energy values of feed ingredients with or without addition of enzymes complex in growing pigs. *Journal of Animal Science* 90(suppl_4): 209–211.

Decimo, M., M. Quattrini, G. Ricci, et al. 2017. Evaluation of microbial consortia and chemical changes in spontaneous maize bran fermentation. *AMB Express* 7: 205.

Der Bedrosian, M. C., K. E. Nestor, L. Kung. 2012. The effects of hybrid, maturity, and length of storage on the composition and nutritive value of corn silage. *Journal of Dairy Science* 95: 5115–5126.

Devant, M., B. Quintana, A. Aris, A. Bach. 2015. Fattening Holstein heifers by feeding high-moisture corn (whole or ground) ad libitum separately from concentrate and straw. *Journal of Animal Science* 93: 4903–4916.

Diarra, S. S., S. Anand. 2020. Impact of commercial feed dilution with copra meal or cassava leaf meal and enzyme supplementation on broiler performance. *Poultry Science* 99: 5867–5873.

Duvnjak, M., K. Kljak, D. Grbeša. 2016. Effect of hybrid, inoculant and storage time on whole plant nitrogen compounds and grain total zein content in maize silage. *Journal of Animal and Feed Sciences* 25: 174–178.

Duvnjak, M., K. Kljak, D. Grbeša. 2019. Response of common silage corn hybrids to inoculant application: Fermentation profile, carbohydrate fractions, and digestibility during ensiling. *Animal Production Science* 59: 1696–1704.

Engberg, R. M., M. Hammershøj, N. F. Johansen, M. S. Abousekken, S. Steenfeldt, B. B. Jensen. 2009. Fermented feed for laying hens: Effects on egg production, egg quality, plumage condition and composition and activity of the intestinal microflora. *British Poultry Science* 50: 228–239.

Fabiszewska, A. U., K. J. Zielińska, B. Wróbel. 2019. Trends in designing microbial silage quality by biotechnological methods using lactic acid bacteria inoculants: A minireview. *World Journal of Microbiology and Biotechnology* 35: 1–8.

FAO. 2012. *Biofuel Co-products as Livestock Feed: Opportunities and Challenges*, ed. H. P. S. Makkar. Rome: FAO.

FAO. 2022. *Food Outlook – Biannual Report on Global Food Markets*. Rome: FAO.

Filya, I. 2003. Nutritive value of whole crop wheat silage harvested at three stages of maturity. *Animal Feed Science and Technology* 103: 85–95.

Filya, I. 2004. Nutritive value and aerobic stability of whole crop maize silage harvested at four stages of maturity. *Animal Feed Science and Technology* 116: 141–150.

Filya, I., E. Socu. 2007. The effect of bacterial inoculants and a chemical preservative on the fermentation and aerobic stability of whole-crop cereal silages. *Asian-Australasian Journal of Animal Sciences* 20: 378–384.

Firkins, J. L., M. L. Eastridge, N. R. St-Pierre, S. M. Noftsger. 2001. Effects of grain variability and processing on starch utilization by lactating dairy cattle. *Journal of Animal Science* 79: E218–E238.

Giuberti, G., A. F. Masoero, L. F. Ferraretto, P. C. Hoffman, R. D. Shaver. 2014. Factors affecting starch utilization in large animal food production system: A review. *Starch-Stärke* 66: 72–90.

Giuberti, G., A. Gallo, M. Moschini, F. Masoero. 2015. New insight into the role of resistant starch in pig nutrition. *Animal Feed Science and Technology* 201: 1–13.

Godoi, L. A., B. C. Silva, F. A. S. Silva, et al. 2021. Effect of flint corn processing methods on intake, digestion sites, rumen pH, and ruminal kinetics in finishing Nellore bulls. *Animal Feed Science and Technology* 271: 114775.

Golder, H. M., H. A. Rossow, I. J. Lean. 2019. Effects of in-feed enzymes on milk production and components, reproduction, and health in dairy cows. *Journal of Dairy Science* 102: 8011–8026.

Grant, R. J., L. F. Ferraretto. 2018. Silage review: Silage feeding management: Silage characteristics and dairy cow feeding behavior. *Journal of Dairy Science* 101: 4111–4121.

Grant, R. H., D. R. Mertens. 1992. Influence of buffer pH and raw corn starch addition on in vitro fiber digestion kinetics. *Journal of Dairy Science* 75: 2762–2768.

Hackl, W., B. Pieper, R. Pieper, U. Korn, and A. Zeyner. 2010. Effects of ensiling cereal grains (barley, wheat, triticale and rye) on total and pre-caecal digestibility of proximate nutrients and amino acids in pigs. *Journal of Animal Physiology and Animal Nutrition* 94: 729–735.

Hamad, A. M., M. L. Fields. 1979. Evaluation of the protein quality and available lysine of germinated and fermented cereals. *Journal of Food Science* 44: 456–459.

Hargreaves, A., J. Hill, J. D. Leaver. 2009. Effect of stage of growth on the chemical composition, nutritive value and ensilability of whole-crop barley. *Animal Feed Science and Technology* 152: 50–61.

Heres, L., B. Engel, F. Van Knapen, J. A. Wagenaar, B. A. P. Urlings. 2003a. Effect of fermented feed on the susceptibility for Campylobacter jejuni colonization in broiler chickens with and without concurrent inoculation of Salmonella enteritidis. *International Journal of Food Microbiology* 87: 75–86.

Heres, L., B. Engel, F. Van Knapen, M. C. De Jong, J. A. Wagenaar, H. A. Urlings. 2003b. Fermented liquid feed reduces susceptibility of broilers for Salmonella enteritidis. *Poultry Science* 82: 603–611.

Holding, D. R., B. A. Larkins. 2006. The development and importance of zein protein bodies in maize endosperm. *Maydica* 51: 243–254.

Hoffman, P. C., N. M. Esser, R. D. Shaver, et al. 2011. Influence of ensiling time and inoculation on alteration of the starch-protein matrix in high-moisture corn. *Journal of Dairy Science* 94: 2465–2474.

Humer, E., Q. Zebeli. 2015. Phytate in feed ingredients and potentials for improving the utilization of phosphorus in ruminant nutrition. *Animal Feed Science and Technology* 209: 1–15.

Humer, E., W. Wetscherek, C. Schwarz, K. Schedle. 2013. Effect of maize conservation technique and phytase supplementation on total tract apparent digestibility of phosphorus, calcium, ash, dry matter, organic matter and crude protein in growing pigs. *Animal Feed Science and Technology* 185: 70–77.

Humer, E., W. Wetscherek, C. Schwarz, K. Schedle. 2014. Effects of maize conservation techniques on the apparent total tract nutrient and mineral digestibility and microbial metabolites in the faeces of growing pigs. *Animal Feed Science and Technology* 197: 176–184.

Jacobs, J. L., J. Hill, T. Jenkin. 2009. Effect of stage of growth and silage additives on whole crop cereal silage nutritive and fermentation characteristics. *Animal Production Science* 49: 595–607.

Jacobsen, H. J., T. A. Samuelsen, A. Girons, K. Kousoulaki. 2018. Different enzyme incorporation strategies in Atlantic salmon diet containing soybean meal: Effects on feed quality, fish performance, nutrient digestibility and distal intestinal morphology. *Aquaculture* 491: 302–309.

Jin, Q., L. Yang, N. Poe, H. Hunag. 2018. Integrated processing of plant-derived waste to produce value-added products based on the biorefinery concept. *Trends in Food Science & Technology* 74: 119–131.

Jørgensen, H., D. Sholly, A. Ø. Pedersen, N. Canibe, K. E. Bach Knudsen. 2010. Fermentation of cereals: Influence on digestibility of nutrients in growing pigs. *Livestock Science* 134: 56–58.

Keady, T. W. J. 2005. Ensiled maize and whole crop wheat forages for beef and dairy cattle: Effects on animal performance. In Silage Production and Utilization. Proceedings of the XIVth International Silage Conference, ed. R. S. Park, M. D. Stronge, 65–82. Wageningen: Wageningen Academic Publishers.

Kellems, R. O., D. C. Chuch. 2010. *Livestock Feeds and Feeding*. 6th ed., 58. Upper Saddle Rive: Prentice Hall.

Khan, N. A., P. Yu, M. Ali, J. W. Cone, W. H. Hendriks. 2015. Nutritive value of maize silage in relation to dairy cow performance and milk quality. *Journal of the Science of Food and Agriculture* 95: 238–252.

Kholif, A. E., M. M. Abdo, U. Y. Anele, M. M. El-Sayed, T. A. Morsy. 2017. *Saccharomyces cerevisiae* does not work synergistically with exogenous enzymes to enhance feed utilization, ruminal fermentation and lactational performance of Nubian goats. *Livestock Science* 206: 17–23.

Kiarie, E., L. F. Romero, C. M. Nyachoti. 2013. The role of added feed enzymes in promoting gut health in swine and poultry. *Nutrition Research Reviews* 26(1): 71–88.

Kljak, K., M. Duvnjak, D. Grbeša. 2019. Effect of starch properties and zein content of commercial maize hybrids on kinetics of starch digestibility in an in vitro poultry model. *Journal of the Science of Food and Agriculture* 99: 6372–6379.

Kondratovich, L. B., J. O. Sarturi, C. A. Hoffmann, M. A. Ballou, S. J. Trojan, P. R. Campanili. 2019. Effects of dietary exogenous fibrolytic enzymes on ruminal fermentation characteristics of beef steers fed high- and low-quality growing diets. *Journal of Animal Science* 97: 3089–3102.

Krause, K. M., D. K. Combs, K. A. Beauchemin. 2002. Effects of forage particle size and grain fermentability in midlactation cows. II. Ruminal pH and chewing activity. *Journal of Dairy Science* 85: 1947–1957.

Kung, L., R. D. Shaver, R. J. Grant, R. J. Schmidt. 2018. Silage review: Interpretation of chemical, microbial, and organoleptic components of silages. *Journal of Dairy Science* 101: 4020–4033.

Lardy, G. P., V. L. Anderson. 2016. *Harvesting, Storing, and Feeding High-moisture Corn*. North Dakota State University. Extension AS1484. Available at https://www.ag.ndsu.edu/publications/livestock/harvesting-storing-and-feeding-high-moisture-corn (accessed May 28, 2021).

Lestienne, I., C. Icard-Vernière, C. Mouquet, C. Picq, S. Trèche. 2005. Effects of soaking whole cereal and legume seeds on iron, zinc and phytate contents. *Food Chemistry* 89: 421–425.

Li, Q., X. Peng, E. R. Burrough, et al. 2020. Dietary soluble and insoluble fiber with or without enzymes altered the intestinal microbiota in weaned pigs challenged with enterotoxigenic *E. coli* F18. *Frontiers in Microbiology* 11: 1110.
Lizardi-Jiménez, M. A., R. Hernández-Martínez. 2017. Solid state fermentation (SSF): Diversity of applications to valorize waste and biomass. *3 Biotech* 7: 44.
Lopez, H. W., A. Ouvry, E. Bervas, C. Guy, A. Messager, C. Demigne, C. Remesy. 2000. Strains of lactic acid bacteria isolated from sour doughs degrade phytic acid and improve calcium and magnesium solubility from whole wheat flour. *Journal of Agricultural and Food Chemistry* 48: 2281–2285.
Łukasiewicz, M., D. Pietrzak, J. Niemiec, J. Mroczek, M. Michalczuk. 2012. Application of dried distillers grains with solubles (DDGS) as a replacer of soybean meal in broiler chickens feeding. *Archives Animal Breeding* 55: 496–505.
Mari, L. J., R. J. Schmidt, L. G. Nussio, C. M. Hallada, L. Kung Jr. 2009. An evaluation of the effectiveness of *Lactobacillus buchneri* 40788 to alter fermentation and improve the aerobic stability of corn silage in farm silos. *Journal of Dairy Science* 92: 1174–1176.
Martinez Tuppia, C., V. Atanasova-Penichon, S. Chéreau, N. Ferrer, G. Marchegay, J. M. Savoie, F. Richard-Forget. 2017. Yeast and bacteria from ensiled high moisture maize grains as potential mitigation agents of fumonisin B1. *Journal of the Science of Food and Agriculture* 97(8), 2443–2452.
Maas, R. M., M. C. Verdegem, T. L. Stevens, J. W. Schrama. 2020. Effect of exogenous enzymes (phytase and xylanase) supplementation on nutrient digestibility and growth performance of Nile tilapia (*Oreochromis niloticus*) fed different quality diets. *Aquaculture* 529: 735723.
Mendoza, G. D., O. Loera-Corral, F. X. Plata-Pérez, P. A. Hernández-García, M. Ramírez-Mella. 2014. Considerations on the use of exogenous fibrolytic enzymes to improve forage utilization. *The Scientific World Journal* 2014: 247437.
Mertens, D. R., J. R. Loften. 1980. The effect of starch on forage fiber digestion kinetics in vitro. *Journal of Dairy Science* 63: 1437–1446.
Missotten, J. A. M., J. Michiels, J. Degroote, S. De Smet. 2015. Fermented liquid feed for pigs: An ancient technique for the future. *Journal of Animal Science and Biotechnology* 6: 1–9.
Mo, W. Y., W. M. Choi, K. Y. Man, M. H. Wong. 2020. Food waste-based pellets for feeding grass carp (*Ctenopharyngodon idellus*): Adding baker's yeast and enzymes to enhance growth and immunity. *Science of the Total Environment* 707: 134954.
Moran, C. A., R. H. Scholten, J. M. Tricarico, P. H. Brooks, M. W. Verstegen. 2006. Fermentation of wheat: Effects of backslopping different proportions of pre-fermented wheat on the microbialand chemical composition. *Archives of Animal Nutrition* 60: 158–169.
Mul, A. J., F. G. Perry. 1994. The role of fructooligosaccharides in animal nutrition. In *Recent Advances in Animal Nutrition*, ed. P. C. Garnsworthy, D. J. A. Cole, 57–79. Nottingham: Nottingham University Press.
Musigwa, S., P. Cozannet, N. Morgan, R. A. Swick, S. B. Wu. 2021. Multi-carbohydrase effects on energy utilization depend on soluble non-starch polysaccharides-to-total non-starch polysaccharides in broiler diets. *Poultry Science* 100: 788–796.
Nadeau, E. 2007. Effects of plant species, stage of maturity and additive on the feeding value of whole crop cereal silage. *Journal of the Science of Food and Agriculture* 87: 789–801.
Nagaraja, T. G., E. C. Titgemeyer. 2007. Ruminal acidosis in beef cattle: The current microbiological and nutritional outlook. *Journal of Dairy Science* 90(Supplement 1): E17–E38.
Nietner, T., M. Pfister, B. Brakowiecka-Sassy, M. A. Glomb, C. Fauhl-Hassek. 2015. Screening for sulfate in distillers dried grains and solubles by FT–IR spectroscopy. *Journal of Agricultural and Food Chemistry* 63: 476–484.

Niven, S. J., C. Zhu, D. Columbus, J. R. Pluske, C. F. M. De Lange. 2007. Impact of controlled fermentation and steeping of high moisture corn on its nutritional value for pigs. *Livestock Science* 109: 166–169.

Nkhata, S. G., E. Ayua, E. H. Kamau, J.-B. Shingiro. 2018. Fermentation and germination improve nutritional value of cereals and legumes through activation of endogenous enzymes. *Food Science & Nutrition* 6: 2446–2458.

Nozière, P., I. Ortigues-Marty, C. Loncke, D. Sauvant. 2010. Carbohydrate quantitative digestion and absorption in ruminants: from feed starch and fibre to nutrients available for tissues. *Animal* 4: 1057–1074.

Okeke, C. A., C. N. Ezekiel, C. C. Nwangburuka, et al. 2015. Bacterial diversity and mycotoxin reduction during maize fermentation (steeping) for ogi production. *Frontiers in Microbiology* 6: 1402.

Oltjen, J. W., K. K. Bolsen. 1980. Wheat, barley, oat and corn silages for growing steers. *Journal of Animal Science* 51: 958–965.

O'Meara, F. M., G. E. Gardiner, J. V. O'Doherty, D. Clarke, W. Cummins, P. G. Lawlor. 2020. Effect of wet/dry, fresh liquid, fermented whole diet liquid, and fermented cereal liquid feeding on feed microbial quality and growth in grow-finisher pigs. *Journal of Animal Science* 98: skaa166.

Osman, M. A. 2004. Changes in sorghum enzyme inhibitors, phytic acid, tannins and *in vitro* protein digestibility occurring during Khamir (local bread) fermentation. *Food Chemistry* 88: 129–134.

Oude Elferink, S. J. W. H., F. Driehuis, J. C. Gottschal, S. F. Spoelstra. 2000. Silage fermentation processes and their manipulation. *FAO Plant Production and Protection Papers* 161: 17–30.

Pahlow, G., R. E. Muck, F. Driehuis, S. J. Oude Elferink, S. F. Spoelstra. 2003. Microbiology of ensiling. In *Silage Science and Technology*, volume 42, ed. D. R. Buxton, R. E. Muck, J. H. Harisson, 31–93. Madison: American Society of Agronomy, Inc., Crop Science Society of America, Inc., Soil Science Society of America, Inc.

Pestana, J. M., B. Puerta, H. Santos, et al. 2020. Impact of dietary incorporation of Spirulina (*Arthrospira platensis*) and exogenous enzymes on broiler performance, carcass traits, and meat quality. *Poultry Science* 99: 2519–2532.

Peyrat, J., P. Nozière, A. Le Morvan, A. Férard, P.-V. Protin, R. Baumont. 2014. Effects of ensiling maize and sample conditioning on in situ rumen degradation of dry matter, starch and fibre. *Animal Feed Science and Technology* 196: 12–21.

Pieper, R., W. Hackl, U. Korn, A. Zeyner, W. B. Souffrant, B. Pieper. 2011. Effect of ensiling triticale, barley and wheat grains at different moisture content and addition of *Lactobacillus plantarum* (DSMZ 8866 and 8862) on fermentation characteristics and nutrient digestibility in pigs. *Animal Feed Science and Technology* 164: 96–105.

Poernama, F., T. A. Wibowo, Y. G. Liu. 2021. The effect of feeding phytase alone or in combination with nonstarch polysaccharides-degrading enzymes on broiler performance, bone mineralization, and carcass traits. *Journal of Applied Poultry Research* 30: 100134.

Puntigam, R., J. Slama, D. Brugger, et al. 2021. Fermentation of whole grain sorghum (*Sorghum Bicolor* (L.) Moench) with different dry datter concentrations: Effect on the apparent total tract digestibility of energy, crude nutrients and minerals in growing pigs. *Animals* 11: 1199.

Qin, M. Z., Y.-X. Shen. 2013. Effect of application of a bacteria inoculant and wheat bran on fermentation quality of peanut vine ensiled alone or with corn stover. *Journal of Integrative Agriculture* 12: 556–560.

Radecki, S. V., M. R. Juhl, E. R. Miller. 1988. Fumaric and citric acids as feed additives in starter pig diets: Effect on performance and nutrient balance. *Journal of Animal Science* 66: 2598–2605.

Radosavljević, M., J. Pejin, S. Kocić-Tanackov, D. Mladenović, A. Djukić-Vuković, L. Mojović. 2018. Brewers' spent grain and thin stillage as raw materials in L-(+)-lactic acid fermentation. *Journal of the Institute of Brewing* 124: 23–30.

Radosavljević, M., J. Pejin, M. Pribić, et al. 2019. Utilization of brewing and malting by-products as carrier and raw materials in L-(+)-lactic acid production and feed application. *Applied Microbiology and Biotechnology* 103: 3001–3013.

Ravindran, V. 2013. Feed enzymes: The science, practice, and metabolic realities. *Journal of Applied Poultry Research* 22: 628–636.

Refat, B., D. A. Christensen, J. J. McKinnon, et al. 2018. Effect of fibrolytic enzymes on lactational performance, feeding behavior, and digestibility in high-producing dairy cows fed a barley silage–based diet. *Journal of Dairy Science* 101: 7971–7979.

Rossow, H. A., H. M. Golder, I. J. Lean. 2020. Variation in milk production, fat, protein, and lactose responses to exogenous feed enzymes in dairy cows. *Applied Animal Science* 36: 292–307.

Salami, S. A., G. Luciano, M. N. O'Grady, et al. 2019. Sustainability of feeding plant by-products: A review of the implications for ruminant meat production. *Animal Feed Science and Technology* 251: 37–55.

Salazar-López, N. J., M. Ovando-Martínez, J. A. Domínguez-Avila. 2020. Cereal/grain by-products. In *Food Wastes and By-products: Nutraceutical and Health Potential*, ed. R. Campos-Vega, B. D. Oomah, H. A.Vergara-Castaneda, 1–34. Hoboken: Wiley & Sons.

Saleh, A. A., A. A. Kirrella, S. E. Abdo, et al. 2019. Effects of dietary xylanase and arabinofuranosidase combination on the growth performance, lipid peroxidation, blood constituents, and immune response of broilers fed low-energy diets. *Animals* 9: 467.

Salim, A. A., S. Grbavčić, N. Šekuljica, et al. 2017. Production of enzymes by a newly isolated *Bacillus* sp. TMF-1 in solid state fermentation on agricultural by-products: The evaluation of substrate pretreatment methods. *Bioresource Technology* 228: 193–200.

Salovaara H. 1998. Lactic acid bacteria in cereal-based products. In *Lactic Acid Bacteria: Microbiology and Functional Aspects*, ed. S. Salminen, A. von Wright, 2nd ed, 115–137. New York: Marcel Dekker Inc.

Sauvant, D., J. M. Perez, G. Tran. 2004. Tables *of Composition and Nutritional Value of Feed Materials: Pigs, Poultry, Cattle, Sheep, Goats, Rabbits, Horses and Fish*. Wageningen: Wageningen Academic Publishers.

Scholten, R. H. J., C. M. C. den van der Peet-Schwering, L. A. den Hartog, M. Balk, J. W. Schrama, M. W. A. Verstegen. 2002. Fermented wheat in liquid diets: effects on gastrointestinal characteristics in weanling piglets. *Journal of Animal Science* 80: 1179–1186.

Selle, P. H., R. Velmurugu. 2008. Phytate-degrading enzymes in pig nutrition. *Livestock Science* 113: 99–122.

Shen, Y., R. Abeynayake, X. Sun, et al. 2019. Feed nutritional value of brewers' spent grain residue resulting from protease aided protein removal. *Journal of Animal Science and Biotechnology* 10: 1–10.

Sholly, D. M., H. Jørgensen, A. L. Sutton, B. T. Richert, K. E. Bach Knudsen. 2011. Effect of fermentation of cereals on the degradation of polysaccharides and other macronutrients in the gastrointestinal tract of growing pigs. *Journal of Animal Science* 89: 2096–2105.

Skrede, G., O. Herstad, S. Sahlstrøm, A. Holck, E. Slinde, A. Skrede. 2003. Effects of lactic acid fermentation on wheat and barley carbohydrate composition and production performance in the chicken. *Animal Feed Science and Technology* 105: 135–148.

Sugiharto, S., S. Ranjitkar. 2019. Recent advances in fermented feeds towards improved broiler chicken performance, gastrointestinal tract microecology and immune responses: A review. *Animal Nutrition* 5: 1–10.
Svihus, B., O. Herstad, C. W. Newman. 1997. Effect of high-moisture storage of barley, oats, and wheat on chemical content and nutritional value for broiler chickens. *Acta Agriculturae Scandinavica A: Animal Sciences* 47: 39–47.
Swiatkiewicz, S., A. Arczewska-Wlosek, D. Jozefiak. 2014. Feed enzymes, probiotic, or chitosan can improve the nutritional efficacy of broiler chicken diets containing a high level of distillers dried grains with solubles. *Livestock Science* 163, 110–119.
Swiatkiewicz, S., M. Swiatkiewicz, A. Arczewska-Wlosek, D. Jozefiak. 2016. Efficacy of feed enzymes in pig and poultry diets containing distillers dried grains with solubles: A review. *Journal of Animal Physiology and Animal Nutrition* 100: 15–26.
Tanasković, S. J., N. Šekuljica, J. Jovanović, et al. 2021. Upgrading of valuable food component contents and anti-nutritional factors depletion by solid-state fermentation: A way to valorize wheat bran for nutrition. *Journal of Cereal Science* 99: 103159.
Teng, P. Y., C. L. Chang, C. M. Huang, S. C. Chang, T. T. Lee. 2017. Effects of solid-state fermented wheat bran by *Bacillus amyloliquefaciens* and *Saccharomyces cerevisiae* on growth performance and intestinal microbiota in broiler chickens. *Italian Journal of Animal Science* 16: 552–562.
Tona, G. O. 2018. Current and future improvements in livestock nutrition and feed resources. In *Animal Husbandry and Nutrition*, eds B. Yücel, T. Taşkin, 147–169. Rijeka: IntechOpen.
Torbica, A., M. Belović, Lj. Popović, J. Čakarević, M. Jovičić, J. Pavličević. 2021. Comparative study of nutritional and technological quality aspects of minor cereals. *Journal of Food Science and Technology* 58: 311–322.
Torres-Pitarch, A., E. G. Manzanilla, G. E. Gardiner, J. V. O'Doherty, P. G. Lawlor. 2019. Systematic review and meta-analysis of the effect of feed enzymes on growth and nutrient digestibility in grow-finisher pigs: Effect of enzyme type and cereal source. *Animal Feed Science and Technology* 251: 153–165.
Torres-Pitarch, A., G. E. Gardiner, P. Cormican, et al. 2020. Effect of cereal fermentation and carbohydrase supplementation on growth, nutrient digestibility and intestinal microbiota in liquid-fed grow-finishing pigs. *Scientific Reports* 10: 1–17.
van Winsen, R. L., B. A. P. Urlings, L. J. A. Lipman, et al. 2001. Effect of fermented feed on the microbial population of the gastrointestinal tracts of pigs. *Applied and Environmental Microbiology* 67: 3071–3076.
Verni, M., C. G. Rizzello, R. Coda. 2019. Fermentation biotechnology applied to cereal industry by-products: Nutritional and functional insights. *Frontiers in Nutrition* 6: 42.
Vötterl, J. C., Q. Zebeli, I. Hennig-Pauka, B. U. Metzler-Zebeli. 2019. Soaking in lactic acid lowers the phytate-phosphorus content and increases the resistant starch in wheat and corn grains. *Animal Feed Science and Technology* 252: 115–125.
Ward, N. E. 2021. Debranching enzymes in corn/soybean meal-based poultry feeds: A review. *Poultry Science* 100: 765–775.
Wilkerson, V. A., B. P. Glenn, K. R. McLeod. 1997. Energy and nitrogen balance in lactating cows fed diets containing dry or high moisture corn in either rolled or ground form. *Journal of Dairy Science* 80: 2487–2496.
Wilkinson, J. 2011. Re-defining efficiency of feed use by livestock. *Animal* 5: 1014–1022.
Wilkinson, J. M., R. E. Muck. 2019. Ensiling in 2050: Some challenges and opportunities. *Grass and Forage Science* 74: 178–187.
Xu, B., Z. Li, C. Wang, J. Fu, Y. Zhang, Y. Wang, Z. Lu. 2020. Effects of fermented feed supplementation on pig growth performance: A meta-analysis. *Animal Feed Science and Technology* 259: 114315.

Zachariasova, M., Z. Dzuman, Z. Veprikova, et al. 2014. Occurrence of multiple mycotoxins in European feedingstuffs, assessment of dietary intake by farm animals. *Animal Feed Science and Technology* 193: 124–140.

Zain, M. E. 2011. Impact of mycotoxins on humans and animals. *Journal of Saudi Chemical Society* 15: 129–144.

Zebeli, Q., M. Tafaj, H. Steingass, B. Metzler, W. Drochner. 2006. Effects of physically effective fiber on digestive processes and milk fat content in early lactating dairy cows fed total mixed rations. *Journal of Dairy Science* 89: 651–668.

Zebeli, Q., J. R. Aschenbach, M. Tafaj, J. Boguhn, B. N. Ametaj, W. Drochner. 2012. Role of physically effective fiber and estimation of dietary fiber adequacy in high-producing dairy cattle. *Journal of Dairy Science* 95: 1041–1056.

Zou, J., P. Zheng, K. Zhang, X. Ding, S. Bai. 2013. Effects of exogenous enzymes and dietary energy on performance and digestive physiology of broilers. *Journal of Animal Science and Biotechnology* 4: 1–9.

Index

A

AAS, *see* Atomic absorption spectroscopy
ABC transporter-dependent pathway, 241, 242
Acetate kinase (AK), 162
Acetic acid bacteria, 251
Acid detergent fiber (ADF), 395
Acidic environments, 44, 120, 204, 236, 250
Acidic hydrolysis method, 102
Acidification, 6, 9, 11, 17, 20, 97, 147, 203, 207, 245, 275, 339, 356, 374
Acid phosphatase, 179
Acid production, 17–18, 350
Acrylamide, 9, 10, 207, 269, 352, 381
Adenosine triphosphate (ATP), 150, 218
ADF, *see* Acid detergent fiber
ADG, *see* Average daily gain
ADI, *see* Arginine deiminase
ADI pathway enzymes, 181–183
ADP-glucose phosphorylase (AGPase), 127
Aerobic phase, 394
Aflatoxin, 394
Agar plate technique, 177
Agricultural waste, 89, 101–102
Agri-food bioresources, 134
AK, *see* Acetate kinase
Aldehydes, 130, 217, 291
Alkenylresorcinols, 42
Alkylresorcinols (ARs), 42, 48, 50, 98, 156
Alpha-amylase, 149, 206
α-Amylase-trypsin-inhibitors (ATIs), 270
Alternansucrase (Asr), 172
Alzheimer's disease, 280
Amino acids, 37, 39, 154
Aminopeptidase (AP), 173, 204
AMP, *see* Antimicrobial peptide
Amylases, 128, 206–207
Amylase/trypsin inhibitors (ATI), 289
Amyloglucosidase, 206
Amylolytic enzymes, 279
Amylopectin, 127
Ando-phthaldialdehyde (OPA), 173
ANFs, *see* Anti-nutritional factors
Angiotensin converting enzyme-inhibitory peptides (ACEIP), 175
Angiotensin I-converting enzyme (ACE), 344
Anthocyanins, 75
Anti-aflatoxigenic activities, 270
Anti-bacillus activity, 348
Antibiotic reutericyclin, 184
Anticancer chemotherapy drugs, 381
Anticarcinogenic effects, 50, 129
Antidiabetic activity, 8
Antifungal activity, 18, 97, 270, 279, 286, 347–348, 378–379
Anti-inflammatory activity, 8
Antimicrobial activity, 8, 74, 278–283, 378–379
Antimicrobial peptide (AMP), 50
Anti-nutritional compounds, 10, 15, 67, 69–71, 289
Anti-nutritional factors (ANFs), 133–134, 392
Antioxidant activity, 8, 42, 74–76, 246, 281, 290–291, 345–346
Anti-staling agents, 347
Apilactobacillus kunkeei, 218
Arabinoxylanoligosaccharides (AXOS), 248
Arabinoxylans (AX), 15, 97, 99, 168, 215, 287, 392
Arginine deiminase (ADI), 131, 169, 182, 218
Arginine metabolism, 182
Aroma compounds, 9, 43, 313, 315, 328, 355
ARs, *see* Alkylresorcinols
Asiatic diet, 291
Aspartic proteases, 203
Aspergillus sp.
  A. ficuum, 209
  A. nidulans, 278
  A. niger, 7, 168, 209
  A. oryzae, 7
Atomic absorption spectroscopy (AAS), 178
ATP, *see* Adenosine triphosphate
ATP-binding cassette (ABC), 129
Autoimmune diseases, 32, 50
Avenanthramides, 42
Average daily gain (ADG), 405
AX, *see* Arabinoxylans

B

BA, *see* Biogenic amines
Bacillus sp.
  B. amyloliquefaciens, 209, 275
  B. cereus, 277, 280, 344, 348
  B. coagulans, 277
  B. licheniformis, 275, 348
  B. megaterium, 344
  B. pseudomesenteroides, 348
  B. subtilis, 73, 74, 209, 275, 277, 290, 348
  B. velezensis, 275
Backslopping, 396

435

Bacterial homoexopolysaccharides, 349
Bacteriocin-like inhibitory substances (BLIS), 278, 378
Bacteriocins, 126, 184, 278, 281–283, 315, 370, 378
Bacteroidetes, 49
Baker's yeast, 418
Bakery products
  LAB and yeasts, 271
  metabolic activity, 4
  nutritional properties, 4–12
  organic acids, 3
  prebiotics, 248–251
  sensory properties
    acid production, 17–18
    grain type and flour composition, 12–17
    sourdough bread, 18–21
    sourdough fermentation, 4–5
  starter cultures, 356
Barley, 32, 273
Bavaricin, 282
BCAAs, *see* Branched-chain amino acids
Beta-amylase, 206
β-galactosidases, 249
Bifidobacterium sp.
  B. aquikefiri, 126
  B. infantis, 178
  B. pseudocatenulatum, 178
Bioaccessibility, 8, 10, 370
Bioactive compounds, 184–185, 374–381
Bioactive food packaging, 385
Bioactive peptides, 6, 8–9, 76–77, 119, 162, 175–176, 205, 281–283, 377
Bioavailability, 8, 11, 14, 99, 208
Biocatalysts, 119, 199
Biochar, 325
Bio-conservations, 370
Bioconversions
  anti-nutritional factors, 133–134
  carbon-based compounds
    carbohydrates, 127–129
    EPS, 129–130
    lipids, 130–131
  nitrogen-based compounds, 131–132
  phenolic compounds, 132–133
Biodegradation, 323
Biodiversity, 67, 371
Biofuel production, 398–400
Biogenic amines (BA), 10
Bioinformatics, 185, 385
Biological acidification, 6
Biotechnology
  acidic environment, 120
  in vitro and *in vivo* health benefits, 121
  LAB, 121–123
  metabolic properties, 121

  microbial strains, 120
  symbiosis, 120
  unconventional probiotic starter cultures, 125–126
  yeasts, 124–125
BLAST analysis, 170
BLIS, *see* Bacteriocin-like inhibitory substances
Bran, 397
Branched-chain amino acids (BCAAs), 49
Bread composition, 312
Bread-making, 13, 14, 148–149, 246, 267–275, 312, 316–319, 325–330
Brewer's spent grain (BSG), 93, 96, 100, 397
Brewing industries, 93–96
Brewing products, 383
Broa sourdough, 277
BSG, *see* Brewer's spent grain
Burger buns, 355
B vitamins, 50, 232–240

C

Calcium alginate, 321, 328
Candida sp.
  C. albicans, 290
  C. glabrata, 124
  C. humilis, 124, 272, 313, 338, 381
  C. krusei, 150
  C. milleri, 150, 313
  C. parapsilosis, 280
  C. pelliculosa, 313
  C. pulcherrima, 280
  C. tropicalis, 179
Capsular polysaccharides (CPS), 240
Carbamate kinase (CK), 182
Carbohydrate metabolism
  coenzymes, 51
  enzymes, 162–169
  EPS producer enzymes, 169–173
  flours' composition, 15
  heterofermentative lactobacilli, 148, 150
  pentosan swelling, 18
  phenolic compounds, 346
Carbohydrates, 14, 67–69, 127–129, 150–151, 162, 371, 373
Carbonaceous materials, 325
Carbon-based compounds
  carbohydrates, 127–129
  EPS, 129–130
Carbon isotope labeling, 170
Carotenoids, 48, 71
CDM, *see* Chemically defined medium
Cd–ninhydrin method, 174
Celiac disease, 32, 246–247, 345, 353, 376
Cell aggregation, 321–322

… Index 437

Cell-envelope-associated serine proteinase (CEP), 154
Cell immobilization
  advantages and disadvantages, 322–323
  aggregation/flocculation, 321–322
  bread making, 325–330
  catalytic activity, 320
  containment behind barriers, 322
  entrapment, 321
  microbial cells, 320
  support types, 323–325
  surface attachment, 320–321
  techniques, 320
Cereal agricultural systems, 88
Cereal by-products, 88, 397; *see also* Pseudocereal by-products
  bioactive compounds, 94–95
  extraction, 98–102
  harvest, 89
  malting and brewing industries, 93–96
  milling, 89–93
  utilization, sourdough fermentation, 97–98
Cereals
  antioxidant activity, 290–291
  by-products, *see* Cereal by-products
  consumption, 46–51
  endogenous enzymes, 6, 149
  flours, 13, 15, 271–275
  food consumption, 87
  grain and chemical composition
    applications, 33
    cell layers, 33
    dietary fiber, 37, 38
    lipids, 40
    minerals and vitamins, 40–42
    phytochemicals, 40, 43
    proteins and amino acids, 37, 39
    total antioxidant activity, 42
  lipid oxidation, 32
  nutrients, 52
  production in Europe, 32
  products, 382
  sourdough potential, 43–46
  straws, 89
  valorization, 286
  wheat, 31
Chemically defined medium (CDM), 124
Chia, 64, 71
Chickpea, 64
Chinese steamed bread (CSB), 272
Chronic diseases, 292
Cicer arietinum, *see* Chickpea
CK, *see* Carbamate kinase
Clostridium sp.
  C. sporogenes, 283
  C. tyrobutyricum, 283

Cobalamin, 239–240
Companilactobacillus sp.
  C. alimentarius, 175, 314, 339
  C. farciminis, 181, 203
  C. kimchii, 338
  C. paralimentarius, 150–151, 338
Corn, 395
  bran, 397
  dry grind biorefinery, 399
  wet milling biorefinery, 399
Corn-soybean based fish feed mixtures, 418
CP, *see* Crude protein
CPS, *see* Capsular polysaccharides
Crohn's disease, 97
Cronobacter sakazakii, 290
Crude protein (CP), 393
CSB, *see* Chinese steamed bread
^{14}C-sucrose radioisotope method, 172
Ctenopharyngodon idellus, 418
Cyclic peptides, 278

D

Dairy products, 383–384
DBE, *see* Debranching enzymes
DDGS, *see* Dried distillers' grains with solubles
Debranching enzymes (DBE), 127
Dekkera bruxellensis, 126
Dextran-producing strains, 172
Dextrans, 243–247, 284
Dextransucrases, 171, 244, 249
DFE, *see* Dietary folate equivalent
D-glucopyranosyl residues, 249
DGS, *see* Distillers' grain with solubles
DIAAS, *see* Digestible Indispensable Amino Acid Score
Dietary fibers, 7–8, 37, 38, 52, 68, 154, 287, 374–376
Dietary folate equivalent (DFE), 234
Dietary products, 50
Digestible Indispensable Amino Acid Score (DIAAS), 76
2,6–Dimethoxybenzoquinone (2,6-DMBQ), 381
Dipeptidase activity, 174
Disproportionating enzymes (DPE), 127
Distillation, 398–400
Distillers' grain with solubles (DGS), 397
DM, *see* Dry matter
DNA-based methods, 122
DNA sequencing techniques, 385
Dough composition, 13
Dried distillers' grains with solubles (DDGS), 397
Dry grind fermentation, 398, 399
Dry matter (DM), 392
Dry milling, 90, 397
Dry pea, 64

E

Edible coating, 385
EFSA, see European Food Safety Authority
Ehrlich pathway, 132
Electromicroscopic techniques, 240
Ellagic/gallic acids, 290
Embden-Meyerhof-Parnas (EMP), 128, 150, 339
Endomycopsis fibuliger, 313
Endopeptidases (EP), 173, 203
Endoxylanases, see Xylanases
Energy metabolism, 47
Ensiling, 393–394
Enterococcus sp.
  E. durans, 121, 207
  E. faecalis, 280, 290, 344
  E. mundtii, 344
  E. pseudoavium, 204, 338
Environmental pollution, 101
Enzymatic activities, 341–343
Enzyme-linked immunosorbent assay (ELISA), 173
Enzymes
  applications, food industries, 201
  bioactive compounds, 184–185
  biological catalyzers, 199
  bread making, 148–149
  carbohydrate metabolism, 162–173
  carbohydrates, 150–151
  cereal grains and flours, 149
  characterization, 163–167
  classes, 148
  EPS, 169–173
  fibers, 154–157
  functional foods, 200
  functional metabolites, 315–320
  glycemic index, 148
  high industrial interest, 316–320
  industrial applications, 201
  manufacturers, 200
  metabolisms, 202
  phenolic metabolism, 176–177
  phytase, 177–180
  production, 200, 202–218
  protein metabolism, 173–176
  proteolytic, 151–154
  shelf life and healthy attributes, 218
  sources, bakery industry, 200
  sourdough fermentation, 147
  sourdough microbiota, 149–150
  spectrophotometry and chromatography, 185
  starter cultures, 157
  strain-dependent, 157
  stress metabolism, 180–184
Enzymology
  anti-nutritional factors, 120

baked goods, 118
biotechnological process, 117, 120
circular economy, 118
fermentation processes, 118
microbial enzymes, 127–134
microbiota, 119
microorganisms, 120–126
probiotic microorganisms, 118
raw materials, 119
EP, see Endopeptidases
EPRS, see European Parliamentary Research Service
EPS, see Exopolysaccharides
ER, see Estrogen receptor
Escherichia coli, 209, 280, 281, 285, 290, 384
Esterase, 216
Estrogen receptor (ER), 50
Ethanol production, 87
European Food Information Council, 237
European Food Safety Authority (EFSA), 293
European Parliamentary Research Service (EPRS), 31
Eurostat, 31
Exogenous enzymes, 424
Exogenous phytase, 404
Exopeptidases, 203
Exopolysaccharides (EPS), 8, 12, 77–79, 97, 126, 129–130, 154–155, 161, 169–173, 216, 232, 240–242, 269, 270, 283–285, 341, 348–350, 379
Extracellular enzymes, 199
Ex-vivo antioxidant activity, 77

F

FAAs, see Free amino acids
Faba bean, 64
FAO, see Food and Agriculture Organization
Fatty acid (FA), 156
Favism, 376
Feed efficiency, 391
Feed enzymes
  monogastric animals and fish, 408–418
  ruminant feed mixtures, 418–422
  SSF, 407
Feed-out phase, 394
Feed processing, 391
Fermentable oligo-, di-, monosaccharides and polyols (FODMAP), 32, 69, 97, 128, 217–218, 270, 312, 345, 374
Fermentation, lactic acid bacteria
  biofuel production, 398–400
  cereal by-products, 397
  distillation, 398–400
  DM, 392–393
  ensiling, 393–394

Index

grain silage, 395–396
liquid feed, 396–397
plant by-products, 397
whole-plant silages, 394–395
Fermentation quotient (FQ), 280
Fermented feeds, animal nutrition
  advances, 423
  cattle, 405–407
  future trends, 422–423
  pigs and poultry, 400, 404, 405
  ruminant and monogastric animal, 401–403
Fermented grain, 382, 395, 400
Ferulic acid (FA), 52, 176, 216, 270, 290
Fiber degrading enzymes
  baking process, 154
  chemical and physical properties, 154
  EPS, 154–155
  hydration, 154
  lipases, 156
  lipoxygenase, 155–156
  phenolic compounds, 156–157
  resistant starches, 154
Fiber-rich foods, 50
Fibrolytic enzymes, 422
Firmicutes, 49
Flavin mononucleotide (FMN), 238
Flippase–polymerase complex, 242
Flocculation, 321–322
Flour composition, 12–17
FODMAP, *see* Fermentable oligo-, di-, monosaccharides and polyols
Folate, 234–237
Folic acid, 50
Food and Agriculture Organization (FAO), 31, 87, 285, 315
Food and Drug Administration (FDA), 316
Food-grade strategies, 234
Food industry, 292–293, 356
Food production, 87
FOS, *see* Fructooligosaccharides
Fourier-transform infrared spectroscopy (FTIR), 122, 125
Free amino acids (FAAs), 204, 272, 377
Frozen dough, 327
Fructansucrases, 129, 216–217, 284
Fructilactobacillus sp.
  F. fructivorans, 203
  F. lindneri, 203
  F. sanfranciscensis, 149, 150, 162, 168, 170, 175, 176, 179, 203, 205, 233, 249, 250, 268, 272, 275, 314, 338, 339
Fructooligosaccharides (FOS), 248, 249, 287
Fructose, 129, 162, 272, 345
Fructose intolerance, 51
Fructosyltransferases (Ftfs), 129, 169, 249

FTIR, *see* Fourier-transform infrared spectroscopy
Functional cereal products, 287–289
Functional foods, 102, 385
Fungistatic activity, 275
Furfurilactobacillus rossiae, 45, 168, 239, 338
Fusarium poae, 278

G

GABA, *see* Gamma-aminobutyric acid
GAD, *see* Glutamic acid decarboxylase
Galactooligosaccharides (GOS), 248, 249
Gallate decarboxylase, 215
Gallic acid, 291
Gamma-aminobutyric acid (GABA), 72–74, 132, 161, 180, 270, 281, 343–344, 380, 384
Gastrointestinal proteases, 282
Gastrointestinal system, 6
Gastrointestinal tract (GIT), 248, 284
Gelatinizes, 39
Gel permeation chromatography (GPC), 170
Generally recognized as safe (GRAS), 11, 248, 284, 316, 384
Genetic diversity, 202
Genomic methodologies, 183
German Nutrition Society, 52
German Research Institute for Food Chemistry, 39
GIP, *see* Glucose-dependent insulin trophic polypeptide
GIT, *see* Gastrointestinal tract
Glucagon-like peptide (GLP)-1, 6
Glucansucrases, 129, 216–217
Glucoamylase, 206
Gluconobacter oxydans, 339
Glucose-dependent insulin trophic polypeptide (GIP), 6
Glucose oxidase, 199, 329
Glucosyltransferase (GTF), 129, 173
Glutamate metabolism, 162
Glutamic acid decarboxylase (GAD), 73, 131, 180–181
Glutaminase, 181–183
Glutamine, 6, 76, 131, 181–183
Glutathione reductase, 183–184
Gluten, 246–247, 270
Gluten degradation, 6–7, 344–345
Gluten-free bread, 353–354
Gluten-free (GF) products, 6, 7
Gluten-free pseudocereals, 134
Gluten intolerance, 32, 345
Gluten proteins, 153
Gluten sensitivity, 376
Glycemic index (GI), 5, 67, 148, 246, 312, 346–347

Glycosidase, 215–216, 288
Glycoside hydrolases, see Glycosidase
Glycosyl hydrolases, 8
Glycosyltransferases (GTF), 154, 169, 241
GMP, see Guanosine monophosphate
GOS, see Galactooligosaccharides
GPC, see Gel permeation chromatography
G protein-coupled receptor, 48
Grain milling, 90
Grain silage, 395–396
Grain type, 12–17
Granules, 68
GRAS, see Generally recognized as safe
Grass carp, 418
Green chemistry, 102
GTF, see Glycosyltransferases
Guanosine monophosphate (GMP), 377
Gut microbiome, 47

H

Hansenula sp.
   H. anomala, 150, 313
   H. polymorpha, 209
Health-promoting effects, 126
Helicobacter pylori, 285
Hemagglutinins, 70
Hemolytic anemia disease, 376
HePS, see Heteropolysaccharides
Heteroexopolysaccharides, 12
Heterofermentative lactobacilli, 6, 169
Heterofermentative metabolism, 19, 20, 43
Heteropolysaccharides (HePS), 78, 155, 240
HFA, see Hydroxy fatty acids
High-fat diet, 48
High moisture corn (HMC), 395, 406
High-performance anion exchange
   chromatography with pulsed
   amperometric detection (HPAEC-PAD), 169
High-performance liquid chromatography
   (HPLC)-refractive index
   system, 168
High-value compounds
   cereal by-products, 88–102
   challenges, 102–105
   extraction, 88
   innovative technologies, 88
   principles, 87
   pseudocereal by-products, 89–96
HMC, see High moisture corn
HMF, see Hydroxymethylfurfural
Homoexopolysaccharides, 12
Homofermentative lactobacilli, 151
Homofermentative metabolism, 19, 20, 43
Homopolysaccharides (HoPS), 78, 155, 240

HPAEC-PAD, see High-performance anion
   exchange chromatography with pulsed
   amperometric detection
Hulls, 397
Human intestinal health, 292
Human nutrition, 294
Hydrogen peroxide, 329
Hydroxybenzoic acids, 133, 176, 281
Hydroxycinnamic acids, 133, 176, 281, 291
Hydroxy fatty acids (HFA), 217
Hydroxymethylfurfural (HMF), 10
Hyperglycemia, 346
Hypotensive compounds, 344

I

Immobilization systems, 312
IMO, see Isomaltooligosaccharides
Inactivated/dead/non-viable microbial cells of
   probiotics, see Paraprobiotics
Indole-3-propionic acid (IPA), 280
Inductively coupled plasma, 178
Industrial enzymes, 199, 200, 202
Inflammatory bowel syndrome (IBS), 345
Inorganic carriers, 325
In silico analyses, 178
In situ fortification, 233
International Association for Cereal Science and
   Technology (ICC), 173
International Scientific Association for Probiotics
   and Prebiotics (ISAPP), 248
Intestinal mucosa, 46, 246
Intracellular enzymes, 199
Inulin, 248
Inulosucrase (IslA), 171
In vitro starch digestibility (IVSD), 5
IPA, see Indole-3-propionic acid
Irritable bowel syndrome (IBS), 51, 69, 97,
   217, 376
ISAPP, see International Scientific Association
   for Probiotics and Prebiotics
Isoamylases (ISA), 127
Isoflavones, 126
Isomaltooligosaccharides (IMO), 248, 285
Isomaltulose, 248
Issatchenkia orientalis, 124
IVSD, see *In vitro* starch digestibility

J

Jiaozi, 272

K

Kazachstania sp.
   K. bulderi, 176, 313

K. *exigua*, 124, 313, 338, 381
K. *humilis*, 176, 250, 313
K. *servazzii*, 383
K. *unispora*, 398
Kidney bean, 64
Klebsiella sp., 209
K. *oxytoca*, 280
Kluyveromyces sp.
  Kl. *aestuarii*, 339
  Kl. *lactis*, 339
  Kl. *marxianus*, 19, 210, 339
Krebs cycle, 73

L

LAB, *see* Lactic acid bacteria
Lactic acid (LA), 5, 279, 280
Lactic acid bacteria (LAB)
  acrylamide, 9
  antimicrobial substances, 315
  bakery products, 147, 271
  baking properties, 267
  B-group vitamins, 233–240
  carbohydrate metabolism, 162
  carbon flow, 279
  cheese whey, 330
  cobalamin production, 239–240
  dextrans, 243–247
  enzymatic equipment, 16
  EPS, 240–242, 283–285, 350
  flours, air, and water, 45
  folate production, 234–237
  food ecosystem, 66
  glycolytic activity, 97
  metabolic products, 123, 232
  metabolism pathways, 372
  organic acids, 3
  organoleptic properties, 233
  phenolic compounds, 8
  phytase-active, 342–343
  prebiotics, bakery products, 248–251
  prokaryotes, 251
  raffinose, 10
  riboflavin production, 237–238
  sourdoughs, 269
  wheat, rye and maize flours, 273
  yeasts, 233, 312, 370
Lacticaseibacillus sp.
  L. *casei*, 45, 155, 178, 272, 328, 338
  L. *pantheris*, 339
  L. *paracasei*, 329, 338
  L. *rhamnosus*, 155
Lactilactubacillus sakei, 314
Lactiol, 248
Lactiplantibacillus sp.
  L. *paracasei*, 76

L. *paraplantarum*, 338
L. *pentoses*, 339
L. *pentosus*, 314
L. *plantarum*, 45, 71–73, 75, 78, 149, 151, 155, 173, 176, 180, 236, 268, 271, 272, 314, 338, 339, 393
L. *plantarum* subsp. *plantarum*, 203
L. *xiangfangensis*, 176
Lactobacillus sp.
  L. *acidifarinae*, 121
  L. *acidophilus*, 19, 203, 250, 393
  L. *alimentarius*, 122
  L. *amylovorus*, 378
  L. *brevis*, 20, 381, 382
  L. *bulgaricus*, 329
  L. *casei*, 422
  L. *crispatus*, 168, 217
  L. *crustorum*, 121
  L. *delbrueckii*, 233
  L. *delbrueckii* subsp. *delbrueckii*, 203–204
  L. *equigenerosi*, 314
  L. *fermentum*, 15, 169, 371, 381, 382
  L. *frumenti*, 15, 121
  L. *gallinarum*, 122, 338
  L. *garviae*, 121
  L. *graminis*, 122
  L. *hammesii*, 121, 377
  L. *helveticus*, 16, 155, 379
  L. *kefiranofaciens*, 122
  L. *kefiri*, 122
  L. *lactis*, 380
  L. *mindensis*, 121
  L. *musae*, 314
  L. *namurensis*, 121
  L. *nantensis*, 121
  L. *nodensis*, 121
  L. *panis*, 382
  L. *paracasei*, 180
  L. *paralimentarius*, 272, 314
  L. *pentosus*, 122
  L. *plantarum*, 16, 20, 131, 371, 376, 378–382
  L. *pontis*, 122
  L. *reuteri*, 15, 118
  L. *rhamnosus*, 374, 379
  L. *rossiae*, 150
  L. *sakei*, 19, 122, 249
  L. *sanfranciscensis*, 7, 15, 20, 73, 125, 131, 371, 381, 382
  L. *secaliphilus*, 121
  L. *siliginis*, 121
  L. *songhuajiangensis*, 121
  L. *uvarum*, 338
  L. *vaccinostercus*, 122, 314
  L. *vaginalis*, 122
Lactococcus sp.
  L. *fermentum*, 238, 239

L. garvieae, 207
L. lactis, 131, 238, 276, 339, 384, 385
L. mesenteroides, 238, 339
L. plantarum, 238, 239
L. raffinolactis, 339
Lactulose, 248
Latilactobacillus sp.
   L. curvatus, 176, 204, 272, 314, 338
   L. sakei, 176, 282, 339
Lectins, 70
Legumes, *see also* Pseudocereals
   biotechnological process, 64
   gluten-free pseudocereals, 134
   metabolites and functional properties, 67–79
   PHA, 381
   pulses, 64
Lens culinaris, *see* Lentil
Lentil, 64
Lentilactobacillus sp.
   L. buchneri, 393
   L. farraginis, 338
   L. hilgardii, 126, 175
Leptin, 46, 47, 49
Leuconostoc sp.
   Leuc. carnosum, 241
   Leuc. citreum, 121, 274, 276, 339
   Leuc. holzapfelii, 122
   Leuc. mesenteroides, 121, 271, 272, 274, 314, 338
   Leuc. monocytogenes, 384
   Leuc. pseudomesenteroides, 338, 339
Levilactobacillus sp.
   L. brevis, 72, 73, 175, 180, 203, 240, 268, 271, 314, 338, 339, 398
   L. freudenreichii, 72, 240
   L. hammesii, 176, 180
   L. senmaizukei, 180
   L. spicheri, 338
Lexicon, 21
L-glutamate, 73
L-glutamic acid, 380
Lignans, 37, 51
Lignocellulosic biomass, 88, 89
Lima bean, 64
Limosilactobacillus sp.
   L. fermentum, 150, 174, 203, 268, 271, 314, 338
   L. pontis, 150, 205, 314, 338
   L. reuteri, 150, 169, 233, 314
   L. fermentum, 236
Linear glucan chains, 127
Lipases, 156, 199
Lipidomics, 130
Lipids, 40, 130–131, 156, 377
Lipoxygenase, 149, 155–156, 217
Liquid feed, 396–397
Liquid fermentation, 400, 404
Liquid state fermentation (LSF), 73
Liquorilactobacillus nagelii, 126

Listeria sp.
   L. innocua, 283
   L. monocytogenes, 74, 130, 281–283
Livestock feeds, 394
Loigolactobacillus coryniformis, 180, 239, 338
LSF, *see* Liquid state fermentation
Lunasin production, 380

M

Maceration, 88
Macronutrients, 287, 346, 374, 375
MAE, *see* Microwave-assisted extraction
Maillard reaction products, 9–10
Maize flour, 273
MALDI-TOF-MS, *see* Matrix assisted laser desorption/ionization – time of flight–mass spectrometry
Malting, 93–96, 397
Maltose, 250, 272
Malt sprouts, 93, 96
Mannanase supplementation, 417
Mannitol, 248, 250, 251
Mannitol dehydrogenase (MDH), 169
Masa Agria, 273
Matrix assisted laser desorption/ionization – time of flight–mass spectrometry (MALDI-TOF-MS), 125, 184
MDH, *see* Mannitol dehydrogenase
Meat products, 384–385
Metabolic disorders, 49
Metabolic engineering, 233
Metabolomics, 120
Metagenetics, 183
Metagenomics, 49, 120, 183, 236
Metal ions, 16
Meta-metabolomics, 183
Metaproteomics, 120, 183
Metatranscriptomics, 183
2-Methoxybenzoquinone (2-MBQ), 380–381
Meyerozyma guilliermondii, 125
MIC, *see* Minimum inhibitory concentration
Microbial biodiversity, 234, 370
Microbial diversity, 268
Microbial fermentation, 248, 397
Microbial metabolites, 4
   antimicrobial activity, 278–283
   bacteriocins and BLIS, 281–283
   LAB and EPS production, 283–285
   low-molecular-weight, 278–281
   trace elements, 278
   volatile organic compounds, 278
Microbial phytases, 209
Microbial production, 199
Microbial starter cultures, 270
Microbiota
   bioprocesses, 119

cereal flours and bread making, 271–275
ingredients, 268
sourdough fermentation, 275–278
sourdough metabolism, 149–150
wheat flour, 13
Microcosmos, 162
Micro-ecosystem, 121
Microencapsulation, 322
Microflora, 338, 341
  classification, 373
Micronutrients, 232, 346, 374, 375
Microorganisms, 67, 119, 120–126, 134, 147, 161, 174, 180, 199, 202, 219
  anti-nutritional factors, 370
  bread making, 267–271, 317–319
  endogenous factors, 338
  exogenous factors, 338
  food applications, 352–356
  food products, 381–385
  immobilization, 312
  importance, 341
  isolated sourdough, 313–314, 340
  metabolites, 285–292, 352
  microflora, 338
  nutrition and health, 341–347
  probiotics, 315
  rye, 276
  sensorial properties, 349–351
  shelf life of products, 347–349
  sourdough bacteria, 370–374
Microwave-assisted extraction (MAE), 88, 96, 100–104
Milk fermentation, 384
Milk kefir grains, 126
Millet, 273, 274
Milling by-products, 89–93
Milling process, 14
Mineral availability, 342–343
Mineral bioavailability, 16, 377–378
Minerals, 40, 41
Minimum inhibitory concentration (MIC), 280
Molecular exclusion chromatography, 179
Monogastric animals, 395
Monosaccharides, 12
Monounsaturated fatty acids, 130
Mucilage, 45
Mycobacterium tuberculosis, 280
Mycotoxins, 269, 394, 398
  elimination, 379

N

NADP, *see* Nicotinamide adenine dinucleotide phosphate
Nano-Liquid chromatography/electrospray ionization mass spectrometry (nano-LC/ESI-MS), 176

Nanomaterials, 325
National Center for Biotechnology Information (NCBI), 183
Natural fermentation (NF), 73
NCBI, *see* National Center for Biotechnology Information
NCDs, *see* Non-communicable diseases
NCGS, *see* Non-celiac gluten sensitivity
NCWS, *see* Non-celiac wheat sensitivity
Nelson-Somogyi method, 172
NF, *see* Natural fermentation
N-glutaryl l-phenyl-alanine-p-NA, 174
Nicotinamide adenine dinucleotide phosphate (NADP), 184–185
Nile tilapia, 418
Nitrogen-based compounds, 131–132
Nitrogen metabolism, 371
Non-celiac gluten sensitivity (NCGS), 376
Non-celiac wheat sensitivity (NCWS), 97, 270, 289, 345
Non-cereal grains, 79
Non-communicable diseases (NCDs), 46
Nonstarch polysaccharides (NSP), 392
NSP-degrading enzymes, 417
NSP-hydrolyzing enzymes, 418
N-succinyl l-phenyl-alanine-p-NA, 173
Nutraceuticals, 102, 385

O

Oat, 273, 274
Obesity, 46
Oenococcus aquikefiri, 126
Oligosaccharides, 215, 248, 249
Olive cake-based lamb diets, 422
OM, *see* Organic matter
Omega-3 fatty acids, 64
Oreochromis niloticus, 418
Organic chemistry, 316
Organic matrices, 325
Organic matter (OM), 394
Organic solvents, 102
Ornithine transcarbamoylase (OTC), 182
Oxalates, 70
Oxyntomodulin, 6

P

Pancreatic polypeptides, 6
Paraprobiotics, 118
Pasta, 354–355
Paucilactobacillus vaccinostercus, 339
PCR, *see* Polymerase chain reaction
PCs, *see* Phenolic compounds
PDO, *see* Protected designation of origin
PE, *see* Phytoestrogens
Pediococcus, 314

P. acidilactici, 121, 204, 327, 338
P. damnosus, 284
P. pentosaceous, 78
P. pentosaceus, 76, 121, 204, 271, 276, 329, 338
peNDF, *see* Physically effective fiber
Penicilium funiculosum, 278
Pentosanases, 210
Pentosans, 44
Pentose phosphate pathway (PPP), 128
Peptidases, 7, 151, 173–175
Peptide transport systems, 154
Peptidomics, 157
pH, 17–18, 171
PHA, *see* Phytohemagglutinin
Phaseolus lunatus, *see* Lima bean
Phaseolus vulgaris, *see* Kidney bean
Phenolic acids, 42, 156, 176, 216
Phenolic compounds (PCs), 119, 132–133, 156–157, 162, 176, 290–291
Phenolic metabolism, 176–177
Phenylmethylsulfonyl fluoride, 179
Phosphoenolpyruvate phosphotransferase system (PTS), 287
Phosphorylases (PHO), 127
Phosphotransferase system (PTS), 172
Physically effective fiber (peNDF), 406
Phytase, 177–180, 207–210
Phytase-active LAB, 342–343
Phytase activity, 11–12
Phytases, 208
Phytate acid activities, 155
Phytic acid, 10, 40, 50, 70, 98, 155, 161, 208, 218, 377, 392
Phytochemicals, 8, 40, 42, 48, 74–76
Phytoestrogens (PE), 51, 126, 291–292
Phytohemagglutinin (PHA), 381
Pichia sp.
  P. anomala, 124
  P. fermentans, 383
  P. kudriavzevii, 124, 125, 179, 313, 338
  P. norvegensis, 150
Pisum sativum, *see* Dry pea
Pizza, 355–356
Plantaricin, 282
Plant by-products, 397
p-nitroaniline (p-NA), 173
p-nitrophenyl-β-D-xylopyranoside (pNPX), 169
p-nitrophenyl phosphate (p-NPP), 179
Polyethylenimine, 322
Polymerase chain reaction (PCR), 125, 282
Polyols, 8, 248
Polyphenols, 74, 290
Polysaccharides, 78
Polyunsaturated fatty acids, 130
Postbiotics, 118
PPP, *see* Pentose phosphate pathway

Probiotic microorganisms, 118
Probiotics, 293, 315, 346
Progesterone receptor (PR), 50
Proline, 6
Propionate, 12
Propionibacterium sp.
  P. acidipropionici, 283
  P. freudenreichii, 72, 98, 246, 349
Protease enzyme, 206
Proteases, 149, 203–206
Protected designation of origin (PDO), 272, 314
Proteinases, 151, 173–175, 199
Protein digestibility, 6, 76–77, 246
Protein Digestibility Corrected Amino Acid Score (PDCAAS), 76
Protein metabolism
  ACE inhibitor activity, 175
  bioactive peptide-forming activities, 175–176
  proteinase and peptidases, 173–175
Proteins, 15, 37, 39, 162, 376–377
Proteolysis, 6, 203
Proteolytic activity, 205
Proteolytic enzymes, 151–154, 394
Proteomics, 157
Providencia stuartii, 280
Pseudocereal by-products, 89–96; *see also* Cereal by-products
Pseudocereals, 13, 14, 274, 286; *see also* Legumes
  agronomic conditions, 64
  dicotyledonous species, 63
  dietary diversity, 63
  fermentation, 66–67
  metabolites, 65
  products, 382
Pseudogenes, 185
Pseudomonas aeruginosa, 290
PTS, *see* Phosphotransferase system
Pullulanases (PUL), 127
Pyrano-3-deoxyanthocyanidins, 379

Q

Qualified presumption of safety (QPS), 270
Quinoa seeds, 71

R

Raffinose oligosaccharides (RFOs), 248
Rapid digestible starch (RDS), 5
RDA, *see* Recommended dietary allowance
RDS, *see* Rapid digestible starch
Real time-polymerase chain reaction (RT-qPCR), 169
Recommended dietary allowance (RDA), 234
Redox balance, 150
Resistant starch (RS), 5, 287, 404

Index

Response surface methodology, 101
Reverse phase-HPLC, 176
RFOs, *see* Raffinose oligosaccharides
Rhodotorula sp.
   R. gracilis, 209
   R. mucilaginosa, 280
Riboflavin, 237–238
Rice, 273
Rice grains, 89
Ropy phenotype, 240
Rovabio Excel AP commercial enzyme, 417
RS, *see* Resistant starch
RT-qPCR, *see* Real time-polymerase chain reaction
Rumen acidosis, 407
Rye dough, 43, 45

S

Saccharomyces sp.
   S. cerevisiae, 20, 124, 150, 176, 208, 217, 238, 249, 269, 272, 313, 338, 381, 383, 418
   S. exiguus, 124, 150
Salmonella sp.
   S. enterica, 280
   S. enteritidis, 130, 290
Salmonellosis, 405
Salvia hispanica L., *see* Chia
Saponins, 70
Satiety perception, 6
Saturated fatty acids, 40
Savory/nonvolatile compounds, 19
SCFAs, *see* Short-chain fatty acids
Schleiferilactobacillus perolens, 338
Schwanniomyces sp.
   S. castellii, 209
   S. occidentalis, 209
SDS, *see* Slow digestible starch
SDS-PAGE, *see* Sodium dodecyl sulfate polyacrylamide gel electrophoresis
Secale cereale, 272
Secondary metabolites, 8, 394
Serratia marcescens, 290
Serum leptin, 48
Short-chain fatty acids (SCFAs), 47, 119, 240, 248, 280
Slow digestible starch (SDS), 5
Small-chain bioactive polypeptides, 377
Sodium dodecyl sulfate polyacrylamide gel electrophoresis (SDS-PAGE), 168
Solid-state fermentation (SSF), 73, 407
Soluble fiber content, 46
Sorghum, 273, 274, 354, 395
Sourdough bakery products, 11
Sourdough bread, 18–21
Soxhlet extraction, 88
Spectrophotometer, 168

Spent brewer's yeast (SBY), 93, 96
Spontaneous fermentation, 398
SSF, *see* Solid-state fermentation
Staphylococcus aureus, 280, 281, 283, 290
Starch branching enzymes (SBE), 127
Starch digestibility, 5–6, 396, 400
Starch–protein matrix, 5
Starch synthases (SS), 127
Starter culture, 356
Steam distillation, 88
Steamed bread/bun, 355
Streptomyces sp.
   S. nataliensis, 286
   S. werraensis, 176
Stress metabolism
   GAD, 180–181
   glutaminase and ADI pathway enzymes, 181–183
   glutathione reductase, 183–184
   sourdough-starter strains, 180
Sucrase proteins, 242
Sugar-alcohols, 355
Sustainability, 88, 102
Sustainable agriculture, 87
Synthase-dependent pathway, 242

T

Tannase, 215
TCA, *see* Tricarboxylic acid
Thin layer chromatography (TLC), 171
Torulaspora delbrueckii, 124, 313, 338
Torulopsis sp.
   T. holmii, 313
   T. stellate, 313
Total energy, 8
Total titratable acidity (TTA), 280
Total-tract digestibility, 404
Trace elements, 51
Tricarboxylic acid (TCA), 150
Trichoderma reesei, 422
Trinitrobenzenesulfonic acid method, 173
Tripeptidase activity, 174
Triticum sp.
   T. aestivum, 272
   T. durum, 32, 272
   T. vulgare, 32
TTA, *see* Total titratable acidity
Type-2 diabetes, 68, 280

U

Ultra-performance liquid chromatography–MS, 177
Ultrasound-assisted extraction (UAE), 88, 96, 98–100
Ultrasound-microwave-assisted extraction (UMAE), 100, 105

Undesirable compounds, 9–10
Unsaturated fatty acids, 40

V

Vasodilator bradykinin, 380
Vegetable-based product, 246
Vicia faba, see Faba bean
Vitamins, 8–9, 40, 42, 71–72, 119, 378
Vitamin stability, 343
Volatile aroma compounds, 315
Volatile compounds, 19, 21

W

Water/salt soluble extracts (WSE), 77
Water-soluble arabinoxylans, 215
Water-soluble carbohydrates (WSC), 393
Weissella sp.
   W. cibaria, 121, 171, 172, 238, 243, 274, 339, 382
   W. confusa, 78, 121, 171, 244, 245, 274, 276, 285, 349
   W. koreensis, 353
   W. paramesenteroides, 121, 204, 338–339
   W. viridescens, 339
Wet milling, 90
Wheat bran, 397
White biotechnologies, 285

Whole-genome microarray, 181
Whole grain bread, 353
Whole grain dietary fiber, 49
Whole-plant silages, 394–395
Wickerhamomyces anomalus, 124, 313, 338, 344, 378
Winemaking, 321
World Health Organization (WHO), 46, 315
WSC, *see* Water-soluble carbohydrates
WSE, *see* Water/salt soluble extracts
Wxz/Wzy-dependent pathway, 129, 241–242

X

Xylanases, 210, 215, 342, 417
Xylitol, 248
Xylooligosaccharides (XOS), 248, 249, 287

Y

Yarrowia spp., 124
Yeast metabolism pathways, 372
Yeast microbiota, 66
Yeasts, 16, 119, 124–125, 151, 268, 271, 370, 396

Z

Zein proteins, 395
Zymographic analysis, 178

Printed in the United States
by Baker & Taylor Publisher Services